T0140595

Nanotechnology in the Life Sciences

Series Editor

Ram Prasad
School of Environmental Science and Engineering,
Sun Yat-Sen University, Guangzhou, China
Amity Institute of Microbial Technology,
Amity University, Noida, Uttar Pradesh, India

Nano and biotechnology are two of the 21st century's most promising technologies. Nanotechnology is demarcated as the design, development, and application of materials and devices whose least functional make up is on a nanometer scale (1 to 100 nm). Meanwhile, biotechnology deals with metabolic and other physiological developments of biological subjects including microorganisms. These microbial processes have opened up new opportunities to explore novel applications, for example, the biosynthesis of metal nanomaterials, with the implication that these two technologies (i.e., thus nanobiotechnology) can play a vital role in developing and executing many valuable tools in the study of life. Nanotechnology is very diverse, ranging from extensions of conventional device physics to completely new approaches based upon molecular self-assembly, from developing new materials with dimensions on the nanoscale, to investigating whether we can directly control matters on/in the atomic scale level. This idea entails its application to diverse fields of science such as plant biology, organic chemistry, agriculture, the food industry, and more.

Nanobiotechnology offers a wide range of uses in medicine, agriculture, and the environment. Many diseases that do not have cures today may be cured by nano-technology in the future. Use of nanotechnology in medical therapeutics needs adequate evaluation of its risk and safety factors. Scientists who are against the use of nanotechnology also agree that advancement in nanotechnology should continue because this field promises great benefits, but testing should be carried out to ensure its safety in people. It is possible that nanomedicine in the future will play a crucial role in the treatment of human and plant diseases, and also in the enhancement of normal human physiology and plant systems, respectively. If everything proceeds as expected, nanobiotechnology will, one day, become an inevitable part of our every-day life and will help save many lives.

More information about this series at http://www.springer.com/series/15921

Ram Prasad • Anal K. Jha • Kamal Prasad
Editors

Exploring the Realms
of Nature for Nanosynthesis

 Springer

Editors
Ram Prasad
Amity Institute of Microbial Technology
Amity University
Noida, Uttar Pradesh, India

Anal K. Jha
Aryabhatta Centre for Nanoscience
and Nanotechnology
Aryabhatta Knowledge University
Patna, Bihar, India

Kamal Prasad
Department of Physics
Tilka Manjhi Bhagalpur University
Bhagalpur, Bihar, India

ISSN 2523-8027   ISSN 2523-8035   (electronic)
Nanotechnology in the Life Sciences
ISBN 978-3-030-07612-2   ISBN 978-3-319-99570-0   (eBook)
https://doi.org/10.1007/978-3-319-99570-0

This Springer imprint is published by the registered company Springer Nature Switzerland AG
The registered company address is: Gewerbestrasse 11, 6330 Cham, Switzerland

Preface

This book discusses the diversity in nature in all forms as far as the natural cohort-based nanomaterials synthesis protocols are concerned. Be it microbes, plants, wastes or any other allied source, the dictum stands tall—nanoscale remains the scale of nature! Nowadays, bionanotechnology is considered as one of the most exciting and upcoming branches of nanotechnology where possibilities seem endless. Actually, physics, chemistry and biology have equipped us with the knowledge of molecular structure, properties and functions which have to be smartly engineered at nano-level to realise various technological applications. Chapter 1 details and emphasises this very fact by delineating different examples and possibly involved thermodynamic mechanisms. Microbes, the minuscule nano-assemblies of nature, have been explored for the purpose of synthesis as they best communicate and deliver as per their cellular level organisation which is very well presented in Chaps. 2 and 3. Chapters 4–9 emphasise the potential promises of different phyto cohorts and their uses in the immediate future. Chapter 10 delineates the use of biogenic silver and gold towards mitigation of dye-related pollution in our aquatic ecosystems. An epoch approach towards waste utilisation has been suggested in Chap. 11 which is bound to have a long-term impact in terms of our knowledge. Chapters 12–14 describe the inherent protocols being utilised by the morphons, broadly unknown to us. While graphene-based biomedical applications have been suggested in Chap. 15, two other chapters (Chaps. 16 and 17) detail the diagnostic prodigality of nanomaterials towards protection of crop plants from microbial annihilation and crystallisation of cellulose nanoparticles, respectively.

This comprehensive book covers the realm of nature and should enlighten readers everywhere.

Noida, Uttar Pradesh, India   Ram Prasad
Patna, Bihar, India   Anal K. Jha
Bhagalpur, Bihar, India   Kamal Prasad

Contents

Contributors

Sidra Abbas Microbiology and Biotechnology Research Lab, Department of Environmental Sciences, Fatima Jinnah Women University, Rawalpindi, Pakistan

Sandra Pérez Álvarez Instituto Politécnico Nacional, CIIDIR Unidad Sinaloa, Depto. de Biotecnología Agrícola, Guasave, Sinaloa, Mexico

Eduardo Fidel Héctor Ardisana Facultad de Ingeniería Agronómica, Universidad Técnica de Manabí, Portoviejo, Ecuador

Rajender Boddula CAS Key Laboratory for Nanosystem and Hierarchical Fabrication, National Center for Nanoscience and Technology, Beijing, P. R. China

Sai Swaroop Dalli Department of Chemistry and Material Sciences, Lakehead University, Thunder Bay, ON, Canada

Ratan Kumar Dey Centre for Applied Chemistry, Central University of Jharkhand, Ranchi, Jharkhand, India

Priyanka Dubey Department of Textile Technology, Indian Institute of Technology Delhi, New Delhi, India

Anila Fariq Microbiology and Biotechnology Research Lab, Department of Environmental Sciences, Fatima Jinnah Women University, Rawalpindi, Pakistan

Saurabh Gautam Department of Cellular Biochemistry, Max Planck Institute of Biochemistry, Martinsried, Germany

Anal K. Jha Aryabhatta Centre for Nanoscience and Nanotechnology, Aryabhatta Knowledge University, Patna, India

Babita Jha Aryabhatta Centre for Nanoscience and Nanotechnology, Aryabhatta Knowledge University, Patna, India

Tabeer Khan Microbiology and Biotechnology Research Lab, Department of Environmental Sciences, Fatima Jinnah Women University, Rawalpindi, Pakistan

Joonseok Koh Department of Organic and Nano System Engineering, Konkuk University, Seoul, Republic of Korea

Santosh Kumar Department of Organic and Nano System Engineering, Konkuk University, Seoul, Republic of Korea

Niraj Kumari Aryabhatta Centre for Nanoscience and Nanotechnology, Aryabhatta Knowledge University, Patna, India

Priti Kumari Aryabhatta Centre for Nanoscience and Nanotechnology, Aryabhatta Knowledge University, Patna, India

Jesús Alicia Chávez Medina Instituto Politécnico Nacional, CIIDIR Unidad Sinaloa, Depto. de Biotecnología Agrícola, Guasave, Sinaloa, Mexico

Ramyakrishna Pothu College of Chemistry and Chemical Engineering, Hunan University, Changsha, P. R. China

Kamal Prasad Department of Physics, Tilka Manjhi Bhagalpur University, Bhagalpur, Bihar, India

S. Rajeshkumar Department of Pharmacology, Saveetha Dental College and Hospitals, Saveetha Institute of Medical and Technical Sciences, Chennai, Tamil Nadu, India

Sudip Kumar Rakshit Department of Chemistry and Material Sciences, Lakehead University, Thunder Bay, ON, Canada

Department of Biotechnology, Lakehead University, Thunder Bay, ON, Canada

Mugdha Rao Aryabhatta Centre for Nanoscience and Nanotechnology, Aryabhatta Knowledge University, Patna, Bihar, India

Sumit K. Roy Department of Physics, St. Xavier's College, Ranchi, India

Mahdieh Samavi Department of Biotechnology, Lakehead University, Thunder Bay, ON, Canada

R. V. Santhiyaa Nanotherapy Laboratory, School of Bio-Sciences and Technology, Vellore Institute of Technology, Vellore, Tamil Nadu, India

Aditya Saran Department of Microbiology, Marwadi University, Rajkot, India

G. P. Singh Centre for Nanotechnology, Central University of Jharkhand, Ranchi, Jharkhand, India

K. P. Singh University Department of Zoology, Vinoba Bhave University, Hazaribag, India

Neha Singh Centre for Applied Chemistry, Central University of Jharkhand, Ranchi, Jharkhand, India

Radhika Singh Department of Chemistry, Dayalbagh Educational Institute, Agra, Uttar Pradesh, India

Marco Antonio Magallanes Tapia Instituto Politécnico Nacional, CIIDIR Unidad Sinaloa, Depto. de Biotecnología Agrícola, Guasave, Sinaloa, Mexico

Bijaya Kumar Uprety Department of Biotechnology, Lakehead University, Thunder Bay, ON, Canada

P. Veena Nanotherapy Laboratory, School of Bio-Sciences and Technology, Vellore Institute of Technology, Vellore, Tamil Nadu, India

María Esther González Vega Instituto Nacional de Ciencias Agrícolas (INCA), Carretera a Tapaste, San José de las Lajas, Mayabeque, Cuba

Mohammad Y. Wani Faculty of Science, Chemistry Department, University of Jeddah, Jeddah, Kingdom of Saudi Arabia

Azra Yasmin Microbiology and Biotechnology Research Lab, Department of Environmental Sciences, Fatima Jinnah Women University, Rawalpindi, Pakistan

Sabiha Zamani Aryabhatta Centre for Nanoscience and Nanotechnology, Aryabhatta Knowledge University, Patna, India

About the Editors

Ram Prasad is associated with Amity Institute of Microbial Technology, Amity University, Uttar Pradesh, India, since 2005. His research interest includes plant-microbe interactions, sustainable agriculture and microbial nanobiotechnology. Dr. Prasad has more than hundred publications to his credit, including research papers, review articles and book chapters, and five patents issued or pending and edited or authored several books. Dr. Prasad has 12 years of teaching experience, and he has been awarded the Young Scientist Award (2007) and Prof. J.S. Datta Munshi Gold Medal (2009) by the International Society for Ecological Communications; FSAB fellowship (2010) by the Society for Applied Biotechnology; the American Cancer Society UICC International Fellowship for Beginning Investigators, USA (2014); Outstanding Scientist Award (2015) in the field of microbiology by Venus International Foundation; BRICPL Science Investigator Award (ICAABT-2017); and Research Excellence Award (2018). Previously, Dr. Prasad served as visiting assistant professor at Whiting School of Engineering, Department of Mechanical Engineering at Johns Hopkins University, USA, and presently, working as research associate professor at School of Environmental Science and Engineering, Sun Yat-Sen University, Guangzhou, China.

Anal K. Jha, MPhil (Cantab), PhD, FCCS (Cambridge, UK), is working as assistant professor in Aryabhatta Centre for Nanoscience and Nanotechnology, Aryabhatta Knowledge University, Patna, India. He has over 60 publications to his credit including more than ten of book chapters and an X-ray diffraction reference data in JCPDS-ICDD, USA. His current research interests include embryonic stem cells, molecular nanomedicine and synthesis and characterisations of advanced nano-materials for different industrial applications using bion-anotechnology approaches. Dr. Jha is currently working as participant scientist on nanosilicon glass-based online water purifier in order to eliminate the foul odour, hard-ness and salinity (Newton Bhabha Foundation of the Royal Society, London, and DST, India). He is an active peer of internationally recognised journals of different publishing houses. He has delivered invited lectures in dozens of national and international conferences.

Kamal Prasad, PhD, is professor at the University Department of Physics, T.M. Bhagalpur University, Bhagalpur, India. He has 25 years of teaching experi-ence in various organisations, such as SLIET, Longowal (Punjab), Central University of Jharkhand Ranchi and Aryabhatta Knowledge University, Patna—a technical university. Prof. Prasad has successfully guided 12 PhD and 2 MTech students and is working as the editorial board member of several journals including *Colloids and Surface B*. He has over 150 publications to his credit including a book, 12 book chapters and 15 X-ray dif-fraction reference data in JCPDS-ICDD, USA. His cur-rent research interests include synthesis and characterisations of eco-friendly ferroelectric/piezoelec-tric ceramics as well as ceramic-polymer composites and advanced nanomaterials for different industrial applications. He is an active peer of internationally rec-ognised journals. He has delivered more than 30 invited lectures in national and international conferences/ workshops.

Chapter 1
Mechanistic Plethora of Biogenetic Nanosynthesis: An Evaluation

Anal K. Jha and Kamal Prasad

> *"You got to accept Nature the way she is not the way you want her to be....."*
>
> *Richard Feynman.*

1.1 Introduction

The Mother Nature seems a superlative manifestation of nano-sized assemblies at its different level of organizations and in general obeying the principles of supramolecular chemistry and thermodynamics. Right from the primitive prokaryotes to the advanced and better organized eukaryotes, inorganic materials have proved their prodigality with time. Nature commenced its organizational endeavor taking use of inorganic materials for various purposes and this indeed led to the development of a myriad of organic structures and subsequently the living systems. This speaks at large of it regarding the time survived interaction of inorganic materials and biological systems. Stress has been a testing cue for the exponential process of evolution which provided various organisms to evaluate and subsequently modulate their metabolic weaponry in order to ensure their survival on this planet. These weapons of adaptability have helped the humanity by all possible means. Right from nutrition to therapeutics, primary or secondary metabolites have proved their utility.

Over the last few years, the scientific and engineering communities have been witnessing an impressive progress in the field of nanoscience and nanotechnology. It is observed that the reduction of materials' dimension has pronounced effects on

A. K. Jha
Aryabhatta Centre for Nanoscience and Nanotechnology, Aryabhatta Knowledge University, Patna, India

K. Prasad (✉)
Department of Physics, Tilka Manjhi Bhagalpur University, Bhagalpur, Bihar, India

© Springer Nature Switzerland AG 2018
R. Prasad et al. (eds.), *Exploring the Realms of Nature for Nanosynthesis*,
Nanotechnology in the Life Sciences, https://doi.org/10.1007/978-3-319-99570-0_1

the physical properties that may be significantly different from the corresponding bulk material. Various physical properties which different nanomaterials are being exhibited are due to: large surface atom, large surface energy, spatial confinement, reduced imperfections, etc. One key aspect of nanotechnology concerns the development of rapid and reliable experimental protocols for the synthesis of nanomaterials over a range of chemical compositions, sizes, high monodispersity, and large-scale production because the synthetic protocols and their processing parameters have large effects on the ultimate properties or performances. Accordingly, many synthetic procedures have emerged with time for the preparation of inorganic nanomaterials most of them capital, time, labor and/or more alarmingly the environmental concerns. Eyebrows were raised genuinely with respect to such protocols as they could not shrug off the fate of the participating chemicals in our immediate milieu. Promulgated laws and framed legislations like WEEE and RoHS restrict the use of such protocol which threatens our survival on this planet. Therefore, synthesis of plethora of nanomaterials ranging from metals and alloys to oxides and chalcogenides, etc. using green protocols is of the major concern nowadays. The objective of this chapter is to address the possible involved synthetic mechanisms in different nanotransformations.

1.2   Synthesis Accomplished by Soft Chemicals

Natural cohorts are very rich in metabolites of a primary or secondary type, and primary ones often take lead in the procedure of nanotransformation for being comparatively labile and regenerative. Organic acids are proven and time tested chelators and often do act as redox agents broadly banking upon the working environment. They indeed act as soft chemicals and help in synthesis of nanomaterials. In order to exemplify, ceramic perovskite with ABO_3 type structure is considered here, which are technologically very important and find applications in multilayer capacitors, ferroelectric, piezoelectric, pyroelectric, memories, solid oxide fuel cells, energy harvesting, microwave, etc. When considered a cubic structure, the oxygen atoms form a cubic lattice of corner-sharing octahedra with B-cations at their centers while A-cations form a second interpenetrating cubic sub-lattice located at 12-fold coordinated sites between the octahedra. The different principles of chemistry which can be combined (viz. ionic radii, valence state, tolerance factor, etc.) to obtain numerous complex perovskite oxides with the mixed-cation formula such as $(A'A''\ldots)BO_3$, $A(B'B''\ldots)O_3$, or $(A'A''\ldots)(B'B''\ldots)O_3$ have a variety of interesting properties, designed for many electronic and/or microelectronic devices (Rödel et al. 2009; Damjanovic et al. 2010; Prasad et al. 2014). Additionally, it has been observed that modifications either at A- or B-site play an important role in tailoring different properties of complex perovskites as the material's properties largely depend upon the size difference of pseudo-cation $(A'A''\ldots)^{2+}$ and/or $(B'B''\ldots)^{4+}$ and on the difference on their valance states. With the intension to realize the optimal properties of such complex perovskite oxides, synthesis of single-phase compound

is the foremost requirement. It is known that the processing technique and/or conditions play very important role in deciding the material's properties. Accordingly, the
synthetic mechanisms of ABO_3 using two soft chemical techniques namely citrate
gel and tartrate gel methods are being considered here keeping in view of the fact
that these organic acids are found in many plants and fruits too.

Organic acids (R-COOH) often make a non-equivalent resonating structure,
whereas RCOO− efficiently makes an equivalent one. In this arrangement, the resonance energy of stabilization of R-COOH goes higher than that of RCOO⁻ and this
energy of dissociation probably remains available all along in the incubation
medium for nanotransformation of both metal and oxide system, as the case may be
(Morrison and Boyd 1983; Jha et al. 2011; Kumar et al. 2013). In case of citrate gel
method, citric acid and citrate ions are considered readily inter-convertible requir-
ing and/or releasing an appreciable amount of free energy (~47 kJ/mol) in aqueous
medium, which is probably sufficient to accomplish nanotransformation (Fig. 1.1).
This may be a very common case with citric acid but with higher one like tartaric
acid the case may be often much amenable. It is found that tartaric acid (H2Tart) is
energetically astir molecule, aptly displaying an interesting pattern of optical isom-
erism. The mean value for heat of combustion for tartaric acid was calculated to be
1124.5 kJ/mol recently (Kochergina et al. 2006). The value of heat of dissociation
was estimated to be of 27.17 kJ/mol for the temperature above 50 °C (Bates and
Canham 1951). It clearly indicates that this energy is in all probabilities sufficient to
accomplish a nanotransformation (Fig. 1.2). Further, this procedure involves com-

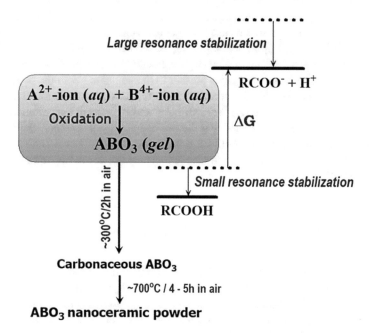

Fig. 1.1 Schematic for the synthesis of nanocrystalline ABO_3 powder using citric acid gel method

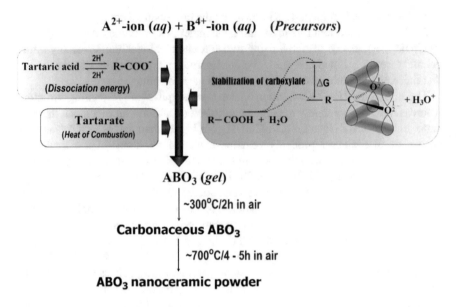

Fig. 1.2 Schematic for the synthesis of nanocrystalline ABO_3 powder using tartaric acid gel method

plexation of metal ions (A^{2+} and B^{4+} in this case) by polyfunctional carboxylic acids, such as tartaric acid having one hydroxyl group. Upon heating the mixture, the solvent (water) parches culminating into heightened viscosity. After complete parching of water, the mixture turns into a gel form and its constituents are mingled at atomic level. The gel upon heating at higher temperature (~700 °C, 4–5 h) produces the nanomaterial of our choice (e.g., nanoceramic like (ABO_3)) which might be due to the combustion of organics present in the gel, along with the evolution of gases during calcination (Kumar et al. 2015). In the present soft-chemical methods, the interactions of citric or tartaric acid with A^{2+} and B^{4+} ions in stoichiometric ratio could make it possible to lower down the surface energy. Therefore, the effect of such soft-chemical ambiance could have made the reaction to occur more easily. Higher congeners may also be employed for this purpose depending upon the material to be synthesized.

1.3   Biosynthesis of Inorganic Nanoparticles

Bionanotechnology, an upcoming and exponentially promising branch of Nanoscience & Technology, refers to the materials and processes at nanometer scale that are based on biological, biomimetic, or biologically inspired molecules. The procedures of preparation of nanoparticles from different biological sources using bionanotechnology approaches could be highly diverse and broadly dependent upon the individual source being employed and its cellular constitution which are discussed in detail in different chapters of this book.

1.4   Biosynthetic Mechanisms

The process tangled in the synthetic cues of nanoparticles, whether metals or oxide, seemingly appears heaped in the responsive culmination of the biological system, congener morph on being taken into use and its immediate metabolite ambiance along with basic bustles of metabolism or its total biochemical and/or metabolite content.

1.5   Microbial Systems

1.5.1   Prokaryotes

As per their nomenclature, prokaryotic systems have obliterated nuclear assembly and cluttered cytoplasmic organization. Metabolite content is also poor in every respect compared to their other advanced congeners (eukaryotes) yet they face challenges of nature in most adroit manner by all viable means. Under metallic stress, their first line response originates from the membrane itself in the mode of a negative electro-kinetic potential, that helps to attract metal ions, and this step subsequently acts as pivot for initiation of nanomaterials synthesis. Further addition of a regular nourishment carbon source such as glucose which works as a soft reducing agent tends to lower the redox potential. The oxidation-reduction potential is asserted to be quantitative character as a degree of aerobiosis and is denoted by rH_2. This can be perceived as a negative logarithm of gaseous hydrogen as its partial pressure. Let us consider the representative example of *Lactobacilli*, they are Gram-positive akin to generate lactic acid through homo and/or hetero-fermentative carbohydrate metabolism. They constantly tend to retort to undulating environmental conditions, which affect their physiological state and consequently their growth rate. Periods of slow growth and non-growing states are often followed by short intervals of fast growth. Mostly, cells of the bacteria remain in a slow or sluggish state to escape cell mortality and lysis under limitation of nutrients (C, N, P, metal ions) and other energy source (Gallegos et al. 1997; Hantke 2001). The *Lactobacilli* are physiologically compelled to constantly regulate the cytoplasmic or intracellular pH. The cells incapable of maintaining the same tend to lose function and viability. Recently, an approach to understanding the pH regulation in cytoplasm has emerged despite pH homeostasis in *lactobacilli* being very important (Hutkins and Nannen 1993).

The principal metabolic product lactic acid is one of the smallest molecules with an asymmetrical carbon in the location of the carboxylic function. This unique feature allows its existence with two defined stereo-isometric forms called L(+) and D(−). In addition to its optical activity, its coetaneous carboxylic and hydroxyl groups concede it to react either as an acid or as an alcohol. This dual functional approach initiates the spontaneous formation on intermolecular esters (oligomers) with a discharge of water, which leads to a thermodynamic equilibrium dependent on the solution concentration and age. The pace of this reaction of condensation is aligned to the temperature and the possible presence of catalyst species (Lewis acid), and to the equilibrium gap of the solution under consideration. As for every esterification reaction, it is reversible by adding water (hydrolysis of formed oligomers). Thermodynamically, heat of dissociation (ΔH amounts to 63 cal/mol (at 25 °C), and the equivalent of free energy of dissociation (ΔF amounts to 5000 cal/mol for lactic acid (www.lactic.com/index.php/lacticacid 2009). This probably ensures ready availability of energy in the culture medium, which along with other crux metabolic fluxes, leads to a successful nanotransformation. Earlier works on molecular level have successfully deciphered the innards of stress response in *Lactobacilli*. Dual integral systems, or histidine-asparagine phosphorylase (HAP) schemes, denote machinery for signal transduction that is extensively suffused among bacteria. Generally, however, these systems have two components: a histidine protein kinase (Hpk) with a response regulator (RR). The Hpk has an N terminal extracellular sensor domain, while the C-terminal portion is made up of protein remains in the cytoplasm. The RR is cytoplasmatic and bears a receiver domain and binds to the DNA component (Robinson et al. 2000). The Hpk acts as a sensor and is extracellular. Once a sudden alteration is found in the ambience, it signals in the form of phosphorylation, via a conserved histidine in the Hpk, to an asparagine residue in receiver segments of the RR (Robinson et al. 2000). Immediately upon phosphorylation, the RR is activated and the DNA-adhering region adheres to the DNA, as a transcriptional enhancer or repressor, hence changing the gene assertion of the cell. As such, 16 two elemental signal transduction could be tangled for regulating the phenomena such as genetic competence (Tortosa and Dubnau 1999), an adaptation to various stresses (Mizuno and Mizushima 1990; Morel-Deville et al. 1998), etc. *Lactobacilli* in general lack a two-component system. *L. acidophilus* encodes nine systems (Altermann et al. 2005) and *L. plantarum* 13 (Kleerebezem 2004).

Under many situations, the transduction of a signal is often more complex than described above. Deciphering gene regulation will increase our knowledge as to how this organism acclimates and interacts in its habitat or against metallic stress. Cells of a probiotic microorganism like *L. sporogenes* are well equipped for undertaking any shear created by toxins or metal ions. This could be accomplished at defined echelons beginning from molecular (genomic/transcriptomic and proteomic levels) through the much trusted metabolomic level to the cytoplasmic one (Urban and Kuthan 2004). Plasmids are commonly found in many members of the *lactobacilli* at the molecular level (Gasson and Shearman 2003). Most of these plasmids are cryptic (Shareck et al. 2004), but their bestowal and significance to the harboring strain remains obscure or unclear. Meanwhile, in *L. plantarum* WCFS1, three plasmids have been functionally deciphered. Two among them had no other function except replication, while the largest plasmid pWCFS103 (36 kb) negotiating resistance to arsenate/arsenite was found to be conjugative (van Kranenburg et al. 2005). The metabolic detoxification takes place using glutathione and thioredoxin (Penninckx 2000; Sundquist and Fahey 1989; Holmgren 1985) and the oxidizing environment ensures biosynthesis of nanoscale oxides or chalcogenides. Further, in the cytoplasm the metal ions may undergo complexation once a suitable candidate ligand is available or finally through the vacuolar retention (Serrano 2008).

1.5.2 Eukaryotes

The ambience of living cohorts exponentially tends to fluctuate; therefore, resilience to the shears of the environment is necessary for organisms to persist. Chemical-stress-inducing factors like pollutants (organic or inorganic) might have different pattern of action but the annihilation stands the same. Along with this, certain natural shearing factors like radiation are also assorted with oxidative damage in cells (Avery 2001; Limon-Pacheco and Gonsebatt 2009). Encountered by metal/metalloid toxicity, fungal members switch on response at manifold levels emanating from cell wall to nuclei that can be understood as follows:

(a) Production of nanoparticles using filamentous fungi has some influence over other organisms. Filamentous fungi are handy, want simple raw materials, and have raised wall-binding capacity (Ray et al. 2011). Members of fungi also harbor a good number of hydroxyl/methoxy derivatives of quinones such as benzoquinones and toluquinones, which are secreted especially by the members of the lower fungi (for example, in *Penicillium* and *Aspergillus* species) in retort to a potential metallic stress. Presence of these metabolites often starts a redox reaction due to tautomerization that leads to synthesis of a nanomaterial (Jha et al. 2009a). Cell membranes are commonly entrusted with like peptides and proteins having small molecular mass and those are cardinally important elements of almost all metal/metalloid detoxification processes, and hence the caliber of extracellular and cytosolic chelation reactions cannot be disparaged (Tamás and Martinoia 2005; Azcón-Aguilar et al. 2009; Wysocki and Tamás 2010; Prasad et al. 2016).

The oxidoreductases, which are either membrane based or cytosolic, could have an indispensable role in the procedure. The pH sensitive oxidoreductases always tend to work in an alternative manner. A lower pH triggers oxidases while a higher value tends to activate the reductase (Jha et al. 2009a). Amino acids and amino acid-derived molecules have a high significance in plants to adapt in heavy metal assault conditions. Nitrogen-containing primary metabolite mainly proline is frequently observed to be incorporated once a heavy metal assault caused by cd, cu, Ni, or Zn is encountered. Proline is bestowed with three main functions in metal detoxification such as binding with metals, signalling, and antioxidative defense (Shanti and Karl 2006). Elaboration of inorganic metal chelators illustriously documented along numerable members of fungi and especially among brown-rot and white rot fungi and this process seems to be stimulated under cu(II) and cd(II) stress (Clausen and Green 2003; Jarosz-Wilkołazka and Gadd 2003). The substantive formation of water-insoluble metal-oxalate crystals is unequivocally an apt way to circumvent toxic metal ions, which tend to enter fungal cells (Jarosz-Wilkołazka and Gadd 2003). In addition to this, oxalate is important to preserve lignolytic system of white rot basidiomycetes (Clausen and Green 2003). A vast range of fungi has been established to produce extracellular mucilaginous materials (ECMM or emulsi-fier) with excellent toxic metal binding capabilities (Paraszkiewicz et al. 2007; Paraszkiewicz et al. 2010; Paraszkiewicz and Długónski 2009). The production of various metabolites like citric acid, homogeneous proteins, heterogeneous proteins, and peroxidases by fungi make them quite effective for detoxification of heavy metals from industry-based effluents. White rot fungi are very widely spread among genera and their enzyme-producing activity makes them effective decolorizers to eliminate toxic metals by biosorption ultimately comprehending the effluents more eco-amenable (Tripathi et al. 2007). The level of oxalic acid is significantly high in comparison to other acids like malonic acid, citric acid, shikimic acid, lactic acid, acetic acid, propionic acid, fumaric acid, formic acid, iso-butyric acid, and butyric acid among different acids production that are present in variable concentrations (Ulla et al. 2000). In addition, *Fusarium Oxysporum*, *Aspergillus niger*, *Aspergillus fumigatus*, and *Penicillium brevi-compactum* being eukaryotes have been bestowed with a much better organized cellular system compared to the bacteria. Therefore, a drastic reduction in par-ticle size of nanomaterials, which is being reported, is natural. *Fusarium* and other members are a benefit to the kingdom fungi. Their adaptability is prodigal, which is manifested as metabolic treasures from membrane-bound cellulases, nitrate reductases, galactosides, and quinones to cytosolic oxidoreductases like cytochrome P450 (Durán et al. 2005), and an active FAD-containing form of nitroalkane oxidase is also present (Gadda and Fitzpatrick 1998).

(b) Cells of eukaryotes, after encountering a metallic stress, may trigger their metabolomic machinery and that can be displayed as intracellular chelation and compartmentalization. Among the usual assaults could be fluctuating osmolar-ity, swinging level of nutrients, availability of virulent molecules, and inclement temperature. While multi-cellular, locomotory morphons can share such assaults but their uni-cellular congeners like yeast are submitted to the clem-ency of their milieu and either have to respond or cease (Morano et al. 2012).

Metal ions enter the cells through multiple transporters and tend to annihi-late the normal course of metabolism and compel the cell to trigger its stress shearing mechanisms and because of such cues, nanomaterials are synthesized. However, the deliberate elimination of the few plasma membrane channels and transporters is noted to confer metal tolerance to metal/metalloid exposed *S. cerevisiae* cells (Tamás and Martinoia 2005; Wysocki and Tamás 2010).

Though, the cells also harbor other chelation sources such as GSH complexes and phytochelatin complexes for the different ionic species such as cd(II), as(III), hg(II), and Pb(II). The actual mechanism could be intricate as it comprises multiple molecular level transporters and seems something beyond the scope of this chapter.

However, the main metabolite contributing towards detoxification procedures among cells of yeast is reckoned to be glutathione (GSH) along with two other groups of metal adhering ligands like metallothioneins and phytochelatins, both very well deciphered. Bearing a structure of c-Glu-Cys-Gly, a tripeptide, they are aptly tangled with enumerable metabolic bustles among bacteria, yeasts, plants, and animals. The *sui generis* properties of an excellent nucleophile places this molecule as a potential detoxifier undertaking bio-reduction with utmost agility and thereby rendering defense against annihilation of free radicals and xenobiotics. GSH also makes a structural unit in molecules of phytochelatin that is one of its major activities. Copied from mRNA translation, metallothioneins are low molecular weight metal binding proteins, very rich in cysteine (Kagi and Schaffer 1988). A considerable number of cysteine fragments bind different metals by S–S bonds. Metallothioneins are classified according to the arrangement of these residues (Butt and Ecker 1987). Bearing a specific sequence of defined amino acids like cysteine along with a low content of other aromatic amino acids, the acumen to adhere metal ions is quite common with metallothioneins (Mehra and Winge 1991). A dedicated clan of genes encodes such a significant protein. Albeit, a model for Cu^{2+} ions detoxification has been created but it suffices other metals in yeast (Mehra et al. 1988). The assault of Cu^{2+} and Zn^{2+} is assorted cadmium toxicity among animals and there are evidences that this significant natural chelator may play similar role among a few plants and yeast (in *S. cerevisiae* and *C. glabrata*) (Murasugi et al. 1983; Grill et al. 1985).

First discovered in the yeast *Schizosaccharomyces pombe* (Cobbett 2000), but found to quite common among most of the plants (Cobbett 2000; Rauser 1995) and a few animal cohorts (Kondo et al. 1983). Initially, though principally described as cadmium-binding peptides (Cobbett 2000; Rauser 1995), synthesis of phytochelatin is triggered by large number of metal ions such as Cd^{2+}, Pb^{2+}, Zn^{2+}, Sb^{3+}, Ag^+, Ni^{2+}, Hg^{2+}, $HAsO^{2-}$, Cu^{2+}, Sn^{2+}, SeO^{2-}, Au^+, Bi^{3+}, Te^{4+}, and W^{6+}, whenever supplemented to the medium (Huang et al. 2007). Phytochelatins bearing a common arrangement $(c\text{-Glu-Cys})n\text{-Gly}$, where $n = 2\text{--}11$, with a multitude of structural variants that have been detailed in the literature (Grill et al. 1989; Rauser 1995; Zenk 1996) also make a promising detoxification system among plants. They are structurally akin to the GSH, performing the function of their substrate during their enzyme biosynthesis, the fact that has broadly been authenticated by physiological, biochemical, and genetic persuasions (Grill et al. 1989; Rauser 1995; Zenk 1996). The synthesis of phytochelatin commences within very small span of time once yeast cells are bared to cadmium ions and is regulated by enzyme action in the company of metal ions. Among the best-reckoned activators are cadmium ions, followed by ag, Bi, Pb, Zn, Cu, Hg, and au (Clemens et al. 1999; Ha et al. 1999; Vatamaniuk et al. 1999). In plants and yeast, Cd-phytochelatin complexes are fabricated in the cytosol but are ameliorated in vacuoles (Salt and Wagner 1993). Specific studies of the fission yeast *S. pombe* deciphered that nearly the whole cadmium and phytochelatin amounts are localized in vacuoles (Mehra and Mulchandani 1995). In comparison to metallothioneins, phytochelatins harbor many appreciable advantages attributable to their structure and specifically the repeated *c*-Glu-Cys units and hence have a better metal binding capacity (Mehra and Mulchandani 1995). Further, phytochelatins can accommodate high quantity of inorganic sulfur culminating into an increased capacity of binding to cadmium (Mehra et al. 1994). The procedure of bio-mineralization of cadmium by the yeast congeners *S. pombe* and *C. glabrata* is a metal-induced biotransformation, in which metal ions are there upon chelated with small choosy peptides and coprecipitated with inorganic sulfur, resulting in nontoxic monodisperse CdS clusters with an average diameter of 2 nm. The crystal lattice consists of 85 CdS pairs covered by approximately 30 (*c*-Glu-Cys)*n*-Gly peptides, with $n = 3$–5 (Dameron and Winge 1990). Vacuoles are the primary sites of intracellular metal/metalloid sequestration and storage in fungi (Tamás and Martinoia 2005; Wysocki and Tamás 2010).

(c) As and when exposed to toxic metal/metalloid annihilation, fungal cells face oxidative cell injuries (Avery 2001). Fungi have well-prepared weaponry of antioxidants to circumvent different kinds of such stresses. GSH-independent and GSH-dependent enzyme activities are capable of overcoming reactive oxygen species with remarkable efficiency (Perego and Howell 1997). While elevated temperature represents the primary assault like heat shock (as described above), among major secondary consequences is production of reactive oxygen species (ROS). All organisms are vulnerable to ROS during the process of normal aerobic metabolism or punctuated by a disclosure to radical-forming compounds (Lewinska and Bartosz 2007). Normally, the molecular oxygen is relatively non-reactive and benign but under partial reduction, they may produce a number of ROS that includes superoxide anion and hydrogen peroxide ($H_2O_2^-$) and can further react to produce the highly sharp hydroxyl radical. Reactive oxygen species (ROS) are toxic to the cellular components culminating into lipid peroxidation, protein oxidation, and DNA modification (genetic damage). *S. cerevisiae* React to an oxidative shear using a number of cellular reactions, which ensure the existence of the cell following disclosure to potentially threatening oxidants including defense systems that detoxify ROS, reduce their rate of production, and repair the damage being created by them. Many responses are ROS-specific, but there are also general stress responses that are typically invoked as a reflex to diverse stress conditions (Bun-Ya et al. 1992). Among metabolically labile morphons, ROS are constantly generated and in *S. cerevisiae* the mechanism to shear such assaults is very well developed in order to maintain a reducing intracellular environment. An oxidative shear is generally understood to have taken place when ROS overcome these countering procedures, resulting in genetic degeneration and physiological impairments, leading eventually to cell mortality. Antioxidant defenses encompass an integer of protective enzymes that are found in different subcellular compartments and may be upregulated in response to ROS exposure. Nonenzymatic defenses typically consist of small molecules that can actually function as free radical scavengers; to date, only ascorbic acid and GSH have been extensively characterized in yeast (Bun-Ya et al. 1992). Antioxidants, like vitamin C, can reduce transition metals giving these compounds a pro-oxidant character (Hiltunen et al. 2003). The different conglomerate of enzymes convoluted in the process of conservation against ROS exposures among fungi may briefly be introduced as follows:

Catalases. Catalases are ubiquitous heme-containing enzymes that catalyze the disalteration of H_2O_2 into H_2O and O_2. Yeast has two parallel enzymes: The peroxisomal catalase a encoded by CTA1, and the cytosolic catalase T encoded by CTT1 (Hiltunen et al. 2003).

Superoxide dismutases. Superoxide dismutases (SODs) convert the superoxide anion to hydrogen peroxide, which can eventually be reduced to water by catalases or peroxidases (Hiltunen et al. 2003). SODs are pervasive antioxidants, which vary in their intracellular locale and metal cofactor urgencies among different organisms (Dean et al. 1997).

Methionine sulfoxide reductase. Amino acids are vulnerable to oxidation by ROS (Le et al. 2009). Most organisms contain methionine sulfoxide reductases (MSRs), which protect against oxidation of methionine through catalysis of thiol depending on the reduction of the oxidized residues of met.

Thioredoxins. S. cerevisiae, Like most eukaryotic morphons, has a cytoplasmic thioredoxin system, which operates in rendering protection against oxidative stress. This comprises two thioredoxins (TRX1 and TRX2) and a thioredoxin reductase (TRR1) (Gan 1991). As in most cohorts, yeast thioredoxins are agile as antioxidants and are critical in protecting against oxidative stress induced by various ROS (Kuge and Jones 1994; Pedrajas et al. 1999).

Peroxiredoxins. Peroxiredoxins (Prx) have multiple roles in stress protection, acting as antioxidants, molecular chaperones, and in the regulation of signal transduction (Pedrajas et al. 1999).

The glutathione system. The oxidation of sulfhydryl groups is one of the earliest detectable advent during ROS conciliated damage. This directionalizes the signification of GSH (g-glutamylcysteinylglycine), which is typically found as the most copious low molecular weight sulfhydryl compound (mM concentrations) in most organisms. Many roles have been advised for GSH in a miscellany of cellular procedures including amino acid transport; synthesis of nucleic acids and proteins; inflection of enzyme activity; and metabolism of carcinogens, xenobiotics, and ROS (Grant et al. 1996a). Not uncommonly, therefore, GSH is an indispensable metabolite in eukaryotes, and for example, mice that are trailing in GSH biosynthesis expire rapidly (Vina 1990). Similarly, GSH is an indispensable metabolite in yeast where it is seemingly essential as a reductant during natural growth conditions (Grant et al. 1996b).

Glutaredoxins. Glutaredoxins (GrX) are heat stable and small oxido-reductase reported firstly from the *E. coli* in which they were found to be a potential hydrogen donor to ribonucleotide reductase (Holmgren 1989).

Glutathione peroxidases. Eukaryotic glutathione peroxidases are the main defense against oxidative assault caused by hydroperoxides. They reduce hydrogen peroxide and other organic hydroperoxides, like fatty acid hydroperoxides, to the analogous alcohol, using their reducing power provided by GSH (Inoue et al. 1999).

- *Glutathione transferases.* Glutathione transferases (GSTs) are the main family of proteins, which are muddled in the curing of many xenobiotic compounds (Sheehan et al. 2001). They catalyze the conjugation of electrophilic substrates to GSH prior to their removal from cells via glutathione conjugate pumps. Two genes encoding functional GSTs, cognominated GTT1 and GTT2 have been identified in yeast (Choi et al. 1998).
- *Ascorbic acid.* Ascorbic acid is a water-soluble antioxidant, which commonly acts in a redox couple with glutathione in many eukaryotes (Winkler et al. 1994). However, the relevance of ascorbate to the yeast oxidative stress reply is obscure since yeast contains a 5-carbon analog, erythroascorbate that can have limited importance as an antioxidant. Using these facts the detailed discussions on the synthesis of different metals, oxides, or chalcogenides using microbes have been undertaken (Jha et al. 2007, 2008, 2009b; Jha and Prasad 2010a, b, c, 2014a; Prasad and Jha 2009; Prasad et al. 2007, 2010).

1.6 Plant Systems and Food Beneficiaries

The deluge of plant-mediated transformation potential is quite vast or seemingly exponential. Seemingly, each member of the plant kingdom in nature is ready to negotiate a nanotransformation as either a metal or an oxide depending upon its own metabolic treasure—the sole factor obligated for survival against the gamut of environmental rigors. Recently, enumerable research has been presented regarding plant-assisted biosynthesis of metallic nanoparticles and the chargeable phytochemical candidates have broadly been identified to be flavones, ketones, aldehydes, amides terpenoids (citronellol and geraniol), and carboxylic acid in the vigil of detailed IR studies. A quick reduction of Ag ions during a few recent investigations might have culminated due to involvement of water-soluble phytochemicals like flavones, quinones, soluble sugars, and organic acids (such as oxalic, malic, tartaric, protocatechuic) abundantly found in plant tissues (Prasad 2014). Phyllanthin and hypophyllanthin (lignan) adjudicated synthesis of Ag and gold (Au) nanoparticles has been noted recently (Kasthuri et al. 2009). Using *Eclipta* leaf extract, Ag nanoparticles were synthesized and a candidate phytochemical was accounted to be flavonoids (Jha et al. 2009c). *Cycas* leaves have been found to contain amentoflavone and hinokiflavone as characteristic biflavonyls, which have been found to synthesize Ag nanoparticles (Jha and Prasad 2010d). Synthesis of Au nanoparticles negotiated by Bael (*Aegle marmelos*) leaves has been noted recently (Rao and Savithramma 2011; Jha and Prasad 2011). Higher plants belonging to the xerophyte, mesophyte, and hydrophyte categories were recently assayed for their nanotransformation promises and all were found suitable for biosynthesis, meeting their inherent metabolic obligations (Jha et al. 2011). Earlier, by using *Aloe vera* (Sathishkumar et al. 2010), Neem (*Azadirachta indica*) (Chandran et al. 2006), *Emblica* (Ghule et al. 2006), and *Avena* (Narayanan and Sakthivel 2008) leaf broth metallic Ag was synthesized. A split-second biosynthesis of Ag nanoparticles in the above reported investigations is net culmination of reductive properties of water-soluble phytochemicals such as flavones, quinines, and organic acids that are generously present in the parenchymatous tissue of leaves (Chandran et al. 2006). Synthesis of nanoscale Ag was reported using mature chili fruit (*Capsicum annuum* L.) (Ghule et al. 2006) and orange juice (Jha et al. 2011). Here the synthesis of nanoscale Ag resulted by dint of redox reactions principally involving different phytochemicals like polyphenols, ascorbic acid, capsaicinoids, etc. Dissociation of citric acid also effectively contributed to the procedure by ensuring the presence of required free energy for nanotransformation. Therefore, where microbes like bacteria or fungi want a comparatively longer duration of incubation in the growth media for reducing a metal ion, plant extracts due to the presence of water-soluble phytochemicals perform in a tick. Therefore, in comparison to microbes, cells of the plants are handy and amenable for biosynthesis of inorganic nanoparticles. Similarly, Ag and Au nanoparticles were synthesized using *Cinnamomum camphora* (Armendariz et al. 2004) leaf biomass and broth. Recently, coriander leaf extract (Awwad et al. 2013) and *Cinnamon zeylanicum* bark extract and powder (Rao and Paria 2013) and quite recently *Trianthema decandra* root extract (Singh et al. 2010) were used to biosynthesize Ag nanoparticles. Both coriander and cinnamon are rich in aromatic phytochemicals such as terpenoids like coriantrol and cinnamomum along with cinnamic acid, vitamin C, tannin, and oxalic acid, which are quite labile in negotiating a nanotransformation as found earlier (Chandran et al. 2006) while the roots of *Trianthema* might contain a rich treasure of catechin and hydroxyflavones (Singh et al. 2010). Earlier, nitrogenase-rich *Gliricidia sepium* (Jacq.) was also employed for preparing Ag nanoparticles (Raut et al. 2009). Besides, few papers have recently been poured in showing the formation of different nanomaterials based on the phytochemicals, organic acids, etc. present in the plant and/or plant parts (Jha and Prasad 2013b, 2014b, 2015, 2016a, b). In addition, the trail continues. A general schematic representation along with the generation of free energy in the medium for the various fabrication of inorganic nanoparticles (metal/oxide/chalcogenide) using plant and/or plant parts have been illustrated in Fig. 1.2 considering the above facts.

1.7   Animal Systems

Like other cohorts of nature, animals too require metals to ensure their physiology and metabolism. Once the limit of this requirement exceeds, the system undergoes stress. It is suggested that heavy metals played an important role in the origin of life itself and so have endothermic among birds and mammals (Arruda et al. 1995; Clarke and Pörtner 2010). Nature has its own orchestration against any assault and that is available among animal cohorts. Like other animals, insects also have affiliated to heavy metals as essential elements in their normal metabolism. The preservation of internal homeostasis though is achieved through diffusion regularly but more meaningfully by ameliorating a specific structure called spherocrystals—originating from the ER-Golgi complex where elements precipitate upon a glycosaminoglycan nucleus in a fragile peripheral stratum. Quite a few of the spherocrystals may store compounds of minerals, commonly phosphates, while others may store waste organic chemicals such as urates. Whenever additional metals are abundant in the environment, insects such as cockroaches and ants are apt enough to stay alive and to trap the metals (e.g., Pb or Cd) in the boundary layers of spherocrystals; the chemistry of cytoplasm remains unaltered. Insects seemingly have the vigor to circumvent the high levels of toxic metals probably due to their odd habitats. In addition, lysosomes are aptly able to retain the potentially toxic heavy metals like Cd or Hg within metallothionein-like proteins (Ballan-Dufrançais 2002).

The body walls (exoskeleton) of most of the insects (like cockroach, ants, banana flies or mosquitoes, etc.) are made up of chitin (polymeric N-acetylglucosamine). Concurrent presence of both acetyl and amino group itself speaks in volumes regarding adaptive vibes among insects and that becomes a boon while one undertakes the pursuit of any nanomaterial synthesis. The broth behaves mildly acidic and is very amenable towards pH modulation, thereby ensuring any nanomaterial of choice. Further, cockroaches contain the proteins that initiate allergic alerts among humans, identified as tropomyosin. The thermodynamically less studied actin-binding protein controls actin mechanics. It has four woven alpha helices—A, B, C, and D—making a quaternary structure and is responsible for muscle contraction (Asturias et al. 1999). Along with its regular enrapturing thermodynamic vibes, this promising protein also provides redial encapsulation to the burgeoning nanomaterials.

Further, a comprehensive collusion of Bla g 2 with the structures of other well-known aspartic proteases suggests that ligand binding could be tangled in the activity of this allergen (Arruda et al. 1995). The loss of balance of metals due to genetic impairments or environmental pollution is a worldwide problem assorted with negative health effects. An exhaustive study uncovered an appreciable number of genes and their regulatory cues tangled in maintaining heavy metal homeostasis involving the metals tested in the study, cadmium, copper and zinc, instigates a wide array of responses and that includes triggering of genes coding for metallothioneins, transporters, glutathione-mediated detoxification pathway components, antimicrobial peptides, ubiquitin conjugating enzymes, heat shock proteins, and cytochrome P450 enzymes. The generic induction of metallothionein by the metals such as copper, zinc, and cadmium is well aligned with their metal scavenger function and protective role against metal toxicity through the downregulation by copper depletion but remarkable disparities in terms of response may arise among metallothionein congeners. MtnA is induced in response to a copper assault, while MtnB and MtnC are more strongly induced by zinc and cadmium, respectively. A recent study noted regarding metallothionein's preference towards different metallic assaults and it was found that MtnA played a crucial defending role against excessive copper, while MtnB got triggered against zinc and cadmium, and other congeners MtnC and MtnD played a sub-ordinative role. So, in complete terms, MtnA elaboration was suggested to be the major hunter of metals at Drosophila larval stage (Yepiskoposyan et al. 2006; Egli et al. 2006).

Fish, like other animals, need heavy metals in their metabolism in traces while most of them act as micronutrients for both plants and animals, which are essential constituents of enzymes and hormones that indeed make them indispensable for an array of metabolic reactions. Among fish, transportation of metals occurs through the blood where the metal ions are usually attached to proteins. Further, when metals are brought into contact with different organs and tissues of the fish, they accumulated there (Asturias et al. 1999). Therefore, metals are usually present in high concentrations in the organs like gills, intestines, and digestive glands. Although fish intestines are seldom consumed, it usually accumulates more heavy metals in the study and this might represent good biomonitors of metals occurring in the surrounding environment. Metallothionein genes are triggered under heavy metal assault by specific transcription factor like MTF-1, which attaches to a short DNA sequence known as metal response elements (MREs) (Ayandiran et al. 2009). At the genomic level, as a homolog to the mammalian MTF-1, the expression of MTs is regulated transcriptionally by metal-responsive transcription factor 1 (MTF-1). MTF-1 binds to the short DNA motifs termed metal response elements (MREs), upon metal assault, in the MT promoter, which is quite significant and capable of mediating the transcriptional rejoinder to heavy metals (Selvaraj et al. 2005; Stuart et al. 1984).

Chicken like other animals have blood as carrier where metals usually are attached to proteins. Many heavy metals are used as trace elements and feed additives in poultry feed. These metals are common in our environment, some of these (iron, copper, manganese, zinc, etc.) are essential for good health, however; other (arsenic, mercury, lead, cadmium, etc.) are poisonous and deleterious for health (Jadhav et al. 2007). Arsenic is most important and usually found in the environment in organic and inorganic forms with different bioavailability. Arsenic can compel antibiotic resistance and this is also true among bacteria. There is trend of using arsenic with antibiotics as feed additive among poultry farmers. However, vitamin C is found to partially mitigate the toxicity of arsenic among broiler birds (Suganya et al. 2016). Similarly, lead, mercury, cadmium, etc. could also have annihilating effect on chicken. It was noticed earlier that functionalities of superoxide dismutase (SOD) and catalase (CAT) are enhanced even under sub-acute toxicity of cadmium while the glutathione (GSH) tends to decline appreciably among the erythrocytes of

Cd exposed birds (Yadav and Khandelwal 2006). A recent exhaustive study suggests the role of ascorbic acid (vitamin C) as an anti-stress among poultries (Ahmadu et al. 2016) and it may be due to their role in glutathione metabolism. Recently, two dead lower animals' broth of banana fly (*Drosophila* sp.) and cockroach (*Periplaneta americana*), respectively, were used to synthesizes CdS and gold nanoparticles (Jha and Prasad 2012, 2013a), and discards of fish, principally intestines, were taken into use for preparing Ag nanoparticles (Jha and Prasad 2014c) as well as chicken and goat meat processing discards for synthesizing silver and zinc oxide nanoparticles (Jha and Prasad 2016c, 2018) which still harbor promising metabolites that are otherwise wasted by festering. Finally, the cessation of cells tends to liberate the potential metabolites that help in nanomaterial fabrication.

1.8 Thermodynamic Considerations

Proteins are naturally bestowed biological ligands, because they contain amino acids, which act as multidentate chelate ligands, providing an amenable spatial fixation and act as a medium themselves due to defined dielectric properties. Initial works particularly by Hsien Wu and others (Anfinsen 1973; Anfinsen and Scheraga 1975; Edsall 1995) established the concept related to the denaturation of soluble proteins involved in transitions between a relatively compact and tidy structure to a whippier, muddled open chain of polypeptide. Tropomyosin (Tm) was discovered more than 65 years ago (Edsall 1995; Bailey 1946). This actin-binding, coiled protein is an enrapturing biological molecule and its stability, structural and functional properties remain yet to be fully understood (Bailey 1948).

Widely distributed among all eukaryotic cell types it has as many as 40 isomeric forms within an individual cell speaking itself regarding functional prodigality of this protein macromolecule (Perry 2001; Gunning et al. 2008). Enumerable experiments for ascertaining thermodynamic parameters of protein transition rest upon approximation of two states behavior for the system itself. For proteins, the native/folded may be represented as A, unfolded form may be denoted as U, respectively, and this transition may be temperature and pH dependent or may broadly depend upon the concentration of denaturant. Free energy difference between native and unfolded (ΔG unf) is reported to be in the range of +20 to 60 kJ mol^{-1}. Experimentally, it is noticed that the temperature craving of ΔG unf displays that for most proteins the folded form is, not unreasonably perhaps, most stout in the physiological temperature range. The relatively small amount of free energy of expansion is due to much larger and much more temperature dependent enthalpy and entropy contributions. Massive molecules like tropomyosin principally framed by tightly held random coils, interpenetrated and tangled with each other (Cooper 1999).

Tropomyosin is quite a large biopolymer in comparison to the solvents. These contain very tightly held random coil units, entangled and interpenetrating all through. The degree of cohesive and other forces of attractions are variable between the molecular units and neighboring coils. The molecules of solvents take little time in order to overcome such forces and to liberate the individual units from the polymeric phase. There lies a clear difference of the procedure of dissolution and a molecule having low molecular weight. The interaction of a polymer molecule with the solvent is an interesting cue of events. The forces of attraction or dispersion commences and emerges between them and at a point solvent–solute interaction overpowers the solute–solute interactions leading to the weakening of the polymer holding forces and subsequently the salvation takes place. This is an overall slow process as the abovementioned interactions take time. The principle of thermodynamics governing dissolution of molecules having high or low molecular weight remains interestingly the same. State functions are quantities that address principally the state of a system and not the procedure or path being taken or followed to reach this state. The quantity V is broadly dependent upon its initial and final value of V, irrespective of the path being used for reaching V_f from V_i. Identically, temperature and pressure are also in the same category. As per the first law of thermodynamics, irrespective of the heat being absorbed or released, a change in internal energy or the work being done on or by the system is not of much significance. The total internal energy of a system is a state function. The transition between one state to another does not involve exchange of heat or work or in other words; heat may or may not be exchanged and work may or may not be done. Therefore, heat and work do not qualify as state functions. However, a change in temperature can be noticed upon absorption of heat but the work done by a system does qualify for the final state. A chemical reaction in solution is being investigated under varied conditions having variable pH, ionic strength and type of solvent, etc. The amount of free energy of activation tends to remain constant or show a tendency towards little change (Holwill and Silvester 1967).

The value of Partial Molar Volume (PMV) remains an important thermodynamic quantity harboring pivotal information regarding solvent–solute interactions along with solute behavior/structure in solvated state. It remains to be the most imperative deal while analyzing the effect of pressure on a defined chemical reaction. The biological sources (cultures, extracts, or broths) are nothing but mere conglomerates of different metabolites, which are released abreast entity due to heat. Bio-legends and metal ions make coordination complexes of broadly biological relevance because metal ions are the sites of Lewis acids capable of accepting lone pairs of electrons donated by the ligand, which behaves as Lewis base. This is well illustrated in metallothionein where 30–35% of its constituent amino acid cysteine carries soft SH-groups. The primordial life activity of metallothioneins is to preserve the cells from assaults of heavy metal toxicity. Secondly, due to a favorable chelate effect, metal chelate complexes are quite stable, because there is always a favorable entropic factor accompanying the release of non-chelating ligands. Similarly, pK_a values of coordinate ligands also play significant role dependent upon tuning of redox potential. Finally, different biopolymers affect the thermodynamic stoutness of a metal center, since it can control, through its three-dimensional structure such as stereochemistry, available ligands for coordination, local hydrophilicity and or hydrophobicity, steric blockage of coordination sites, and hydrogen bonding formation. This enrapturing thermodynamic interaction/ interdependence keeps molecules agile and ensures the formation of nanomaterials is dependent upon the modulation of experimental cues. The mechanism of such nanotransformation could be understood using the nucleation and growth theory in which the overall free energy change (ΔG) must be overcome. Here the case of spherical nanoparticles is considered. ΔG is the prognostic of sum of the free energy due to the formation of a new volume and new surface being created (Jha and Prasad 2013b; Li et al. 2007): $\Delta G = -\dfrac{4}{V}\pi r^3 k_B T \ln(S) + 4\pi r^2 \gamma$, where V, r, k_B, S, and γ are the molecular volume of the precipitated species, radius of the nucleus, Boltzmann constant, saturation ratio, and surface free energy per unit surface area, respectively. It follows from this equation that a decrease in or an increase in S is helpful in the formation of desired/targeted nanoparticles. In a biochemical reaction involving microbe/plant metabolites, the interactions of broth/culture medium with metal salt make the concentration of metal ions surrounded by different metabolites like flavonoids, terpenoids, organic acids, polypeptides, etc. adhered to the metal nuclei that lead to a lower surface energy of the crystal lattice. So far, we can draw this emphatic conclusion that it is the core molecular wealth/metabolites along with the experimental cues, which decide the fate of fabrication of the type of nanomaterial, and the source/mean/method whether it is chemical or biological is oblivious as the crux lies with might of thermodynamics! All these facts have been illustrated in the Fig. 1.3 that clearly delineates the participation of different pivotal factors in the synthetic cue such as energy of decomposition if we opt to take a soft-chemical

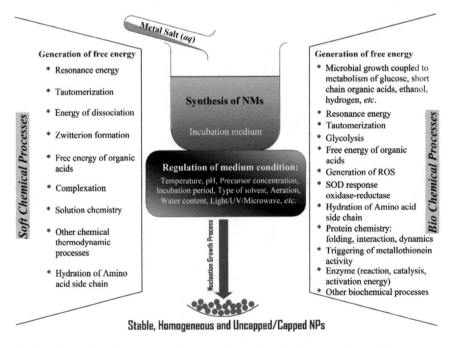

Fig. 1.3 Generation of free energy in the incubation medium for the preparation of different nanomaterials: Comparison between soft-chemical and biochemical methods

route, generation of ROS and subsequent triggering of phytochelatin, metallothionein and GSH along with many other factors leading to a successful nanotransformation.

References

Ahmadu S, Mohammed AA, Buhari H, Auwal A (2016) An overview of vitamin C as an antistress in poultry. Malays J Vet Res 7:9–22

Altermann E, Russell WM, Azcarate-Peril MA, Barrangou R, Buck BL, McAuliffe O, Souther N, Dobson A, Duong T, Callanan M, Lick S, Hamrick A, Cano R, Klaenhammer TR (2005) Complete genome sequence of the probiotic lactic acid bacterium *Lactobacillus acidophilus* NCFM. Proc Natl Acad Sci U S A 102:3906–3912

Anfinsen CB (1973) Principles that govern the folding of protein chains. Science 181:223–229

Anfinsen CB, Scheraga HA (1975) Experimental and theoretical aspects of protein folding. Adv Protein Chem 29:205–300

Armendariz V, Herrera I, Peralta-Videa J, Jose-Yacaman M, Troiani H, Santiago P (2004) Size controlled gold nanoparticles formation by *Avena sativa* biomass: use of plants in nanobiotechnology. J Nanopart Res 6:377–382

Arruda LK, Vailes LD, Mann BJ, Shannon J, Fox JW, Vedvick TS, Haden ML, Chapman MD (1995) Molecular cloning of a major cockroach (*Blattella germanica*) allergen, Bla g 2 sequence homology to the aspartic proteases. J Biol Chem 270:19563–19568

Asturias JA, Gómez-Bayón N, Arilla MC, Martínez A, Palacios R, Sánchez-Gascón F, Martínez J (1999) Molecular characterization of American cockroach tropomyosin (*Periplaneta americana* allergen 7), a cross-reactive allergen. J Immunol 162:4342–4348

Avery SV (2001) Metal toxicity in yeasts and the role of oxidative stress. Adv Appl Microbiol 49:111–142

Awwad AM, Salem NM, Abdeen AO (2013) Biosynthesis of silver nanoparticles using loquat leaf extract and its antibacterial activity. Adv Mater Lett 4:338–342

Ayandiran TA, Fawole OO, Adewoye SO, Ogundiran MA (2009) Bioconcentration of metals in the body muscle and gut of *Clarias gariepinus* exposed to sub-lethal concentrations of soap and detergent effluent. J Cell Anim Biol 3:113–118

Azcón-Aguilar C, Barea JM, Gianinazzi S, Gianinazzi-Pearson V (2009) Mycorrhizas-functional processes and ecological impacts. Springer, Berlin

Bailey K (1946) Tropomyosin: a new asymmetric protein component of muscle. Nature 157:368–369

Bailey K (1948) Tropomyosin: a new asymmetric protein component of the muscle fibril. Biochem J 43:271–279

Ballan-Dufrançais C (2002) Localization of metals in cells of pterygote insects. Microsc Res Tech 56:403–420

Bates RG, Canham RG (1951) pH of solutions of potassium hydrogen d-tartrate from 0° to 60°C. J Res Natl Bur Stand 47:343–438

Bun-Ya M, Harashima S, Oshima Y (1992) Putative GTP-binding protein, Gtr1, associated with the function of the Pho84 inorganic phosphate transporter in Saccharomyces cerevisiae. Mol Cell Biol 12:2958–2966

Butt TR, Ecker DJ (1987) Yeast metallothionein and applications in biotechnology. Microbiol Rev 51:351–364

Chandran SP, Chaudhary M, Pasricha R, Ahmad A, Sastry M (2006) Synthesis of gold nanotriangles and silver nanoparticles using *Aloe vera* plant extract. Biotechnol Prog 22:577–583

Choi JH, Lou W, Vancura A (1998) A novel membrane-bound glutathione S-transferase functions in the stationary phase of the yeast *Saccharomyces cerevisiae*. J Biol Chem 273:29915–29922

Clarke A, Pörtner HO (2010) Temperature, metabolic power and the evolution of endothermy. Biol Rev 85:703–727

Clausen CA, Green F (2003) Oxalic acid overproduction by copper-tolerant brown-rot basidiomycetes on southern yellow pine treated with copper-based preservatives. Int Biodeter Biodegr 51:139–144

Clemens S, Kim EJ, Neumann D, Schroeder JI (1999) Tolerance to toxic metals by a gene family of phytochelatin synthases from plants and yeast. EMBO J 18:3325–3333

Cobbett CS (2000) Phytochelatins and their roles in heavy metal detoxification. Plant Physiol 123:825–832

Cooper A (1999) In: Geoffrey A (ed) Protein: a comprehensive treatise, vol 2. JAI Press, Stamford, pp 217–270

Dameron CT, Winge DR (1990) Peptide mediated formation of quantum semiconductors. Trends Biotechnol 8:3–6

Damjanovic D, Klein N, Li J, Porokhonskyy V (2010) What can be expected from lead-free piezoelectric materials? Funct Mater Lett 3:5–13

Dean RT, Fu S, Stocker R, Davies MJ (1997) Biochemistry and pathology of radical-mediated protein oxidation. Biochem J 324:1–18

Durán N, Marcato PD, Alves OL, De Souza GIH, Esposito E (2005) Mechanistic aspects of biosynthesis of silver nanoparticles by several *Fusarium oxysporum* strains. J Nanobiotechnol 3:1–8. https://doi.org/10.1186/1477315538

Edsall JT (1995) Hsien Wu and the first theory of protein denaturation. Adv Protein Chem 46:1–5

Egli D, Domenech J, Selvaraj A, Balamurugan K, Hua H, Capdevila M, Georgiev O, Schaffner W, Atrian S (2006) The four members of the Drosophila metallothionein family exhibit distinct yet overlapping roles in heavy metal homeostasis and detoxification. Genes Cells 11:647–658

Gadda G, Fitzpatrick PF (1998) Biochemical and physical characterization of the active FAD containing form of nitroalkane oxidase from *Fusarium oxysporum*. Biochemistry 37:6154–6164

Gallegos MT, Schleif R, Bairoch A, Hofmann K, Ramos JL (1997) Arac/XylS family of transcriptional regulators. Microbiol Mol Biol Rev 61:393–410

Gan Z (1991) Yeast thioredoxin genes. J Biol Chem 266:1692–1696

Gasson MJ, Shearman CA (2003) In: BJB W, Warner PJ (eds) Genetics of lactic acid bacteria, vol 3. Kluwer Academic/Plenum Publishers, New York

Ghule K, Ghule AV, Liu JY, Ling YC (2006) Microscale size triangular gold prisms synthesized using Bengal gram beans (*Cicerarietinum* L.) extract and $HAuCl_4 \times 3H_2O$: a green biogenic approach. J Nanosci Nanotechnol 6:3746–3751

Grant CM, Collinson LP, Roe JH, Dawes IW (1996a) Yeast glutathione reductase is required for protection against oxidative stress and is a target gene for yAP-1 transcriptional regulation. Mol Microbiol 21:171–179

Grant CM, MacIver FH, Dawes IW (1996b) Glutathione is an essential metabolite required for resistance to oxidative stress in the yeast *Saccharomyces cerevisiae*. Curr Genet 29:511–515

Grill E, Winnacker EL, Zenk MH (1985) Phytochelatins: the principal heavy-metal complexing peptides of higher plants. Science 230:674–676

Grill E, Loffler S, Winnacker EL, Zenk MH (1989) Phytochelatins, the heavy-metal-binding peptides of plants, are synthesized from glutathione by a specific gamma-glutamylcysteine dipeptidyl transpeptidase (phytochelatin synthase). Proc Natl Acad Sci U S A 86:6838–6842

Gunning P, O'Neill G, Hardeman E (2008) Tropomyosin-based regulation of the actin cytoskeleton in time and space. Physiol Rev 88:1–35

Ha SB, Smith AP, Howden R, Dietrich WM, Bugg S, O'Connell MJ, Goldsbrough PB, Cobbett CS (1999) Phytochelatin synthase genes from arabidopsis and the yeast *Schizosaccharomyces pombe*. Plant Cell 11:1153–1163

Hantke K (2001) Iron and metal regulation in bacteria. Curr Opin Microbiol 4:172–177

Hiltunen JK, Mursula AM, Rottensteiner H, Wierenga RK, Kastaniotis AJ, Gurvitz A (2003) The biochemistry of peroxisomal beta-oxidation in the yeast *Saccharomyces cerevisae*. FEMS Microbiol Rev 27:35–64

Holmgren A (1985) Thioredoxin. Annu Rev Biochem 54:237–271

Holmgren A (1989) Thioredoxin and glutaredoxin systems. J Biol Chem 264:13963–13966

Holwill MEJ, Silvester NR (1967) Thermodynamic aspects of flagellar activity. J Exp Biol 47:249–265

http://www.lactic.com/index.php/lacticacid (2009)

Huang J, Li Q, Sun D, Lu Y, Su Y, Yang X, Wang H, Wang Y, Shao W, He N, Hong J, Chen C (2007) Biosynthesis of silver and gold nanoparticles by novel sundried *Cinnamomum camphora* leaf. Nanotechnology 18:105104–105115

Hutkins RW, Nannen NL (1993) pH homeostasis in lactic acid bacteria. J Dairy Sci 76:2354–2365

Inoue Y, Matsuda T, Sugiyama KI, Izawa S, Kimura A (1999) Genetic analysis of glutathione peroxidase in oxidative stress response of *Saccharomyces cerevisiae*. J Biol Chem 274:27002–27009

Jadhav SH, Sarkar SN, Patil RD, Tripathi HC (2007) Effects of sub-chronic exposure via drinking water to a mixture of eight water-contaminating metals: a biochemical and histopathological study in male rats. Arch Environ Contam Toxicol 53:667–677

Jarosz-Wilkołazka A, Gadd GM (2003) Oxalate production by wood- rotting fungi growing in toxic metal-amended medium. Chemosphere 52:541–547

Jha AK, Prasad K (2010a) Biosynthesis of CdS nanoparticles: an improved green and rapid procedure. J Colloid Interface Sci 342:68–72

Jha AK, Prasad K (2010b) Ferroelectric $BaTiO_3$ nanoparticles: biosynthesis and characterization. Colloids Surf B Biointerfaces 75:330–334

Jha AK, Prasad K (2010c) Synthesis of $BaTiO_3$ nanoparticles: a new sustainable green approach. Integr Ferroelectr 117:49–54

Jha AK, Prasad K (2010d) Green synthesis of silver nanoparticles using *Cycas* leaf. Int J Green Nanotechnol Phys Chem 1:P110–P117

Jha AK, Prasad K (2011) Biosynthesis of gold nanoparticles using bael (*Aegle marmelos*) leaf: mythology met technology. Int J Green Nanotechnol Phys Chem 3:92–97

Jha AK, Prasad K (2012) Banana fly (*Drosophila* sp.) synthesizes CdS nanoparticles! J Bionanosci 6:99–103

Jha AK, Prasad K (2013a) Can animals too negotiate nano transformations? Adv Nano Res 1:35–42

Jha AK, Prasad K (2013b) Rose (*Rosa sp.*) petals assisted green synthesis of gold nanoparticles. J Bionanosci 7:245–250

Jha AK, Prasad K (2014a) Green synthesis and characterization of $BaFe_{0.5}Nb_{0.5}O_3$ nanoparticles. J Chin Adv Mater Soc 2:294–302

Jha AK, Prasad K (2014b) Green synthesis of silver nanoparticles and its activity on SiHa cervical cancer cell line. Adv Mater Lett 5:501–505

Jha AK, Prasad K (2014c) Synthesis of silver nanoparticles employing fish processing discard: an eco-amenable approach. J Chin Adv Mater Soc 2:179–185

Jha AK, Prasad K (2015) Facile green synthesis of metal and oxide nanoparticles using papaya juice. J Bionanosci 9:311–314

Jha AK, Prasad K (2016a) Green synthesis and antimicrobial activity of silver nanoparticles onto cotton fabrics: an amenable option for textile industries. Adv Mater Lett 7:42–46

Jha AK, Prasad K (2016b) Aquatic fern (*Azolla* sp.) assisted synthesis of gold nanoparticles. Int J Nanosci 15:1650008–1650012

Jha AK, Prasad K (2016c) Synthesis of ZnO nanoparticles from goat slaughter waste for environmental protection. Int J Curr Eng Technol 6:147–151

Jha AK, Prasad K (2018) Nanomaterials from biological and pharmaceutical wastes – a step towards environmental protection. Mater Today: Proc, in press

Jha AK, Prasad K, Kulkarni AR (2007) Microbe mediated nano transformation: cadmium. Nano 2:239–242

Jha AK, Prasad K, Kulkarni AR (2008) Yeast mediated synthesis of silver nanoparticles. Int J Nanosci Nanotechnol 4:17–21

Jha AK, Prasad K, Prasad K (2009a) Biosynthesis of Sb_2O_3 nanoparticles: a low cost green approach. Biotechnol J 4:1582–1585

Jha AK, Prasad K, Kulkarni AR (2009b) Synthesis of TiO_2 nanoparticles using microorganisms. Colloids Surf B Biointerfaces 71:226–229

Jha AK, Prasad K, Kumar V, Prasad K (2009c) Biosynthesis of silver nanoparticles using *Eclipta* leaf. Biotechnol Prog 25:1476–1479

Jha AK, Kumar V, Prasad K (2011) Biosynthesis of metal and oxide nanoparticles using orange juice. J Bionanosci 5:162–166

Kagi JHR, Schaffer A (1988) Biochemistry of metallothionein. Biochemistry 27:8509–8515

Kasthuri J, Kathiravan K, Rajendiran N (2009) Phyllanthin assisted synthesis of silver and gold nanaoparicles;a novel biological approach. J Nanopart Res 11:1075–1085

Kleerebezem M (2004) Quorum sensing control of lantibiotic production; nisin and subtilin auto-regulate their own biosynthesis. Peptides 25:1405–1414

Kochergina LA, Volkov AV, Krutov DV, Krutova ON (2006) The standard enthalpies of formation of citric and tartaric acids and their dissociation products in aqueous solutions. Russ J Phys Chem A 80:1029–1033

Kondo N, Isobe M, Imai K, Goto T, Murasugi A, Hayashi Y (1983) Structure of cadystin, the unit-peptide of cadmium-binding peptides induced in a fission yeast, *Schizosaccharomyces pombe*. Tetrahedron Lett 24:925–928

van Kranenburg R, Golic N, Bongers R, Leer RJ, de Vos WM, Siezen RJ, Kleerebezem M (2005) Functional analysis of three plasmids from *Lactobacillus plantarum*. Appl Environ Microbiol 71:1223–1230

Kuge S, Jones N (1994) YAP1 dependent activation of TRX2 is essential for the response of *Saccharomyces cerevisiae* to oxidative stress by hydroperoxides. EMBO J 13:655–664

Kumar S, Sahay LK, Jha AK, Prasad K (2013) Synthesis and characterization of nanocrystalline $Al_{0.5}Ag_{0.5}TiO_3$ powder. Adv Nano Res 1:211–218

Kumar S, Jha AK, Prasad K (2015) Green synthesis and characterization of $(Ag_{1/2}Al_{1/2})TiO_3$ nanoc-eramics. Mater Sci-Pol 33:59–72

Le DT, Lee BC, Marino SM, Zhang Y, Fomenko DE, Kaya A, Hacioglu E, Kwak GH, Koc A, Kim HY, Gladyshev VN (2009) Functional analysis of free methionine-R-sulfoxide reductase from *Saccharomyces cerevisiae*. J Biol Chem 284:4354–4364

Lewinska A, Bartosz G (2007) Protection of yeast lacking the Ure2 protein against the toxicity of heavy metals and hydroperoxides by antioxidants. Free Radic Res 41:580–590

Li S, Qui L, Shen Y, Xie A, Yu X, Zhang L, Zhang Q (2007) Green synthesis of silver nanoparticles using *Capsicum annuum* L. extract. Green Chem 9:852–858

Limon-Pacheco J, Gonsebatt ME (2009) The role of antioxidants and antioxidant-related enzymes in protective responses to environmentally induced oxidative stress. Mutat Res 674:137–147

Mehra RK, Mulchandani P (1995) Glutathione-mediated transfer of cu(I) into phytochelatins. Biochem J 307:697–705

Mehra RK, Winge DR (1991) Metal ion resistance in fungi: molecular mechanisms and their regu-lated expression. J Cell Biochem 45:30–40

Mehra RK, Tarbet EB, Gray WR, Winge DR (1988) Metal-specific synthesis of two metallo-thioneins and gamma-glutamyl peptides in *Candida glabrata*. Proc Natl Acad Sci U S A 85:8815–8819

Mehra RK, Mulchandani P, Hunter TC (1994) Role of CdS quantum crystallites in cadmium resis-tance in *Candida glabrata*. Biochem Biophys Res Commun 200:1193–1200

Mizuno T, Mizushima S (1990) Signal transduction and gene regulation through the phosphoryla-tion of two regulatory components: the molecular basis for the osmotic regulation of the porin genes. Mol Microbiol 4:1077–1082

Morano KA, Grant CM, Moye-Rowley WS (2012) The response to heat shock and oxidative stress in Saccharomyces cerevisiae. Genetics 190:1157–1195

Morel-Deville F, Fauvel F, Morel P (1998) Two-component signal-transducing systems involved in stress responses and vancomycin susceptibility in *Lactobacillus sakei*. Microbiology 144:2873–2883

Morrison RT, Boyd RN (1983) Advanced organic chemistry. Allyn and Bacon, Boston

Murasugi A, Wada C, Hayashi Y (1983) Occurrence of acid-labile sulfide in cadmium-binding peptide 1 from fission yeast. J Biochem 93:661–664

Narayanan KB, Sakthivel N (2008) Coriander leaf mediated biosynthesis of gold nanoparticles. Mater Lett 62:4588–4590

Paraszkiewicz K, Długónski J (2009) Effect of nickel, copper, and zinc on emulsifier production and saturation of cellular fatty acids in the filamentous fungus *Curvularia lunata*. Int Biodeter Biodegr 63:100–105

Paraszkiewicz K, Frycie A, Słaba M, Długónski J (2007) Enhancement of emulsifier production by *Curvularia lunata* in cadmium,zinc and lead presence. Biometals 20:797–805

Paraszkiewicz K, Bernat P, Naliwajski M, Długónski J (2010) Lipid peroxidation in the fungus *Curvularia lunata* exposed to nickel. Arch Microbiol 192:135–141

Pedrajas JR, Kosmidou E, Miranda-Vizuete A, Gustafsson JA, Wright AP, Spyrou G (1999) Identification and functional characterization of a novel mitochondrial thioredoxin system in *Saccharomyces cerevisiae*. J Biol Chem 274:6366–6373

Penninckx M (2000) A short review on the role of glutathione in the response of yeasts to nutritional, environmental and nutritive stresses. Enzym Microb Technol 26:737–742

Perego P, Howell SB (1997) Molecular mechanisms controlling sensitivity to toxic metal ions in yeast. Toxicol Appl Pharmacol 147:312–318

Perry SV (2001) Vertebrate tropomyosin: distribution, properties and function. J Muscle Res Cell Motil 22:5–49

Prasad R (2014) Synthesis of silver nanoparticles in photosynthetic plants. J Nanoparticles Article ID 963961. https://doi.org/10.1155/2014/963961

Prasad K, Jha AK (2009) ZnO nanoparticles: synthesis and adsorption study. Nat Sci 1:129–135

Prasad K, Jha AK, Kulkarni AR (2007) *Lactobacillus* assisted synthesis of titanium nanoparticles. Nanoscale Res Lett 2:248–250

Prasad K, Jha AK, Prasad K, Kulkarni AR (2010) Can microbes mediate nano-transformation. Indian J Phys 84:1355–1360

Prasad K, Priyanka ANK, Chandra KP, Kulkarni AR (2014) Dielectric relaxation in Ba(Y$_{1/2}$Nb$_{1/2}$)O$_3$-BaTiO$_3$ ceramics. J Mater Sci Mater Electron 25:4856–4866

Prasad R, Pandey R, Barman I (2016) Engineering tailored nanoparticles with microbes: quo vadis. Wiley Interdiscip Rev Nanomed Nanobiotechnol 8:316–330. https://doi.org/10.1002/wnan.1363

Rao KJ, Paria S (2013) Green synthesis of silver nanoparticles from aqueous *Aegle marmelos* leaf extract. Mater Res Bull 48:628–634

Rao ML, Savithramma N (2011) Biological synthesis of silver nanoparticles using *Svensonia hyderabadensis* leaf extract and evaluation of their antimicrobial efficacy. J Pharm Sci Res 3:1117–1121

Rauser WE (1995) Phytochelatins and related peptides structure, biosynthesis, and function. Plant Physiol 109:1141–1149

Raut WR, Lakkakula JR, Kolekar NS, Mendhulkar VD, Kashid SB (2009) Phytosynthesis of silver nanoparticles using *Gliricidia sepium* (Jacq.). Curr Nanosci 5:117–121

Ray S, Sarkar S, Kundu S (2011) Extracellular biosynthesis of silver nanoparticles using the mycorrhizal mushroom *Tricholoma crassum* (BERK.) SACC: its antimicrobial activity against pathogenic bacteria and fungus, including multidrug resistant plant and human bacteria. Dig J Nanomater Biostruct 6:1289–1299

Robinson VL, Buckler DR, Stock AM (2000) A tale of two components: a novel kinase and a regulatory switch. Nat Struct Biol 7:626–633

Rödel J, Jo W, Seifert KTP, Anton EM, Granzow T, Damjanovic D (2009) Perspective on the development of lead-free piezoceramics. J Am Ceram Soc 92:1153–1177

Salt DE, Wagner GJ (1993) Cadmium transport across tonoplast of vesicles from oat roots. Evidence for a Cd^{2+}/H$^+$ antiport activity. J Biol Chem 268:12297–12302

Sathishkumar M, Krishnamurthy S, Yun YS (2010) Immobilization of silver nanoparticles synthesized using *Curcuma longa* tuber powder and extract on cotton cloth for bactericidal activity. Bioresour Technol 101:7958–7965

Selvaraj A, Balamurugan K, Yepiskoposyan H, Zhou H, Egli D, Georgiev O, Thiele DJ, Schaffner W (2005) Metal-responsive transcription factor (MTF-1) handles both extremes, copper load and copper starvation, by activating different genes. Genes Dev 19:891–896

Serrano LM (2008) Oxidative stress response in *Lactobacillus plantarum* WCFS1: a functional genomics approach. Ph.D. Thesis, Wageningen University and Research Centre, The Netherlands

Shanti SS, Karl JD (2006) The significance of amino acids and amino acid derived molecules in plant responses and adaptation to heavy metal stress. J Exp Bot 57:711–726

Shareck J, Choi Y, Lee B, Miguez CB (2004) Cloning vectors based on cryptic plasmids isolated from lactic acid bacteria: their characteristics and potential applications in biotechnology. Crit Rev Biotechnol 24:155–208

Sheehan D, Meade G, Foley VM, Dowd CA (2001) Structure, function and evolution of glutathione transferases: implications for classification of non-mammalian members of an ancient enzyme superfamily. Biochem J 360:1–16

Singh A, Jain D, Upadhyay MK, Khandelwal N, Verma HN (2010) Green synthesis of silver nanoparticles using *Argemone mexicana* leaf extract and evaluation of their activity. Dig J Nanomater Biostruct 5:483–489

Stuart GW, Searle PF, Chen HY, Brinster RL, Palmiter RD (1984) A 12-base-pair DNA motif that is repeated several times in metallothionein gene promoters confers metal regulation to a heterologous gene. Proc Natl Acad Sci U S A 81:7318–7322

Suganya T, Senthilkumar S, Deepa K, Muralidharan J, Sasikumar P, Muthusamy N (2016) Metal toxicosis in poultry – a review. Int J Sci Environ Technol 5:515–524

Sundquist AR, Fahey RC (1989) Evolution of antioxidant mechanisms: thiol-dependent peroxidases and thioltransferase among procaryotes. J Mol Evol 29:429–435

Tamás MJ, Martinoia E (2005) Molecular biology of metal homeostasis and detoxification: from microbes to man. Springer, Heidelberg

Tortosa P, Dubnau D (1999) Competence for transformation: a matter of taste. Curr Opin Microbiol 2:588–592

Tripathi AK, Harsh NSK, Gupta N (2007) Fungal treatment of industrial effluents: a mini review. Life Sci J 4:78–81

Ulla AJ, Patrick AWV, Ulla SL, Roger DF (2000) Organic acids produced by mycorrhizal *Pinus sylvestris* exposed to elevated aluminium and heavy metal concentrations. New Phytol 146:557–567

Urban PL, Kuthan RT (2004) Application of probiotics in the xenobiotic detoxification therapy. Nukleonika 49(suppl 1):S43–S45

Vatamaniuk OK, Mari S, Lu YP, Rea PA (1999) AtPCS1, a phytochelatin synthase from arabidopsis: isolation and in vitro reconstitution. Proc Natl Acad Sci U S A 96:7110–7115

Vina J (ed) (1990) Glutathione: metabolism and physiological functions. CRC Press, Boca Raton

Winkler BS, Orselli SM, Rex TS (1994) The redox couple between glutathione and ascorbic acid: a chemical and physiological perspective. Free Radic Biol Med 17:333–349

Wysocki R, Tamás MJ (2010) How *Saccharomyces cerevisiae* copes with toxic metals and metalloids. FEMS Microbiol Rev 34:925–951

Yadav N, Khandelwal S (2006) Effect of Picroliv on cadmium-induced hepatic and renal damage in the rat. Hum Exp Toxicol 25:581–591

Yepiskoposyan H, Egli D, Fergestad T, Selvaraj A, Treiber C, Multhaup G, Georgiev O, Schaffner W (2006) Transcriptome response to heavy metal stress in *Drosophila* reveals a new zinc transporter that confers resistance to zinc. Nucleic Acids Res 34:4866–4877

Zenk MH (1996) Heavy metal detoxification in higher plants-a review. Gene 179:21–30

Chapter 2
Microbes: Nature's Cell Factories of Nanoparticles Synthesis

Tabeer Khan, Sidra Abbas, Anila Fariq, and Azra Yasmin

2.1 Introduction

Modern science and technology is working on the fabrication of new yet improved nanodevices and systems possible for a variety of industrial, consumer, and biomedical applications. Theoretical application of nanotechnology was first envisioned by a Physicist Richard Feynman in 1959. Recent developments in nanotechnology has led to the synthesis of nanoparticles from biological sources. The prefix "nano" is a Greek word which means "dwarf" very small or miniature size. These nanoparticles were considered as building blocks for the development of optoelectronic, electronics, various chemical and biochemical sensors (Narayanan and Sakthivel 2010; Faraz et al. 2018). Nanoparticles are defined as the small solid particles of one dimension with the average size of 10–1000 nm. Few examples of nanosized biological materials include DNA (2.5nm) and viruses (100nm) (Wagner et al. 1992; Lyubchenko and Shlyakhtenko 1997).

Nanoparticles possesses unique wavelength that is below the wavelength of light, which makes them appear transparent. Thus metal nanoparticles can also be easily attached to the single strand of DNA. Due to these unique properties, they have gained enormous importance to be used in diverse areas such as electronics, cosmetics, coatings, packaging, and biotechnology by controlling and modifying their size shape and physicochemical behavior. Basically, there are two approaches to synthesize nanoparticles: bottom up approach and top down approach. In bottom up approach, basic building blocks were synthesized and assembled into the final nanostructure by chemical and biological procedures. Top down approach starting larger material is reduced into smaller one by physical and chemical means. Major

T. Khan · S. Abbas · A. Fariq · A. Yasmin (✉)
Microbiology and Biotechnology Research Lab, Department of Environmental Sciences, Fatima Jinnah Women University, Rawalpindi, Pakistan
e-mail: azrayasmin@fjwu.edu.pk

© Springer Nature Switzerland AG 2018
R. Prasad et al. (eds.), *Exploring the Realms of Nature for Nanosynthesis*, Nanotechnology in the Life Sciences, https://doi.org/10.1007/978-3-319-99570-0_2

drawback of both the approaches is that they require huge amount of energy, toxic chemicals, and formation of hazardous by-products (Thakkar et al. 2009; Prasad et al. 2016). Also physically and chemically synthesized nanoparticles are less biocompatible and due to their toxic nature their use is limited. So there is need to develop ecofriendly and economical method for the synthesis of nanoparticles (Hosseini and Sarvi 2015).

Recently, biological synthesis emerged as an alternative to conventional methods of nanoparticle synthesis. It involves the use of microorganisms and plants for synthesis. Biological entities present on this planet have been in regular touch with its surroundings. Microbes are found to be the small living entities and microbial mediated synthesis of nanoparticles established a new field of research nano-biotechnology. With the passage of time organisms evolve certain mechanisms to alter the chemical nature of toxic metal to survive extreme metal concentrations in environment. So, one can say that nanoparticles are the by-products of reduction mechanisms by organisms (Durán et al. 2007). Microorganisms produce different types of enzymes that are involved in the reduction of toxic and hazardous metals and convert them into nanoparticles. Nanoparticles synthesis through biological entities offers a clean, nontoxic and environmental friendly way with wide range of size, shape, composition, and physicochemical properties (Sastry et al. 2005; Mohanpuria et al. 2008; Prasad et al. 2016). Extensive studies have been conducted on metal bioremediation and metal–microbe interactions but new dimensions in microbial nano-biotechnology research are yet to be explored (Narayanan and Sakthivel 2010).

2.2   Traditional Sources of Nanoparticles Synthesis

Most common mode of nanoparticles synthesis includes physical and chemical methods. Both processes follow two approaches for synthesis of nanomaterial which are top down and bottom up. In bottom up approach, atom and molecular constituents are made up of single molecule. These molecules or atoms are attached via covalent bonding making them far stronger. A lot of information could be stored in devices made from bottom up approach. Chemical and physical techniques that use bottom up approach are AFM, liquid phase techniques based on inverse micelles, sol-gel processing, chemical vapor deposition (CVD), laser pyrolysis, etc. *Top down* method includes synthesis of smaller and smaller particles by using bulk materials. This approach includes cutting, carving, and molding of materials but there is still need to develop highly advanced nanodevices through this process. Methods that follow top down for nanomaterial synthesis include laser ablation, milling, nano-lithography, hydrothermal technique, physical vapor deposition, electrochemical method, etc. (Nikalje 2015; Rajput 2015).

Depending on target material to be fabricated, almost every element of periodic table is utilized for nanomaterial synthesis. These nanomaterials have applications from nanoelectronics to nanomedicines. Nanotechnology revolutionized material manufacturing of desired shape and design without use of machining, metals, polymers, and ceramics. Furthermore it also provides benefits to chemical catalysis from fuel cell to catalytic converters and photocatalytic devices due to their large surface to volume ratio. Some of the examples of physical and chemical methods of nanomaterial synthesis are;

2.2.1 Gas Condensation

To synthesize nanocrystalline metals and alloys gas condensation was used. General process involves vaporization of inorganic or metallic material by thermal evaporation. In this process, ultrafine particle of 100 nm of size is synthesized by gas phase collision (Tissue and Yuan 2003). This method had some limitations such as maintenance of temperature ranges, source-precursor incompatibility, and dissimilar evaporation rates in an alloy. Besides thermal evaporation, alternative techniques such as sputtering electron beam heating and plasma methods were used for production of ultrafine nanoparticles. Variety of clusters of Ag, Fe, and Si were produced via sputtering (Hasany et al. 2012).

2.2.2 Vacuum Evaporation

In vacuum evaporation or deposition and vaporization, compounds to be fabricated are vaporized and deposited in a vacuum. Thermal processes are involved in the vaporization of material at a pressure of 0.1 Pa in vacuum level of 10–0.1 MPa. During vapor phase nucleation atoms are passed through gas where they are collided and fine particles are synthesized of size range from 1 to 100 nm. This method is economical and deposition rate is also high (Winterer et al. 2003).

2.2.3 Chemical Vapor Deposition and Chemical Vapor Condensation

In chemical vapor deposition (CVD) method, chemical deposition vapor or gas phase is converted into solid which in turn deposited onto the heated surface via a chemical reaction. Temperature above 900 °C is required for activation of thermal CVD reaction. Nanocomposite powders had been prepared by CVD. CVC (Chemical Vapor condensation) reactions were proceeded by activation energy generated by several methods. Particles of ZrO_2, Y_2O_3 and nanowhiskers had been produced by CVC method (Konrad et al. 2001; Winterer et al. 2003).

2.2.4   Mechanical Attrition

Mechanical attrition involves synthesis of nanomaterial by structural decomposition of coarser grained structures. Ceramic nanocomposite and magnesium oxide had been fabricated through this method. Mechanical alloying method was advantageous as it can be carried out on room temperature. This process can also be performed on high energy mills, centrifugal type mills, vibratory and low energy type mills (Sharma et al. 2009).

2.2.5   Chemical Precipitation

Formation of nanomaterial by chemical precipitation involved reaction between material and suitable solvent. Size of nanomaterial to be synthesized may be controlled by precipitation, by avoiding the physical changes, and by aggregation of crystals. Surfactants are added to avoid clustering of particles (Konrad et al. 2001; Sharma et al. 2009).

2.2.6   Sol-Gel Methods

Among all the other methods, sol-gel methods are extensively used recently. Precursors for synthesizing colloids are ions of metal alkoxides and alkoxysilanes. Furthermore these colloids are synthesized by a network in continuous liquid phase. Sol-gel process is carried out in four steps which includes hydrolysis, condensation, and growth of particles and agglomeration of particles (Lu and Jagannathan 2002; Morita et al. 2004).

All these methods are robust but not economical, require a lot of energy and a large amount of toxic by-products are generated. To reduce these risks associated with the physical and chemical method of synthesis, focus on less toxic methods is increasing.

2.3   Green Routes of Nanoparticles Biosynthesis

Besides conventional methods of nanoparticles synthesis, increasing amount of research is being performed on the green synthesis of nanoparticles. Green synthesis of nanoparticles utilizes plants and plants extracts which have advantage over other methods as they are nontoxic and possess natural capping or stabilizing agents. Moreover green synthesized nanoparticles have diverse nature, more stability and process is one step procedure. General phenomenon of nanoparticles biosynthesis include bioreduction of metals upon interaction of plant extracts with metals. Plants metabolites like sugars, terpenoids, polyphenols, alkaloids, phenolic acids, and proteins play an important role in metal ions reduction into nanoparticles (Prasad 2014).

Lot of researches had been conducted on green synthesis of silver and gold nanoparticles. Silver and gold nanoparticles synthesis was carried out by utilizing *Geranium* extract, *Aloe vera* plant extracts, and sundried *Cinnamomum camphora* (Shankar et al. 2004a, b; Chandran et al. 2006; Huang et al. 2007). A homogeneous solution of AgNPs was obtained simply by reacting the *Jatropha curcas* extract with silver nitrate solution. AgNPs obtained were of 10–20 nm in size (Bar et al. 2009). Another inexpensive synthesis of bimetallic Ag and Au nanoparticles had been synthesized with the aid of *Plumeria rubra* plant latex (Patil et al. 2012). As discussed earlier that plant metabolites act as reducing agents in the reduction of metal ions into nanoparticles. Study reported by Christensen et al. (2011) explained the presence of natural reducing agents eugenol and carbazoles in the leaf extracts of *Murraya koenigii* (Christensen et al. 2011). Silver nanoparticles also possess effective antimicrobial property against *Escherichia coli* and *Vibrio cholera* at a concentration of 10 µg/mL (Krishnaraj et al. 2010). Besides silver nanoparticles plant also synthesizes gold nanoparticles. Neem plant, *Azadirachta indica* was utilized for the synthesis of silver, gold, and bimetallic AuAgNPs. Terpenoids and flavanone components of plant seemed to act as stabilizing agents of nanoparticles (Shankar et al. 2004a, b). Similarly leaves of *Mirabilis jalapa* were exploited for the synthesis of gold (Au) nanoparticles (Vankar and Bajpai 2010). Only drawback of plant-based synthesis over microbial synthesis is huge biomass with prolonged synthesis time. Otherwise, it would be a cheap and green synthesis of nanoparticles.

2.4 Mechanisms of Biosynthesis of Nanoparticles

Nanoparticles have attracted attention of the researchers all around the world due to their size ranging from 1 to 100 nm, unusual and captivating properties, and applications in numerous fields (Daniel and Astruc 2004; Kato 2011). There are various methods such as physical, chemical, and biological available for the synthesis of nanoparticles (Liu et al. 2011; Luechinger et al. 2010; Tiwari et al. 2008; Mohanpuria et al. 2008). Among these, physical and chemical methods have gained popularity due to bulk production of nanoparticles in a relatively short period of time. However, due to high costs, use of toxic chemicals, and formation of toxic by-products during their production have limited their use in clinical applications (Bhattacharya and Mukherjee 2008).

Use of enzymes for nanoparticles has reduced the cost of the process as expensive chemicals are no longer required. This method is more acceptable as it is more environmental friendly and not as energy intensive as chemical method (Simkiss and Wilbur 1989; Bhattacharya and Mukherjee 2008). Biological method reduces the detrimental effects produced as a result of physical and chemical methods.

Biosynthesis of nanoparticles has been exploited throughout the world because of its low cost, high efficiency, and nontoxic by-products. Microbes like bacteria, algae, actinomycetes, yeast, and fungi have gained attention of the researchers due to their high adaptability to toxic environment and ability to synthesize nanoparticles at mild temperature, pH, and pressure (Li et al. 2011; Prasad et al. 2016).

2.4.1 Intracellular and Extracellular Biosynthesis

The exact mechanism of nanoparticles using microorganism is not yet fully comprehended because different microorganisms interact differently with the metal ions. Many organisms have the ability to produce nanoparticles from inorganic materials either intracellularly or extracellularly (Mukherjee et al. 2001a, b). However, the mechanism of intracellular or extracellular production is unique to each organism. The intracellular mechanism requires a special ion transport system and cell wall plays important role in this process. During the process there is an electrostatic interaction of positively charged metal ions with the negatively charged cell wall, thereby converting the metal ion into elemental form via various enzymes. The nanoparticles diffuse off through the cell wall (Fig. 2.1). Extracellular biosynthesis of nanoparticles is mediated via enzymes either releases in the growth medium or by enzymes in cell membrane (Fig. 2.2) (Li et al. 2011).

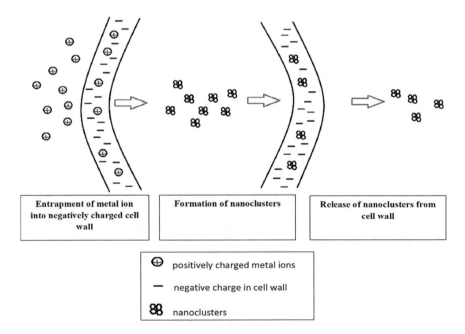

| Entrapment of metal ion into negatively charged cell wall | Formation of nanoclusters | Release of nanoclusters from cell wall |

⊕ positively charged metal ions

— negative charge in cell wall

88 nanoclusters

Fig. 2.1 Intracellular synthesis of nanoparticles

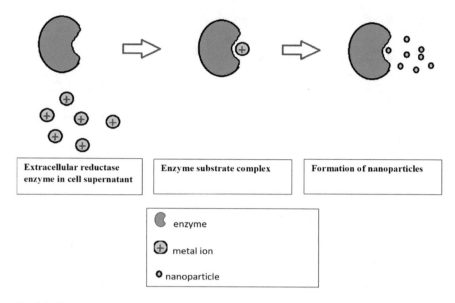

Fig. 2.2 Extracellular synthesis of nanoparticles

2.5 Microbes as Tiny Cell Factories of Nanoparticles Biosynthesis

Biological entities like microbes were exploited for nanoparticles synthesis. Among microbes, bacteria, yeast, and fungi were preferred due to their fast growth rate, easy cultivation and their ability to grow at ambient physiological conditions. A brief overview of microbe mediated synthesis of different nanoparticles has been tabulated in Table 2.1.

2.5.1 Nanoparticles Synthesis from Bacteria

Due to difference in nature/properties of colloidal and bulk solution, gold nanoparticles are synthesized due to their excessive antioxidant potential. Microbial mechanism of gold nanoparticle synthesis involves solubilization to promote oxidation of gold to create ions. To stabilize these gold ions, ligands were secreted by microorganisms to avoid complex formation (Reith et al. 2009). Alkalotolerant *Rhodococcus* sp. was used for the synthesis of monodispersed gold nanoparticles (Ahmad et al. 2003a, b). Ability of high radiation and oxidant resistant bacterium *Deinococcus radiodurans* was explored to synthesize AuNPs. Data suggested that Au(III) may be reduced to Au(I), and further to Au (0) with the capping groups to stabilize the AuNPs (Li et al. 2016).

Table 2.1 Microbial mediated synthesis of different nanoparticles

Microorganisms	Type of nanoparticle synthesized	Size (nm)	References
Bacteria			
E. coli	Cu, Ag	3–9	Ruparelia et al. (2008)
Bacillus subtilis	Cu, Ag	3–9	Ruparelia et al. (2008)
Staphylococcus aureus	Cu, Ag	3–9	Ruparelia et al. (2008)
Vibrio alginolyticus	Ag	50–100	Rajeshkumar et al. (2014a, b)
Bacillus sp.	Ag	65–70	Malarkodi et al. (2013)
Proteus mirabilis	Ag	5–45	Al-Harbi et al. (2014)
Corynebacterium glutamicum	Ag	5–50	Sneha et al. (2010)
B. amyloliquefaciens	Ag	10–100	Behera et al. (2013)
Thermomonospora	Au	50–100	Kasthuri et al. (2008)
Rhodopseudomonas capsulata	Au	10–20	He et al. (2007)
Delftia acidovorans	Au	50	Johnston et al. (2013)
Pseudomonas stutzeri	Cu	11	Varshney et al. (2010)
Pseudomonas fluorescens	Cu	49	Shantkriti and Rani (2014)
Morganella psychrotolerans	Cu	2–5	Ramanathan et al. (2011a, b)
Rhodobacter sphaeroides	CdS	4.3	Bai et al. (2009)
E. coli, Klebsiella pneumoniae	Cd	5–200	Mousavi et al. (2012)
Bacillus licheniformis	Cd	20–40	Shivashankarappa and Sanjay (2015)
Bacillu subtilis	FeO	60–80	Sundaram et al. (2012)
Magnetospirillum	Magnetite	47	Elblbesy et al. (2014)
Klebsiella oxytoca	Magnetite	Not detected	Arčon et al. (2012)
Alcaligenes faecalis	$FeSO_4$, Magnetite	43.60, 12.30	Kaul et al. (2012)
Shewanella oneidensis	Magnetite	40–50	Perez-Gonzales et al. (2010)
Clostridium sp.	Magnetite	2–10	Kim and Roh (2014)
Thermoanaerobacter sp.	Magnetite	13	Moon et al. (2010)
Lactobacillus plantarum	ZnO	7–1	Selvarajan and Mohanasrinivasan (2013)
Aeromonas hydrophila	ZnO	57.72	Jayaseelan et al. (2012)
Pseudomonas sp.	Pd	Not detected	Schlüter et al. (2014)
Bacillus megaterium	Ag, Pb, Cd	10–20	Prakash et al. (2010)
Fungus			
Phanerochaete chrysosporium	Ag	25	Vigneshwaran et al. (2006)
Fusarium oxysporum	Ag, Au	20–50	Vilchis-Nestor et al. (2008)
Mucor hiemalis	Ag	5–15	Aziz et al. (2016)

(continued)

Table 2.1 (continued)

Microorganisms	Type of nanoparticle synthesized	Size (nm)	References
Phanerochaete chrysosporium	Ag	100	Mondal et al. (2011)
Aspergillus flavus	Ag	7–10	Evanoff Jr and Chumanov (2005)
Aspergillus fumigatus	Ag	5–25	Vigneshwaran et al. (2007)
Coriolus versicolor	Ag	350–600	Merga et al. (2007)
Fusarium solani	Ag	5–35	Hu et al. (2008)
Aspergillus terreus	Ag	13.80–20.0	Velhal et al. (2016)
Aspergillus fumigatus	ZnO	1.2–6.8	Raliya and Tarafdar (2013)
Pestalotia sp.	Ag	10–40	Raheman et al. (2011)
Aspergillus oryzae	$FeCl_3$	10–24.6	Raliya and Tarafdar (2013)
Aspergillus terreus	Mg	48–98	Raliya and Tarafdar (2013)
Yeast			
Rhizopus oryzae	Au	10	Das et al. (2009)
Candida glabrata	CdS	–	Kowshik et al. (2002)
Yarrowia lipolytica	Au	–	Pimprikar et al. (2009)

Besides gold nanoparticles, different bacterial species are known to produce silver nanoparticles. Saifuddin et al. (2009) reported synthesis of silver nanoparticles from *Bacillus subtilis* with a size of about 5–50 nm. Extracellular synthesis of silver nanoparticles was conducted with five psychrophilic bacteria *Phaeocystis antarctica, Pseudomonas proteolytica, Pseudomonas meridiana, Arthrobacter kerguelensis,* and *Arthrobacter gangotriensis* and two mesophilic bacteria *Bacillus indicus* and *Bacillus cecembensis* (Shivaji et al. 2011). Physiological factors like pH, temperature, and salt concentration play important role in nanoparticle synthesis. *Bacillus* sp. was reported to synthesize silver nanoparticle with 1 mM silver nitrate, room temperature within 24 h (Das et al. 2014). Along with environmental factors extracellular enzymes were recognized in biological synthesis. Activity of nitroreductase in bioreduction of silver was studied in *Enterobacteriaceae* and *K. pneumoniae*. Overall reaction was quite fast and within few minutes of silver metal exposure to organisms; silver nanoparticles were synthesized (Mokhtari et al. 2009). Bacteria also possess ability to utilize other metal nanoparticles like *E.coli* (Singh et al. 2010), *M. psychrotolerans* and *M. morganii* (Ramanathan et al. 2011a, b), *Pseudomonas* sp. (Majumder 2012) and *Pseudomonas stutzeri* (Varshney et al. 2010), *Serratia* (Hasan et al. 2008), *Streptomyces* (Usha et al. 2010) had reported for the synthesis of copper nanoparticles. A bacterial synthesis mechanism is illustrated in Fig. 2.3, as an example.

About one-third of the consumer product market utilized metal oxide nanoparticles which possessed wide range of applications (Maynard et al. 2006). These oxide nanoparticles had small size and high density which is why they have different physical and chemical properties. As per their unique physicochemical

Fig. 2.3 Bacterial
extracellular synthesis of
gold nanoparticle

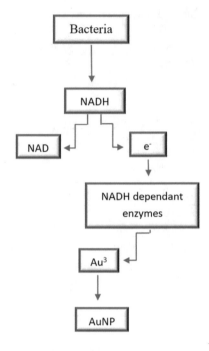

Fig. 2.3 Bacterial extracellular synthesis of gold nanoparticle

properties, oxide nanoparticles have wide applications in different industries includ-ing synthesis of biosensors, ceramics, absorbents, and/or catalysts (Garcíaa and Rodriguezb 2007). Biosynthesis utilizing bacterial species gained much interest due to low cost, easy reproducibility, and confined particle size. In biological system, extracellular enzymes possess great redox potential. The potential of producing reducing agents helps microorganisms to reduce metal ions into nanoparticles (Baker and Tatum 1998). Biosynthesis of zinc oxide nanoparticles had been reported in numerous researches as simple and low-cost procedure. As antibacterial and anti-fungal agent zinc oxide nanoparticles gained much interest in preparation of per-sonal care products. Studies reported the bacterial mediated synthesis of zinc oxide nanoparticles using *Aeromonas hydrophila* and *Acinetobacter schindleri* along with their antimicrobial activities (Jayaseelan et al. 2012; Busi et al. 2016).

Some oxide forms of metal nanoparticles are difficult to synthesize like copper oxide due to its excessive susceptibility to oxidation. So, to cope with this problem biological synthesis of copper nanoparticles gained much importance. As in case of bacterial synthesis, oxidation of copper nanoparticles was prevented by presence of thiol group which act as capping agent. Biosynthesis of copper nanoparticles offers unique optical, electrical, and antimicrobial applications (Yang et al. 2006; Hasan et al. 2008). Kirthi et al. (2011) reported *Bacillus subtilis* mediated biosynthesis of titanium oxide nanoparticles with an average size of 66–77 nm.

Apart from oxide nanoparticles, sulfide nanoparticles had great appliance in applied research due to their novel electronic and optical properties. These are used in the synthesis of quantum dot fluorescent biomarkers and as cell labeling agents, etc. (Yang and Holloway 2004). *Desulfobacteraceae* and *Rhodobacter sphaeroides* synthesized zinc sulfide nanoparticles with the average size of 2–8 nm and spherical in shape (Labrenz et al. 2000; Bai et al. 2006). Similarly, *Escherichia coli* and *Rhodopseudomonas palustris* were reported for the synthesis of cadmium sulfide nanoparticles with crystalline structure (Sweeney et al. 2004; Bai et al. 2009). Biologically synthesized nanoparticles had been stabilized by the capping of biomolecular compounds. Cadmium sulfate synthesized by the marine bacteria was capped by biomolecular compounds that were involved in the reduction of cadmium sulfate to sulfide nanoparticles. They also hold significant antimicrobial potential against pathogenic bacteria (Rajeshkumar et al. 2014a, b).

With the emerging trend in use of fluorescent biological labels, semiconductor nanocrystals had been widely used for the purpose. Biological cadmium nanocrystal was synthesized utilizing *E. coli* (Rozamond et al. 2004). Bacteria synthesize greater nanocrystals in stationary phase probably due to the presence of lipids and poly-phosphates as capping agents that stabilized the synthesized nanostructure (Blattner 1996; Rao and Kornberg 1996). *Serratia marcescens* was capable of synthesizing antimony sulfide nanoparticles. The transmission electron micrograph of the nanoparticles revealed that they were small, regular, and non-aggregated particles with size less than 35 nm (Bahrami et al. 2012). On the other hand, mercury is the highly toxic metal accumulated in the environment through anthropogenic activities. Microorganisms detoxify mercury through their metabolic processes and convert it into nontoxic form. Mercuric nanoparticles were biosynthesized using *Enterobacter* sp. Uniform sized intracellular mercury nanoparticles were synthesized at alkaline pH and with lower concentration of mercury (Sinha and Khare 2011). Another bacterium *Bacillus cereus* MRS-1 isolated from industrial effluents was assessed for its potential to produce mercury sulfide nanoparticles. Strain extracellularly synthesizes mercury sulfide nanoparticles with the size range (10–100 nm) (Sathyavanthi et al. 2013). *Bacillus megaterium* obtained from north Bihar, India, could synthesize silver, lead, and cadmium nanoparticles (Prakash et al. 2010).

There is an emerging development of luminescent quantum dots and semiconductors for biological detection and cell imaging. Bacterial species *Clostridium thermoaceticum* had the ability to precipitate $CdCl_2$ in to CdS in the presence of cysteine hydrochloride at the surface of cell as well as in the medium (Cunningham and Lundie 1993). Another study reported use of *E. coli* for synthesis of CdS nanocrystals using cadmium chloride and cadmium sulfide with a size distribution of 2–5 nm. Nanocrystal formation not only depends on genetic conditions but also on physiological conditions as evidenced by *E. coli*. CdS nanocrystal formation increases 20-fold when in stationary phase as compared to late logarithmic phase (Sweeney et al. 2004). In the biofilm overriding *Desulfobacteraceae* family, spherical ZnS (Sphalerite) particles (2–5 nm) were synthesized (Labrenz et al. 2000).

2.5.2 Nanoparticle Synthesis from Fungus

Compared to bacteria, fungi have an advantage in large-scale production as it produces large amounts of enzymes involved in silver nanoparticle synthesis (Prasad et al. 2016; Aziz et al. 2016). Moreover, fungal culture is simpler to grow both in the laboratory and at industrial scale. Many fungal species were exploited for the synthesis of silver nanoparticles. Biosynthesis of silver nanoparticle from filamentous fungi *Verticillium* sp. and different species of *Fusarium* was utilized (Mukherjee et al. 2001a, b; Ahmad et al. 2003a, b; Durán et al. 2005). Most of the fungal species produce naphthoquinones and anthraquinones, which act as reducing agents (Siddiqi and Husen 2016). Also enzymes like nitroreductases were involved in the bioreduction of metals. Nitrate reductase is essential for ferric ion reduction to iron nanoparticles (Durán et al. 2005). To identify the possible mechanism involved in nanoparticle synthesis via fungus, different enzymes and metabolites were analyzed. In this regard α-NADPH-dependent nitrate reductase produced by *Rhizopus stolonifer*, as well as phytochelatin, was utilized to form silver NPs successfully (Binupriya et al. 2010). Another report also suggested the possible mechanism of bioreduction from *F. oxysporum* of H2PtCl6 and PtCl2 into platinum NPs by means of a filtered hydrogenase enzyme (Govender et al. 2009).

Vast majority of researches were performed with silver and gold nanoparticles but a little is known about the biological synthesis of platinum NPs. Intracellular or extracellular mode of synthesis and effect of environmental conditions on synthesis of platinum nanoparticles were explored using fungal species *Neurospora crassa* and *F. oxysporum*. *N. crassa* utilized intracellular pathway while *F. oxysporum* follows extracellular pathway of synthesis (Riddin et al. 2006; Castro-Longoria et al. 2012). Similarly *F. oxysporum* was exploited for the synthesis of different sulfide nanoparticles. Studies showed that fungus could synthesize cadmium sulfide (CdS), lead sulfide (PbS), zinc sulfide (ZnS), and molybdenum sulfide (MoS) nanoparticles, when the appropriate salt is added to the growth medium (Ahmad et al. 2002).

Potential application of nanoparticles synthesized from fungal species was limited due to their pathogenicity towards humans. Antimicrobial activity of fungal mediated silver NPs was evaluated against certain bacterial and fungal pathogens (Jaidev and Narasimha 2010; Musarrat et al. 2010). *Rhizopus oryzae* mediated synthesized nano gold bio-conjugate was utilized as antimicrobial agent against pathogenic bacteria such as *P. aeruginosa*, *E. coli*, *B. subtilis*, *S. aureus*, *Salmonella* sp. and the yeasts *S. cerevisiae* and *C. albicans* (Das et al. 2009).

2.5.3 Nanoparticles Synthesis from Yeast

Various studies had revealed the ability of yeast to accumulate significant amount of heavy metals. This characteristic was further explored by the researchers for production of nanoparticles either by intracellular or extracellular enzymes. Due to ease of handling and mass production at laboratory scale, yeast had more advantage over bacterial mediated synthesis of nanoparticles. Metallic nanoparticles were synthesized by investigating different yeast strains. Silver nanoparticles were synthesized using yeast strain which possess good antimicrobial activity (Kumar et al. 2011). Silver-tolerant yeast strain MKY3 was utilized for extracellular synthesis of silver nanoparticles. Nanoparticles obtained were hexagonal in shape with 2–5 nm in size (Kowshik et al. 2003). *Yarrowia lipolytica* mediated synthesis of gold nanoparticles showed that nanoparticles synthesized were of varying shape. Varying salt concentrations are used to effect the size of NPs (Pimprikar et al. 2009). Similarly *Schizosaccharomyces pombe* was exploited to study the synthesis of CdS NPs and its route intracellular or extracellular pathway of synthesis (Dameron et al. 1989; Kowshik et al. 2002; and Krumov et al. 2007). Along with yeast *Hansenula anomala* amine-terminated polyamidoamine dendrimer as stabilizer was added to aid the synthesis of gold nanoparticles (Kumar et al. 2008). Extracellular synthesis of silver and gold nanoparticles was performed by using *Saccharomyces cerevisiae*. Few examples of nanosized biological materials include DNA (2.5nm) and viruses (100nm) (Lim et al. 2011). Similar finding was obtained when *Candida guilliermondii* was used for synthesis of silver and gold NPs. These nanoparticles also possess significant antimicrobial activity against bacterial pathogen *S. aureus* (Mishra et al. 2011).

2.5.4   Nanoparticles Synthesis by Algae

Green alga *Chlorella vulgaris* was successfully investigated for the production of gold nanoparticles. Gold was bound to algae upon addition of dried algal cell suspension to hydro-tetrachloro-aurate supplemented solution. The rate and amount of gold bound to algal cell was increased with time (Hosea et al. 1986). Three cyanobacteria *P. valderianum*, *P. tenue*, *M. chthonoplastes* and four green algae *R. fontinale*, *U. intestinalis*, *C. zeylanica*, *P. oedogoniana* were screened to their gold nanoparticle production after exposing to hydrogen tetrachloro-aurate supplemented medium. It was observed that all three cyanobacteria were able to produce intracellular gold nanoparticles. Whereas, among four green algae, *R. fontinale* and *U. intestinalis* were able to produce intracellular gold nanoparticles. X-ray diffractometry showed that gold metal ions were successfully reduced. Sizes and shapes of nanoparticles were observed using transmission electron microscopy, and UV-Vis spectrometry. At pH 9 and 7 *P. valderianum* produced hexagonal, triangular, and spherical gold nanoparticles. At acidic pH 5 gold nanorods and spheres were produced by *P. valderianum*, so selection of algae is important as not all species have the ability to produce nanoparticles (Parial et al. 2012). Filamentous cyanobacterium *Plectonema boryanum* was successfully used for biosynthesis of Ag nanoparticles in solution supplemented with silver nitrate at 25 °C (Lengke et al. 2007). In another study, rapid extracellular production of gold nanoparticles was achieved by marine *Sargassum wightii* (Singaravelu et al. 2007). Aziz et al. (2014, 2015) also revealed that Ag nanoparticles synthesized from *Scenedesmus abundans* and *Chlorella pyrenoidosa* could be used as antimicrobial agents.

2.5.5 Nanoparticle Synthesis by Actinomycetes

Like other microorganisms, actinomycetes and yeast are also well known for synthesis of nanoparticles. However, secondary metabolites resulting from actinomycetes are comparatively less explored for synthesis of metal nanoparticles. There are several studies indicating actinomycetes as suitable candidates for intra- as well as extracellular biosynthesis of metal nanoparticles. Actinomycetes produced stable nanoparticles effective against many pathogens (Golinska et al. 2014). Silver nanoparticles were produced intracellularly by actinomycetes as a result of electrostatic bonding between silver ions to the negatively charged carboxylic group of enzymes located on the mycelia wall. Enzymes present in the cell wall reduce the silver ion into silver nuclei. These silver nuclei accumulate together into silver nanoparticles. Intracellular synthesis occurs on the surface of the mycelia due to the electrostatic binding of Ag^+ ions to the negatively charged carboxylate groups in the enzyme present on the cell wall of mycelia. The Ag^+ ions are then reduced by the enzymes in the cell wall forming silver nuclei. The accumulation of the silver nuclei leads to formation of nanoscale silver particles (Sunitha et al. 2013). In another study, *Streptomyces* species was used to synthesize metal oxide nanoparticles extracellularly. Zinc nitrate was readily reduced to nanoparticles when exposed to *Streptomyces* species. This reduction was most likely due to reductase enzyme.

2.6 Advantages of Microbes over Other Routes for Synthesis of Nanoparticles

Besides conventional methods nanoparticles synthesis by using microorganisms gained much interest due to their small size, safe to use, ecofriendly, and economical. Biological methods of nanoparticle synthesis follow single step process, thus eliminating the ruthless physiological conditions. Most of the microbes use their physiological condition, like pH, temperature, etc. for synthesis of nanoparticles. Due to incredible properties, microbial synthesized nanoparticles have turned into noteworthy in many fields in the recent years, such as energy, health care, environment, and agriculture.

2.7 Applications

2.7.1 Microbial Nanoparticles in Agriculture

Phytopathogens are one of the major causes of crop diseases. Biologically synthesized silver nanoparticles act as strong fungicidal against these phytopathogens and control spread of plant diseases (Abd-Elsalam and Prasad 2018). Fungicidal activity of biologically synthesized silver nanoparticles was assessed in various researches which could be beneficial in protecting plants against these pathogens (Jaidev and Narasimha 2010; Mala et al. 2012; Gopinath and Velusamy 2013; Mishra et al. 2014). Biologically synthesized silver and gold nanoparticles had promising applications in life sciences but very little research had been conducted on their applications in agriculture. Mishra and coworkers synthesized silver nanoparticles using *Stenotrophomonas* sp. with the size of 12 nm. Antifungal effect and inhibitory effect of AgNPs on condial germination were observed. Furthermore reduced sclerotia germination, phenolic acids induction, altered lignification, and H_2O_2 production were observed which could be the possible mechanisms providing protection to chickpea against *Sclerotium rolfsii*. So these biologically synthesized silver nanoparticles could possibly be used in crop protection (Mishra et al. 2017). Another study reported to check the effect of *Aspergillus fumigatus* mediated ZnO NPs cluster bean (*Cyamopsis tetragonoloba* L.) germination and phosphorous mobilizing enzymes. After application of ZnO NPs, improvement in cluster bean biomass, shoot length, root length, root area, chlorophyll content, total soluble leaf protein, rhizospheric microbial population, acid phosphatase, alkaline phosphatase, and phytase activity was observed (Raliya and Tarafdar 2013). These findings suggest that nanoparticles synthesized from microorganisms could potentially be used in agriculture sector to protect plant against phytopathogens and to enhance the germination of plants (Prasad et al. 2014, 2017a). Further research in this field is required to strengthen the findings.

2.7.2 Microbial Nanoparticles in Environment

Large amount of industrial and agricultural waste is discarded in environment each year which is toxic for human and animal consumption/health (Sangeetha et al. 2017a, b, c). Bio-nanoparticles can also be utilized for environmental remediation. For degradation of organic compounds, microbial based nanoparticles have successfully been utilized. Bacterium *Clostridium* was reported to successfully reduce palladium, Pd (II) ions to form metallic Pd nanoparticles. They efficiently degrade organic azo dyes, methyl orange, and Evans blue. This method could be used as alternative to conventional waste water and ground water treatment (Johnson et al. 2013). In conventional methods for synthesis of nanomaterials like quantum dots, a lot of energy, elevated temperatures and pressure is required while biological synthesis has more advantage over conventional methods. Highly luminescent quantum dots of CdS could be synthesized at room temperature from fungus *Fusarium oxysporum* (Kumar et al. 2007). Quantum dots synthesized by conventional methods were used for the detection of pathogenic bacteria in the environment (Zhu et al. 2004). These quantum dots had applications in biomedical research, while their applications in environmental research are emerging. Mono dispersed and different sizes of nanoparticles could be synthesized by fungal species *Fusarium oxysporum, Aspergillus fumigatus*, and *F. semitectum* (Dhillon et al. 2012; Shedbalkar et al. 2014).

2.7.3 Microbial Nanoparticles in Biomedicine

(a) Antimicrobial Activity

Concerns over antibiotics resistance focused the researcher to identify alternate methods. Promising alternative is use of silver which is known for its antimicrobial activities (Goodman et al. 2013). Silver nanoparticles can change the functional properties such as solubility and surface adhesion property of bacteria more difficult (Campoccia et al. 2013, Romanò et al. 2013). The synthesis and antimicrobial activity of cadmium sulfide and zinc sulfide nanoparticles against oral pathogens was tested. Size of CdS is from 10 to 25 and that of ZnS was 65 nm. During formation of sulfide nanoparticles, proteins which contain amine groups played their role in reduction process. The antimicrobial activity assessed against oral pathogens such as *Streptococcus* sp., *Staphylococcus* sp., *Lactobacillus* sp., and *Candida albicans*, confirmed microbicidal activity of sulfide nanoparticles (Malarkodi et al. 2014). Fungus *Trichoderma viride* was used to produce silver nanoparticles. Silver metal ions were reduced to stable nanoparticles ranging in size from 5 to 40 nm. Antimicrobial activity of these nanoparticles along with antibiotics was assessed against variety of Gram negative and Gram positive bacteria. It was observed that silver nanoparticles increased the efficiency of ampicillin, kanamycin, erythromycin, and chloramphenicol. It could be concluded that antibiotics along with nanoparticles have better antimicrobial activity (Fayaz et al. 2010; Aziz et al. 2016). Textile fabric fused with silver nanoparticles from fungi *Fusarium oxysporum* can be used as anti-infection agent against pathogens such as *S. aureus* (Durán et al. 2007).

(b) Drug Delivery System

Due to small size, nanoparticles have shown to be an excellent candidate for drug delivery to the target as they can easily surpass the blood–brain barriers and epithelial joints of the skin. Moreover, their increase surface to volume ratio showed better distribution, and as a result of their high surface area to volume ratio, nanocarriers have shown improved distribution of the particular drug exhibiting maximum therapeutically impact (Häfeli et al. 2009; Prasad et al. 2017b).

In a study biocompatibility of magnetic nanoparticles Fe_3O_4 and Fe_2O_3 from *Magnetospirillum gryphiswaldense* was explored. It was concluded that magnetosomes can be used as carrier for novel drug or gene at the cancer tissue sites in hyperthermia cancer therapy, medical imaging such as MRI, and investigation of DNA (Xiang et al. 2007).

Felfoul and fellow researchers studied drug delivery to the target site using magnetotactic bacteria (MTB) with magnetosomes. They used magnetotaxis to change to course of the bacteria and its flagella to direct it in the blood vessels was also used as drug delivery agent (Felfoul et al. 2007). Another group of researchers reported utilization of magnetotactic bacteria nanoparticles (MTB-NPs) for delivering gene to the target site. PEI-associated MTB-NPs were successfully employed to deliver β-galactosidase plasmids both in vitro and in vivo. They managed to use PEI-associated MTB-NPs to deliver β-galactosidase plasmids, at both in vitro and in vivo levels (Xie et al. 2009). Drug and gene delivery potential of gold nanoparticles over conventional carriers has been explored. Study suggested that gold nanoparticles can be used as drug or gene carriers (Giljohann et al. 2010).

(c) As Anticancer Agents

Cancer is one of the leading causes of death per year all over the world. Conventional methods of treatment of cancer involve surgery followed by chemotherapy/radiotherapy, which had a lot of side effects. Early diagnosis and targeted drug delivery to eliminate the risk of other organ failure is still not well developed. So, nanotechnology offers better alternative in cancer treatment with the development of nanomedicine (Fariq et al. 2017). Silver nanoparticles were extensively studied as potential antimicrobial agents and disinfectants. Recently silver nanoparticles have been used as anticancerous agents. Gurunathan et al. (2013) reported that AgNPs inhibited the growth of human breast cancer MDA-MB-231 cells through activation of lactate dehydrogenase, caspase-3, generation of reactive oxygen species that lead to induction of apoptosis (Gurunathan et al. 2013). MCF-7 breast cancer cells were treated against AgNPs and their cytotoxicity confirmed the possible pathway of induction of apoptosis. These silver nanoparticles inhibit Bcl-2 and activate Bax which further induce cytochrome C release from mitochondria and in turn induce apoptosis (Kulandaivelu and Gothandam 2016). Beside silver nanoparticles, other metal nanoparticles also possess anticancerous activity such as selenium. Selenium nanoparticles from *Streptomyces bikiniensis* had anticancer property against MCF-7 and Hep-G2 human cancer cells. They induce cell death by mobilization of chromatin-bound copper which in turn generate oxidized species resulted in apoptosis of cells (Ahmad et al. 2015). Further study on immune response and toxicity of biologically synthesized nanoparticles prior to commercialization and their application in cancer diagnosis and treatment is desired.

2.7.4   Microbial Nanoparticles in Biotechnology as Biosensors

Nanoparticles having unique optical and electrochemical feature are best suited for use in biosensor devices. Nanoparticles have appealing optical and electronic features and can be utilized in biosensor applications (Faraz et al. 2018). Zheng and coworkers used gold-silver alloy nanoparticles produced extracellularly by yeast in construction of vanillin biosensors. It was concluded that electrochemical response increased many times in Au-Ag alloy based vanillin biosensors. This modified vanillin sensors were efficacious in determination of vanillin from variety of vanilla samples (Zheng et al. 2010).

Bacillus subtilis was used to produce selenium nanoparticles. Nanoparticles produced were of spherical shape with size ranging from 50 to 400 nm. These selenium nanoparticles after 24 h at room temperature were transformed to one-dimensional structure. It was suggested that selenium nanoparticles have the potential to be used as enhancing and settled material in production of horseradish peroxide biosensor. Prasad and coworkers successfully used microbial selenium nanoparticles from *Bacillus pumilus* sp. for the synthesis of H_2O_2 biosensor (Wang et al. 2010; Prasad et al. 2015).

2.8   Challenges and Future Prospects

Biosynthesis and potential applications of microorganism-based nanoparticles have been focus of researchers for past several years with incredible developments in this field. However, need of the hour is to improve the efficiency of synthesis as nanoparticles synthesized from microorganisms require long incubation/contact time from several hours to sometimes days compared to physical and chemical processes. Particle size, its morphology and mono-dispersity are vital in nanoparticle synthesis and also important challenges, frequently encountered by researchers which require further investigation. It was observed in several studies that nanoparticles formed from microorganisms undergo decomposition or lose stability after sometime. There is a need to focus on stability of the nanomaterials produced biologically. Focus to optimize conditions for large-scale production of nanoparticles at industrial level is also necessary. Genomic and proteomic analysis is another focus point, which will provide the insight about the possible genetic pathways involved in the reduction process. By altering the genetic and proteomic pathways, time span and amount of synthesis may be enhanced.

References

Abd-Elsalam K, Prasad R (2018) Nanobiotechnology applications in plant protection. Springer, Berlin https://www.springer.com/us/book/9783319911601

Ahmad A, Mukherjee P, Mandal D, Senapati S, Khan MI (2002) Enzyme mediated extracellular synthesis of CdS nanoparticles by the fungus, *Fusarium oxysporum*. J Am Chem Soc 124:12108–12109

Ahmad A, Mukherjee P, Senapati S, Mandal D, Khan MI, Kumar R, Sastry M (2003a) Extracellular biosynthesis of silver nanoparticles using the fungus *Fusarium oxysporum*. Colloids Surf B Biointerfaces 28:313–318

Ahmad A, Senapati S, Khan MI, Kumar R, Ramani R, Sirinivas V, Sastry M (2003b) Intracellular synthesis of gold nanoparticles by a novel alkalotolerant actinomycete, Rhodococcus species. Nanotechnology 14:824–828

Ahmad MS, Yasser MM, Sholkamy EN, Ali AM, Mehanni MM (2015) Anticancer activity of biostabilized selenium nanorods synthesized by *Streptomyces bikiniensis* strain. Int J Nanomed 10:3389. https://doi.org/10.2147/IJN.S82707

Al-Harbi MS, El-Deeb BA, Mostafa N, Amer SAM (2014) Extracellular biosynthesis of AgNPs by the bacterium *Proteus mirabilis* and its toxic effect on some aspects of animal physiology. Adv Nanoparticles 3:83–91

Arčon I, Piccolo O, Pagnelli S, Baldi F (2012) XAS analysis of a nanostructured iron polysaccharide produced anaerobically by a strain of *Klebsiella oxytoca*. Biometals. https://doi.org/10.1007/s10534-012-9554-6

Aziz N, Fatma T, Varma A, Prasad R (2014) Biogenic synthesis of silver nanoparticles using *Scenedesmus abundans* and evaluation of their antibacterial activity. J Nanoparticles Article ID 689419. https://doi.org/10.1155/2014/689419

Aziz N, Faraz M, Pandey R, Sakir M, Fatma T, Varma A, Barman I, Prasad R (2015) Facile algae-derived route to biogenic silver nanoparticles: synthesis, antibacterial and photocatalytic properties. Langmuir 31:11605–11612. https://doi.org/10.1021/acs.langmuir.5b03081

Aziz N, Pandey R, Barman I, Prasad R (2016) Leveraging the attributes of *Mucor hiemalis*-derived silver nanoparticles for a synergistic broad-spectrum antimicrobial platform. Front Microbiol 7:1984. https://doi.org/10.3389/fmicb.2016.01984

Bahrami K, Nazari P, Sepehrizadeh Z, Zarea B, Shahverdi AR (2012) Microbial synthesis of antimony sulfide nanoparticles and their characterization. Ann Microbiol 62:1419–1425

Bai H, Zhang Z, Gong J (2006) Biological synthesis of semiconductor zinc sulfide nanoparticles by immobilized *Rhodobacter sphaeroides*. Biotechnol Lett 28(14):1135–1139

Bai H, Zhang Z, Guo Y, Jia W (2009) Biological synthesis of size-controlled cadmium sulfide nanoparticles using immobilized *Rhodobacter sphaeroides*. Nanoscale Res Lett 4:717–723. https://doi.org/10.1007/s11671-009-9303-0

Baker RA, Tatum JH (1998) Novel anthraquinones from stationary cultures of *Fusarium oxysporum*. J Ferment Bioeng 85:359–361

Bar H, Bhui DK, Sahoo GP, Sarkar P, De Sankar P et al (2009) Green synthesis of silver nanoparticles using latex of *Jatropha curcas*. Colloids Surf A Physicochem Eng Aspects 339:134–139

Behera SS, Jha S, Arakha M, Panigrahi TK (2013) Synthesis of silver nanoparticles from microbial source-a green synthesis approach, and evaluation of its antimicrobial activity against *Escherichia coli*. Int J Eng Res Appl 3(2):58–62

Bhattacharya R, Mukherjee P (2008) Biological properties of "naked" metal nanoparticles. Adv Drug Deliv Rev 60(11):1289–1306

Binupriya AR, Sathishkumar M, Yun SI (2010) Biocrystallization of silver and gold ions by inactive cell filtrate of *Rhizopus stolonifer*. Colloids Surf B Biointerfaces 79:531–534

Blattner FR (1996) *E. coli* genome project. http://www.genome.wisc.edu/

Busi S, Rajkumari J, Pattnaik S, Parasuraman P, Hnamte S (2016) Extracellular synthesis of zinc oxide nanoparticles using *Acinetobacter schindleri* SIZ7 and its antimicrobial property against foodborne pathogens. J Microbiol Biotechnol Food Sci 5(5):407–411

Campoccia D, Montanaro L, Arciola CR (2013) A review of the biomaterials technologies for infection-resistant surfaces. Biomaterials 34(34):8533–8554. https://doi.org/10.1016/j.biomaterials.2013.07.089

Castro-Longoria E, Moreno-Velázquez SD, Vilchis-Nestor AR, Arenas Berumen E, Avalos-Borja M (2012) Production of platinum nanoparticles and nano-aggregates using *Neurospora crassa*. J Microbiol Biotechnol 22:1000–1004

Chandran SP, Chaudhary M, Pasricha R, Ahmad A, Sastry M (2006) Synthesis of gold nanotriangles and silver nanoparticles using *Aloe vera* plant extract. Biotechnol Prog 22:577–583

Christensen L, Vivekanandhan S, Misra M, Mohanty A (2011) Biosynthesis of silver nanoparticles using *murraya koenigii* (curry leaf): An investigation on the effect of broth concentration in reduction mechanism and particle size. Adv Mater Lett 2:429–434

Cunningham DP, Lundie LL (1993) Precipitation of cadmium by *Clostridium thermoaceticum*. Appl Environ Microbiol 59:7–14

Dameron CT, Reese RN, Mehra RK, Kortan AR, Carroll PJ, Steigerwald ML, Brus LE, Winge DR (1989) Biosynthesis of cadmium sulphide quantum semiconductor crystallites. Nature 338:596–597

Daniel MC, Astruc D (2004) Gold nanoparticles: assembly, supramolecular chemistry, quantum-size-related properties, and applications toward biology, catalysis, and nanotechnology. Chem Rev 104(1):293–346

Das SK, Das AR, Guha AK (2009) Gold nanoparticles: microbial synthesis and application in water hygiene management. Langmuir 25:8192–8199

Das VL, Thomas R, Varghese RT, Soniya EV, Mathew J, Radhakrishnan EK (2014) Extracellular synthesis of silver nanoparticles by the *Bacillus strain CS 11* isolated from industrialized area. 3 Biotechnology 4:121–126. https://doi.org/10.1007/s13205-013-0130-8

Dhillon GS, Brar SK, Kaur S, Verma M (2012) Green approach for nanoparticle biosynthesis by fungi: current trends and applications. Crit Rev Biotechnol 32:49–73

Durán N, Marcato PD, Alves OL, de Souza GIH, Esposito E (2005) Mechanistic aspects of biosynthesis of silver nanoparticles by several *Fusarium oxysporum* strains. J Nanobiotechnol 3. https://doi.org/10.1186/1477-3155-3-8

Durán N, Marcato PD, De Souza GIH, Alves OL, Esposito E (2007) Antibacterial effect of silver nanoparticles produced by fungal process on textile fabrics and their effluent treatment. J Biomed Nanotechnol 3(2):203–208

Elblbesy MAA, Madbouly AK, Hamdan TAA (2014) Bio-synthesis of magnetite nanoparticles by bacteria. Am J Nano Res Appl 2(5):98–103. https://doi.org/10.11648/j.nano.20140205.12

Evanoff DD Jr, Chumanov G (2005) Synthesis and optical properties of silver nanoparticles and arrays. Chem Phys Chem 6:1221–1231

Faraz M, Abbasia A, Naqvia FK, Khare N, Prasad R, Barman I, Pandey R (2018) Polyindole/CdS nanocomposite based turn-on, multi-ion fluorescence sensor for detection of Cr3+, Fe3+ and Sn2+ ions. Sens Actuators B 269:195–202 https://doi.org/10.1016/j.snb.2018.04.110

Fariq A, Khan T, Yasmin A (2017) Microbial synthesis of nanoparticles and their potential applications in biomedicine. J Appl Biomed 15(2017):241–248

Fayaz AM, Balaji K, Girilal M, Yadav R, Kalaichelvan PT, Venketesan R (2010) Biogenic synthesis of silver nanoparticles and their synergistic effect with antibiotics: a study against gram-positive and gram-negative bacteria. Nanomed Nanotechnol Biol Med 6(1):e103–e109

Felfoul O, Mohammadi M, Martel S (2007) Magnetic resonance imaging of Fe$_3$O$_4$ nanoparticles embedded in living magnetotactic bacteria for potential use as carriers for in vivo applications. In: Proceedings of the 29th annual international conference of the IEEE Engineering in Medicine and Biology Society (EMBS '07), pp 1463–1466

Garcíaa MF, Rodriguezb A (2007) Handbook of Research on Nanoscience, Nanotechnology, and Advanced Materials. Brookhaven national laboratory -79479-2007-BC

Giljohann DA, Seferos DS, Daniel WL, Massich MD, Patel PC, Mirkin CA (2010) Gold nanoparticles for biology and medicine. Angew Chem Int 49:3280–3294

Golinska P, Wypij M, Ingle AP, Gupta I, Dahm H, Rai M (2014) Biogenic synthesis of metal nanoparticles from actinomycetes: biomedical applications and cytotoxicity. Appl Microbiol Biotechnol 98(19):8083–8097

Goodman SB, Yao Z, Keeney M, Yang F (2013) The future of biologic coatings for orthopaedic implants. Biomaterials 34(13):3174–3183

Gopinath V, Velusamy P (2013) Extracellular biosynthesis of silver nanoparticles using *Bacillus* sp. GP-23 and evaluation of their antifungal activity towards *Fusarium oxysporum*. Spectrochim Acta A Mol Biomol Spectrosc 106:170–174

Govender Y, Riddin T, Gericke M, Whiteley CG (2009) Bioreduction of platinum salts into nanoparticles: a mechanistic perspective. Biotechnol Lett 31:95–100

Gurunathan S, Han JW, Eppakayala V, Jeyaraj M, Kim JH (2013) Cytotoxicity of biologically synthesized silver nanoparticles in MDA-MB-231 human breast cancer cells. BioMed Res Int. https://doi.org/10.1155/2013/535796

Häfeli UO, Riffle JS, Harris-Shekhawat L, Carmichael-Baranauskas A, Mark F, Dailey JP, Bardenstein D (2009) Cell uptake and in vitro toxicity of magnetic nanoparticles suitable for drug delivery. Mol Pharm 6:1417–1428

Hasan S, Singh S, Parikh RY, Dharne MS, Patole MS, Prasad BL, Shouche YL (2008) Bacterial synthesis of copper/copper oxide nanoparticles. J Nanosci Nanotechnol 8(6):3191–3196

Hasany SF, Ahmad I, Ranjan J, Rehman A (2012) Systematic review of the preparation techniques of iron oxide magnetic nanoparticles. Nanosci Nanotechnol 2(6):148–158

He S, Guo Z, Zhang Y, Zhang S, Wang J et al (2007) Biosynthesis of gold nanoparticles using the bacteria *Rhodopseudomonas capsulata*. Mater Lett 61:3984–3987

Hosea M, Greene B, Mcpherson R, Henzl M, Dennis MDA, Darnall W (1986) Accumulation of elemental gold on the alga *Chlorella vulgaris*. Inorg Chim Acta 123(3:161–165

Hosseini MR, Sarvi MN (2015) Recent achievements in the microbial synthesis of semiconductor metal sulfide nanoparticles. Mater Sci Semicond Process 40:293–301

Hu B, Wang SB, Wang K, Zhang M, Yu SH (2008) Microwave assisted rapid facile 'green' synthesis of uniform silver nanoparticles: self-assembly into multilayered films and their optical properties. J Phys Chem 112:11169–11174

Huang J, Li Q, Sun D, Lu Y, Su Y et al (2007) Biosynthesis of silver and gold nanoparticles by novel sundried *Cinnamomum* camphora leaf. Nanotechnology 18:11–15

Jaidev LR, Narasimha G (2010) Fungal mediated biosynthesis of silver nanoparticles, characterization and antimicrobial activity. Colloids Surf B Biointerfaces 81(2):430–433

Jayaseelan C, Rahuman AA, Kirthi AV, Marimuthu S, Santhoshkumar T, Bagavan A, Gaurav K, Karthik L, Rao KVB (2012) Novel microbial route to synthesize ZnO nanoparticles using *Aeromonas hydrophila* and their activity against pathogenic bacteria and fungi. Spectrochim Acta A 90:78–84

Johnson A, Merilis G, Hasting J, Palmer EM, Fitts JP, Chidambaram D (2013) Reductive degradation of organic compounds using microbial nanotechnology. J Electrochem Soc 160:G27–G31

Johnston CW, Wyatt MA, Li X, Ibrahim A, Shuster J et al (2013) Gold biomineralization by a metallophore from a gold-associated microbe. Nat Chem Biol 9:241–243

Kasthuri J, Kathiravan K, Rajendiran N (2008) Phyllanthin-assisted biosynthesis of silver and gold nanoparticles: a novel biological approach. J Nanopart Res 11:1075–1085

Kato H (2011) In vitro assays: tracking nanoparticles inside cells. Nat Nanotechnol 6(3):139–140

Kaul RK, Kumar P, Burman U, Joshi P, Agrawal A, Raliya R, Tarafdar JC (2012) Magnesium and iron nanoparticles production using microorganisms and various salts. Mater Sci 30(3):254–258

Kim Y, Roh Y (2014) Effects of microbial growth conditions on synthesis of magnetite nanoparticles by indigenous Fe (III)-reducing bacteria. In: The 2014 World congress on advances in civil, environmental and materials research, Busan, Korea

Kirthi AV, Rahuman AA, Rajakumar G, Marimuthu S, Santhoshkumar T, Jayaseelan C, Elango G, Zahir AA, Kamaraj C, Bagavan A (2011) Biosynthesis of titanium dioxide nanoparticles using bacterium *Bacillus subtilis*. Mater Lett 65:2745–2747

Konrad A, Herr U, Tidecks R, Samwer F (2001) Luminescence of bulk and nanocrystalline cubic yttria. J Appl Phys 90(7):516–3523

Kowshik M, Deshmukh N, Vogel W, Urban J, Kulkarni SK, Paknikar KM (2002) Microbial synthesis of semiconductor CdS nanoparticles, their characterization, and their use in the fabrication of an ideal diode. Biotechnol Bioeng 78:583–588

Kowshik M, Ashtaputre S, Kulkani SK, Parknikar KMM (2003) Extracellular synthesis of silver nanoparticles by a silver-tolerant yeast strain MKY3. Nanotechnology 14:95–100

Krishnaraj C, Jagan EG, Rajasekar S, Selvakumar P, Kalaichelvan PT et al (2010) Synthesis of silver nanoparticles using *Acalypha indica* leaf extracts and its antibacterial activity against water borne pathogens. Colloids Surf B Biointerfaces 76:50–56

Krumov N, Oder S, Perner-Nochta I, Angelov A, Posten C (2007) Accumulation of CdS nanoparticles by yeasts in a fed-batch bioprocess. J Biotechnol 132:481–486

Kulandaivelu, B., & Gothandam, K. M. (2016). Cytotoxic effect on cancerous cell lines by biologically synthesized silver nanoparticles. Brazilian Archives of Biology and Technology, 59.

Kumar SA, Ayoobul AA, Absar A, Khan MI (2007) Extracellular biosynthesis of CdSe quantum dots by the fungus, *Fusarium oxysporum*. J Biomed Nanotechnol 3:190–194

Kumar SK, Peter YA, Nadeau JL (2008) Facile biosynthesis, separation and conjugation of gold nanoparticles to doxorubicin. Nanotechnology 19:495101

Kumar D, Karthik L, Kumar G, Roa KB (2011) Biosynthesis of silver nanoparticles from marine yeast and their antimicrobial activity against multidrug resistant pathogens. Pharmacol Online 3:1100–1111

Labrenz M, Druschel GK, Thomsen-Ebert T, Gilbert B, Welch SA, Kemner KM, Logan GA, Summons RE, Stasio GD, Bond PL, Lai B, Kelly SD, Banfield JF (2000) Formation of sphalerite (ZnS) deposits in natural biofilms of sulfate-reducing bacteria. Science 290:1744–1747

Lengke MF, Fleet ME, Southam G (2007) Biosynthesis of silver nanoparticles by filamentous Cyanobacteria from a silver(I) nitrate complex. Langmuir 23(5):2694–2699. https://doi.org/10.1021/la0613124

Li X, Xu H, Chen ZS, Chen G (2011) Biosynthesis of nanoparticles by microorganisms and their application. J Nanometer 8. https://doi.org/10.1155/2011/270974

Li J, Li Q, Ma X, Tian B, Li T, Yu J, Dai D, Weng Y, Hua Y (2016) Biosynthesis of gold nanoparticles by the extreme bacterium *Deinococcus radiodurans* and an evaluation of their antibacterial properties. Int J Nanomedicine 11:5931–5944

Lim HA, Mishra A, Yun SI (2011) Effect of pH on the extracellular synthesis of gold and silver nanoparticles by *Saccharomyces cerevisae*. J Nanosci Nanotechnol 11:518–522

Liu J, Qiao SZ, Hu QH, Lu GQ (2011) Magnetic nanocomposites with mesoporous structures: synthesis and applications. Small 7(4):425–443

Lu CH, Jagannathan J (2002) Cerium-ion-doped yttrium aluminum garnet nanophosphors prepared through sol-gel pyrolysis for luminescent lighting. Appl Phys Lett 80(19):3608–3610

Luechinger NA, Grass RN, Athanassiou EK, Stark WJ (2010) Bottom-up fabrication of metal/metal nanocomposites from nanoparticles of immiscible metals. Chem Mater 22(1):155–160

Lyubchenko YL, Shlyakhtenko LS (1997) Visualization of supercoiled DNA with atomic force microscopy in situ. Proc Natl Acad Sci U S A 94:496–501

Majumder BR (2012) Bioremediation: copper nanoparticles from electronic-waste. Int J Eng Sci Technol 4:4380

Mala R, Arunachalam P, Sivsankari M (2012) Synergistic bactericidal activity of silver nanoparticles and ciprofloxacin against phytopathogens. J Cell Tissue Res 12:3249–3254

Malarkodi C, Rajeshkumar S, Paulkumar K, Gnanajobitha G, Vanaja M, Annadurai G (2013) Bacterial synthesis of silver nanoparticles by using optimized biomass growth of *Bacillus* sp. Nanosci Nanotechnol Int J 3(2):26–32

Malarkodi C, Rajeshkumar S, Paulkumar K, Vanaja M, Gnanajobitha G, Annadurai G (2014) Biosynthesis and antimicrobial activity of semiconductor nanoparticles against oral pathogens. Bioinorg Chem Appl 2014. https://doi.org/10.1155/2014/347167

Maynard AD, Aitken RJ, Butz T, Colvin V, Donaldson K, Oberdörster G et al (2006) Safe handling of nanotechnology. Nature 444:267–269

Merga G, Wilson R, Geoffrey L, Milosavljevic BH, Meisel D (2007) Redox catalysis on "naked" silver nanoparticles. J Phys Chem 111:12220–12226

Mishra A, Tripathy S, Yun SI (2011) Bio-synthesis of gold and silver nanoparticles from *Candida guilliermondii* and their antimicrobial effect against pathogenic bacteria. J Nanosci Nanotechnol 1:243–248

Mishra S, Singh BR, Singh A, Keswani C, Naqvi AH, Singh HB (2014) Biofabricated silver nanoparticles act as a strong fungicide against *Bipolaris sorokiniana* causing spot blotch disease in wheat. PLoS One 9(5):e97881

Mishra S, Singh BR, Naqvi AH, Singh HB (2017) Potential of biosynthesized silver nanoparticles using *Stenotrophomonas sp*. BHU-S7 (MTCC 5978) for management of soil-borne and foliar phytopathogens. Sci. Rep. 27(7):45154. https://doi.org/10.1038/srep45154

Mohanpuria P, Rana NK, Yadav SK (2008) Biosynthesis of nanoparticles: technological concepts and future applications. J Nanopart Res 10(3):507–517

Mokhtari, N., Daneshpajouh, S., Seyedbagheri, S., Atashdehghan, R., Abdi, K., Sarkar, S., ... & Shahverdi, A. R. (2009). Biological synthesis of very small silver nanoparticles by culture supernatant of *Klebsiella pneumonia*: The effects of visible-light irradiation and the liquid mixing process. Materials research bulletin, 44(6), 1415–1421.

Mondal AK, Mondal S, Samanta S, Mallick S (2011) Synthesis of ecofriendly silver nanoparticles from plant latex used as an important taxonomic tool for phylogenetic interrelationship. Adv Bioresour 2(1):122–133

Moon JW, Rawn CJ, Rondinone AJ, Love LJ, Roh Y, Everett SM, Lauf RJ, Phelps TJ (2010) Large-scale production of magnetic nanoparticles using bacterial fermentation. J Ind Microbiol Biotechnol 37:1023–1031

Morita M, Rau D, Kajiyama S, Sakurai T, Baba M, Iwamura M (2004) Luminescence properties of nano-phosphors: metal-ion doped sol-gel silica glasses. Mater Sci-Pol 22(1):5–15

Mousavi RA, Sepahy AA, Fazeli MR (2012) Biosynthesis, purification and characterization of Cadmium sulfide nanoparticles using *Enterobacteriaceae* and their application. Proc Int Conf Nanomater Appl Prop 1(1):5

Mukherjee P, Ahmad A, Mandal D, Senapati S, Sainkar SR, Khan MI, Parishcha R, Ajaykumar PV, Alam M, Kumar R et al (2001a) Fungus-mediated synthesis of silver nanoparticles and their immobilization in the Mycelial matrix: a novel biological approach to nanoparticle synthesis. Nano Lett 1:515–519

Mukherjee P, Ahmad A, Mandal D, Senapati S, Sainkar SR, Khan MI, Ramani R, Parischa R, Ajayakumar PV, Alam M, Sastry M, Kumar R (2001b) Bioreduction of AuCl(4)(-) ions by the fungus, *Verticillium* sp. and surface trapping of the gold nanoparticles formed. Angew Chem Int Ed Engl 40(19):3585–3588

Musarrat J, Dwivedi S, Singh BR, Al-Khedhairy AA, Azam A, Naqvi A (2010) Production of antimicrobial silver nanoparticles in water extracts of the fungus *Amylomyces rouxii* strain KSU-09. Bioresour Technol 101:8772–8776

Narayanan KB, Sakthivel N (2010) Biological synthesis of metal nanoparticles by microbes. Adv Colloid Interface Sci 156:1–13. https://doi.org/10.1016/j.cis.2010.02.001

Nikalje AP (2015) Nanotechnology and its applications in medicine. Med Chem 5:2

Parial D, Patra HK, Dasgupta AK, Pal R (2012) Screening of different algae for green synthesis of Gold nanoparticles. Eur J Phycol 47(1):22–29

Patil CD, Patil SV, Borase HP, Salunke BK, Salunkhe RB (2012) Larvicidal activity of silver nanoparticles synthesized using *Plumeria rubra* plant latex against *Aedes aegypti* and *Anopheles stephensi*. Parasitol Res 110:1815–1822

Perez-Gonzales T, Jimenez-Lopez C, Neal AL, Rull-Perez F, Rodriguez-Navarro A, Fernandez-Vivas A, Ianez-Pareja E (2010) Magnetite biomineralization induced by *Shewanella oneidensis*. Geochem Cormochem Acta 74:967–979

Pimprikar PS, Joshi SS, Kumar AR, Zinjarde SS, Kulkarni SK (2009) Influence of biomass and gold salt concentration on nanoparticle synthesis by the tropical marine yeast *Yarrowia lipolytica* NCIM 3589. Colloids Surf B Biointerfaces 74:309–316

Prakash A, Sharma S, Ahmad N, Ghosh A, Sinha P (2010) Bacteria mediated extracellular synthesis of metallic nanoparticles. Int Res J Biotechnol 6:71–79

Prasad R (2014) Synthesis of silver nanoparticles in photosynthetic plants. J Nanoparticles Article ID 963961. https://doi.org/10.1155/2014/963961

Prasad R, Kumar V, Prasad KS (2014) Nanotechnology in sustainable agriculture: present concerns and future aspects. Afr J Biotechnol 13(6):705–713

Prasad KS, Vaghasiya JV, Soni SS, Patel J, Patel R, Kumari M, Jasmani F, Selvaraj K (2015) Microbial Selenium nanoparticles (SeNPs) and their application as a sensitive hydrogen peroxide biosensor. Appl Biochem Biotechnol 177(6):1386–1393. https://doi.org/10.1007/s12010-015-1814-9

Prasad R, Pandey R, Barman I (2016) Engineering tailored nanoparticles with microbes: quo vadis. Wiley Interdiscip Rev Nanomed Nanobiotechnol 8:316–330. https://doi.org/10.1002/wnan.1363

Prasad R, Bhattacharyya A, Nguyen QD (2017a) Nanotechnology in sustainable agriculture: recent developments, challenges, and perspectives. Front Microbiol 8:1014. https://doi.org/10.3389/fmicb.2017.01014

Prasad R, Pandey R, Varma A, Barman I (2017b) Polymer based nanoparticles for drug delivery systems and cancer therapeutics. In: Kharkwal H, Janaswamy S (eds) Natural polymers for drug delivery. CAB International, Wallingford, pp 53–70

Raheman F, Deshmukh S, Ingle A, Gade A, Rai M (2011) Silver nanoparticles: Novel antimicrobial agent synthesized from an endophytic fungus *Pestalotia* sp. isolated from leaves of *Syzygium cumini* (L). Nano Biomed Eng 3(3):174–178. https://doi.org/10.5101/nbe.v3i3.p174-178

Rajeshkumar S, Malarkodi C, Sivakumar V, Paulkumar P, Vanaja M (2014a) Biosynthesis of silver nanoparticles by using marine bacteria *Vibrio alginolyticus*. Int Res J Pharm Biosci 1(1):19–23

Rajeshkumar S, Ponnanikajamideen M, Malarkodi C, Malini M, Annadurai G (2014b) Microbe-mediated synthesis of antimicrobial semiconductor nanoparticles by marine bacteria. J Nanostruct Chem 4:96. https://doi.org/10.1007/s40097-014-0096

Rajput N (2015) Methods of preparation of nanoparticles-a review. Int J Adv Eng Technol 7(4):1806–1811

Raliya R, Tarafdar JC (2013) ZnO nanoparticle biosynthesis and its effect on phosphorous-mobilizing enzyme secretion and gum contents in Clusterbean (*Cyamopsis tetragonoloba* L.). Agric Res 2:48–57

Ramanathan R, Bhargava SK, Bansal V (2011a) Biological synthesis of copper/copper oxide nanoparticles. In Rose Amal (ed.) CHEMECA 2011 - "Engineering A Better World", Sydney, Australia, 18-21 September, pp. 1-8. Chem Conf 466

Ramanathan R, O'Mullane AP, Parikh RY, Smooker PM, Bhargava SK, Bansal V (2011b) Bacterial kinetics-controlled shape-directed biosynthesis of silver nanoplates using *Morganella psychrotolerans*. Langmuir 27(2):714–719

Rao NN, Kornberg A (1996) Inorganic polyphosphate supports resistance and survival of stationary-phase *Escherichia coli*. J Bacteriol 178:1394–1400

Reith F, Etschmann B, Grosse C, Moors H, Benotmane MA, Monsieurs P et al (2009) Mechanisms of gold biomineralization in the bacterium *Cupriavidus metallidurans*. Proc Natl Acad Sci 106:17757–17762

Riddin TL, Gericke M, Whiteley CG (2006) Analysis of the inter- and extracellular formation of platinum nanoparticles by *Fusarium oxysporum f. sp. lycopersici* using response surface methodology. Nanotechnology 17:3482–3489

Romanò CL, Toscano M, Romanò D, Drago L (2013) Antibiofilm agents and implant-related infections in orthopaedics: where are we? J Chemother 25(2):67–80. https://doi.org/10.1179/1973947812Y.0000000045

Rozamond Y, Sweeney C, Mao X, Gao JLB, Angela MB, Georgiou G, Brent LI (2004) Bacterial biosynthesis of cadmium sulfide, nanocrystals. Chem Biol 11:1553–1559

Ruparelia JP, Chatterjee AK, Duttagupta SP, Mukherji S (2008) Strain specificity in antimicrobial activity of silver and copper nanoparticles. Acta Biomater 4(3):707–716. https://doi.org/10.1016/j.actbio.2007.11.006

Saifuddin N, Wong CW, Nur Yasumira AA (2009) Rapid biosynthesis of silver nanoparticles using culture supernatant of bacteria with microwave irradiation. Eur J Chem 6(1):61–70

Sangeetha J, Thangadurai D, Hospet R, Harish ER, Purushotham P, Mujeeb MA, Shrinivas J, David M, Mundaragi AC, Thimmappa AC, Arakera SB, Prasad R (2017a) Nanoagrotechnology for soil quality, crop performance and environmental management. In: Prasad R, Kumar M, Kumar V (eds) Nanotechnology. Springer, Singapore, pp 73–97

Sangeetha J, Thangadurai D, Hospet R, Purushotham P, Karekalammanavar G, Mundaragi AC, David M, Shinge MR, Thimmappa SC, Prasad R, Harish ER (2017b) Agricultural nanotechnology: concepts, benefits, and risks. In: Prasad R, Kumar M, Kumar V (eds) Nanotechnology. Springer, Singapore, pp 1–17

Sangeetha J, Thangadurai D, Hospet R, Purushotham P, Manowade KR, Mujeeb MA, Mundaragi AC, Jogaiah S, David M, Thimmappa SC, Prasad R, Harish ER (2017c) Production of bionanomaterials from agricultural wastes. In: Prasad R, Kumar M, Kumar V (eds) Nanotechnology. Springer, Singapore, pp 33–58

Sastry M, Ahmad A et al (2005) Microbial nanoparticle production. In: Niemeyer PDCM, Mirkin PDCA (eds) Nanobiotechnology. Wiley, Chichester, pp 126–135

Sathyavanthi S, Manjula A, Rajendhran J, Gunasekaran P (2013) Biosynthesis and characterization of mercury sulphide nanoparticles produced by *Bacillus cereus* MRS-1. Indian J Exp Biol 51:973–978

Schlüter M, Hentzel T, Suarez C, Koch M, Lorenz WG et al (2014) Synthesis of novel palladium(0) nanocatalysts by microorganisms from heavy-metalinfluenced high-alpine sites for dehalogenation of polychlorinated dioxins. Chemosphere 117C:462–470

Selvarajan E, Mohanasrinivasan V (2013) Biosynthesis and characterization of ZnO nanoparticles using *Lactobacillus plantarum VITES07*. Mater Lett 112:180–182

Shankar SS, Rai A, Ahmad A, Sastry M (2004a) Biosynthesis of silver and gold nanoparticles from extracts of different parts of the Geranium plant. Appl Nanotechnol 1:69–77

Shankar SS, Rai A, Ahmad A, Sastry M (2004b) Rapid synthesis of Au, Ag, and bimetallic Au core-Ag shell nanoparticles using neem *(Azadirachta indica)* leaf broth. J Colloid Interface Sci 275:496–502

Shantkriti S, Rani P (2014) Biological synthesis of copper nanoparticles using *Pseudomonas fluorescens*. Int J Curr Microbiol App Sci 3(9):374–383

Sharma AB, Sharma M, Pandey RK (2009) Synthesis, properties and potential applications of semiconductor Quantum particles. Asian J Chem 21(10):S033–S038

Shedbalkar U, Singh R, Wadhwani S, Gaidhani S, Chopade BA (2014) Microbial synthesis of gold nanoparticles: current status and future prospects. Adv Colloid Interface Sci 209:40–48

Shivaji S, Madhu S, Singh S (2011) Extracellular synthesise of antibacterial silver nanoparticles using psychrophilic bacteria. Process Biochem 49:830–837

Shivashankarappa A, Sanjay KR (2015) Study on biological synthesis of cadmium sulfide nanoparticles by *Bacillus licheniformis* and its antimicrobial properties against food borne pathogens. Nanosci Nanotechnol Res 3(1):6–15

Siddiqi KS, Husen A (2016) Fabrication of metal nanoparticles from fungi and metal salts: scope and application. Nanoscale Res Lett 11:98. https://doi.org/10.1186/s11671-016-1311-2

Simkiss K, Wilbur KM (1989) Biomineralization. Academic, New York

Singaravelu G, Arockiamary JS, Kumar VG, Govindaraju K (2007) A novel extracellular synthesis of monodisperse gold nanoparticles using marine alga, *Sargassum wightii* Greville. Colloids Surf B Biointerfaces 57(1:97–101

Singh V, Patil R, Ananda A, Milani P, Gade W (2010) Biological synthesis of copper oxide nano particles using *Escherichia coli*. Curr Nanosci 6(4):365–369

Sinha A, Khare SK (2011) Mercury bioaccumulation and simultaneous nanoparticle synthesis by *Enterobacter sp.* cells. Bioresour Technol 102:4281–4284

Sneha K, Sathishkumar M, Mao J, Kwak IS, Yun YS (2010) *Corynebacterium glutamicum*-mediated crystallization of silver ions through sorption and reduction processes. Chem Eng J 162:989–996

Sundaram PA, Augustine R, Kannan M (2012) Extracellular biosynthesis of iron oxide nanoparticles by *Bacillus subtilis* strains isolated from rhizosphere soil. Biotechnol Bioprocess Eng 17:835–840. https://doi.org/10.1007/s12257-011-0582-9

Sunitha A, Isaac RSR, Geo S, Sornalekshmi S, Rose A, Praseetha PK (2013) Evaluation of antimicrobial activity of biosynthesized iron and silver nanoparticles using the fungi *Fusarium oxysporum* and *Actinomycetes sp.* on human pathogens. Nano Biomed Eng 5:39–45

Thakkar KN, Mhatre SS, Parikh RY (2009) Biological synthesis of metallic nanoparticles. Nanomed Nanotechnol Biol Med. https://doi.org/10.1016/j.nano.2009.07.002

Tissue BM, Yuan HB (2003) Structure particle size and annealing gas phase-condensed $Eu^{3+}:Y_2O_3$ nanophosphors. J Solid State Chem 171:12–18

Tiwari DK, Behari J, Sen P (2008) Timeanddose-dependent antimicrobial potential of Ag nanoparticles synthesized by top-down approach. Curr Sci 95(5):647–655

Usha R, Prabu E, Palaniswamy M, Venil CK, Rajendran R (2010) Synthesis of metal oxide nanoparticles by *Streptomyces* sp. for development of antimicrobial textiles. Glob J Biotechnol Biochem 5(3):153–160

Vankar PS, Bajpai D (2010) Preparation of gold nanoparticles from *Mirabilis jalapa* flowers. Indian J Biochem Biophys 47:157–160

Varshney R, Bhadauria S, Gaur MS, Pasricha R (2010) Characterization of copper nanoparticles synthesized by a novel microbiological method. J Miner Metals Mater Soc 62(12):102–104

Velhal SG, Kulkarni SD, Latpate RV (2016) Fungal mediated silver nanoparticle synthesis using robust experimental design and its application in cotton fabric. Int Nano Lett 6:257–264

Vigneshwaran N, Kathe AA, Varadarajan PV, Nachane RP, Balasubramanya RH (2006) Biomimetics of silver nanoparticles by white rot fungus *Phaenerochaete chrysosporium*. Colloids Surf B Biointerfaces 53:55–59

Vigneshwaran N, Ashtaputre NM, Varadarajan PV, Nachane RP, Paralikar KM, Balasubramanya RH (2007) Biological synthesis of silver nanoparticles using the fungus *Aspergillus flavus*. Mater Lett 61:413–1418

Vilchis-Nestor AR, Sánchez-Mendieta V, Camacho-López MA, Gómez-Espinosa RM, Camacho-López MA, Arenas Alatorre JA (2008) Solventless synthesis and optical properties of Au and Ag nanoparticles using *Camellia sinensis* extract. Mater Lett 62(17):3103–3105

Wagner E, Plank C, Zatloukal K, Cotton M, Birnstiel ML (1992) Influenza virus hemagglutinin HA-2 N-terminal fusogenic peptides augment gene transfer by transferrin-polylysine-DNA complexes: toward a synthetic virus-like gene transfer vehicle. Proc Natl Acad Sci 89:7934–7938

Wang T, Yang L, Zhang B, Liu J (2010) Extracellular biosynthesis and transformation of selenium nanoparticles and application in H_2O_2 biosensor. Colloids Surf B 80(1):94–102

Winterer M, Hahn H, Metallkd Z (2003) Chemical vapor synthesis of nanocrystalline powders. Nanoceramics Chem Vapor Synthesis 94:1084–1090

Xiang L, Wei J, Jianbo S, Guili W, Feng G, Ying L (2007) Purified and sterilized magnetosomes from *Magnetospirillum gryphiswaldense MSR-1* were not toxic to mouse fibroblasts in vitro. Lett Appl Microbiol 45:75–81

Xie J, Chen K, Chen X (2009) Production, modification and bio-applications of magnetic nanoparticles gestated by magnetotactic bacteria. Nano Res 2(4):261–278

Yang H, Holloway PH (2004) Efficient and photostable ZnS-passivated CdS:Mn luminescent nanocrystals. Adv Funct Mater 14:152–156

Yang JG, Zhou YL et al (2006) Preparation of oleic acid-capped copper nanoparticles. Chem Lett 35(10):1190–1191

Zheng D, Hu C, Gan T, Dang X, Hu S (2010) Preparation and application of a novel vanillin sensor based on biosynthesis of Au-Ag alloy nanoparticles. Sens Actuators B Chem 148:247–252

Zhu L, Ang S, Liu WT (2004) Quantum dots as a novel immunofluorescent detection system for *Cryptosporidium parvum* and *Giardia lamblia*. Appl Environ Microbiol 70:597–598

Chapter 3
Myco-Nanoparticles: A Novel Approach for Inhibiting Amyloid-β Fibrillation

Aditya Saran, Rajender Boddula, Priyanka Dubey, Ramyakrishna Pothu, and Saurabh Gautam

3.1 Introduction

The requirement of eco-friendly and green technology in the area of material science is of a great interest due to its distinct biological applications. "Nanotechnology" belongs to one of the imperative areas of modern material science due to its key importance in the development of drug delivery, gene therapy, antibacterial agents, electronics, magnetic resonance imaging, and separation science, to name a few (Sonvico et al. 2005; Wilkinson 2003; Nie et al. 2007). Therefore, the advancement in the synthesis of nanoparticles with varying shape, size, chemical composition, and morphology is indispensable. There are various protocols described for the synthesis of different types of nanoparticles such as chemical, physical, biological, and hybrid methods (Mohanpuria et al. 2008; Luechinger et al. 2009). The synthesis of nanoparticles via chemical and physical channel has gained a huge attention recently (Iravani et al. 2014). However, these methodologies are complicated, expensive, and

A. Saran
Department of Microbiology, Marwadi University, Rajkot, India

R. Boddula
CAS Key Laboratory for Nanosystem and Hierarchical Fabrication, National Center for Nanoscience and Technology, Beijing, P. R. China

P. Dubey
Department of Textile Technology, Indian Institute of Technology Delhi, New Delhi, India

R. Pothu
College of Chemistry and Chemical Engineering, Hunan University, Changsha, P. R. China

S. Gautam (✉)
Department of Cellular Biochemistry, Max Planck Institute of Biochemistry, Martinsried, Germany

© Springer Nature Switzerland AG 2018
R. Prasad et al. (eds.), *Exploring the Realms of Nature for Nanosynthesis*, Nanotechnology in the Life Sciences, https://doi.org/10.1007/978-3-319-99570-0_3

require the use of different types of toxic reagents resulting in restricted biomedical applications. As a result, there is a considerable and urgent need for nontoxic, reliable, eco-friendly, and "green" route techniques for the synthesis of such nanoparticles, especially for biomedical applications. In order to achieve this, the utilization of biological methods using different microorganisms for the synthesis of nanoparticles has emerged as a unique field of research (Li et al. 2011; Prasad et al. 2016). Based on the site of synthesis of nanoparticles by the microorganisms, the process can be classified into two types: intracellular and extracellular synthesis (Talham 2002). The intracellular synthesis is defined as the transportation of metal ions (building block of respective nanoparticles) in the presence of enzymes into the microbial cell (Otari et al. 2015). However, when the metal ions are absorbed at the surface of the microbial cell, forming nanoparticles, it is known as extracellular synthesis of nanoparticles (Zhang et al. 2011). In the following sections, we have described in detail the biological modes for the synthesis of different metallic nanoparticles.

3.2 Biosynthesis of Nanoparticles Using Microorganisms

3.2.1 Gold Nanoparticles (AuNPs)

Michael Faraday for the first time observed that the colloidal gold solutions have different properties from bulk gold (Stephen and Macnaughtont 1999). The synthesis of different dimensions (1D, 2D, and 3D shapes) and hollow structures of AuNPs offers an imperative biological application (Kowalczyk et al. 2010). AuNPs are the nontoxic carriers in the area of gene and drug delivery (Ghosh et al. 2008). Their ability to interact with thiol group offers a selective means of controlled intracellular release. The extracellular synthesis of AuNPs has been reported using *Fusarium oxysporum* (fungus) and *Thermomonospora* sp. (actinomycete) (Southam and Beveridge 1996). Moreover, the intracellular synthesis of AuNPs by fungi *Verticillium* sp. has been reported by the same group (Beveridge and Murray 1980). The bacterium *Rhodopseudomonas capsulata* was also used for the synthesis of the AuNPs with different shapes and sizes by modulating the pH (Zhou et al. 2007). The alkalotolerant actinomycete (*Rhodococcus* sp.) has also been reported in the intracellular synthesis of AuNPs (Ahmad et al. 2003b). Alkalotolerant *Rhodococcus* sp. has been known to produce AuNPs under alkaline conditions (Klaus et al. 1999). Nanocrystals and nanoalloys synthesis has been reported using *Lactobacillus* (Mukherjee et al. 2001). The synthesis of AuNPs with distinct shapes (cubic, spherical, and octahedral) has been reported from filamentous cyanobacteria (Klaus-Joerger et al. 2001). Synthesis of AuNPs from blue green alga *Spirulina platensis* has also been reported recently (Suganya et al. 2015).

3.2.2 Silver Nanoparticles (AgNPs)

Silver nanoparticles are gaining significant attention due to their potent antibacterial activity against Gram-positive and Gram-negative bacteria (Fayaz et al. 2010; Aziz et al. 2014, 2015). They have also been reported to possess anticancer and antioxidant properties (Mohanta et al. 2017). AgNPs are capable of physical interaction with the cell surface of various bacteria (Dakal et al. 2016). Studies have reported that AgNPs can break the cellular membranes of microorganisms which further results in an enhanced anti-microbial activity (Franci et al. 2015). The synthesis of AgNPs by bacteria has been shown to be influenced by silver-resistant genes, c-type cytochromes, cellular enzymes (nitrate reductase), peptides, and reducing cofactors (Hamedi et al. 2017). The synthesis of the AgNP film from fungi such as *Fusarium oxysporum*, *Verticillium, Aspergillus flavus,* or *Mucor hiemalis* has been demonstrated by various workers (Ahmad et al. 2003a; Bhainsa and D'souza 2006, Aziz et al. 2016). AgNPs synthesis using aqueous extract of the seaweed *Sargassum muticum* has also been reported recently (Madhiyazhagan et al. 2015). Another study has reported that when *Pseudomonas stutzeri* AG259 bacterium was kept in the concentrated aqueous solution of silver nitrate, it led to the reduction of the Ag^+ ions and subsequently the formation of AgNPs in the periplasmic space of the bacteria (Pugazhenthiran et al. 2009).

3.2.3 Alloy Nanoparticles

Alloy nanoparticles are of enormous interest for in situ applications due to their flexible properties (Oezaslan et al. 2011). Alloy nanoparticles are contributing to different fields of science and technology such as optical materials, catalysis, electronics, and coatings (Cortie and McDonagh 2011; Ferrando et al. 2008). An alloy nanoparticle of gold and silver has been used for the catalytic reduction of methylene blue (Tripathi et al. 2015). *Trichoderma harzianum* has been used as a stabilizing and reducing agent in the synthesis of gold-silver alloy (Tripathi et al. 2015). *F. oxysporum* has been used for the synthesis of gold and silver alloy (Senapati et al. 2005). The biosynthesis of alloy of gold and silver nanoparticles under microwave irradiation by their metal precursors and *Jasminum sambac* leaves extract has also been illustrated earlier (Yallappa et al. 2015). Another study has demonstrated the synthesis of gold and silver alloy nanoparticles using yeast cells via extracellular routes (Zheng et al. 2010).The synthesis of gold and silver allloy nanoparticles from fungal strain *Fusarium semitectum* was also reported which was demonstrated to be stable for many weeks (Sawle et al. 2008).

3.3 Nanoparticles Synthesis Using Fungi

As compared to various other groups of microorganisms, fungi are an important group of microorganisms responsible for the synthesis of large amount of metal nanoparticles (Syed et al. 2013; Aziz et al. 2016). This could be due to their capability of synthesizing monodisperse nanoparticles with well-defined shape and size (Mukherjee et al. 2002). Fungal strains consist of large number of enzymes and proteins facilitating the formation of nanoparticles; moreover, these strains are also easy to handle (Mohanpuria et al. 2008; Prasad et al. 2016). A study has reported wherein fungus *Verticillium luteoalbum* was used to synthesize gold nanoparticles using extracellular enzymes (Gericke and Pinches 2006). Another study has reported the synthesis of silver nanoparticle when fungus *Aspergillus niger* was incubated with silver nitrate solution (Gade et al. 2008). The enzymes and proteins produced by fungus facilitated the stability of nanoparticles, as reported in that study. The nanoparticles were shown to have a size distribution between 5 and 25 nm and were monodispersed in nature. In another study, *Fusarium oxysporum* was also used to synthesize CdS nanoparticles (Shakibaie et al. 2010). The authors concluded that the release of reductase enzymes in the presence of *F. oxysporum* led to the formation of CdS nanoparticles. *Aspergillus fumigates* is another fungal strain known for synthesis of metal nanoparticles at wide level (Bhainsa and D'souza 2006). The presence of reductase enzymes facilitates the biosynthesis of metal nanoparticles of different chemical compositions (Adelere and Lateef 2016). Mukherjee *et al.* have reported the synthesis of gold nanoparticles by the intracellular reduction of metal ions in *Verticillium* sp. (Mukherjee et al. 2001). Another study reported the biosynthesis of Ag, Au, Zn, and, Ag-Au nanoparticles using fungi *Volvariella volvacea* (Philip 2009). Another species *Trichothecium* sp. was used for the biosynthesis of gold nanoparticle (Ahmad et al. 2005). Authors showed that the shaking phase results in the formation of intracellular nanoparticles; however, an stationary phase (without shaking) also leads to the formation of extracellular nanoparticles (Ahmad et al. 2005). *Penicillium brevicompactum* was reported for the biosynthesis of silver nanoparticles using nitrate reductase enzyme (Hemath Naveen et al. 2010). An interesting study reported the biosynthesis of monodispersed silver nanoparticles using *Aspergillus fumigates* within 10 min (Bhainsa and D'souza 2006). This biosynthesis method for nanoparticles has been reported to be significantly rapid as compared to chemical and physical methods (Bhainsa and D'souza 2006). Based on abovementioned findings by various workers, it can be deduced clearly that biosynthesis of nanoparticles using fungal-based method is a rational, green, cost-effective, faster, and nontoxic approach. This is primarily due to the capability of fungi for reduction of the metal ions by an enzymatic process.

3.4 Introduction to Protein Aggregation

The specific functional and structural properties of protein aggregates play a key role in many routine physiological activities (Alsberg et al. 2006). However, the misfolding during protein assembly and aggregation could lead to several neurological disorders and diseases such as Huntington's, Alzheimer's, Parkinson's, and, prion diseases to name a few (Lashuel et al. 2002). The protein aggregates defined as amyloids found in the patients with these disorders were characterized histologically more than 150 years ago (Aguzzi and O'Connor 2010).

3.4.1 Types of Protein Aggregates

Based on morphological analysis, protein aggregates can be categorized into two classes: ordered aggregates (amyloid) and disordered aggregates (amorphous) (Fink 1998). Amyloid fibrils are most commonly found extracellularly; however, there are various reports showing the presence of intracellular amyloid fibrils as well (Takahashi et al. 1989; Friedrich et al. 2010). Amyloids are known to have ordered structure with a high proportion of β-sheets (cross-β structure) and are known to demonstrate apple-green birefringence upon binding with dye Congo red (Fink 1998; Rambaran and Serpell 2008; Tycko 2011; Chiti and Dobson 2006). On the other hand, amorphous aggregates lack any ordered secondary structure (Fink 1998; Baneyx and Mujacic 2004; Villaverde and Carrio 2003).

3.4.2 Mechanism of Protein Aggregation

The mechanism of protein aggregation has been categorized into three different types such as nucleation dependent protein aggregation (seeded polymerization), templated assembly model, and nucleated conformational conversion mechanism (Philo and Arakawa 2009; Invernizzi et al. 2012; Idicula-Thomas and Balaji 2007; Gsponer and Vendruscolo 2006). Nucleation dependent polymerization consists of a rate limiting step, termed as lag phase, wherein oligomers form a critical nucleus and increase its size by the formation of the larger aggregates known as amyloids (Jarrett and Lansbury 1993; Eichner and Radford 2011). However, in template assembly model the protein aggregate functions as a template for straight addition of monomers (Griffith 1967). Nucleated conformational conversion mechanism consists of both the aforementioned mechanisms (Chatani and Yamamoto 2018). The latter aggregation pathway initially includes the formation of amorphous nuclei, which is followed by template assembly aggregation mechanism (Serio et al. 2000). The amyloid aggregates are known to follow the nucleation dependent polymerization pathway (Morris et al. 2009; Fink 1998; Eichner and Radford 2011).

3.4.3 Therapeutic Intervention of Protein Aggregation Diseases

Proteins tend to misfold and form aggregates under the specific stress conditions leading to the various human pathological disorders (Ross and Poirier 2004). Therefore, in order to control this, efforts are ongoing globally on controlling and/ or reversing the formation of amyloid aggregates responsible for various neurode-generative disorders (Sharma et al. 2012; Wang et al. 2013; Uversky 2007). Different approaches have been reported in the literature to deal with protein aggregation such as small peptides (Xiong et al. 2015), osmolytes (Macchi et al. 2012), antibiotics such as rifampicin (Tomiyama et al. 1994), naturally derived polyphenols (Nedumpully-Govindan et al. 2016), and compounds based on amyloid binding dyes (Hasegawa et al. 2007). However, we will discuss about the nanoparticles as therapeutic agents for protein aggregation diseases.

3.4.4 Nanoparticles as Therapeutic Agents

Various studies have been reported on the effect of different nanoparticles on amyloid aggregation (Brambilla et al. 2011; Kogan et al. 2006; Araya et al. 2008). Gold and silver nanoparticles have been utilized in biomedical sciences (cancer treatment, controlled drug delivery, biomedical imaging, etc.) due to their apposite properties such as low toxicity, tunable stability, biocompatibility, small dimensions, and the possibility to interact with a variety of substances (Belanova et al. 2018). Recent studies have demonstrated that gold nanoparticles are capable of crossing the blood–brain barrier as well (Cabuzu et al. 2015; Cheng et al. 2014). Liao et al. have illustrated the effect of gold nanoparticles on amyloid aggregation (Liao et al. 2012). The authors showed that bare gold nanoparticles were able to abolish the formation of large fibrils and giving rise to fragmented fibrils and spherical oligomers (Liao et al. 2012). Authors further demonstrated that due to negative surface potential of gold nanoparticles, these act as nano-chaperones and further redirect and inhibit the amyloid aggregation. Recently a study demonstrated the biosynthesis of gold nanoparticles via extracellular mode using fungus *Fusarium oxysporum* (Mukherjee et al. 2002). The authors demonstrated that the extracellular synthesis of nanoparticles provided various advantages such as homogeneous catalysis when the nanoparticles were synthesized in solution form (Mukherjee et al. 2002). Xiong et al. have designed LVFFARK and LVFFARK-functionalized nanoparticles in order to prevent the amyloid-β protein aggregation (Xiong et al. 2015). A study reported that peptide-gold nanoparticles irradiated with microwave are capable of selective binding to β-amyloid and inhibiting the amyloidogenesis (Araya et al. 2008). Obulesu and Jhansilakshmi (2016) have synthesized redox nanoparticles such as nitroxide radical, 2,2,6,6 tetramethylpiperidinyl-N-oxyl. They concluded that piperine in association with redox nanoparticles confers improved prevention against Alzheimer's disease *in vitro*.

Another interesting study has attempted to interfere with amyloid-β aggregation growth and distantly re-dissolve these deposits via local heat dissipation of gold nanoparticles by selective binding to the aggregates (Kogan et al. 2006). Association with aggregates also allowed both, noninvasive exploration and dissolution of molecular aggregates. Therefore, based on abovementioned finding a conclusion can be drawn that a promising therapeutic approach for controlling the protein aggregation can be developed using different aspects of nanotechnology.

3.5   Future Prospect

Based on a recent study, it is estimated that Alzheimer's disease caused by amyloid plaques has affected about 44 million people worldwide (Alzheimer's Association 2015). This figure is expected to rise to 115 million by 2050 (Alzheimer's Association 2015). Presently, there is no competent therapy available for treating the patients with Alzheimer's disease. Moreover, another disadvantage of many of the available drug candidates is the incapability to cross the blood–brain barrier (Patel and Patel 2017). In that scenario, nanotechnology has emerged as a potential therapy in order to control the Alzheimer's disease due to the capability of various nanoparticles to cross the blood–brain barrier (Saraiva et al. 2016) and control the production and aggregation of amyloid plaques (Fig. 3.1). However, there are certain issues which have to be addressed to improve their efficacy. For example, the drug-loaded nanoparticles in comparison with the free drug, are currently limited quantitatively into the brain. Therefore, more focus should be paid on the development of systems transporting pharmacologically significant quantity of drugs into the brain. Moreover, a complete understanding of mechanisms of Alzheimer's disease will contribute towards the accurate application of targeted nanoparticles.

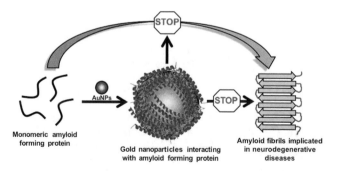

Fig. 3.1 Representative diagram showing the interaction of gold nanoparticles with the amyloid forming protein (e.g. amyloid β-peptide) in order to prevent its aggregation

References

Adelere I, Lateef A (2016) A novel approach to the green synthesis of metallic nanoparticles: the use of agro-wastes, enzymes, and pigments. Nanotechnol Rev 5(6):567–587

Aguzzi A, O'Connor T (2010) Protein aggregation diseases: pathogenicity and therapeutic perspectives. Nat Rev Drug Discov 9(3):237–248. https://doi.org/10.1038/nrd3050

Ahmad A, Mukherjee P, Senapati S, Mandal D, Khan MI, Kumar R, Sastry M (2003a) Extracellular biosynthesis of silver nanoparticles using the fungus Fusarium oxysporum. Colloids Surf B Biointerfaces 28(4):313–318

Ahmad A, Senapati S, Khan MI, Kumar R, Ramani R, Srinivas V, Sastry M (2003b) Intracellular synthesis of gold nanoparticles by a novel alkalotolerant actinomycete, Rhodococcus species. Nanotechnology 14(7):824

Ahmad A, Senapati S, Khan MI, Kumar R, Sastry M (2005) Extra-/intracellular biosynthesis of gold nanoparticles by an alkalotolerant fungus, Trichothecium sp. J Biomed Nanotechnol 1(1):47–53

Alsberg E, Feinstein E, Joy MP, Prentiss M, Ingber DE (2006) Magnetically-guided self-assembly of fibrin matrices with ordered nano-scale structure for tissue engineering. Tissue Eng 12(11):3247–3256. https://doi.org/10.1089/ten.2006.12.3247

Alzheimer's Association (2015) 2015 Alzheimer's disease facts and figures. Alzheimers Dement 11(3):332

Araya E, Olmedo I, Bastus NG, Guerrero S, Puntes VF, Giralt E, Kogan MJ (2008) Gold nanoparticles and microwave irradiation inhibit beta-amyloid amyloidogenesis. Nanoscale Res Lett 3(11):435

Aziz N, Fatma T, Varma A, Prasad R (2014) Biogenic synthesis of silver nanoparticles using Scenedesmus abundans and evaluation of their antibacterial activity. J Nanoparticles Article ID 689419. https://doi.org/10.1155/2014/689419

Aziz N, Faraz M, Pandey R, Sakir M, Fatma T, Varma A, Barman I, Prasad R (2015) Facile algae-derived route to biogenic silver nanoparticles: synthesis, antibacterial and photocatalytic properties. Langmuir 31:11605–11612. https://doi.org/10.1021/acs.langmuir.5b03081

Aziz N, Pandey R, Barman I, Prasad R (2016) Leveraging the attributes of Mucor hiemalis-derived silver nanoparticles for a synergistic broad-spectrum antimicrobial platform. Front Microbiol 7:1984. https://doi.org/10.3389/fmicb.2016.01984

Baneyx F, Mujacic M (2004) Recombinant protein folding and misfolding in Escherichia coli. Nat Biotechnol 22(11):1399–1408. https://doi.org/10.1038/nbt1029

Belanova AA, Gavalas N, Makarenko YM, Belousova MM, Soldatov AV, Zolotukhin PV (2018) Physicochemical properties of magnetic nanoparticles: implications for biomedical applications in vitro and in vivo. Oncol Res Treat 41(3):139–143. https://doi.org/10.1159/000485020

Beveridge T, Murray R (1980) Sites of metal deposition in the cell wall of Bacillus subtilis. J Bacteriol 141(2):876–887

Bhainsa KC, D'souza S (2006) Extracellular biosynthesis of silver nanoparticles using the fungus Aspergillus fumigatus. Colloids Surf B Biointerfaces 47(2):160–164

Brambilla D, Le Droumaguet B, Nicolas J, Hashemi SH, Wu L-P, Moghimi SM, Couvreur P, Andrieux K (2011) Nanotechnologies for Alzheimer's disease: diagnosis, therapy, and safety issues. Nanomed Nanotechnol Biol Med 7(5):521–540

Cabuzu D, Cirja A, Puiu R, Mihai Grumezescu A (2015) Biomedical applications of gold nanoparticles. Curr Top Med Chem 15(16):1605–1613

Chatani E, Yamamoto N (2018) Recent progress on understanding the mechanisms of amyloid nucleation. Biophys Rev 10(2):527–534. https://doi.org/10.1007/s12551-017-0353-8

Cheng Y, Dai Q, Morshed RA, Fan X, Wegscheid ML, Wainwright DA, Han Y, Zhang L, Auffinger B, Tobias AL (2014) Blood-brain barrier permeable gold nanoparticles: an efficient delivery platform for enhanced malignant glioma therapy and imaging. Small 10(24):5137–5150

Chiti F, Dobson CM (2006) Protein misfolding, functional amyloid, and human disease. Annu Rev Biochem 75:333–366. https://doi.org/10.1146/annurev.biochem.75.101304.123901

Cortie MB, McDonagh AM (2011) Synthesis and optical properties of hybrid and alloy plasmonic nanoparticles. Chem Rev 111(6):3713–3735

Dakal TC, Kumar A, Majumdar RS, Yadav V (2016) Mechanistic basis of antimicrobial actions of silver nanoparticles. Front Microbiol 7:1831. https://doi.org/10.3389/fmicb.2016.01831

Eichner T, Radford SE (2011) A diversity of assembly mechanisms of a generic amyloid fold. Mol Cell 43(1):8–18. https://doi.org/10.1016/j.molcel.2011.05.012

Fayaz AM, Balaji K, Girilal M, Yadav R, Kalaichelvan PT, Venketesan R (2010) Biogenic synthesis of silver nanoparticles and their synergistic effect with antibiotics: a study against gram-positive and gram-negative bacteria. Nanomed Nanotechnol Biol Med 6(1):103–109

Ferrando R, Jellinek J, Johnston RL (2008) Nanoalloys: from theory to applications of alloy clusters and nanoparticles. Chem Rev 108(3):845–910

Fink AL (1998) Protein aggregation: folding aggregates, inclusion bodies and amyloid. Fold Des 3(1):R9–R23. https://doi.org/10.1016/S1359-0278(98)00002-9

Franci G, Falanga A, Galdiero S, Palomba L, Rai M, Morelli G, Galdiero M (2015) Silver nanoparticles as potential antibacterial agents. Molecules 20(5):8856–8874

Friedrich RP, Tepper K, Ronicke R, Soom M, Westermann M, Reymann K, Kaether C, Fandrich M (2010) Mechanism of amyloid plaque formation suggests an intracellular basis of Abeta pathogenicity. Proc Natl Acad Sci U S A 107(5):1942–1947. https://doi.org/10.1073/pnas.0904532106

Gade A, Bonde P, Ingle A, Marcato P, Duran N, Rai M (2008) Exploitation of Aspergillus niger for synthesis of silver nanoparticles. J Biobaased Mater Bioenergy 2(3):243–247

Gericke M, Pinches A (2006) Microbial production of gold nanoparticles. Gold Bull 39(1):22–28

Ghosh P, Han G, De M, Kim CK, Rotello VM (2008) Gold nanoparticles in delivery applications. Adv Drug Deliv Rev 60(11):1307–1315

Griffith JS (1967) Self-replication and scrapie. Nature 215(5105):1043–1044

Gsponer J, Vendruscolo M (2006) Theoretical approaches to protein aggregation. Protein Pept Lett 13(3):287–293

Hamedi S, Ghaseminezhad M, Shokrollahzadeh S, Shojaosadati SA (2017) Controlled biosynthesis of silver nanoparticles using nitrate reductase enzyme induction of filamentous fungus and their antibacterial evaluation. Artif Cells Nanomed Biotechnol 45(8):1588–1596

Hasegawa T, Sato Y, Okada T, Shibukawa M, Li C, Orbulescu J, Leblanc RM (2007) Inhibition of aggregation of a biomimic peptidolipid Langmuir monolayer by Congo red studied by UV-vis and infrared spectroscopies. J Phys Chem B 111(51):14227–14232. https://doi.org/10.1021/jp0759269

Hemath Naveen K, Kumar G, Karthik L, Bhaskara Rao K (2010) Extracellular biosynthesis of silver nanoparticles using the filamentous fungus Penicillium sp. Arch Appl Sci Res 2(6):161–167

Idicula-Thomas S, Balaji PV (2007) Protein aggregation: a perspective from amyloid and inclusion-body formation. Curr Sci 92(6):758–767

Invernizzi G, Papaleo E, Sabate R, Ventura S (2012) Protein aggregation: mechanisms and functional consequences. Int J Biochem Cell Biol 44(9):1541–1554. https://doi.org/10.1016/j.biocel.2012.05.023

Iravani S, Korbekandi H, Mirmohammadi SV, Zolfaghari B (2014) Synthesis of silver nanoparticles: chemical, physical and biological methods. Res Pharm Sci 9(6):385–406

Jarrett JT, Lansbury PT Jr (1993) Seeding "one-dimensional crystallization" of amyloid: a pathogenic mechanism in Alzheimer's disease and scrapie? Cell 73(6):1055–1058

Klaus T, Joerger R, Olsson E, Granqvist C-G (1999) Silver-based crystalline nanoparticles, microbially fabricated. Proc Natl Acad Sci 96(24):13611–13614

Klaus-Joerger T, Joerger R, Olsson E, Granqvist C-G (2001) Bacteria as workers in the living factory: metal-accumulating bacteria and their potential for materials science. Trends Biotechnol 19(1):15–20

Kogan MJ, Bastus NG, Amigo R, Grillo-Bosch D, Araya E, Turiel A, Labarta A, Giralt E, Puntes VF (2006) Nanoparticle-mediated local and remote manipulation of protein aggregation. Nano Lett 6(1):110–115

Kowalczyk B, Lagzi I, Grzybowski BA (2010) "Nanoarmoured" droplets of different shapes formed by interfacial self-assembly and crosslinking of metal nanoparticles. Nanoscale 2(11):2366–2369. https://doi.org/10.1039/c0nr00381f

Lashuel HA, Hartley D, Petre BM, Walz T, Lansbury PT (2002) Neurodegenerative disease: amyloid pores from pathogenic mutations. Nature 418(6895):291–291

Li X, Xu H, Chen Z-S, Chen G (2011) Biosynthesis of nanoparticles by microorganisms and their applications. J Nanomater 2011:270974

Liao YH, Chang YJ, Yoshiike Y, Chang YC, Chen YR (2012) Negatively charged gold nanoparticles inhibit Alzheimer's amyloid-β fibrillization, induce fibril dissociation, and mitigate neurotoxicity. Small 8(23):3631–3639

Luechinger NA, Grass RN, Athanassiou EK, Stark WJ (2009) Bottom-up fabrication of metal/metal nanocomposites from nanoparticles of immiscible metals. Chem Mater 22(1):155–160

Macchi F, Eisenkolb M, Kiefer H, Otzen DE (2012) The effect of osmolytes on protein fibrillation. Int J Mol Sci 13(3):3801–3819. https://doi.org/10.3390/ijms13033801

Madhiyazhagan P, Murugan K, Kumar AN, Nataraj T, Dinesh D, Panneerselvam C, Subramaniam J, Kumar PM, Suresh U, Roni M (2015) Sargassum muticum-synthesized silver nanoparticles: an effective control tool against mosquito vectors and bacterial pathogens. Parasitol Res 114(11):4305–4317

Mohanpuria P, Rana NK, Yadav SK (2008) Biosynthesis of nanoparticles: technological concepts and future applications. J Nanopart Res 10(3):507–517

Mohanta YK, Panda SK, Jayabalan R, Sharma N, Bastia AK, Mohanta TK (2017) Antimicrobial, antioxidant and cytotoxic activity of silver nanoparticles synthesized by leaf extract of Erythrina suberosa (Roxb.). Front Mol Biosci 4:14. https://doi.org/10.3389/fmolb.2017.00014

Morris AM, Watzky MA, Finke RG (2009) Protein aggregation kinetics, mechanism, and curve-fitting: a review of the literature. Biochim Biophys Acta 1794(3):375–397. https://doi.org/10.1016/j.bbapap.2008.10.016

Mukherjee P, Ahmad A, Mandal D, Senapati S, Sainkar SR, Khan MI, Ramani R, Parischa R, Ajayakumar P, Alam M (2001) Bioreduction of AuCl4– ions by the fungus, Verticillium sp and surface trapping of the gold nanoparticles formed. Angew Chem Int Ed 40(19):3585–3588

Mukherjee P, Senapati S, Mandal D, Ahmad A, Khan MI, Kumar R, Sastry M (2002) Extracellular synthesis of gold nanoparticles by the fungus Fusarium oxysporum. Chembiochem 3(5):461–463

Nedumpully-Govindan P, Kakinen A, Pilkington EH, Davis TP, Chun Ke P, Ding F (2016) Stabilizing off-pathway oligomers by polyphenol nanoassemblies for IAPP aggregation inhibition. Sci Rep 6:19463. https://doi.org/10.1038/srep19463

Nie S, Xing Y, Kim GJ, Simons JW (2007) Nanotechnology applications in cancer. Annu Rev Biomed Eng 9:257–288

Obulesu M, Jhansilakshmi M (2016) Neuroprotective role of nanoparticles against Alzheimer's disease. Curr Drug Metab 17(2):142–149

Oezaslan M, Heggen M, Strasser P (2011) In situ observation of bimetallic alloy nanoparticle formation and growth using high-temperature XRD. Chem Mater 23(8):2159–2165

Otari SV, Patil RM, Ghosh SJ, Thorat ND, Pawar SH (2015) Intracellular synthesis of silver nanoparticle by actinobacteria and its antimicrobial activity. Spectrochim Acta A Mol Biomol Spectrosc 136(pt B):1175–1180. https://doi.org/10.1016/j.saa.2014.10.003

Patel MM, Patel BM (2017) Crossing the blood-brain barrier: recent advances in drug delivery to the brain. CNS Drugs 31(2):109–133. https://doi.org/10.1007/s40263-016-0405-9

Philip D (2009) Biosynthesis of Au, Ag and Au–Ag nanoparticles using edible mushroom extract. Spectrochim Acta A Mol Biomol Spectrosc 73(2):374–381

Philo JS, Arakawa T (2009) Mechanisms of protein aggregation. Curr Pharm Biotechnol 10(4):348–351

Prasad R, Pandey R, Barman I (2016) Engineering tailored nanoparticles with microbes: quo vadis. WIREs Nanomed Nanobiotechnol 8:316–330. https://doi.org/10.1002/wnan.1363

Pugazhenthiran N, Anandan S, Kathiravan G, Prakash NKU, Crawford S, Ashokkumar M (2009) Microbial synthesis of silver nanoparticles by Bacillus sp. J Nanopart Res 11(7):1811

Rambaran RN, Serpell LC (2008) Amyloid fibrils: abnormal protein assembly. Prion 2(3):112–117

Ross CA, Poirier MA (2004) Protein aggregation and neurodegenerative disease. Nat Med 10(suppl):S10–S17. https://doi.org/10.1038/nm1066

Saraiva C, Praca C, Ferreira R, Santos T, Ferreira L, Bernardino L (2016) Nanoparticle-mediated brain drug delivery: Overcoming blood-brain barrier to treat neurodegenerative diseases. J Control Release 235:34–47. https://doi.org/10.1016/j.jconrel.2016.05.044

Sawle BD, Salimath B, Deshpande R, Bedre MD, Prabhakar BK, Venkataraman A (2008) Biosynthesis and stabilization of Au and Au–Ag alloy nanoparticles by fungus, Fusarium semi-tectum. Sci Technol Adv Mater 9(3):035012

Senapati S, Ahmad A, Khan MI, Sastry M, Kumar R (2005) Extracellular biosynthesis of bimetal-lic Au–Ag alloy nanoparticles. Small 1(5):517–520

Serio TR, Cashikar AG, Kowal AS, Sawicki GJ, Moslehi JJ, Serpell L, Arnsdorf MF, Lindquist SL (2000) Nucleated conformational conversion and the replication of conformational information by a prion determinant. Science 289(5483):1317–1321

Shakibaie M, Forootanfar H, Mollazadeh-Moghaddam K, Bagherzadeh Z, Nafissi-Varcheh N, Shahverdi AR, Faramarzi MA (2010) Green synthesis of gold nanoparticles by the marine microalga Tetraselmis suecica. Biotechnol Appl Biochem 57(2):71–75

Sharma AK, Pavlova ST, Kim J, Finkelstein D, Hawco NJ, Rath NP, Kim J, Mirica LM (2012) Bifunctional compounds for controlling metal-mediated aggregation of the abeta42 peptide. J Am Chem Soc 134(15):6625–6636. https://doi.org/10.1021/ja210588m

Sonvico F, Dubernet C, Colombo P, Couvreur P (2005) Metallic colloid nanotechnology, applica-tions in diagnosis and therapeutics. Curr Pharm Des 11(16):2091–2105

Southam G, Beveridge TJ (1996) The occurrence of sulfur and phosphorus within bacterially derived crystalline and pseudocrystalline octahedral gold formed in vitro. Geochim Cosmochim Acta 60(22):4369–4376

Stephen JR, Macnaughtont SJ (1999) Developments in terrestrial bacterial remediation of metals. Curr Opin Biotechnol 10(3):230–233

Suganya KU, Govindaraju K, Kumar VG, Dhas TS, Karthick V, Singaravelu G, Elanchezhiyan M (2015) Blue green alga mediated synthesis of gold nanoparticles and its antibacterial efficacy against Gram positive organisms. Mater Sci Eng C 47:351–356

Syed A, Saraswati S, Kundu GC, Ahmad A (2013) Biological synthesis of silver nanoparticles using the fungus Humicola sp. and evaluation of their cytoxicity using normal and cancer cell lines. Spectrochim acta A Mol Biomol Spectrosc 114:144–147. https://doi.org/10.1016/j.saa.2013.05.030

Takahashi M, Yokota T, Kawano H, Gondo T, Ishihara T, Uchino F (1989) Ultrastructural evidence for intracellular formation of amyloid fibrils in macrophages. Virchows Arch A Pathol Anat Histopathol 415(5):411–419

Talham DR (2002) Biomineralization: principles and concepts in bioinorganic materials chemistry Stephen Mann. Oxford University Press, New York (2001 ACS Publications)

Tomiyama T, Asano S, Suwa Y, Morita T, Kataoka K, Mori H, Endo N (1994) Rifampicin pre-vents the aggregation and neurotoxicity of amyloid beta protein in vitro. Biochem Biophys Res Commun 204(1):76–83. https://doi.org/10.1006/bbrc.1994.2428

Tripathi RM, Gupta RK, Bhadwal AS, Singh P, Shrivastav A, Shrivastav B (2015) Fungal bio-molecules assisted biosynthesis of Au–Ag alloy nanoparticles and evaluation of their catalytic property. IET Nanobiotechnol 9(4):178–183

Tycko R (2011) Solid-state NMR studies of amyloid fibril structure. Annu Rev Phys Chem 62:279–299. https://doi.org/10.1146/annurev-physchem-032210-103539

Uversky VN (2007) Neuropathology, biochemistry, and biophysics of alpha-synuclein aggrega-tion. J Neurochem 103(1):17–37. https://doi.org/10.1111/j.1471-4159.2007.04764.x

Villaverde A, Carrio MM (2003) Protein aggregation in recombinant bacteria: biological role of inclusion bodies. Biotechnol Lett 25(17):1385–1395

Wang Q, Yu X, Patal K, Hu R, Chuang S, Zhang G, Zheng J (2013) Tanshinones inhibit amyloid aggregation by amyloid-beta peptide, disaggregate amyloid fibrils, and protect cultured cells. ACS Chem Nerosci 4(6):1004–1015. https://doi.org/10.1021/cn400051e

Wilkinson J (2003) Nanotechnology applications in medicine. Med Device Technol 14(5):29–31

Xiong N, Dong X-Y, Zheng J, Liu F-F, Sun Y (2015) Design of LVFFARK and LVFFARK-functionalized nanoparticles for inhibiting amyloid β-protein fibrillation and cytotoxicity. ACS Appl Mater Interfaces 7(10):5650–5662

Yallappa S, Manjanna J, Dhananjaya B (2015) Phytosynthesis of stable Au, Ag and Au–Ag alloy nanoparticles using J. sambac leaves extract, and their enhanced antimicrobial activity in presence of organic antimicrobials. Spectrochim Acta A Mol Biomol Spectrosc 137:236–243

Zhang X, Yan S, Tyagi R, Surampalli R (2011) Synthesis of nanoparticles by microorganisms and their application in enhancing microbiological reaction rates. Chemosphere 82(4):489–494

Zheng D, Hu C, Gan T, Dang X, Hu S (2010) Preparation and application of a novel vanillin sensor based on biosynthesis of Au–Ag alloy nanoparticles. Sens Actuators B 148(1):247–252

Zhou BR, Liang Y, Du F, Zhou Z, Chen J (2007) Mixed macromolecular crowding accelerates the oxidative refolding of reduced, denatured lysozyme. Implications for protein folding in intracellular environments (vol 279, pg 55109, 2004). J Biol Chem 282(37):27556–27556

Chapter 4
Synthesis and Characterization of Selenium Nanoparticles Using Natural Resources and Its Applications

S. Rajeshkumar, P. Veena, and R. V. Santhiyaa

4.1 Introduction

Nanoparticles are the particles with nano size of less than 100 nm and one or more dimension. Nanoparticles have a higher surface area which gives larger target interaction. It has many important properties such as low melting point, catalytic activity, high photoconductivity, and high semiconductivity (Presentato et al. 2018; Prasad et al. 2016). This property of nanoparticles is used in the medical application (Palomo-Siguero et al. 2016). Nanoparticles are the very major part of nanotechnology playing an important role in the lot of applications related to biomedical engineering, chemical technology, environmental engineering, polymer technology, biotechnology, food science, agricultural development, etc. (Prasad et al. 2014, 2016, 2017). The nanoparticles used in various applications are gold, silver, silver oxide, copper, copper oxide, copper sulfide, zinc, zinc oxide, zinc sulfide, cadmium sulfide, titanium oxide, zirconium, cerium oxide, selenium, chitosan, cellulose, silica, iron, iron oxide, antimony trioxide, zirconium dioxide, platinum, and palladium (Mandal et al. 2006; Arvizo et al. 2010; Zhang et al. 2011; Dizaj et al. 2014; Rajeshkumar and Bharath 2017).

The different types of chemical and physical methods and materials are used in the synthesis of different nanoparticles using citrate, ammonia, sodium borohydride and radiation-chemical, microwave irradiation, electrochemical process, photochemical, ionic liquids, sonochemical and laser ablation (Dahl et al. 2007; Bhaduri et al. 2013). The biological synthesis of nanoparticles is developed vigorously

S. Rajeshkumar (✉)
Department of Pharmacology, Saveetha Dental College and Hospitals, Saveetha Institute of Medical and Technical Sciences, Chennai, Tamil Nadu, India

P. Veena · R. V. Santhiyaa
Nanotherapy Laboratory, School of Bio-Sciences and Technology, Vellore Institute of Technology, Vellore, Tamil Nadu, India

© Springer Nature Switzerland AG 2018
R. Prasad et al. (eds.), *Exploring the Realms of Nature for Nanosynthesis*,
Nanotechnology in the Life Sciences, https://doi.org/10.1007/978-3-319-99570-0_4

nowadays (Fig. 4.1) with microorganisms such as bacteria like *Bacillus subtilis,* sulfate-reducing bacteria *Serratia marcescens, E. coli, B. licheniformis, Aeromonas* sp. *SH10, B. megatherium D01, Enterobacter cloacae, Klebsiella pneumoniae, Brevibacterium casei, Rhodobacter sphaeroides, Gluconacetobacter xylinus, Rhodopseudomonas palustris, Acetobacter xylinum, Klebsiella aerogenes,* marine bacteria *Enterococcus* sp., *Vibrio alginolyticus, S. oneidensis MR-1, Actinobacteria* sp., thermophilic bacteria *Thermoanaerobacter ethanolicus (TOR-39), Sulfurospirillum barnesii, B. selenitireducens, Morganella* sp., *Thermophilic bacteria TOR-39, Geobacter metallireducens GS-15, P. boryanum UTEX 485, Lactobacillus* sp., *Sulfurospirillum barnesii* (Narayanan and Sakthivel 2010; Malarkodi et al. 2013a, b; Rajeshkumar et al. 2014b; Rajeshkumar 2016b). The different types of fungus are involved in the synthesis of various metal nanoparticles through intracellular and extracellular methods. Some of the fungus involved in the nanoparticles synthesis are *Penicillium fellutanum, Aspergillus flavus, Fusarium oxysporum Colletotrichum* sp., *Trichothecium* sp., *Cladosporium cladosporioides, Penicillium brevicompactum, Phanerochaete chrysosporium, Trichoderma asperellum, Trichoderma viride, Phoma glomerata, Aspergillus niger, Coriolus versicolor, Mucor hiemalis,* and *Fusarium solani* (Lengke et al. 2011; Sharma et al. 2015; Aziz et al. 2016; Soumya et al. 2017).

Fig. 4.1 Green synthesis of nanoparticles

The marine macro algae such as *Sargassum wightii, Sargassum longifolium, Turbinaria conoides,* and *Padina tetrastromatica* are used in the synthesis of silver and gold nanoparticles synthesis and its applications on antibacterial, antifungal, and anticancer activities (Rajeshkumar et al. 2012, 2013a, b, 2014a, Sangeetha et al. 2017a, b). Apart from this some other natural resources like viruses tobacco mosaic virus (TMV) used in the synthesis of SiO_2, Fe_2O_3, CdS, and PbS, and M13 bacteriophage used in the semiconductor nanoparticles synthesis. The yeast *Candida glubrata, Schizosaccharomyces cerevisiae, Yarrowia lipolytica, Torulopsis* sp., *Pichia jadinii,* and *S. pombe* also involved in the biosynthesis of many metals and metal oxide nanoparticles (Mandal et al. 2006; Narayanan and Sakthivel 2010; Thakkar et al. 2010; Lengke et al. 2011; Moses 2014).

Finally the plants, one of the major resources used in the green synthesis of different types of nanoparticles and a lot of phytochemicals present in the plant responsible for the nanoparticles production. Some important plants involved in the green synthesis of nanoparticles are *Pongamia pinnata, Passiflora* sp., *Coleus aromaticus, Andrographis paniculata, Solanum Xanthocarpum, Capsicum annuum, Coriandrum Sativum, Euphorbia hirta, Gliricidia sepium, Ficus carica, Lippia citriodora, Paederia foetida, Rosa rugosa, Catharanthus roseus, Elaeagnus Indica, Arbutus unedo, Allium cepa,* Fissidens minutes, *Elaeagnus latifolia, Ocimum bacillicum, Anogeissus latifolia,* and *Vitis vinifera* (Vanaja and Annadurai 2012; Gnanajobitha et al. 2013; Vanaja et al. 2013, 2014; Prasad 2014; Paulkumar et al. 2014; Rajeshkumar 2016a; Rajeshkumar and Bharath 2017). The biomolecules present in the bacteria, fungus, algae, yeast, plant, and virus are major agents responsible for the synthesis of metal oxide, metal and metal sulfide nanoparticles synthesis and characters. The Fourier transform infrared spectroscopy is the major technique used in the identification of functional group involved in the synthesis of nanoparticles. Some of the biomolecules involved in the green synthesis of nanoparticles are proteins, carbohydrates, sulfated polysaccharides, nitrate reductases, carotenoids and NADPH-dependent enzymes 90-kDa protein, 66- and 10-kDa protein, membrane-bound quinine or membrane-bound/cytosolic pH-dependent oxidoreductase, glutamate and aspartate on the surface, peptidoglycan reducing sugars, α-NADPH-dependent sulfite reductase (35.6 kDa), and phytochelatin (Narayanan and Sakthivel 2010; Paulkumar et al. 2013, 2014; Prasad et al. 2016). A possible green synthesis of various nanoparticles using plant and/or plants' part has been illustrated in Fig. 4.2.

4.2 Selenium Nanoparticles

Selenium is one of the important trace elements which is required up to 40–300 µg for human body every day. It helps in regulating the function of the human body. It helps in protecting cardiovascular health, regulating thyroid hormones and immune response, and preventing progression of cancer (Gautam et al. 2017). The only small amount is required for maintaining the function, and a large amount of

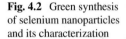

Fig. 4.2 Green synthesis of selenium nanoparticles and its characterization

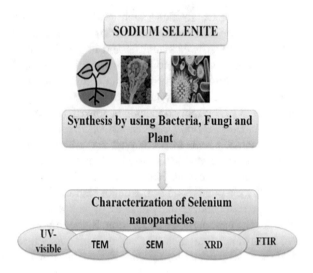

selenium may be harmful to the human body (Srivastava and Mukhopadhyay 2015a). Selenium has codons in mRNA which form seleno-cysteine by inserting as selenoprotein. Thioredoxin reductase and glutathione peroxidize are the enzymes comprising selenoprotein that function as antioxidant and detoxificant (Srivastava and Mukhopadhyay 2015b). Vitamins, ATP, and other capping agent stabilize and control the size, and they also increase the circulation and uptake of selenium nanoparticles (Zhu et al. 2017). Selenium also acts as an alternate for gold in alcohol beverage industry (Kalishwaralal et al. 2015). Selenium is less toxicity and high activity of selenium nanoparticles are used in many medical applications such as anticancer (Huang et al. 2013; Ramamurthy et al. 2013), antimicrobial (Bartůněk et al. 2015), drug delivery. The human body is able to degrade selenium nanoparticles naturally; the residues of the selenium nanoparticles act as the selenium nutrient source and are not toxic to the human body. High surface area and the low particle size of selenium nanoparticles enhance the biological activity. Selenium nanoparticles actively work against the antibiotic-resistant bacteria which has been increased recently (Bartůněk et al. 2015). The average size of the selenium nanoparticle is 80 nm and the property and activity of the selenium nanoparticles depend on the size, shape, and the biomolecule's nature (Husen and Siddiqi 2014).

Selenium nanoparticles can be synthesized by various methods. Usually, SeNPs is synthesized by the biological method and chemical method. Biological method of synthesis of selenium nanoparticle is by using living organisms such as bacteria, yeast, fungi, and plants (Husen and Siddiqi 2014; Ezhuthupurakkal et al. 2017; Hamza et al. 2017). Selenium nanoparticles synthesized using biological method have many applications such as antibacterial, antifungal, and also against pathogenic yeast (Dutta et al. 2014; Fernández-Llamosas et al. 2016). Chemical method of synthesis of selenium nanoparticles is by using chemicals as reducing agent for reducing selenite to selenium nanoparticles (Panahi-Kalamuei et al. 2014). Many other methods are also followed to synthesize selenium nanoparticles such as chemical vapor deposition, template free, template-assisted and electrodeposition (Chen et al. 2011).

4.3   Chemical Synthesis of Selenium Nanoparticles

Biopolymers such as gum acacia, glutathione, sodium alginate, and carboxymethyl cellulose are used for the synthesis of selenium nanoparticles (Tayebee et al. 2005). Several studies were carried out with the synthesis and application of the selenium nanoparticles. Different sizes of selenium nanoparticles are synthesized by facile reduction method by using ascorbic acid as reducing agent and to control the reaction kinetics polysorbate 20 surfactant is used. TEM is used to analyze the size and quality of the selenium nanoparticles. AAS is used to assess the presence of selenium. XRD is used to measure the structure of the nanoparticles. Three bacterial strains are used to test the antibacterial property of the synthesized nanoparticles. Selenium nanoparticles showed activity against common gram-positive bacteria *Staphylococcus epidermidis* and *Staphylococcus aureus* (Bartůněk et al. 2015). Selenium plays an important role in human health. The cells are protected by selenium from free radical. Bovine serum albumin and keratin are used to synthesize two different sized selenium nanoparticles. The characteristics of synthesized selenium nanoparticles are analyzed using SEM, XRD, and EDX. In vivo and in vitro antioxidative property of the selenium nanoparticles and sodium selenite is studied. The results show that the selenium nanoparticles can be effectively used in the alcoholic beverage industry as an additive (Kalishwaralal et al. 2015) by using beta-lactoglobulin as a stabilizer and ascorbic acid as reducing agent. TEM, dynamic light scattering (DLS), ultraviolet visible spectrophotometry, and FTIR are used to analyze the size, shape, structure, morphology, biological activity, and stability. Bls- SeNPs is highly stable in higher and lower pH. The synthesized nanoparticles are spherical in shape and the toxicity of the selenium nanoparticles is lower than that of selenite (Zhang et al. 2017). Lentinan is naturally occurring β-glucan which is used as stabilizing selenium nanoparticles. Transmission electron microscopy (TEM), X-ray diffraction (XRD), energy dispersive X-Ray (EDX), dynamic light scattering (DLS), and ultraviolet visible spectrophotometry are used to analyze the characteristics of the selenium nanoparticles. By controlling the concentration of the sodium selenite and the reducing agent, different size of the selenium nanoparticles are obtained. The anticancer activity of the selenium nanoparticles is based on the distribution of size (Jia et al. 2015). The shape of the selenium nanoparticles is controlled by designing the chemical structure of the stabilizing agent by a self-assembly process. Three hundred nanometer, the cubic structure of the selenium nanoparticles are synthesized using the folic acid-gallic acid-*N, N, N*-trimethyl chitosan as a stabilizing agent. Two hundred nanometer, spherical shaped particles are produced using unmodified chitosan. Better uptake of breast cancer cells and lesser toxicity towards noncancer cells are achieved by using cubic selenium nanoparticles (Luesakul et al. 2016).

4.4   Biological Synthesis of Selenium Nanoparticles

Synthesis of selenium nanoparticles by using bacterial biomass has gained huge interest due to its low cost, high effect, and environment-friendly nature. Several types of bacteria, yeast, fungi, and parts of plants are used based on its tolerance to selenite in order to convert it into selenium nanoparticles. Actinomycete, *Rhodococcus aetherivorans* BCP1 is anaerobic bacterium which is used to convert selenite into selenium nanoparticles. The strain has 500 mM of MHC and has high tolerance towards selenite. Two states of cells are used in bioconversion, conditioned and unconditioned. Bioconversion process is more effective in conditioned cells (Presentato et al. 2018). Fish gills were used to synthesize the selenium nanoparticles and used in thermal tolerance enhancement of *Pangasius hypophthalmus* high temperature and lead concentration (Kumar et al. 2017). A fast-growing bacterium *Vibrio natriegens* is used in the bioconversion of selenite into selenium nanoparticles. *Vibrio natriegens* is easily engineered with the gene of interest and can be grown with any carbon source. *Vibrio natriegens* has a high tolerance to selenite at an extreme concentration of 100 mM and effective growth is observed in 15 mM. X-ray spectroscopy and an electron microscope are used to analyze the characteristics of synthesized selenium nanoparticles and the growth of *Vibrio natriegens* on selenite. Culture is grown in LB media; spherical shaped selenium nanoparticles with 100–400 nm size are synthesized. Reduction of selenium is seen in the exponential growth phase of the cell and the synthesis of selenium nanoparticles is seen in cell lysis. *Vibrio natriegens* is used to enhance the speed of the production of selenium nanoparticles (Fernández-Llamosas et al. 2017). Development of an appropriate method for synthesis and retrieval of the nanoparticle is necessary. Selenium nanoparticles are synthesized using *Klebsiella pneumonia* strain from selenium chloride. Culture broth is sterilized at 121 °C for 20 min at 17 psi. Synthesized nanoparticles are 100–550 nm. Wet heat sterilization is carried out since no chemical changes are obtained (Fesharaki et al. 2010). Proteins also play an important role in the synthesis of the selenium nanoparticles. It involves in the production of selenium nanoparticles by reducing selenium oxyanions. Proteins involve in reduction of selenite to selenium and also they are involved in the stabilization of the selenium nanoparticles. The structure of the biosynthesized nanoparticles is different with the chemically synthesized nanoparticles. Their structure and characteristics are also analyzed (Tugarova and Kamnev 2017). Table 4.1 depicts the green synthesis of selenium nanoparticles along with their shapes and sizes.

Selenium nanoparticles synthesized using actinobacteria possess several other properties such as antioxidant, antiviral, and anticancer. *Streptomyces minutiscleroticus* M10A62 are able to synthesize selenium nanoparticles extracellularly. The different types of analysis such as XRD, FTIR, HR-TEM, and UV visible spectroscopy are carried out to study the characteristics of the selenium nanoparticles. The nanoparticles synthesized are of 100–250 nm with identical sphere structure. The selenium NPs synthesized from actinobacteria has activity against dengue virus (Ramya et al. 2015). *Enterococcus faecalis* are able to synthesize selenium NPs from

Table 4.1 Green synthesis of selenium nanoparticles

Source	Scientific name	UV-visible (nm)	FTIR	Size (in nm)	Shape	References
Bacteria	*Streptomyces minutiscleroticus*	510	FTIR shows that the protein secondary structure binds with the NPs	10–250	Spherical	Ramya et al. (2015)
Bacteria	*Enterococcus faecalis*	200–1000	–	29–195	Spherical	Shoeibi and Mashreghi (2017)
Bacteria	*Azospirillum brasilense*		FTIR shows that the surface of selenium NPs is capped by the polysaccharides and proteins of biomacromolecules	50–100	Nanosphere	Kamnev et al. (2017)
Bacteria	*Zooglea ramigera*	330	–	30–150	Spherical	Srivastava and Mukhopadhyay (2013)
Bacteria	*Alcaligenes* sp.	300–800	–	90–120	Rod-like structure	Mesbahi-Nowrouzi and Mollania (2018)
Bacteria	*Idiomarina* sp. PR58-8	320	–	150–350	Spherical	Srivastava and Kowshik (2016)
Bacteria	*Pseudomonas aeruginosa*	300–900	FTIR peaks show that the bacterial protein involves in the synthesis and stabilization of NPs	47–165	Spherical	Kora and Rastogi (2016)
Bacteria	*Stenotrophomonasmaltophilia*		Carbohydrates, proteins, and lipids act as the capping agent	160–250	Spherical	Lampis et al. (2017)
Bacteria	*Azospirillum thiophilum*		FTIR shows that the surface of selenium NPs is capped by the polysaccharides and proteins of biomacromolecules	160–250	Nanosphere shape	Tugarova et al. (2018)
Proteobacteria	*Burkholderia fungorum*		–	170–200	Spherical	Khoei et al. (2017)

(continued)

Table 4.1 (continued)

Source	Scientific name	UV-visible (nm)	FTIR	Size (in nm)	Shape	References
Protozoa	*Tetrahymena thermophila* SB210Yin	240–600	–	50–500	Amorphous sphere	Cui et al. (2016)
Yeast	*Yarrowia lipolytica*	220	Presence of the amine and carboxylic group in nanoparticle synthesis			Hamza et al. (2017)
Yeast	*Pichia pastoris*		–	180	Spherical	Elahian et al. (2017)
Fungi	*Cordyceps sinensis*	200–800	The result shows that surface atom of SeNPs and OH group of EPS has strong bonding	80–125	Homogenous sphere	Xiao et al. (2017)
Plant	*Azadirachta indica* leaf	600	The FTIR result shows that the selenium nanoparticles and the biomolecule of the plant leaf have the physical interaction	35	Spherical	Tareq et al. (2018)
Plant	*Allium sativum*	200–800	Peaks in FTIR confirm that the formation of SeNPs is due to the organic acid, polymeric agent of *Allium sativum* and they also act as the stabilizing agent	205	Crystalline sphere	Ezhuthupurakkal et al. (2017), Satgurunathan et al. (2017)
Plant	Lemon leaf	395	Hydroxyl, amine, carboxyl are involved in SeNPs synthesis	60–80	Sphere	Prasad et al. (2013)
Plant	Fenugreek seed	200–400	FTIR shows that the selenious acid reduction is due to the bond stretching and bending of functional group	50–150	Oval	Ramamurthy et al. (2013)
Plant	*Vitis vinifera*		The result shows that the selenium nanoparticles are synthesized using the reducing agent present in the *V. vinifera.*	3–18	Sphere	Husen and Siddiqi (2014)
Plant	*Bougainvillea spectabilis* wildflower	590	Flavonoids present in the *B. spectabilis* act as the reducing agent in the synthesis of SeNPs	25	Hollow crystals	Ganesan (2015)

Source	Scientific name	UV-visible (nm)	FTIR	Size (in nm)	Shape	References
Plant	*Catharanthus roseus*	325	Reduction of sodium selenite is due to the vibration by stretching of ketone	17–34	Spherical	Deepa and Ganesan (2015)
Plant	*Peltophorum pterocarpum*	325	–	21–42	Spherical	Deepa and Ganesan (2015)
Plant	*Diospyros Montana*	270–350	Alkaloids, steroids, and flavonoids from leaf act as the capping agent	4–16	Spherical	Kokila et al. (2017)

sodium selenite. The characteristics of the NPs are analyzed using TEM and UV-spectroscopy. TEM reveals that the size of the NPs is 29–195 nm and is spherical in shape (Shoeibi and Mashreghi 2017). *Idiomarina* sp. PR58-8 is a bacteria used in the process of reduction of sodium selenite to selenium nanoparticles. The nanoparticles are 150–350 nm in size and spherical in shape. The synthesized nanoparticles are highly anti-neoplastic (Srivastava and Kowshik 2016). In vivo synthesis of selenium nanoparticles by protozoa is environment-friendly without complicated devices. Several other genes are also regulated with the nanoparticles producing group. The synthesized nanoparticles are spherical in shape with diameter of 50–500 nm (Cui et al. 2016). A biomass of rhizobacterium azospirillum brasilence is used to synthesize selenium nanoparticles by selenite reduction. FTIR and TEM are used to analyze the structure and morphology which reveals that the nanoparticles are homogenous with average diameter 50–200 nm (Kamnev et al. 2017). The bacterial protein of *Zooglea ramigera* is responsible for the reduction of selenium nanoparticles. The prepared nanoparticles are hexagonal phase nanocrystals. XRD, TEM, SEM, SAED, and DLS are carried out to test the characteristics of the nanoparticles which reveals that the SeNPs is 30–150 nm and spherical in shape (Srivastava and Mukhopadhyay 2013). Gram-negative bacteria *Pseudomonas aeruginosa* is used in the reduction of selenite to selenium nanoparticle. Presence of selenium nanoparticles is confirmed by UV-spectra peaks and the color change to red in broth.

Cordyceps sinensis is a fungus that is used to synthesize EPS conjugated selenium nanoparticles. Characteristics are analyzed and the results reveal that the size is uniform and the shape is a 80–125 nm homogenous sphere (Xiao et al. 2017). *Azadirachta indica* leaf is used in the reduction of selenium ion to selenium nanoparticle. By conducting different analysis it is found that the selenium nanoparticles are spherical in shape and have average size of 35 nm. The synthesized nanoparticles are antibacterial and antioxidant (Tareq et al. 2018). Eco-friendly preparation of selenium nanoparticles is done by using an aqueous extract of *Allium sativum* which has good interaction with the DNA. This property is used in chemotherapy for DNA targeted (Ezhuthupurakkal et al. 2017). Selenium nanoparticles synthesized using lemon leaf extract are used to treat the DNA damage caused by UV exposure (Prasad et al. 2013). Fenugreek seed extract is also used for reduction of the selenious acid to selenite nanoparticles in an eco-friendly process. The synthesized nanoparticles are loaded with doxorubicin which acts as an anticancer agent. Flower of *Bougainvillae spectabilis* will be used for reducing sodium selenite to selenium nanoparticles. The synthesis is easy, cost-effective, and fast (Ganesan 2015). Using the aqueous extract of the Diospyros Montana leaf, selenium nanoparticles are synthesized by the eco-friendly process. The synthesis is a simple precipitation process and different analysis is made for characterization of the synthesized nanoparticles (Kokila et al. 2017). Detailed applications of selenium nanoparticles have been summarized in Table 4.2.

Yarrowia lipolytica is marine yeast that shows tolerance for the sodium selenite and is able to form selenium nanoparticles. *Artemia salina* is fed with the selenium nanoparticles rich biomass and the size of the larvae is found to be larger (Hamza et al. 2017). Applications of selenium nanoparticles in various fields have been illustrated in Fig. 4.3.

Table 4.2 Applications of selenium nanoparticles

Field	Application	Remarks	References
Antibacterial	Selenium NPs act against *Staphylococcus aureus* which is used in catheterization and prosthesis	8 mm zone is obtained in disc diffusion test	Shoeibi and Mashreghi (2017), Huang et al. (2016)
	Multidrug-resistant superbugs	Quercetin and acetylcholine decorated on the surfaces of the Se NPs act as antibacterial against MRSA	
Drug delivery	HepG2 hepatocellular carcinoma, breast adenocarcinoma cells are tested with doxorubicin	Cell apoptosis is promoted by activating p53 pathway	Huang et al. (2013)
	Human breast carcinoma cells, murine colon carcinoma cells are inhibited by epirubicin	Human breast carcinoma cells, murine colon carcinoma cells are inhibited by epirubicin	Jalalian et al. (2018)
Cancer	Therapeutic agent	Several biomacromolecules, polymers, and metals are conjugated with selenium NPs to achieve maximum efficacy for human cancer	Fernandes and Gandin (2015)
Dye degradation	Degradation of model dye trypan blue	ROS generated by the selenium doped zinc nanoparticles are efficient in degrading trypan blue	Dutta et al. (2014)
Inhibits biofilm	Staphylococcus aureus biofilm formation is inhibited	SeNPs synthesized using Bacillus licheniform is used for inhibiting the growth of Staphylococcus aureus adherence on surfaces	Sonkusre and Singh Cameotra (2015)
Cancer	Chemoradiation for cervical cancer cells, mouse embryonic fibroblast	Selenium NPs with PEG surface decorated possess X-ray responsive property and act as a therapeutic agent for cancer	Yu et al. (2016)
Anti-leishmaniasis	Selenium nanoparticles and selenium dioxide are tested against *Leishmania infantum*	Selenium nanoparticles are more efficient than selenium dioxide	Soflaei et al. (2014)

Selenium nanoparticles synthesized from *Enterococcus faecalis* are active against *Staphylococcus aureus*. Antibacterial test for selenium nanoparticles against gram-positive bacteria *Bacillus subtilis*, *Staphylococcus aureus* and gram-negative *Pseudomonas aeruginosa and Escherichia coli* is done by disc diffusion method. The result shows that selenium nanoparticles are active against *Staphylococcus aureus* with 8 mm zone (Shoeibi and Mashreghi 2017). Multidrug-resistant superbugs are also a great threat to the world. To overcome this selenium nanoparticle is decorated with quercetin and acetylcholine on the surface which permanently damage the cell membrane of methicillin-resistant *Staphylococcus aureus* (Huang et al. 2016).

Fig. 4.3 Applications of
selenium nanoparticles in
various fields

Selenium nanoparticles have great attention towards cancer therapy. Nanoparticles
are widely used for a selected target which reduces the effect on normal cells.
Nanoparticles based drug delivery is widely due to its major advantages. Transferrin-
conjugated selenium nanoparticles is loaded with cancer drug doxorubicin. By activating p53 and MAPKs pathways, transferrin-conjugated SeNPs mediate cell
apoptosis. This design of drug delivery decreases the side effects and increases the
efficiency (Huang et al. 2013). Dye degradation is enhanced by the oxidation process
which is carried out by the selenium doped zinc nanoparticle for degrading trypan
blue dye. The degradation is carried out by selenium doped zinc nanoparticle by ROS
oxidative system (Dutta et al. 2014). Selenium nanoparticles synthesized using
Bacillus licheniform are used for inhibiting the biofilm formation of *Staphylococcus
aureus* on surfaces such as glass, catheter, and polystyrene which can be used to prevent the formation of biofilm on medical devices (Sonkusre and Singh Cameotra
2015). Chemoresistant and radioresistant cancer cells are treated using chemoradia-
tion. Selenium nanoparticles decorated with PEG have an X-ray responsive property
and the growth of cancer is inhibited by cell apoptosis (Yu et al. 2016).

4.5   Conclusion

Selenium nanoparticles are synthesized using living organisms and physical/chemi-
cal method. SeNPs prepared using bacteria have high effect against pathogenic bacteria and yeast. Selenium nanoparticles show less toxicity towards normal cells and
more toxic to cancer cells. This property can be used in drug delivery and for cancer
therapy. Selenium nanoparticles are degraded by the human body naturally and the
selenium acts as a source of nutrients for the function of the human and animal

body. Selenium nanoparticles synthesized using several methods possess many important properties which can be used in different applications such as drug delivery, cancer therapy, and dye degradation. This property enhances the use of selenium nanoparticles in the medical and biological field.

References

Arvizo R, Bhattacharya R, Mukherjee P (2010) Gold nanoparticles: opportunities and challenges in nanomedicine. Expert Opin Drug Deliv 7:753–763. https://doi.org/10.1517/17425241003777010

Aziz N, Pandey R, Barman I, Prasad R (2016) Leveraging the attributes of Mucor hiemalis-derived silver nanoparticles for a synergistic broad-spectrum antimicrobial platform. Front Microbiol 7:1984. https://doi.org/10.3389/fmicb.2016.01984

Bartůněk V, Junková J, Šuman J et al (2015) Preparation of amorphous antimicrobial selenium nanoparticles stabilized by odor suppressing surfactant polysorbate 20. Mater Lett 152:207–209. https://doi.org/10.1016/j.matlet.2015.03.092

Bhaduri GA, Little R, Khomane RB, Lokhande SU (2013) Green synthesis of silver nanoparticles using sunlight. J Photochem Photobiol A Chem 258:1–9. https://doi.org/10.1016/j.jphotochem.2013.02.015

Chen H, Yoo JB, Liu Y, Zhao G (2011) Green synthesis and characterization of Se nanoparticles and nanorods. Electron Mater Lett 7:333–336. https://doi.org/10.1007/s13391-011-0420-4

Cui YH, Li LL, Zhou NQ et al (2016) In vivo synthesis of nano-selenium by Tetrahymena thermophila SB210. Enzyme Microb Technol 95:185–191. https://doi.org/10.1016/j.enzmictec.2016.08.017

Dahl JA, Maddux BLS, Hutchison JE (2007) Toward greener nanosynthesis. Chem Rev 107:2228–2269. https://doi.org/10.1021/cr050943k

Deepa B, Ganesan V (2015) Bioinspiredsynthesis of selenium nanoparticles using flowers of Catharanthus roseus(L.) G.Don.and Peltophorum pterocarpum(DC.)Backer ex Heyne – a comparison. Int J ChemTech Res 7:725–733

Dizaj SM, Lotfipour F, Barzegar-Jalali M et al (2014) Antimicrobial activity of the metals and metal oxide nanoparticles. Mater Sci Eng C 44:278–284. https://doi.org/10.1016/j.msec.2014.08.031

Dutta RK, Nenavathu BP, Talukdar S (2014) Anomalous antibacterial activity and dye degradation by selenium doped ZnO nanoparticles. Colloids Surf B Biointerfaces 114:218–224. https://doi.org/10.1016/j.colsurfb.2013.10.007

Elahian F, Reiisi S, Shahidi A, Mirzaei SA (2017) High-throughput bioaccumulation, biotransformation, and production of silver and selenium nanoparticles using genetically engineered Pichia pastoris. Nanomed Nanotechnol Biol Med 13:853–861. https://doi.org/10.1016/j.nano.2016.10.009

Ezhuthupurakkal PB, Polaki LR, Suyavaran A et al (2017) Selenium nanoparticles synthesized in aqueous extract of Allium sativum perturbs the structural integrity of Calf thymus DNA through intercalation and groove binding. Mater Sci Eng C 74:597–608. https://doi.org/10.1016/j.msec.2017.02.003

Fernandes AP, Gandin V (2015) Selenium compounds as therapeutic agents in cancer. Biochim Biophys Acta 1850:1642–1660. https://doi.org/10.1016/j.bbagen.2014.10.008

Fernández-Llamosas H, Castro L, Blázquez ML et al (2016) Biosynthesis of selenium nanoparticles by Azoarcus sp. CIB. Microb Cell Fact 15:1–10. https://doi.org/10.1186/s12934-016-0510-y

Fernández-Llamosas H, Castro L, Blázquez ML et al (2017) Speeding up bioproduction of selenium nanoparticles by using Vibrio natriegens as microbial factory. Sci Rep 7:1–9. https://doi.org/10.1038/s41598-017-16252-1

Fesharaki PJ, Nazari P, Shakibaie M et al (2010) Biosynthesis of selenium nanoparticles using Klebsiella pneumoniae and their recovery by a simple sterilization process. Braz J Microbiol 41:461–466. https://doi.org/10.1590/S1517-83822010000200028

Ganesan V (2015) Biogenic synthesis and characterization of selenium nanoparticles using the flower of Bougainvillea spectabilis Willd. Int J Sci Res 4:690–695

Gautam PK, Kumar S, Tomar MS et al (2017) Selenium nanoparticles induce suppressed function of tumor associated macrophages and inhibit Dalton's lymphoma proliferation. Biochem Biophys Rep 12:172–184. https://doi.org/10.1016/j.bbrep.2017.09.005

Gnanajobitha G, Paulkumar K, Vanaja M et al (2013) Fruit-mediated synthesis of silver nanoparticles using Vitis vinifera and evaluation of their antimicrobial efficacy. J Nanostruct Chem 3:67

Hamza F, Vaidya A, Apte M et al (2017) Selenium nanoparticle-enriched biomass of Yarrowia lipolytica enhances growth and survival of Artemia salina. Enzyme Microb Technol 106:48–54. https://doi.org/10.1016/j.enzmictec.2017.07.002

Huang Y, He L, Liu W et al (2013) Selective cellular uptake and induction of apoptosis of cancer-targeted selenium nanoparticles. Biomaterials 34:7106–7116. https://doi.org/10.1016/j.biomaterials.2013.04.067

Huang X, Chen X, Chen Q et al (2016) Investigation of functional selenium nanoparticles as potent antimicrobial agents against superbugs. Acta Biomater 30:397–407. https://doi.org/10.1016/j.actbio.2015.10.041

Husen A, Siddiqi KS (2014) Plants and microbes assisted selenium nanoparticles: characterization and application. J Nanobiotechnol 12:1–10. https://doi.org/10.1186/s12951-014-0028-6

Jalalian SH, Ramezani M, Abnous K, Taghdisi SM (2018) Targeted co-delivery of epirubicin and NAS-24 aptamer to cancer cells using selenium nanoparticles for enhancing tumor response in vitro and in vivo. Cancer Lett 416:87–93. https://doi.org/10.1016/j.canlet.2017.12.023

Jia X, Liu Q, Zou S et al (2015) Construction of selenium nanoparticles/β-glucan composites for enhancement of the antitumor activity. Carbohydr Polym 117:434–442. https://doi.org/10.1016/j.carbpol.2014.09.088

Kalishwaralal K, Jeyabharathi S, Sundar K, Muthukumaran A (2015) Sodium selenite/selenium nanoparticles (SeNPs) protect cardiomyoblasts and zebrafish embryos against ethanol induced oxidative stress. J Trace Elem Med Biol 32:135–144. https://doi.org/10.1016/j.jtemb.2015.06.010

Kamnev AA, Mamchenkova PV, Dyatlova YA, Tugarova AV (2017) FTIR spectroscopic studies of selenite reduction by cells of the rhizobacterium Azospirillum brasilense Sp7 and the formation of selenium nanoparticles. J Mol Struct 1140:106–112. https://doi.org/10.1016/j.molstruc.2016.12.003

Khoei NS, Lampis S, Zonaro E et al (2017) Insights into selenite reduction and biogenesis of elemental selenium nanoparticles by two environmental isolates of Burkholderia fungorum. N Biotechnol 34:1–11. https://doi.org/10.1016/j.nbt.2016.10.002

Kokila K, Elavarasan N, Sujatha V (2017) Diospyros montana leaf extract-mediated synthesis of selenium nanoparticles and their biological applications. New J Chem. https://doi.org/10.1039/C7NJ01124E

Kora AJ, Rastogi L (2016) Biomimetic synthesis of selenium nanoparticles by Pseudomonas aeruginosa ATCC 27853: an approach for conversion of selenite. J Environ Manage 181:231–236. https://doi.org/10.1016/j.jenvman.2016.06.029

Kumar N, Krishnani KK, Gupta SK, Singh NP (2017) Selenium nanoparticles enhanced thermal tolerance and maintain cellular stress protection of Pangasius hypophthalmus reared under lead and high temperature. Respir Physiol Neurobiol 246:107–116. https://doi.org/10.1016/j.resp.2017.09.006

Lampis S, Zonaro E, Bertolini C et al (2017) Selenite biotransformation and detoxification by Stenotrophomonas maltophilia SeITE02: novel clues on the route to bacterial biogenesis of selenium nanoparticles. J Hazard Mater 324:3–14. https://doi.org/10.1016/j.jhazmat.2016.02.035

Lengke MF, Sanpawanitchakit C, Southam G (2011) Biosynthesis of gold nanoparticles: a review. Met Nanoparticles Microbiol:37–74. https://doi.org/10.1007/978-3-642-18312-6_3

Luesakul U, Komenek S, Puthong S, Muangsin N (2016) Shape-controlled synthesis of cubic-like selenium nanoparticles via the self-assembly method. Carbohydr Polym 153:435–444. https://doi.org/10.1016/j.carbpol.2016.08.004

Malarkodi C, Rajeshkumar S, Paulkumar K et al (2013a) Biosynthesis of semiconductor nanoparticles by using sulfur reducing bacteria Serratia nematodiphila. Adv Nano Res 1:83–91. https://doi.org/10.12989/anr.2013.1.2.083

Malarkodi C, Rajeshkumar S, Paulkumar K et al (2013b) Bactericidal activity of bio mediated silver nanoparticles synthesized by Serratia nematodiphila. Drug Invent Today 5:119–125. https://doi.org/10.1016/j.dit.2013.05.005

Mandal D, Bolander ME, Mukhopadhyay D et al (2006) The use of microorganisms for the formation of metal nanoparticles and their application. Appl Microbiol Biotechnol 69:485–492. https://doi.org/10.1007/s00253-005-0179-3

Mesbahi-Nowrouzi M, Mollania N (2018) Purification of selenate reductase from Alcaligenes sp. CKCr-6A with the ability to biosynthesis of selenium nanoparticle: enzymatic behavior study in imidazolium based ionic liquids and organic solvent. J Mol Liq 249:1254–1262. https://doi.org/10.1016/j.molliq.2017.10.117

Moses V (2014) Biological synthesis of copper nanoparticles and its impact - a review. Int J Pharm Sci Invent 3:2319–6718

Narayanan KB, Sakthivel N (2010) Biological synthesis of metal nanoparticles by microbes. Adv Colloid Interface Sci 156:1–13. https://doi.org/10.1016/j.cis.2010.02.001

Palomo-Siguero M, Gutiérrez AM, Pérez-Conde C, Madrid Y (2016) Effect of selenite and selenium nanoparticles on lactic bacteria: a multi-analytical study. Microchem J 126:488–495. https://doi.org/10.1016/j.microc.2016.01.010

Panahi-Kalamuei M, Salavati-Niasari M, Hosseinpour-Mashkani SM (2014) Facile microwave synthesis, characterization, and solar cell application of selenium nanoparticles. J Alloys Compd 617:627–632. https://doi.org/10.1016/j.jallcom.2014.07.174

Paulkumar K, Rajeshkumar S, Gnanajobitha G et al (2013) Biosynthesis of silver chloride nanoparticles using Bacillus subtilis MTCC 3053 and assessment of its antifungal activity. ISRN Nanomater 2013:1–8. https://doi.org/10.1155/2013/317963

Paulkumar K, Gnanajobitha G, Vanaja M et al (2014) Piper nigrum leaf and stem assisted green synthesis of silver nanoparticles and evaluation of its antibacterial activity against agricultural plant pathogens. ScientificWorldJournal 2014:829894. https://doi.org/10.1155/2014/829894

Prasad R (2014) Synthesis of silver nanoparticles in photosynthetic plants. J Nanoparticles Article ID 963961. https://doi.org/10.1155/2014/963961

Prasad KS, Patel H, Patel T et al (2013) Biosynthesis of Se nanoparticles and its effect on UV-induced DNA damage. Colloids Surf B Biointerfaces 103:261–266. https://doi.org/10.1016/j.colsurfb.2012.10.029

Prasad R, Kumar V, Prasad KS (2014) Nanotechnology in sustainable agriculture: present concerns and future aspects. Afr J Biotechnol 13(6):705–713

Prasad R, Pandey R, Barman I (2016) Engineering tailored nanoparticles with microbes: quo vadis. WIREs Nanomed Nanobiotechnol 8:316–330. https://doi.org/10.1002/wnan.1363

Prasad R, Bhattacharyya A, Nguyen QD (2017) Nanotechnology in sustainable agriculture: recent developments, challenges, and perspectives. Front Microbiol 8:1014. https://doi.org/10.3389/fmicb.2017.01014

Presentato A, Piacenza E, Anikovskiy M et al (2018) Biosynthesis of selenium-nanoparticles and -nanorods as a product of selenite bioconversion by the aerobic bacterium Rhodococcus aetherivorans BCP1. N Biotechnol 41:1–8. https://doi.org/10.1016/j.nbt.2017.11.002

Rajeshkumar S (2016a) Green synthesis of different sized antimicrobial silver nanoparticles using different parts of plants – a review. Int J ChemTech Res 9:197–208

Rajeshkumar S (2016b) Anticancer activity of eco-friendly gold nanoparticles against lung and liver cancer cells. J Genet Eng Biotechnol 14:195–202. https://doi.org/10.1016/j.jgeb.2016.05.007

Rajeshkumar S, Bharath LV (2017) Mechanism of plant-mediated synthesis of silver nanoparticles – a review on biomolecules involved, characterisation and antibacterial activity. Chem Biol Interact 273:219–227. https://doi.org/10.1016/j.cbi.2017.06.019

Rajeshkumar S, Kannan C, Annadurai G (2012) Green synthesis of silver nanoparticles using marine brown Algae turbinaria conoides and its antibacterial activity. Int J Pharm Bio Sci

Rajeshkumar S, Malarkodi C, Gnanajobitha G et al (2013a) Seaweed-mediated synthesis of gold nanoparticles using Turbinaria conoides and its characterization. J Nanostruct Chem 3:44. https://doi.org/10.1186/2193-8865-3-44

Rajeshkumar S, Malarkodi C, Vanaja M et al (2013b) Antibacterial activity of algae mediated synthesis of gold nanoparticles from turbinaria conoides. Der Pharma Chem 5:224–229

Rajeshkumar S, Malarkodi C, Paulkumar K et al (2014a) Algae mediated green fabrication of silver nanoparticles and examination of its antifungal activity against clinical pathogens. Int J Metal 2014:1–8. https://doi.org/10.1155/2014/692643

Rajeshkumar S, Ponnanikajamideen M, Malarkodi C et al (2014b) Microbe-mediated synthesis of antimicrobial semiconductor nanoparticles by marine bacteria. J Nanostruct Chem 4:96. https://doi.org/10.1007/s40097-014-0096-z

Ramamurthy CH, Sampath KS, Arunkumar P et al (2013) Green synthesis and characterization of selenium nanoparticles and its augmented cytotoxicity with doxorubicin on cancer cells. Bioprocess Biosyst Eng 36:1131–1139. https://doi.org/10.1007/s00449-012-0867-1

Ramya S, Shanmugasundaram T, Balagurunathan R (2015) Biomedical potential of actinobacterially synthesized selenium nanoparticles with special reference to anti-biofilm, anti-oxidant, wound healing, cytotoxic and anti-viral activities. J Trace Elem Med Biol 32:30–39. https://doi.org/10.1016/j.jtemb.2015.05.005

Sangeetha J, Gayathri S, Rajeshkumar S (2017a) Antimicrobial assessment of marine brown algae Sargassum whitti against UTI pathogens and its phytochemical analysis. Res. J Pharm Technol. https://doi.org/10.5958/0974-360X.2017.00334.1

Sangeetha J, Gayathri S, Rajeshkumar S (2017b) Antimicrobial assessment of marine brown algae Sargassum whitti against UTI pathogens and its phytochemical analysis. Res J Pharm Technol 10:6–11

Satgurunathan T, Bhavan PS, Komathi S (2017) Green synthesis of selenium nanoparticles from sodium selenite using garlic extract and its enrichment on Artemia nauplii to feed the freshwater prawn Macrobrachium rosenbergii post-larvae. Res J Chem Environ 21:1–12

Sharma D, Kanchi S, Bisetty K (2015) Biogenic synthesis of nanoparticles: a review. Arab J Chem. https://doi.org/10.1016/j.arabjc.2015.11.002

Shoeibi S, Mashreghi M (2017) Biosynthesis of selenium nanoparticles using Enterococcus faecalis and evaluation of their antibacterial activities. J Trace Elem Med Biol 39:135–139. https://doi.org/10.1016/j.jtemb.2016.09.003

Soflaei S, Dalimi A, Abdoli A et al (2014) Anti-leishmanial activities of selenium nanoparticles and selenium dioxide on Leishmania infantum. Comp Clin Pathol 23:15–20. https://doi.org/10.1007/s00580-012-1561-z

Soumya M, Venkat Kumar S, Rajeshkumar S (2017) A review on biogenic synthesis of gold nanoparticles, characterization, and its applications. 1643:1–12. https://doi.org/10.1016/j.reffit.2017.08.002 Article In Press

Sonkusre P, Singh Cameotra S (2015) Biogenic selenium nanoparticles inhibit Staphylococcus aureus adherence on different surfaces. Colloids Surf B Biointerfaces 136:1051–1057. https://doi.org/10.1016/j.colsurfb.2015.10.052

Srivastava P, Kowshik M (2016) Anti-neoplastic selenium nanoparticles from Idiomarina sp. PR58-8. Enzyme Microb Technol 95:192–200. https://doi.org/10.1016/j.enzmictec.2016.08.002

Srivastava N, Mukhopadhyay M (2013) Biosynthesis and structural characterization of selenium nanoparticles mediated by Zooglea ramigera. Powder Technol 244:26–29. https://doi.org/10.1016/j.powtec.2013.03.050

Srivastava N, Mukhopadhyay M (2015a) Green synthesis and structural characterization of selenium nanoparticles and assessment of their antimicrobial property. Bioprocess Biosyst Eng 38:1723–1730. https://doi.org/10.1007/s00449-015-1413-8

Srivastava N, Mukhopadhyay M (2015b) Biosynthesis and structural characterization of selenium nanoparticles using Gliocladium roseum. J Clust Sci 26:1473–1482. https://doi.org/10.1007/s10876-014-0833-y

Tareq FK, Fayzunnesa M, Kabir MS, Nuzat M (2018) Mechanism of bio molecule stabilized selenium nanoparticles against oxidation process and Clostridium Botulinum. Microb Pathog 115:68–73. https://doi.org/10.1016/j.micpath.2017.12.042

Tayebee R, Silva PBM, Oliveira KA, Coltro WKT (2005) Short Report. Tetrahedron 16:108–111. https://doi.org/10.7196/AJHPE.2016.v8i1.523

Thakkar KN, Mhatre SS, Parikh RY (2010) Biological synthesis of metallic nanoparticles. Nanomed Nanotechnol Biol Med 6:257–262. https://doi.org/10.1016/j.nano.2009.07.002

Tugarova AV, Kamnev AA (2017) Proteins in microbial synthesis of selenium nanoparticles. Talanta 174:539–547. https://doi.org/10.1016/j.talanta.2017.06.013

Tugarova AV, Mamchenkova PV, Dyatlova YA, Kamnev AA (2018) FTIR and Raman spectroscopic studies of selenium nanoparticles synthesised by the bacterium Azospirillum thiophilum. Spectrochim Acta A Mol Biomol Spectrosc 192:458–463. https://doi.org/10.1016/j.saa.2017.11.050

Vanaja M, Annadurai G (2012) Coleus aromaticus leaf extract mediated synthesis of silver nanoparticles and its bactericidal activity. Appl Nanosci. https://doi.org/10.1007/s13204-012-0121-9

Vanaja M, Rajeshkumar S, Paulkumar K et al (2013) Kinetic study on green synthesis of silver nanoparticles using Coleus aromaticus leaf extract. Pelagia Res Libr 4:50–55

Vanaja M, Paulkumar K, Gnanajobitha G et al (2014) Herbal plant synthesis of antibacterial silver nanoparticles by Solanum trilobatum and its characterization. Int J Metal 2014:1–8. https://doi.org/10.1155/2014/692461

Xiao Y, Huang Q, Zheng Z et al (2017) Construction of a Cordyceps sinensis exopolysaccharide-conjugated selenium nanoparticles and enhancement of their antioxidant activities. Int J Biol Macromol 99:483–491. https://doi.org/10.1016/j.ijbiomac.2017.03.016

Yu B, Liu T, Du Y et al (2016) X-ray-responsive selenium nanoparticles for enhanced cancer chemo-radiotherapy. Colloids Surf B Biointerfaces 139:180–189. https://doi.org/10.1016/j.colsurfb.2015.11.063

Zhang X, Yan S, Tyagi RD, Surampalli RY (2011) Synthesis of nanoparticles by microorganisms and their application in enhancing microbiological reaction rates. Chemosphere 82:489–494. https://doi.org/10.1016/j.chemosphere.2010.10.023

Zhang J, Teng Z, Yuan Y et al (2017) Development, physicochemical characterization and cytotoxicity of selenium nanoparticles stabilized by beta-lactoglobulin. Int J Biol Macromol 107:1406–1413. https://doi.org/10.1016/j.ijbiomac.2017.09.117

Zhu C, Zhang S, Song C et al (2017) Selenium nanoparticles decorated with Ulva lactuca polysaccharide potentially attenuate colitis by inhibiting NF-KB mediated hyper inflammation. J Nanobiotechnol 15:1–15. https://doi.org/10.1186/s12951-017-0252-y

Chapter 5
Nanofabrication by Cryptogams: Exploring the Unexplored

Sabiha Zamani, Babita Jha, Anal K. Jha, and Kamal Prasad

5.1 Introduction

The canvas of nanotechnology recounts diverse fascinating endeavours to control, understand, and manipulate matters at infinitesimal scale. This rapidly emerging and fascinating technology, that deals with nanometre (10^{-9}) sized entities, as a breakthrough among the existent technologies, promises to unravel the nanoscale processes for novel applications. Cells, which is considered as the basic unit of all living organisms, are usually the size of 10 μm while the cell organelles although smaller yet lies in the micron size. The size of proteins, which is roughly around 5 nm, is equivalent to the size of smallest manmade nanoparticles (Taton 2002). This gives an idea as to how small 'nano' is and nanomaterials with dimension lying amidst 1–100 nm are blessed with amazing properties as compared to its bulk counterpart. Novel properties of materials at the nanoscale may be attributed to its altered physicochemical properties and surface to volume ratio (Suman et al. 2010). At nanoscale, the optical (Parak et al. 2003), physical, and magnetic (Pankhurst et al. 2003) properties are amended, and this leads to outspread applications of these entities in variegated sectors, medical sector being one of them.

With the increasing implementation of nanomaterials in diversified sectors the urge for its legitimate fabrication route is of prime concern. Nanofabrication is achieved by two types of amenable approaches, namely top down approach and bottom up approach as depicted in Fig. 5.1. In top down approach, a bulk piece of material is engraved to nano size through processes like mechanical grinding, ball milling, thermal evaporation, explosion process, and different kinds of lithographic cutting

S. Zamani · B. Jha · A. K. Jha
Aryabhatta Centre for Nanoscience and Nanotechnology, Aryabhatta Knowledge University, Patna, India

K. Prasad (✉)
Department of Physics, Tilka Manjhi Bhagalpur University, Bhagalpur, Bihar, India

© Springer Nature Switzerland AG 2018
R. Prasad et al. (eds.), *Exploring the Realms of Nature for Nanosynthesis*,
Nanotechnology in the Life Sciences, https://doi.org/10.1007/978-3-319-99570-0_5

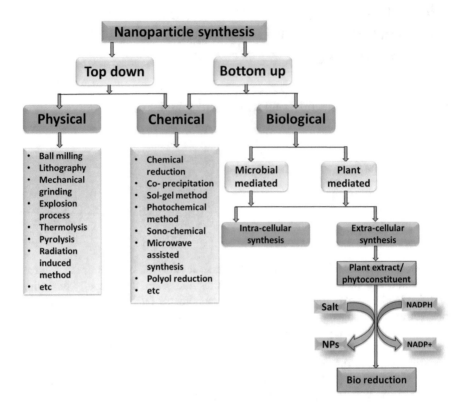

Fig. 5.1 The amenable approaches for nanofabrication

techniques like photo ion beam or X-ray lithography cutting (Li et al. 1999a, b). Top down approaches lead to some imperfections on the surface. This is a major disadvantage of this method because the surface structure plays an important role in physical properties of nanoparticles (Thakkar et al. 2010). Bottom up approaches are basically driven by self-arrangement of atoms to fabricate the nanomaterial through atom-by-atom and molecule-by-molecule interaction. The bottom up synthesis generally depends on chemical and biological methods. Chemical routes involve condition such as use of strong chemicals, reducing agents, capping agents, high temperature, and pressure, with protective agents (e.g. sodium borohydride, sodium citrate, alcohol). These organic solvents along with the reducing and capping agents are toxic, flammable, and their release in environment causes environmental hazards. Occurrence of these harsh chemicals limits the chemical route of fabrication and the further applicability of nanomaterials in areas like medical sector. In order to have biological and environmental safety in nanofabrication, the biological fabrication route applicability was considered. The living world is controlled by most efficient and best eco-friendly nanoscale processes, and this was a clue to the quest for an eco-friendly fabrication route. In contrast to physical and chemical methods, in this route natural biological agents are used for hydrolysing the metals. It is a simple, nontoxic, less labour intensive, low cost, environment-friendly greener approach of nanofabrication with no disposal issues. The biological agents used for fabrication are microorganisms (bacteria, fungi, actinomycetes), yeast, algae, and plants (bryophytes, pteridophytes, angiosperm, gymnosperm). Synthesis based on microorganisms has been commonly reported (Sastry et al. 2003; Gericke and Pinches, 2006; Mohanpuria et al. 2008; Korbekandi et al. 2009; Li et al. 2010; Sanghi and Verma, 2010; Kaler et al. 2011; Luangpipat et al. 2011; Dhillon and Mukhopadhyay 2015; Prasad et al. 2016). Synthesis from microbes can be scaled up in the lab, but as far as production is concerned, it is more expensive than plant extracts. The plant-mediated nanoparticles remain stable in nature for long time, and the rate of synthesis is also faster as compared to the microbes. Synthesis from plant extracts is a one-step, efficient, inexpensive, and environmentally safe method for producing nanoparticles with specified properties. The plant biomolecules besides being involved in bio-reduction also helps in stabilizing the nanoparticles and enhancing its properties. Therefore, plant-mediated nanoparticle fabrication is significant in all prospects (Park et al. 2011; Prasad 2014). The plant diversity is rich and bestowed with members starting from microscopic algae to giant trees. Each member has its important role in maintaining the ecosystem and all are rich in diverse metabolites, capable of nano-transformation. Numerous nanoparticles have been synthesized and reported from phanerogams (angiosperm and gymnosperm) (Prathna et al. 2010), but fewer attempts have made in this direction with lower cryptogamic plants. These cryptogams are rich sources of diverse metabolites and have been used as ethnomedicine since ages. The success story of initial attempts of nanofabrication through these cryptogams has proved them to be strong contenders in the plant world. This chapter attempts to explore these salient members as tools for nanofabrication. It presents an overview about cryptogams, its phylogeny, and role in plant diversity along with its ethnomedicine properties. The focal point of this chapter will be the nanofabrication via the cryptogams (bryophytes and pteridophytes), summarizing all the reports till date along with the characterization tools used and deciphering the fundamental science behind the fabrication. Special attention will be given to the fabricated nanoparticles biomedical applications and effect on human health.

5.2 Cryptogams: Salient Component of Plant Diversity

Cryptogams are integral part of plant diversity and ecosystem, playing important role in ecological balance. These are non-seed-bearing plants reproducing by spores and are divided into four orders, namely fungi, algae, mosses (bryophytes), and ferns (pteridophytes) (Smith, 2014). Nowadays, fungi are specified in separate kingdom, more closely to animals than plants and algae (thallophyte). The order algae are an enlarged group with numerous members and in nanotechnological perspective many members have been exploited for nanofabrication. This chapter tends to explore the other two members of the cryptogam family, namely bryophytes and pteridophytes wherein the penetration of nanotechnology is meagre. In the compass of bryophytes and pteridophytes, there are only few reports of nanofabrication through them. Before getting into the nanofabrication sphere, there is a need to have an overview of distribution and appliance of these members for better understanding as to why they are important as fabricators.

Bryophytes taxonomically exists between pteridophytes and algae, there are approx. 24,000 species in the world. These are cellular (non-vascular) simply organized plant body with a single spore-forming organ. Their species are classified into three main groups: Firstly, Anthocerotophyta (hornworts), which has approximately 300 species, secondly Marchantiophyta (liverworts), which has approximately 6000 species and lastly, Bryophyta (mosses), which has approximately 14,000 species. They have the ability to grow on bare soil and cracks of rocks because they do not depend on roots for their nutrition from soil, but a thin layer of water is mandatory on the surface of the plant for reproduction and survival (Shaw et al. 2011). Therefore, they are found in ranges from chilly to dessert, sea to alpine, and dry to wet rainforest (Everet and Susan 2013). Pteridophytes are richly found in several different habitats of the world. In Carboniferous period, they were leading, hence named as 'Age of Ferns'. Most of the carboniferous ferns are extinct, but some of these ferns have evolved. Around 12,000 species of pteridophytes are existing in the world (Antony et al. 2000). They are vascular plants, with xylem and phloem for nutrition, water and gas movements (Schuettpelz et al. 2016). These ferns are mainly found in humid, moist, well-drained sites, and crack of rock walls for growth. Such wide distribution in almost all the habitats displays that these cryptogams are allotted with rich phyto-constituents that give them the ability to restrain the environmental challenges.

5.2.1   Appliance of Cryptogams

The bryophytes and pteridophytes are bestowed with many rich and treasured phytochemicals that make them valuable and accounts for their diverse applications. Bryophytes besides being used as bioindicators for the presence of heavy metals in soil and air pollution are also used to improve soil conditions, by water retention in air space within soil (Kumar et al. 1999). Some bryophytes have been used as natural pesticides, biochemicals produced by bryophytes act as antifeedants, for example: the liverwort, *Plagiochila* are poisonous to mice (Glime 2007). *Pythium sphagnum* inhibits growth of fungus; this antibiotic property of bryophytes makes them a valuable packaging material and surgical dressing tool (Glime 2007). Although enriched with remarkable quality, bryophytes largely remain unexplored and untouched in many aspects, nanofabrication being one of them. Pharmacological analysis of bryophytes as studied by many researchers revealed out unique and active compounds that have the potential of therapeutic applications. At present, approximately 400 new compounds have been isolated and categorized for their biochemical and antimicrobial properties. Mostly, bryophytes can't be damaged by insects, and other small animals, because of fragrant odours and an intense bitter taste. They are rarely infected by microorganisms because of the presence of metabolites. These biologically active compounds enhance their medicinal importance worldwide (Ilhan et al. 2006). Many bryophytes are being used in medicinal purposes like burns, wounds,and bruises (Cramer et al. 1999). Active biological substances found in bryophytes, with respect to their bio-chemistry, pharmacology, and applications are used in cosmetics and medicinal drugs. Most of the bryophytes are rich in alkaloids (lycopodine, clavatine, nicotine), aromatic compounds (isocoumarins, cinnamates, bibenzyls, bis-bibenzyls, benzoates, naphthalenes, phthalides), flavonoids (apiginine and triterpene), and oil bodies (acetogenins). The presence of these compounds enhances their biological activities (Asakawa et al. 1995, Asakawa 2001) and also suggests their appliance as nanofabricators. The di-terpenoids and aromatic compounds isolated from the bryophytes (Lohlau et al. 2000) are used as antibacterial agent (Banerjee 2001; Scher et al. 2004; Sabovljevic et al. 2006; Mathur et al. 2007; Dulger et al. 2009), antifungal agent (Alam et al. 2011, 2012), anti-HIV, cytotoxic agent, DNA polymerase inhibitor (Rhoades 1999; Asakawa 2001), antioxidant, and antitumor agents (Vats et al. 2012; Sathish et al. 2016). The wide distribution of bryophytes in all habitats and the presence of abundant ethno therapeutic compounds make them a competent candidate for nanofabrication.

There are around 1250 species of pteridophytes found in India; out of these approx. 173 species have been found to be used as medicine, bio-fertilizers, green-manure, food, and flavour and dye (Manickam and Irudayaraj 1992). The presence of essential biological compounds enhances its popularity among plants. The presence of phenolic group attributes effective antioxidants property; flavonoids help to decrease the risk of cancer, cardiovascular disease, and other age-related diseases. Ancient people were using these ferns for ethno-medicinal purposes (Britto et al. 2012). These plants have been effectively used in homeopathic, ayurvedic, and unani system of medicines (Khare 2007). The pteridophytes have been efficiently used as antimicrobial (Dalli et al. 2007; Parihar et al. 2010) as well as an antioxidant agents (Lai and Lim 2011). Dihydrochalcone isolated from *Pityrogramma calomelanos* is a potential antibiotic and anticancer chemotherapeutic agent (Sukumaran and Kuttan 1991). The tribal groups have been using rhizomes, stem, fronds, pinnae, and spores in various ways for the treatment of various ailments (Kumari et al. 2011). The rhizome of fern is a pouch for storage of food as well as variety of secondary metabolites. Although pteridophytes are store house of such important bioactive compounds, sufficient attention has not been paid towards their valuable aspects, especially the phytochemical constituent usage. Pteridophytes belonging to the group of lower plants are rich in metabolites, but because of ignorance it remains unattended and their useful aspects are largely unexplored. Biochemical constituent of these pteridophytes makes it a suitable candidate for the fabrication of nanomaterials.

5.2.2   Phylogeny of Cryptogams

In the plant kingdom, plants are classified on the basis of geographies, their similarities and dissimilarities. In 1883, Eichler classified whole plant domain, known as the traditional phylogenetic system. Eichler divided plant kingdom into two sub-kingdoms (cryptogams and phanerogams). The cryptogams were further subdivided into three divisions: thallophyta, bryophyta, and pteridophyta. Bryophytes parted into three classes: Marchantiophyta, Anthocerotophyta, and Bryophyta. Pteridophytes parted into two classes: lycopodiopsida and polypodiopsida. The phylogeny of cryptogams as represented in Fig. 5.2, provides an ecological relationship, showing similarities and dissimilarities among its members. This knowledge may help in deciphering a common probable mechanistic aspect of nano-transformation by these members.

5.3   Nanofabrication Using Cryptogams

In several papers, researchers have often reported synthesis of nanoparticles from phanerogams (angiosperms and gymnosperms). However, very slight work has been done in the lower groups like bryophytes and pteridophytes (cryptogams), primitive plants with highest assemblage next to angiosperm. If we compare cryptogams with phanerogams plants, in perspective of nanofabrication, cryptogams are found to be more advantageous with simple organization of the plant body. The phytochemical analysis of these primitive plants confirmed that they have a variety of biochemicals and can be used in many ways. Fabrication of metal silver and gold nanoparticles is reported from cryptogams, as portrayed in Fig. 5.3, but the bio-fabrication of platinum nanoparticles (Pt-NPs) and palladium nanoparticles (Pd-NPs) has not been explored yet. Cryptogams are imitated as powerful ion exchanger, generally bryophytes due to simple organization of the tissues have the capability of heavy metal uptake from soil and water (industrial effluents) (Tyler 1990; Zhang et al. 2004). Thus, it has been used as the sorbent for heavy metal ions and further reduction to metallic nanoparticles (Ganji et al. 2005). Cryptogams are bestowed with the capacity to uptake metal ions and converse them to corresponding nano-metallic form (Chefetz et al. 2005). Therefore, researchers have a practical interest to reveal the possibilities of nanofabrication through cryptogams.

5.3.1   Nanofabrication Using Bryophytes

Bryophytes are used as an experimental model system for the explication of biological processes. They exhibit several key characteristics such as ease of growth and maintenance, fast generation time (Giles 1971), DNA sequencing, and

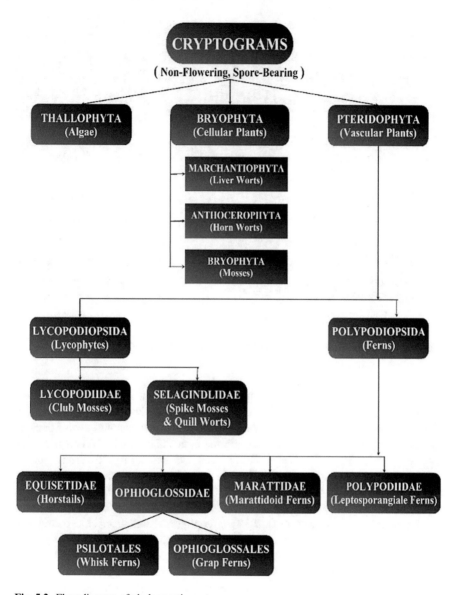

Fig. 5.2 Flow diagram of phylogenetic system

production of transgenic individuals via transformation (Wood et al. 2000). These properties prove it to be powerful model system for study of nanomaterial synthesis and their affects. The reduction of metal ions from aqueous and ethanolic extract of moss led to the formation of metal nanoparticles. Silver nanoparticles (Ag-NPs) synthesis with lower to higher (1–10 M) concentration of silver nitrate ($AgNO_3$) solutions was reported. Biosynthesis of Ag-NPs from a bryophyte was initially

Fig. 5.3 Photograph of some species of cryptogams used in the biosynthesis

proposed by Kulkarni et al. (2011), they demonstrated that 1 mM aqueous solution of silver nitrate with aqueous filtrate of *Anthoceros* when kept at 25 °C on a shaker at 150 rpm in dark led to Ag-NPs fabrication. 1 mM aqueous solution was decided by optimizing the experiment by varying the concentration of silver nitrate from 0.5, 1, 2, 3 to 5 mM (Kulkarni et al. 2011). Later, synthesis of Ag-NPs from aqueous extract of *Fissidens minutus* with 0.5 mM silver nitrate solution at room temperature was testified (Srivastava et al. 2011). The fabrication of silver nanoparticles using aqueous extract of *Riccia* on a shaker at 150 rpm in dark at 25 °C was reported in their experiment (Kulkarni et al. 2012). Bryophytes also reduced chloroauric acid (HAuCl$_4$) solution to gold nanoparticles when 3 mM HAuCl$_4$ was treated with aqueous filtrate of *Taxithelium nepalense* (Acharya and Sarkar 2014). Silver nanoparticles production from *Bryum medianum* plant extract treated with aqueous 1 mM AgNO$_3$ solution at room temperature for 24 h has also been reported (Sathish et al. 2016). Similarly, aqueous filtrate of *C. flexuosus* with 1 mM AgNO$_3$ solution when exposed to sunlight for an hour led to Ag-NPs synthesis (Vimala et al. 2017). The bryophytes mediated synthesized particles show antioxidant property with higher degree of scavenging free radicals activity. These members of cryptogams have the ability to produce diverse types of metal and oxide nanoparticles that inhibited the pathogenic organisms (Sathish et al. 2016). These synthesized nanoparticles may combine with antibiotics to enhance their bactericidal activities (Srivastava et al. 2011).

5.3.2 Nanofabrication Using Pteridophytes

Pteridophytes have also the facility to reduce metal salt solution to metal nanoparticles due to fullness in phytochemicals. First time bio-reduction of silver nanoparticles from aquatic pteridophyte (*Azolla*) was reported by Chefetz et al., they testified that the aquatic plant *Azolla* used as the sorbent for Ag^+, Pb^{2+}, and Ru^{3+} ions could transform them to the corresponding nano-metallic particles. Chefetz, combined two methods, adsorption of metallic ions on aquatic plants and conversion of the adsorbed heavy ions, into metallic nanoparticles. This combined process offers a hopeful approach for the removal of heavy metals from wastewater and the reutilizing of adsorbed metals into beneficiary products of metallic nanoparticles (Chefetz et al. 2005). Later, Kang et al. reported more than 85% reduction of the Ag ions is completed within 12 h, when it is synthesized from pteridophyte extract. Prepared nanoparticles remained stable for 12 months at room temperature, and there were no signs of any aggregation (Kang et al. 2008). Stability, aggregation, shape, size, and nature of nanoparticles are decided by some selected parameters; these known parameters are concentration of metal salt, pH, temperature, nature of plant extract, and reaction time. While synthesis of gold and silver nanoparticles using fern *A. philippense* L., the importance of these parameters were considered by Sant et al. According to Sant and co-workers, the optimum proportion for Au-NPs synthesized to the original plant extract was 1:1 at pH 11 with 5 mM salt concentration (tetrachloroauric acid), whereas for Ag-NPs synthesis 1:1 proportion of original plant extract at pH 12 and 9 mM concentration of silver nitrate yielded the desired results (Sant et al. 2013). The phytochemical examination of plant extract showed *Adiantum* contains terpenoids (filicene, filicenal, adiantone, isoadiantone, 23-hydroxyfernene), triterpenes (adininaonol and adininaneone), flavonoids (rutin, quercetine-3-glucaronyl, and isoquercetin), hentriacontane, 16-hentriacontanone, β-sitosterol, and fernene. The presence of these phytochemicals in *Adiantum* made it a probable candidate for bio-reduction of silver and gold salts. *Adiantum philippense* L. is known for its antioxidant potential (Sawant et al. 2009) and its ethnomedicinally significance. This fern is also used in treatment of diseases like rabies, paralysis, epileptic fits, elephantiasis, dysentery, pimples, wounds, and blood-related diseases (Karthik, et al. 2011) due to its antimicrobial activity (Pan et al. 2011). The presence of medicinal carrier, termed phytochemicals, makes them suitable candidate for biosynthesis of nanoparticles and in treatment of disease. *Adiantum capillus-veneris*-assisted Ag-NPs showed extreme toxicity against multi-drug resistant human pathogenic bacteria (Santhoshkumar and Nagarajan 2014). An aquatic fern *Azolla* sp. had been proved to have the ability to synthesize the metal (Au) as well as oxide (ZnO) nanoparticles (Jha and Prasad 2016; Asha and Francis 2015). *Azolla* has the potential to act as the stabilizing and reducing agent due to richness in phenols, tannins, sugars, flavanoids, steroids, and proteins. The aqueous extract of *Pteris* contains flavonoids (Kaempferol, Kaempferol-3-*o*-D glucopyranoside, Quercetin, and rutin) enzymes, proteins, amino acids, alkaloids, polysaccharides, phenolics, terpenoids, and vitamins, which has the ability to convert the silver ions into silver nanoparticles (Baskaran et al. 2016). *Asplenium scolopendrium* L. mediated Ag-NPs, improved antioxidant activity as compared to frond extract, tested on the root apexes of *Allium cepa by* genotoxic assay. This may be due to adsorption of those phytochemicals, which is responsible for antioxidant activity from the extract onto the surface of the nanoparticles (Şuţan et al. 2016). Combination of nanoparticles with other nanoparticles or conjugating with drugs enhances the property of nanoparticles. Synthesized gold nanoparticles from *Adiantum philippense* extract conjugated with antibiotics were found to have a synergistic activity against microorganisms (Kalita et al. 2016). The outer surface of gold nanoparticles was shielded with β-lactam antibiotic and gold nanoparticles-amoxicillin (GNP-Amox) conjugate enhanced antibacterial activity against Gram-positive as well as Gram-negative bacteria. Besides, *in vitro* and *in vivo* assays of GNP-Amox conjugate lead to effective anti-MRSA (methicillin-resistant *Staphylococcus aureus*) activity. Hence, GNP-Amox conjugate has been used as a promising antibacterial therapeutic agent against MRSA as well as other pathogens (Ghosh et al. 2012; Kalita et al. 2016). Although these members of cryptogam family are rich in active metabolites, there are only few reports of nanoparticle fabrication by bryophytes and pteridophytes, as compared to that of higher plants. The reported fabrications are tabulated in Table 5.1.

5.4 Monitoring of Nanoparticles

Observation and monitoring of a process is paramount for new discoveries, and this is applicable for nanotechnology too. As far as nano entities are concerned one cannot proceed further with their appliance until and unless these are monitored for needful fabrication and properties. Synthesized nanoparticles are confirmed by characterization tools and monitoring of nanoparticles is worthy to understand properties of nanoparticles. Nanoparticles are usually attributed for shape, size, and over all composition. Elucidation of nanoparticles has been resoluted by numerous characterization techniques under the umbrella of characterization of nanomaterials using UV-vis spectrometry, energy dispersive X-ray (EDX), X-ray diffraction (XRD), dynamic light scattering (DLS), Fourier transform infrared spectroscopy (FTIR), scanning electron microscopy (SEM), transmission electron microscopy (TEM), and atomic force microscopy (AFM) as shown in Fig. 5.4. In the case of reported nanofabrication by cryptogams, primarily the change in colour of colloidal solution symbolizes fabrication of nanomaterials. Then after, results of UV-vis spectrometer confirmed the fabrication of nanomaterials by surface plasmon resonance (SPR) peak absorbance value. UV-vis spectral analyses depend on concentration of H^+ ion (pH) (Akhtar et al. 2013), metal salt concentration (Dubey et al. 2010), ratio of reactants (Sheny et al. 2011), temperature and total reaction time, which reveals the shape and size (morphology) and its stability. Sharpness in peak indicates the smaller size and broadening shows chances of agglomeration (Sant et al. 2013). When SPR peak shifted towards red, this indicates the size of particle

Table 5.1 Nanomaterials fabrication from bryophytes and pteridophytes

S. no.	Reducing agent	Nano particle	Characterization	Particle characteristics	Application	References
Bryophytes						
1	Aqueous filtrate of *Anthoceros*	Ag-NPs	UV-Vis SEM EDS	Size 20–50 nm Shape-cub/triangular	Antibacterial action against *E. coli, B. subtilis, K. pneumoniae, P. aeruginosa*	Kulkarni et al. (2011)
2	Aqueous and ethanol filtrate of *Fissidens minutus*	Ag-NPs	UV-Vis SEM EDS	Shape-nearly spherical	Antibacterial action against *E. coli, B. cereus, K. pneumoniae, P. aeruginosa*	Srivastava et al. (2011)
3	Ethanol filtrate of *Riccia*	Ag-NPs	UV-Vis SEM EDS	Shape-cub/triangular	Antibacterial against *P. aeruginosa*	Kulkarni et al. (2012)
4	Ethanol filtrate of *Anthoceros*	Ag-NPs	UV-Vis SEM EDS	Size 20–50 nm Shape-cub/triangular	Antibacterial activity after incorporation into gauze cloth	Kulkarni et al. (2012)
5	Aqueous filtrate of *Taxithelium nepalense*	Au-NPs	UV-Vis, DLS, FTIR, EDX, XRD, TEM	Size range, 42–145 nm. Shape-spherical, triangular, hexagonal	N/A	Acharya and Sarkar (2014)
6	Aqueous filtrate of *Bryum medianum* Mitt.	Ag-NPs	UV-Vis FTIR, FESEM, EDX XRD	Size 85 nm shaped nanocrystals	Antimicrobial action against *Proteus mirabilis, Escherichia coli, Klebsiella pneumonia, Aspergillus fumigatus, Candida albicans,* and *Trichophyton mentagrophytes*	Sathish et al. (2016)
7	Aqueous filtrate of *Campylopus flexuosus*	Ag-NPs	UV-Vis FTIR FESEM Zeta potential XRD	Size 51 nm Shape-spherical	N/A	Vimala et al. (2017)

(continued)

Table 5.1 (continued)

S. no.	Reducing agent	Nano particle	Characterization	Particle characteristics	Application	References
Pteridophytes (fern)						
8	Aqueous filtrate of *Pteridophyta*	Ag-NPs	UV-Vis TEM EDX	Size 20–30 nm Shape-spherical	Ag-Nps are stable for 12 months	Kang et al. (2008)
9	Aqueous filtrate of *Adiantum philippense* L.	Ag-NPs and Au-NPs	UV-Vis FTIR EDS TEM DLS XRD	Size 10–18 nm Shape-anisotropic Structure-FCC	Ag-Nps from medicinally important plants opens spectrum of medical applications	Sant et al. (2013)
10	Aqueous filtrate of *Cheilanthes Forinosa* Forsk leaf	Ag-NPs	UV-Vis SEM XRD	Size ~26.58 nm Shape-spherical Structure-FCC	Antibacterial action against *S. aureus* and *Proteus morgani*	Nalwade et al. (2013)
11	Aqueous filtrate of *Pteris argyraea, Pteris confuse,* and *Pteris biaurita*	Ag-NPs	N/A	N/A	Antibacterial action against *Shigella boydii, Shigella dysenteriae, S. aureus, Klebsiella vulgaris,* and *Salmonalla typhi*	Britto et al. (2014)
12	Aqueous filtrate of *Nephrolepis exaltata* L. fern	Ag-NPs	UV-Vis SEM XRD	Size—avg. 24.76 nm Shape-spherical Structure-FCC	Antibacterial against many human and plant pathogens	Bhor et al. (2014)
13	Aqueous filtrate of *Adiantum capillus-veneris* L.	Ag-NPs	UV-Vis SEM EDX	Size 25–37 nm Cubic structure Shape-spherical	Antimicrobial against human pathogenic bacteria such as *Streptococcus pyogenes, Staphylococcus aereus, Escherichia coli,* and *Klebsiella pneumonia*	Santhoshkumar and Nagarajan (2014)

(continued)

Table 5.1 (continued)

S. no.	Reducing agent	Nano particle	Characterization	Particle characteristics	Application	References
14	Aqueous extract of *Diplazium esculentum* (*retz.*) *sw*	Ag-NPs	UV-Vis FTIR TEM XRD	Size 9.71 nm spherical, oval, and triangular	As catalyst in degradation of methylene blue (MB) and rhodamine B (RhB) dyes under solar light illumination	Paul et al. (2015)
15	Aqueous filtrate of *Pteris tripartita* Sw.	Ag-NPs	UV-Vis SEM TEM, XRD, EDX, FTIR	Size 32 nm hexagonal, spherical, and rod-shaped structures	Antibacterial against 12 different bacteria, antifungal against five different fungus, in vitro antioxidant, in vivo anti-inflammatory activities	Baskaran et al. (2016)
16	Ethanolic extracts of *Asplenium scolopendrium* L.	Ag-NPs	UV-Vis EDXRF FTIR	Size—below 50 nm	Antioxidant activity and genotoxic effects of the extracts on the root apexes of *Allium cepa*	Șuțan et al. (2016)
17	Aqueous extract of *Dicranopteris linearis*	Ag-NPs	UV-vis, FTIR SEM, EDX, XRD, zeta potential, and particle size analysis	Size—about 40–60 nm in diameter spherical in shape	Mosquitocidal assays, Smoke toxicity against *A. aegypti*	Rajaganesh et al. (2016)
Pteridophytes (aquatic fern)						
18	Aqueous solution of *Azolla filiculoides*	Ag-NPs	XRD, TEM	Size 20–30 nm spherical	N/A	Chefetz et al. (2005)
19	Aqueous extract of *Azolla*	ZnO-NPs	UV-vis spectroscopy, XRD, SEM, and FTIR	Size 19 nm spherical in shape	Antioxidant activity Antibacterial agent against *S. aureus*	Asha and Francis (2015)

(continued)

Table 5.1 (continued)

S. no.	Reducing agent	Nano particle	Characterization	Particle characteristics	Application	References
20	Ethanolic extract of *Azolla*	Au-NPs	XRD, TEM	Size 10–50 nm spherical shape	N/A	Jha and Prasad (2016)
21	Aqueous extract of *Salvinia molesta*	Ag-NPs	FESEM, HRTEM, AFM, EDX, XRD, AES	Size 12 nm spherical in shape	Antibacterial agent against both Gram-positive and Gram-negative bacteria	Verma et al. (2016)

increase and blue shift for decrease in size (Asha and Francis 2015). The morphology of fabricated nanomaterial by cryptogams and their particle shape and size were determined by SEM, TEM, and AFM. Spherical Ag-NPs were fabricated by the cryptogams like *Pteris, Cheilanthes forinosa, Diplazium esculentum,* and *Salvinia molesta* (Nalwade et al. 2013; Paul et al. 2015; Baskaran et al. 2016; Verma et al. 2016). Besides this rod-shape, hexagonal, oval shape, triangular and cuboidal shape were also reported from *Pteris, Diplazium esculentum, Anthoceros,* respectively. (Kulkarni et al. 2012; Paul et al. 2015; Baskaran et al. 2016).

AFM are superior over traditional microscopes like SEM and TEM as it enables measurement of three-dimensional structure of NPs. Particles height and volume can also be calculated through AFM. X-ray diffractometer (XRD) reveals molecular, atomic phase identification of crystalline structure of NPs. In literature, cryptogams mediated nanofabrication XRD analysis has been used to predict crystallinity of the particles. EDS or EDAX are used for analysing elemental composition of the nanomaterials. Dynamic light scattering (DLS) rectifies particle size distribution. The surface chemistry and functional (biological) groups that bound distinctively on surface of particles and were involved in the synthesis of these nanoparticles are determined by FTIR spectroscopy. In the present scenario, use of some of these characterization tools for nanoparticles fabricated by cryptogams is yet to be applied. In the near future, the increasing demand for development of fabrication processes through these cryptogams will ensure their applicability. A generalized flow chart predicting the characterization techniques is displayed in Fig. 5.4.

5.5 Probable Biosynthetic Mechanism

Biosynthesis of nanoparticles from cryptogams is currently under exploration. These cryptogams are known as lower plant and shows similarities to higher plants in some of the aspects. In the framework of nanofabrication, understanding the underlying mechanism is of paramount interest. Till date, as per the literature survey there are no reports elucidating this aspect of nanofabrication from bryophytes and

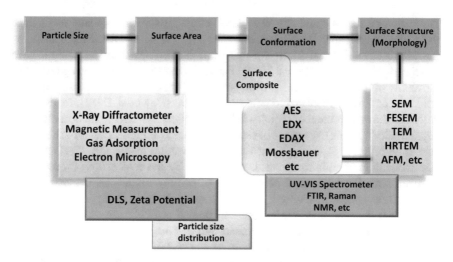

Fig. 5.4 A generalized flow chart predicting the characterization techniques

pteridophytes. The phyto-constituents present in these members of the family cryptogams have been found to be similar to that of higher plants. Hence, herein an attempt has been made to predict the probable mechanism underlying it by virtue of the mechanistic aspect reported for the higher plants. The mechanism process and pathway for nanofabrication are imprecise, even for higher plants. Prediction of exact mechanism faces several problems due to variation in phytochemical composition (Ahmed et al. 2014). However, some researchers reported the role of phenolic acid, flavonoids, terpenoids, proteins, organic acid, essential oils, etc. for synthesis and stabilization (Basha et al. 2010; Tamuly et al. 2014; Prasad 2014). These phytochemicals are also found in bryophytes and pteridophytes (Asakawa 2007; Kumari et al. 2011) rendering them the capability of nanofabrication. Limited reports are existing, where phytochemicals are directly utilized for the metallic nanoparticles synthesis (Singh et al. 2010; Ahmed et al. 2014; Durai et al. 2014; Iram et al. 2014; Dauthal and Mukhopadhyay 2016). These reports provide a clue for future endeavours employing bryophytes and pteridophytes for extraction of phytochemicals for nanofabrication. Flavonoids have the ability to donate hydrogen atom (Pietta 2000; Zhou et al. 2010) through keto-enol conversion. Consequently, these flavonoids compel the reduction of metal ions to corresponding metal nanoparticles (Ahmad et al. 2010). Similarly, hydroxyl groups (OH^-) of flavonoids (quercetin) reacts to carbonyl groups ($C=O$) during bio-reduction (Ghoreishi et al. 2011). Synthesis of Ag-NPs supported by intermediate complex in which phenolic hydroxyl groups consequently undergo oxidation to quinone with subsequent reduction of silver ions to Ag-NPs has been reported (Edison and Sethuraman, 2012). The polyphenolic compounds are capable of chelating many kinds of metallic ions (Pt^{4+}, Au^{3+}, Pd^{2+}, Ag^+, etc.) (Mohsen and Ammar 2009). The phenolic compounds have higher antioxidant potential and it reduces metal ions into stable metal nanoparticles, thus favouring the stabilization of nanoparticles. Terpenoids also play a role as bio-reducing agent for fabrication of Au-NPs and Ag-NPs (Singh et al. 2010). Proteins can also reduce metal silver ions to metallic NPs for example; tryptophan has reductive properties by the release of an electron during conversion of the tryptophan residue (Adyanthaya and Sastry 2004; Si and Mandal 2007). Tan et al. have tested the reducing and binding ability of twenty amino acids with metal ions (Tan et al. 2010). Tyrosine reduces gold ions and forming of stable gold nanoparticles (Roy et al. 2014). Researchers theorized the involvement of proteins in bio-reduction. Metal ions can be reduced by organic acid (alkaloid) by release of reactive hydrogen (Tamuly et al. 2014). Ascorbic acid is an effective reducing agent for the stabilization of Ag-NPs. The keto-enol tautomerization of anthroquinones is conjectured by Jha et al. (2009) for reduction of metal ions. Phytochemical examination of cryptogams reveals the presence of carbohydrates, phenols, alkaloids, steroids, terpenoids, hydroxy-hypnone, catecholamines, tannins, saponins, anthraquinones, coumarins, fats, enzymes (NADP reductase), proteins, functional groups (amines and ketones), essential oils, flavonoids, glycosides, gums, iridoids, and mucilage (Asakawa 2007; Santos et al. 2010; Kumari et al. 2011; Suganya et al. 2011; Rai et al. 2016). As per our assumption, similar

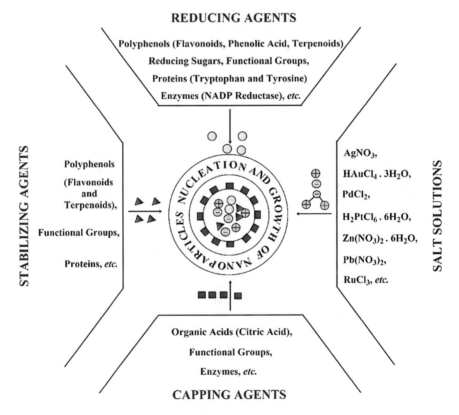

Fig. 5.5 Schematic diagram presenting probable mechanisms behind the formation of nanoparticles

bioactive compounds which are present in the extracts of cryptogams, act as a reducing as well as a stabilizing agent. Flavonoids and terpenoids act as stabilizing agent; proteins, enzymes, and functional groups act as capping agent and protein, reducing sugars, enzymes (NADP reductase), phenolic compounds, flavonoids (quercetin) as reducing agent (Akhtar et al. 2013). Thus, we can conclusively remark the participation of phytochemicals in bio-reduction of metal ions to their corresponding nanoparticles. This probable mechanism is schematized in Fig. 5.5 wherein potential reducing, stabilizing, and capping agents that are present and responsible for nanotransformation by cryptogams are indicated.

5.6   Pharmacological Applications

Nano systems and biological components both exhibiting nanometre dimensions provide probable synergies amidst them. This very fact gives implications of the foreseeable appliance of nano entities for understanding the biological processes and as nanomedicine in biomedical field. Nanoprobes or sensors for disease detection, nano systems with large surface area as drug carriers and nanodrugs as alternative to conventional drugs turn up as a precise, controllable, reliable, cost-effective, and

Table 5.2 Applications of different nanoparticles synthesized from cryptogams

Medium sources	Nanoparticles	Applications	References
(Bryophytes) *Anthoceros, Bryum medianum* Mitt., *Fissidens minutus, Riccia, Campylopus flexuosus* (Pteridophytes) *Pteridophyta, Adiantum philippense* L., *Adiantum capillus-veneris* L., *Cheilanthes forinosa, Pteris argyraea, Pteris confuse, Pteris biaurita, Nephrolepis exaltata* L., *Diplazium esculentum (retz.) sw, Pteris tripartita* Sw., *Asplenium scolopendrium* L, *Dicranopteris linearis, Azolla filiculoides, Salvinia molesta*	Ag-NPs	Antibacterial activity Antifungal activity Antioxidant activity Anti-inflammatory activity Larvicidal Activity Genotoxic activity Catalytic activity	Kulkarni et al. (2012, 2011), Srivastava et al. (2011), Sathish et al. (2016), Vimala et al. (2017), Kang et al. (2008), Sant et al. (2013), Nalwade et al. (2013), Britto et al. (2014), Bhor et al. (2014), Santhoshkumar and Nagarajan (2014), Paul et al. (2015), Baskaran et al. (2016), Şuţan et al. (2016), Rajaganesh et al. (2016), Verma et al. (2016), Chefetz et al. (2005)
Taxithelium nepalense (Bryophytes) *Adiantum philippense* L., *Azolla* (Pterodophytes)	Au-NPs	Potential medical applications	Acharya and Sarkar (2014), Sant et al. (2013), Jha and Prasad (2016)
Azolla	ZnO-NPs	Antioxidant activity Antibacterial activity	Asha and Francis (2015)

rapid diagnostic and treatment solution (Prasad et al. 2014, 2017). In the framework of nanoparticles fabricated by cryptogams, there are several reports wherein these particles have been tested for the pharmacological applications. The results were agreeable and indicative of the fact that these metabolite-rich cryptogams probably add on to pharmacological efficiency of bare nanoparticles. An insight to the pharmacological applications of cryptogam-mediated nanoparticles is illustrated in the following sections and summarized in Table 5.2.

5.6.1 Antimicrobial Activity

According to several scientific reports, it has been evidenced that nanoparticles (NPs) are highly effective against the pathogenic microorganisms like viruses, fungi, Gram-negative, and Gram-positive bacteria (Rai et al. 2012; You et al. 2012; Rizzello and Pompa 2014; Aziz et al. 2014, 2015, 2016; Joshi et al. 2018). It also exhibits bactericidal and inhibitory effects. By virtue of this property, it is widely used as an antimicrobial agent. Zone of inhibition is a very common method, which is being used to assess the antimicrobial activity of NPs (Ortiz et al. 2017). The mechanism of nanoparticles antimicrobial effect is yet to be explored. Among the nanoparticles, silver nanoparticles are widely studied for antimicrobial activity due to its medicinal history. Few researchers testified that Ag-NPs may bound to the surface of the cell membrane and interact with the disulphide bonds of the glycoprotein of microorganisms (viruses, bacteria, and fungi) instigating change in three-dimensional structures of proteins, which causes blocking the functional operations of the microorganism (Amro et al. 2000; Kim et al. 2007; Morones et al. 2005; Rai and Bai 2011; Aziz et al. 2015, 2016). In spite of interacting with the surface membrane, silver nano particles can penetrate inside the bacteria and destruct on genetic level. It may also form reactive oxygen species which interacts and damages the ATP assembly of respiration. Silver nanoparticles can also accumulate inside the cells of microbes causing damage to cell. Hence, by virtue of this mechanism, it is an effective antimicrobial agent (Rai et al. 2009; Sadeghi and Gholamhoseinpoor 2015). Some researchers have also reported that the Gram-positive bacteria are more sensitive to Ag-NPs than the Gram-negative and vice versa (Wrotniak et al. 2014; Perito et al. 2016; Dakal et al. 2016). This is very contradictory part of antibacterial action. This difference in sensitiveness is due to the difference in structure (shape and size) and methodology applies during the synthesis process of Ag-NPs. are more sensitive than spherical followed by rod shape (Xiu et al. 2012). Smaller size of Ag-NPs shows better antibacterial result. It is a well-known fact that resistant to antibiotics is found to be more in Gram-negative bacteria as compared to the Gram-positive. This resistance is mainly due to the differences in their cell wall composition. It is also revealed by the several studies that susceptibility to the cryptogams is found more in Gram-negative bacteria. This susceptibility is mainly because of the presence of broad spectrum of antibiotic compounds present in the ferns. So, it can be said that cryptogams mediated Ag-NPs are susceptible against Gram-negative bacteria.

Fissidens minutus (bryophytes)-assisted Ag-NPs as investigated showed susceptibility against *Bacillus subtilis, Klebsiella pneumoniae, Pseudomonas aeruginosa,* and *Escherichia coli* (Srivastava et al. 2011). *Anthoceros*-assisted Ag-NPs showed positive result against four pathogenic strains: *Klebsiella pneumoniae, Escherichia coli, Bacillus subtilis,* and *Pseudomonas aeruginosa* (Kulkarni et al. 2011). *Bryum medianum*-mediated synthesized Ag-NPs showed greater inhibitory effect on the bacteria *Klebsiella pneumonia* and *Escherichia coli* than *Proteus mirabilis* (Sathish et al. 2016). The growth of *Klebsiella pneumoniae* NCIM 2719, *Proteus morgani* NCIM 2719, *Corynebacterium diphtheriae, Pseudomonas testosteroni* NCIM 5098, *Bacillus subtilis* NCIM 2063, *Escherichia coli,* and *Xanthomonas axonopodis pv. punicae* were highly inhibited by the Ag-NPs synthesized using extract of a fern *Nephrolepis exaltata* L. The highest inhibitory activity was recorded for *Xanthomonas axonopodis* pv. punicae and lowest for *Pseudomonas teststeroni* NCIM 5098 (Bhor et al. 2014). The Ag-NPs obtained from *Adiantum capillus-veneris* also disclosed antibacterial activity against four bacterial strains: *Staphylococcus aureus,* followed by *Streptococcus pyogenes, Escherichia coli,* and *Klebsiella pneumonia* than standard drugs (standard antibiotic erythromycin) (Santhoshkumar and Nagarajan 2014). Up to 3 mM zone of inhibition was found against Gram-positive bacterial strain (*S. aureus*) by *Azolla*-assisted ZnO-NPs (Asha and Francis 2015). Another source of Ag-NPs fabrication aquatic fern: *Salvinia molesta* was also explored to be a very effective antibacterial agent against Gram-negative as well as Gram-positive bacteria (Verma et al. 2016). Ag-NPs, which are produced from *Pteris tripartita Sw.* showed a positive antibacterial action against as much as 12 different bacterial strains. Experiment was accomplished by disc diffusion, time course growth methods, and minimum inhibitory concentration (MIC). The studies regarding in vivo (anti-inflammatory) and in vitro (antimicrobial and antioxidant) are also carried out for Ag-NPs. Ag-NPs also showed anti-inflammatory activity. This activity is well explained through carrageenan-induced paw volume experiment in female Wistar albino rats (Baskaran et al. 2016).

5.6.2 Antifungal Activity

Ag-NPs are found to be effective fungicides against common fungi such as *Candida, Aspergillus, and Saccharomyces* (Yu et al. 2005). *Bryum medianum*-assisted synthesized Ag-NPs also showed antifungal activity; the highest degree of inhibition was observed in *Candida albicans* in comparison to *Aspergillus fumigatus* and *Trichophyton mentagrophytes* (Sathish et al. 2016). Ag-NPs obtained from *Pteris tripartita Sw.* was found to exhibit antifungal activity against four different species. It showed positive antifungal activities against *Fusarium oxysporum, Rhizopus oryzae, Aspergillus niger,* and *Aspergillus flavus* (Baskaran et al. 2016).

5.6.3 Larvicidal Activity

The Ag-NPs are known to show the larvicidal activity by killing the larvae of vectors at an early stage (Buhroo et al. 2017). Many Ag-NPs have shown larvicidal activity against malarial vectors *Aedes aegypti* (Suresh et al. 2014). Metal nanoparticles, especially Ag-NPs, show maximum larvicidal activity. Silver nanoparticles fabricated using leaf extract of *Dicranopteris linearis* were testified that it has larvicidal activity against *A. aegypti* adults, egg hatchability reduced by 100% when treated with 25 ppm of Ag-NPs. Ag-NPs reduced oviposition rates more than 70%. One of the useful applications of *D. linearis*-mediated Ag-NPs is to develop nano-formulated oviposition deterrents, which is effective against dengue vectors (Rajaganesh et al. 2016).

5.6.4 Wound Healing Activity

The Ag-NPs showed wound healing activity. It can penetrate into the fibroblast cells and accrued over time, at the same time it does not interfere with the cell viability. During apoptosis, the mitochondria are activated in the cell to reduce its viability (Garg et al. 2014). The wound-healing application of Ag-NPs has already been tested on animal models (Arunachalam et al. 2013). This fact gives an idea that cryptogam-mediated Ag-NPs can also be used for wound healing activity.

5.6.5 Medicinal Textiles

In the field of Textiles, NPs may enhance the quality of fabric. Kulkarni et al. testified antimicrobial activity against *Pseudomonas aeruginosa* of gauze cloth in disc form incorporated with Ag-NPs, which is synthesized from *Anthoceros* (Kulkarni et al. 2012).

5.6.6 Catalytic Activity

Diplazium esculentum-mediated Ag-NPs were also used as catalyst in degradation of dyes. Rhodamine B (RhB) and methylene blue (MB) dyes degradation under solar light in the presence of Ag-NPs was checked for analysis of its catalytic efficiency. The discolouration of dyes occurred in minutes, indicative of the fact that small-sized Ag-NPs caused structural changes and depletion of chromophoric group from the dye. The pronounced photocatalytic activity of Ag-NPs was attributed to its small size achieved through the bio-reductive process (Paul et al. 2015).

5.6.7 Antioxidant Activity

The presence of free radicals in cells is responsible for the development of many diseases like cancer, cardiovascular Alzheimer's atherosclerosis, diabetes, and ageing (Coyle and Puttfarcken 1993). Therefore, evaluation of the free radical scavenging potential of biosynthesized nanoparticles supports in the treatment of such diseases. *Bryum medianum* moss plant-mediated Ag-NPs antioxidant activity as determined by the DPPH assay showed the higher percentage of inhibition compared to the ascorbic acid. Thus, Ag-NPs obtained from *Bryum medianum* have the potential to scavenge the free radicals (Sathish et al. 2016). Baskaran et al. also reported in vitro antioxidant activities of Ag-NPs derived from *Pteris tripartita* by DPPH radical scavenging method, hydrogen peroxide scavenging method, and azinobis 3-ethylbenzothiazo-line-6-sulfonate (ABTS˙+) assay. (Baskaran et al. 2016). Similarly, Şuţan et al. from ethanolic extracts of *A. scolopendrium* L. obtained Ag-NPs and compared it for the genotoxic effect on the root apexes of *Allium cepa*. The finding showed that Ag-NPs synthesized using the extract of *A. scolopendrium* L. enhanced the antioxidant activity as compared to the extract. Enhancing of antioxidant activity may be attributed to small size and adsorption of antioxidant components from the extract to the surface of the nanoparticles thereby increasing its efficiency (Şuţan et al. 2016).

5.6.8 Cytotoxicity

Cancer is a life-threatening disease and its treatment is lethiferous. Nanotechnology turns up as a novel approach for the treatment of cancer because of its lesser side effects compared to conventional chemotherapy and synthetic drugs. Nanoparticles as reported by several researchers have anticancer potential. It is anti-proliferative (Firdhouse and Lalitha 2013), apoptosis inducer (Mukherjee et al. 2015), anti-metastatic (Karuppaiya et al. 2013), and has cytotoxic effect (Suman et al. 2013). The cytotoxicity studies of green synthesized NPs on different mammalian cell lines are usually convoyed with quantitative cell viability via MTT (3-(4,5-dimethylthiazol-2-y l)-2,5-diphenyltetrazolium bromide) and qualitative viability through death double staining, acridine orange, and ethidium bromide live assays (Sukirthaa et al. 2012; Ramamurthy et al. 2013; Preethi and Padma 2016). The outcome from all those current research is that cytotoxic activity was tremendously sensitive to the size of the nanoparticles and the viability capacities decreased with increasing dosage. Thus, cryptogams mediated nanoparticles could be used as anticancer agent. The potential of such medicinally important lower plants can be used for the development of new anti-cancer drugs.

5.7   Conclusion

The developing era of nanoscience is a famous flair for the development of science all over the world. With the advent of application of nanoparticles in almost every field, the need for an inexpensive, eco-friendly, simple process of fabrication is of core interest. Several studies have been done over the last decade on application of biological entities as nanofabricators starting from microbes to higher plants, but there are still significant openings in our knowledge wherein groups like crypto-gams have not been explored judiciously. These cryptogams have been gifted with immense reservoir of active metabolites with bio-reductive and pharmacological potentials. This chapter summarizes information about the nanofabrication by cryp-togams and its scope in pharmacological actions. Moreover, in order to have better understanding general description about cryptogams has also been included. In framework of nanofabrication, the bryophytes and pteridophytes remain largely unexplored and unattended, with reports in scratches. The chapter describes the various possibilities and benefits with intent to draw attention towards these mem-bers for nanofabrication explorations. Further, it has stressed on the need to deter-mine the phytochemical constituents of the extract, as well as to find out the mechanism responsible for synthesis. Conclusively, this chapter is an attempt to explore the unexplored members of plant diversity, the cryptogams for fabrication of nanoparticles.

References

Acharya K, Sarkar J (2014) Bryo-synthesis of gold nanoparticles. Int J Pharm Sci Rev Res 29:82–86

Adyanthaya SD, Sastry MJ (2004) Water-dispersible tryptophan-protected gold nanoparticles pre-pared by the spontaneous reduction of aqueous chloroaurate ions by the amino acid. J Colloid Interface Sci 269:97–102

Ahmad N, Sharma S, Alam MK, Singh VN, Shamsi SF, Mehta BR, Fatma A (2010) Rapid synthe-sis of silver nanoparticles using dried medicinal plant of basil. Colloids Surf B Biointerfaces 81:81–86

Ahmed KBA, Subramaniam S, Veerappan G, Hari N, Sivasubramanian A, Veerappan A (2014) β-Sitosterol-d-glucopyranoside isolated from Desmostachya bipinnata mediates photoinduced rapid green synthesis of silver nanoparticles. RSC Adv 4:59130–59136

Akhtar MS, Panwar J, Yun YS (2013) Biogenic synthesis of metallic nanoparticles by plant extracts. ACS Sustain Chem Eng 1:591–602

Alam A, Tripathi A, Vats S, Behera KK, Sharma V (2011) In vitro antifungal efficacies of aqueous extract of *Dumortiera hirsute* (Swaegr.) *Nees* against sporulation and growth of postharvest phytopathogenic fungi. Arch Bryol 103:1–9

Alam A, Sharma SC, Sharma V (2012) In vitro antifungal efficacies of aqueous extract of *Targionia hypophylla* L. against growth of some pathogenic fungi. Int J Ayurvedic Herb Med 2:229–233

Amro NA, Kotra LP, Wadu-Mesthrige K, Bulychev A, Mobashery S, Liu G (2000) High-resolution atomic force microscopy studies of the *Escherichia coli* outer membrane: structural basis for permeability. Langmuir 16:2789–2796

Antony RS, Khan AE, Thomas J (2000) Rare, endangered and threatened ferns from Chemunji hills, Kerala. J Econ Taxon Bot 24:413–415

Arunachalam KD, Annamalai SK, Arunachalam AM, Kennedy S (2013) Green synthesis of crystalline silver nanoparticles using *Indigofera aspalathoides*-medicinal plant extract for wound healing applications. Asian J Chem 25:S311–S314

Asakawa Y (2001) Recent advances in phytochemistry of bryophytes acetogenins, terpenoids and bis(bibenzyl)s from selected Japanese, Taiwanese, New Zealand, Argentinean and European liverworts. Phytochemistry 56:297–312

Asakawa Y (2007) Biologically active compounds from bryophytes. Pure Appl Chem 79:557–580

Asakawa Y, Herz W, Kirby G, Moore RE, Steglich W, Tamm C (1995) Chemical constituents of the Bryophytes. In: Progress in the chemistry of organic natural products, vol 65. Springer, Vienna, pp 1–562

Asha PS, Francis J (2015) One pot green synthesis of ZnO nanoparticles using Azolla extract and accessing its biological activities. Int J Curr Res 7:22520–22527

Aziz N, Fatma T, Varma A, Prasad R (2014) Biogenic synthesis of silver nanoparticles using Scenedesmus abundans and evaluation of their antibacterial activity. J Nanoparticles Article ID 689419. https://doi.org/10.1155/2014/689419

Aziz N, Faraz M, Pandey R, Sakir M, Fatma T, Varma A, Barman I, Prasad R (2015) Facile algae-derived route to biogenic silver nanoparticles: synthesis, antibacterial and photocatalytic properties. Langmuir 31:11605–11612. https://doi.org/10.1021/acs.langmuir.5b03081

Aziz N, Pandey R, Barman I, Prasad R (2016) Leveraging the attributes of Mucor hiemalis-derived silver nanoparticles for a synergistic broad-spectrum antimicrobial platform. Front Microbiol 7:1984. https://doi.org/10.3389/fmicb.2016.01984

Banerjee RD (2001) Antimicrobial activities of bryophytes: a review. In: Nath V, Asthana AK (eds) Perspectives in Indian bryology. Bishen Singh Mahendra Pal Singh Publisher, Dehradun, India

Basha SK, Govindaraju K, Manikandan R, Ahn JS, Bae EY, Singaravelu G (2010) Phytochemical mediated gold nanoparticles and their PTP 1B inhibitory activity. Colloids Surf B Biointerfaces 75:405–409

Baskaran X, Antony VGV, Parimelazhagan T, Rao DM, Zhang S (2016) Biosynthesis, characterization, and evaluation of bioactivities of leaf extract-mediated biocompatible silver nanoparticles from an early tracheophyte, *Pteris tripartita* Sw. Int J Nanomed 11:5789–5805

Bhor G, Maskare S, Hinge S, Singh L, Nalwade A (2014) Synthesis of silver nanoparticles by using leaflet extract of *Nephrolepi exaltata* L. and evaluation of antibacterial activity against human and plant pathogenic bacteria. Asian J Pharm Technol Innov 2:23–31

Britto AJD, Herin SGD, Benjamin JRKP (2012) *Pteris biaurita* L.: a potential antibacterial fern against Xanthomonas and Aeromonas bacteria. J Pharm Res 5:678–680

Britto AJD, Gracelin DHS, Kumar PBJR (2014) Antibacterial activity of silver nanoparticles synthesized from a few medicinal ferns. Int J Pharm Res Dev 6:25–29

Buhroo AA, Nisa G, Asrafuzzaman S, Prasad R, Rasheed R, Bhattacharyya A (2017) Biogenic silver nanoparticles from *Trichodesma indicum* aqueous leaf extract against Mythimna separata and evaluation of its larvicidal efficacy. J Plant Prot Res 57(2):194–200

Chefetz B, Sominski L, Pinchas M, Ginsburg T, Elmachliy S, Tel-Or E, Gedanken A (2005) New approach for the removal of metal ions from water: adsorption onto aquatic plants and microwave reaction for the fabrication of nanometals. J Phys Chem B 109:15179–15181

Coyle JT, Puttfarcken P (1993) Oxidative stress, glutamate, and neurodegenerative 22 disorders. Science 262:689–695

Cramer JV, Asakawa Y (1999) Phytochemistry of bryophytes. In: Romeo J (ed) Phytochemicals in human health protection, nutrition, and plant defense. Kluwer Academic/Plenum Publishers, New York

Dakal TC, Kumar A, Majumdar RS, Yadav V (2016) Mechanistic basis of antimicrobial actions of silver nanoparticles, frontiers in microbiology antimicrobial, resistance chemotherapy. Front Microbiol 7:1831, 17p. https://doi.org/10.3389/fmicb.2016.01831

Dalli AK, Saha G, Chakraborty U (2007) Characterization of antimicrobial compounds from a common fern, *pteris biaurita*. Indian J Exp Biol 45:285–290

Dauthal P, Mukhopadhyay M (2016) Noble metal nanoparticles: plant-mediated synthesis, mechanistic aspects of synthesis, and applications. Ind Eng Chem Res 55:9557–9577

Dhillon NK, Mukhopadhyay SS (2015) Nanotechnology and allelopathy: synergism in action. J Crop Weed 11:187–191

Dubey SP, Lahtinen M, Sillanp M (2010) Tansy fruit mediated greener synthesis of silver and goldnanoparticles. Process Biochem 45:1065–1071

Dulger B, Hacıolu N, Uyar G (2009) Evaluation of antimicrobial activity of some mosses from Turkey. Asian J Chem 21:4093–4096

Durai P, Gajendran ACB, Ramar M, Pappu S, Kasivelu G, Thirunavukkarasu A (2014) Synthesis and characterization of silver nanoparticles using crystal compound of sodium para-hydroxybenzoate tetrahydrate isolated from *Vitex negundo* L. leaves and its apoptotic effect on human colon cancer cell lines. Eur J Med Chem 84:90–99

Edison TJI, Sethuraman MG (2012) Instant green synthesis of silver nanoparticles using *Terminalia chebula* fruit extract and evaluation of their catalytic activity on reduction of methylene blue. Process Biochem 47:1351–1357

Everet R, Susan E (2013) Biology of plants, 18th edn. W.H. Freeman, New York

Firdhouse MJ, Lalitha P (2013) Biosynthesis of silver nanoparticles using the extract of *Alternanthera sessilis*-antiproliferative effect against prostate cancer cells. Cancer Nanotechnol 4:137–143

Ganji M, Khosravi M, Rakhshaee R (2005) Biosorption of Pb, Cd, Cu and Zn from the wastewater by treated *Azolla filiculoides* with $H_2O_2/MgCl_2$. Int J Environ Sci Technol 1:265–271

Garg S, Chandra A, Mazumder A, Mazumder R (2014) Green synthesis of silver nanoparticles using *Arnebia nobilis* root extract and wound healing potential of its hydrogel. Asian J Pharm 8:95–101

Gericke M, Pinches A (2006) Microbial production of gold nanoparticles. Gold Bull 39:22–28

Ghoreishi SM, Behpour M, Khayatkashani M (2011) Green synthesis of silver and gold nanoparticles using *Rosa damascena* and its primary application in electrochemistry. Phys E Low Dimens Syst Nanostruct 44:97–104

Giles KL (1971) Dedifferentiation and regeneration in bryophytes: a selective review. N Z J Bot 9:689–694

Glime JM (2007) Economic and ethnic uses of bryophytes, Flora of North American North of Mexico. Bryophytes 27:14–41

Ghosh K, Harish CR, Baghel MS (2012) A preliminary pharmacognostical and physcicochemical assay of *Shunthikhana granules*. Internat Res J Pharmacy 3:170–175

Ilhan S, Savaroglu F, Çolak F, Isçen C, Erdemgil F (2006) Antimicrobial activity of *Palustriella commutata* (Hedw.) Ochyra extracts (Bryophyta). Turk J Biol 30:149–152

Iram F, Iqbal MS, Athar MM, Saeed MZ, Yasmeen A, Ahmad R (2014) Glucoxylan-mediated green synthesis of gold and silver nanoparticles and their phyto toxicity study. Carbohydr Polym 104:29–33

Jha AK, Prasad K (2016) Aquatic fern (*Azolla* sp.) assisted synthesis of gold nanoparticles. Int J Nanosci 15:1650008–1650012

Jha AK, Prasad K, Prasad K, Kulkarni AR (2009) Plant system: Nature's nanofactory. Colloids Surf B Biointerfaces 73:219–223

Joshi N, Jain N, Pathak A, Singh J, Prasad R, Upadhyaya CP (2018) Biosynthesis of silver nanoparticles using *Carissa carandas* berries and its potential antibacterial activities. J Sol Gel Sci Technol. https://doi.org/10.1007/s10971-018-4666-2

Kaler A, Nankar R, Bhattacharyya MS, Banerjee UC (2011) Extracellular biosynthesis of silver nanoparticles using aqueous extract of *Candida viswanathii*. J Bionanosci 5:53–58

Kalita S, Kandimalla R, Sharma KK, Kataki AC, Deka M, Kotoky J (2016) Amoxicillin functionalized gold nanoparticles reverts MRSA resistance. Mater Sci Eng C 61:720–727

Kang KC, Kim SS, Baik MH, Choi JW, Kwon SH (2008) Synthesis of silver nanoparticles using green chemical method. Appl Chem 12:281–284

Karthik V, Raju K, Ayyanar M, Gowrishankar K, Sekar T (2011) Ethno medicinal uses of pteridophytes in Kolli Hills, Eastern Ghats of Tamil Nadu, India. J Nat Prod Plant Resour 1:50–55

Karuppaiya P, Satheeshkumar E, Chao WT, Kao LY, Chen EF, Tsay HS (2013) Anti-metastatic activity of biologically synthesized gold nanoparticles on human fibrosarcoma cell line HT-1080. Colloids Surf B Biointerfaces 110:163–170

Khare CP (2007) Indian medicinal plants. Springer, Berlin

Kim JS, Kuk E, Yu KN, Kim JH, Park SJ, Lee HJ, Kim SH, Park YK, Park YH, Hwang CY, Kim YK, Lee YS, Jeong DH, Haing M (2007) Antimicrobial effects of silver nanoparticles. Nanomed Nanotechnol Biol Med 3:95–101

Korbekandi H, Iravani S, Abbasi S (2009) Production of nanoparticles using organisms. Crit Rev Biotechnol 29:279–306

Kulkarni AP, Srivastava AA, Harpale PM, Zunjarrao RS (2011) Plant mediated synthesis of silver nanoparticles tapping the unexploited resources. J Nat Prod Plant Resour 1:100–107

Kulkarni AP, Srivastava AA, Nagalgaon RK, Zunjarrao RS (2012) Phytofabrication of silver nanoparticles from a novel plant source and its application. Inte J Biol Pharm Res 3:417–421

Kumar K, Singh KK, Asthana AK, Nath V (1999) Ethnotherapeutics of bryophyte *plagiochasma appendiculatum* among the Gaddi Tribes of Kangra Valley, Himachal Pradesh, India. Pharm Biol 37:1–4

Kumari P, Otaghvri AM, Govindpyari H, Bahuguna YM, Uniyal PN (2011) Some ethnomedicinally important pteridophytes of India. Int J Med Arom Plants 1:18–22

Lai HY, Lim YY (2011) Antioxidant Properties of some Malaysian Ferns. In: Proceedings of third international conferene on chemical, biological and environmental engineering, vol 20, pp 8–12

Li CZ, Bogozi A, Huang W, Tao NJ (1999a) Fabrication of stable metallic nanowires with quantized conductance. Nanotechnology 10:221–223

Li Y, Duan X, Qian Y, Li Y, Liao H (1999b) Nanocrystalline silver particles: synthesis, agglomeration, and sputtering induced by electron beam. J Colloid Interface Sci 209:347–349

Li WR, Xie XB, Shi QS, Duan SS, Ouyang YS, Chen YB (2010) Antibacterial effect of silver nanoparticles on *Staphylococcus aureus*. Biol Met 24:135–141

Lohlau EH, Hashimoto T, Asakawa Y (2000) Chemical constituent of the liverwort *Plagiochasma japonica* and *Marchantia tosana*. J Hattori Bot Lab 88:271–275

Luangpipat T, Beattie IR, Chisti Y, Richard GH (2011) Gold nanoparticles produced in amicroalgae. J Nanopart Res 13:6439–6445

Manickam VS, Irudayaraj V (1992) Pteridophyte flora of the Western Ghats, South India. B.I. Publications, New Delhi

Mathur V, Vats S, Jain M, Bhojak J, Kamal R (2007) Antimicrobial activity of bioactive metabolites isolated from selected medicinal plants. Asian J Exp Sci 21:267–272

Mohanpuria P, Rana NK, Yadav SK (2008) Biosynthesis of nanoparticles: technological concepts and future applications. J Nanopart Res 10:507–517

Mohsen SM, Ammar ASM (2009) Total phenolic contents and antioxidant activity of corn tassel extracts. Food Chem 112:595–598

Morones JR, Elechiguerra JL, Camacho A, Holt K, Kouri JB, Yacaman MJ (2005) The bactericidal effect of silver nanoparticles. Nanotechnology 16:2346–2353

Mukherjee S, Dasari M, Priyamvada S, Kotcherlakota R, Bollua VS, Patra CR (2015) A green chemistry approach for the synthesis of gold nanoconjugates that induce the inhibition of cancer cell proliferation through induction of oxidative stress and their in vivo toxicity study. J Mater Chem B 3:3820–3830

Nalwade AR, Badhe MN, Pawale CB, Hinge SB (2013) Rapid biosynthesis of silver nanoparticles using fern leaflet extract and evaluation of their antibacterial activity. Int J Biol Technol 4:12–18

Ortiz EP, Ruiz JHR, Márquez EAH, Esparza JL, Cornejo AD, González JCC, Cristóbal LFE, Lopez SYR (2017) Dose-dependent antimicrobial activity of silver nanoparticles on polycaprolactone fibers against gram-positive and gram-negative bacteria. J Nanomater Article ID 4752314, 9p. https://doi.org/10.1155/2017/4752314

Pan C, Chen YG, Ma XY, Jiang JH, He F, Zhang Y (2011) Phytochemical constituents and pharmacological activities of plants from the *Genus adiantum*: a review. Trop J Pharm Res 10:681–692

Pankhurst QA, Connolly J, Jones SK, Dobson J (2003) Applications of magnetic nanoparticles in biomedicine. J Phys D Appl Phys 36:R167–R181

Parak WJ, Gerion D, Pellegrino T, Zanchet D, Micheel C, Williams CS, Boudreau R, Le Gros MA, Larabell CA, Alivisatos AP (2003) Biological applications of colloidal nanocrystals. Nanotechnology 14:R15–R27

Parihar P, Parihar L, Bohra A (2010) In vitro antibacterial activity of fronds (leaves) of some important pteridophytes. J Microbiol Antimicrob 2:19–22

Park Y, Hongn YN, Weyers A, Kim YS, Linhardt RJ (2011) Polysaccharides and phytochemicals: a natural reservoir for the green synthesis of gold and silver nanoparticles. IET Nanobiotechnol 5:69–78

Paul B, Bhuyan B, Dhar D, Purkayastha DSS (2015) Green synthesis of silver nanoparticles using dried biomass of *Diplazium esculentum* (retz.) sw. and studies of their photocatalytic and anticoagulative activities. J Mol Liq 212:813–817

Perito B, Giorgetti E, Marsili P, Miranda MM (2016) Antibacterial activity of silver nanoparticles obtained by pulsed laser ablation in pure water and in chloride solution Beilstein. J Nanotechnol 7:465–473

Pietta PG (2000) Flavonoids as antioxidants. J Nat Prod 63:1035–1042

Prasad R (2014) Synthesis of silver nanoparticles in photosynthetic plants. J Nanoparticles Article ID 963961. https://doi.org/10.1155/2014/963961

Prasad R, Kumar V, Prasad KS (2014) Nanotechnology in sustainable agriculture: present concerns and future aspects. Afr J Biotechnol 13(6):705–713

Prasad R, Pandey R, Barman I (2016) Engineering tailored nanoparticles with microbes: quo vadis. WIREs Nanomed Nanobiotechnol 8:316–330. https://doi.org/10.1002/wnan.1363

Prasad R, Bhattacharyya A, Nguyen QD (2017) Nanotechnology in sustainable agriculture: recent developments, challenges, and perspectives. Front Microbiol 8:1014. https://doi.org/10.3389/fmicb.2017.01014

Prathna TC, Mathew L, Chandrasekaran N, Ashok MR, Mukherjee A (2010) Biomimetic synthesis of nanoparticles: science, technology and applicability. In: Mukherjee A (ed) Biomimetics learning from nature. Intech Open. https://doi.org/10.5772/8776

Preethi R, Padma PR (2016) Green synthesis of silver nanobioconjugates from Piper betle leaves and its anticancer activity on A549 cells, Asian J Pharm Clin Res 9:252-257

Rai RV, Bai J (2011) Nanoparticles and their potential application as antimicrobials. In: Vilas M (ed) Science against microbial pathogens: communicating current research and technological advances. Formatex, Badajoz, pp 197–209

Rai M, Yadav A, Gade A (2009) Silver nanoparticles as a new generation of antimicrobials. Biotechnol Adv 27:76–83

Rai MK, Deshmukh SD, Ingle AP, Gade AK (2012) Silver nanoparticles: the powerful nanoweapon against multidrug resistant bacteria. J Appl Microbiol 112:841–852

Rai S, Yadav SK, Mathur K, Goyal M (2016) A review article on *Adiantum incisum*. World J Pharm Pharm Sci 5:861–867

Ramamurthy V, Rajeswari DM, Gowri R, Vadivazhagi MK, Jayanthi, G, Raveendran S (2013) Study of the phytochemical analysis and antimicrobial activity of Dodonaea viscosa. J Pure Appl Zool 1:178-184

Rajaganesh R, Muruganan K, Panneerselvam C, Jayashanthini S, Aziz AT, Roni M, Suresh U, Trivedi S, Rehman H, Higuchi A, Nicoletti M, Benelli G (2016) Fern-synthesized silver nanocrystals: towards a new class of mosquito oviposition deterrents? Res Vet Sci 109:40–51

Rhoades FM (1999) A review of lichen and bryophyte elemental content literature with reference to pacific northwest species. United States Department of Agriculture Forest Service Pacific Northwest Region, Bellingham

Rizzello L, Pompa PP (2014) Nanosilver-based antibacterial drugs and devices: mechanisms, methodological drawbacks, and guidelines. Chem Soc Rev 43:1501–1518

Roy B, Mukherjee S, Mukherjee N, Chowdhury P, Babu SPS (2014) Design and green synthesis of polymer inspired nanoparticles for the evaluation of their antimicrobial and antifilarial efficiency. RSC Adv 4:34487–34499

Sabovljevic A, Sokovic M, Sabovljevic M, Grubisic D (2006) Antimicrobial activity of *Bryum argenteum*. Fitoterapia 77:144–145

Sadeghi B, Gholamhoseinpoor F (2015) A study on the stability and green synthesis of silver nanoparticles using *Ziziphora tenuior* (Zt) extract at room temperature. Spectrochim Acta A Mol Biomol Spectrosc 134:310–315

Sanghi R, Verma P (2010) pH dependant fungal proteins in the 'green' synthesis of gold nanoparticles. Adv Mater Let 1:193–199

Sant DG, Gujarathi TR, Harne SR, Ghosh S, Kitture R, Kale S, Chopade BA, Pardesi KR (2013) *Adiantum phillipense* L. frond assisted rapid green synthesis of gold and silver nanoparticles. J Nanoparticles 2013:1–9. https://doi.org/10.1155/2013/182320

Santhoshkumar S and Nagarajan N (2014) Biological synthesis of silver nanoparticles of *Adiantum capillus veneris* L. and their evaluation of antibacterial activity against human pathogenic bacteria. J Pharm Sci Res 5:5511–5518

Santos MG, Kelecom A, Paiva SR, Moraes MG, Rocha L, Garret R (2010) Phytochemical studies in pteridophytes growing in Brazil: A review. Am J Plant Sci Biotechnol 4:113–125

Sastry M, Ahmad A, Khan MI, Kumar R (2003) Biosynthesis of metal nanoparticles using fungi and actinomycete. Curr Sci 85:2–6

Sathish SS, Vimala A, Kanaga A, Murugan M (2016) Antioxidant and antimicrobial studies on biosynthesized silver nanoparticles using *Bryum medianum* mitt. A bryophyte from Kolli Hills, Eastern Ghats of Tamilnadu, India. J Pharm Sci Res 8:704–709

Sawant O, Kadam VJ, Ghosh R (2009) In vitro free radical scavenging and antioxidant activity of *Adiantum lunulatum*. J Herb Med Toxicol 3:39–44

Scher JM, Speakman JB, Zapp J, Becker H (2004) Bioactivity guided isolation of antifungal compounds from the liverwort *Bazzania trilobata* (L.) Gray SF. Phytochemistry 65:2583–2588

Schuettpelz E, Schneider H, Smith AR, Hovenkamp P, Prado J, Rouhan G, Salino A, Sundue M, Almeida TE,Parris B, Sessa EB, Field AR, de Gasper AL, Rothfels CJ, Windham MD, Lehnert M, Dauphin B, Ebihara A,Lehtonen S, Schwartsburd PB, Metzgar J, Zhang LB, Kuo LY, Brownsey PJ, Kato M, Arana MD (2016) A community-derived classification for extant lycophytes and ferns. J Systematics Evolution 54:563–603

Shaw AJ, Péter S, Shaw B (2011) Bryophyte diversity and evolution: windows into the early evolution of land plants. Am J Bot 98:352–369

Si S, Mandal TK (2007) Tryptophan-based peptides to synthesize gold and silver nanoparticles: a mechanistic and kinetic study. Chem A Eur J 13:3160–3168

Singh AK, Talat M, Singh DP, Srivastava ON (2010) Biosynthesis of gold and silver nanoparticles by natural precursor clove and their functionalization with amine group. J Nanopart Res 12:1667–1675

Smith GM (2014) Cryptogamic botany, bryophytes and pteridophytes, vol 1. McGraw-Hill, New York Record No. 20057004954

Srivastava AA, Kulkarni AP, Harpale PM, Zunjarrao RS (2011) Plant mediated synthesis of silver nanoparticles using a bryophyte: *Fissidens minutus* and its anti-microbial activity. Int J Eng Sci Technol 3:8342–8347

Suganya S, Irudayaraj V, Johnson M (2011) Pharmacognostical studies on an endemic Spike-Moss *Selaginella tenera* (Hook. and Grev.) Spring from the Western Ghats, South India. J Chem Pharm Res 3:721–731

Sukirthaa R, Priyankaa KM, Antonya JJ, Kamalakkannana S, Thangamb R, Gunasekaranb P (2012) Cytotoxic effect of green synthesized silver nanoparticles using *Melia azedarach* against in vitro HeLa cell lines and lymphoma mice model. Process Biochem 47:273–279

Sukumaran K, Kuttan R (1991) Screening of 11 ferns for cytotoxic and antitumor potential with special reference to *pityrogramma calomelanos*. J Ethnopharmacol 34:93–96

Suman PR, Jain VK, Varma A (2010) Role of nanomaterials in symbiotic fungus growth enhancement. Curr Sci 99:1189–1191

Suman TY, Rajasree SRR, Kanchana A, Elizabeth SB (2013) Biosynthesis, characterization and cytotoxic effect of plant mediated silver nanoparticles using *Morinda citrifolia* root extract. Colloids Surf B Biointerfaces 106:74–78

Suresh P, Gunasekar PH, Kokila D, Prabhu D, Dinesh D, Ravichandran N, Ramesh B, Koodalingam A, Siva GV (2014) Green synthesis of silver nanoparticles using *Delphinium denundatum* root extract exhibits antibacterial and mosquito larvicidal activities. Spectrochim Acta A Mol Biomol Spectrosc 127:61–66

Şuţan NA, Radu IF, Fierăscub C, Ştefania D, Liliana M, Soarea C (2016) Comparative analytical characterization and in vitro-cytogenotoxic activity evaluation of *Asplenium scolopendrium* L. leaves and rhizome extracts prior to and after Ag nanoparticles phytosynthesis. Ind Crop Prod 83:379–338

Sheny DS, Mathew J, Philip D (2011) Phytosynthesis of Au, Ag and Au-Ag bimetallic nanoparticles using aqueous extract and dried leaf of Anacardium occidentale. Spectrochim Acta A Mol Biomol Spectrosc 79:254–262

Tamuly C, Hazarika M, Bordoloi M, Bhattacharyya PK, Kar R (2014) Biosynthesis of Ag nanoparticles using pedicellamide and its photocatalytic activity: an eco-friendly approach. Spectrochim. Acta A Mol Biomol Spectrosc 132:687–691

Tan YN, Lee JY, Wang DIC (2010) Uncovering the design rules for peptide synthesis of metal nanoparticles. J Am Chem Soc 132:5677–5686

Taton TA (2002) Nanostructures as tailored biological probes. Trends Biotechnol 20:277–279

Thakkar KN, Mhatre SS, Parikh RY (2010) Biological synthesis of metallic nanoparticles. Nanomedicine 6:257–262

Tyler G (1990) Bryophytes and heavy metals: a literature review. Bot J Linn Soc 104:231–253

Vats S, Tiwari R, Alam A, Behera KK, Pareek R (2012) Evaluation of phytochemicals, antioxidant and antimicrobial activity of in vitro culture of *Vigna unguiculata* L. Walp Res 4:70–74

Verma DK, Syed HH, Banika RM (2016) Photo-catalyzed and phyto-mediated rapid green synthesis of silver nanoparticles using herbal extract of *Salvinia molesta* and its antimicrobial efficacy. J Photochem Photobiol B Biol 155:51–59

Vimala A, Sathish SS, Thamizharasi T, Palani R, Vijayakanth P, Kavitha R (2017) Moss (bryophyte) mediated synthesis and characterization of silver nanoparticles from *Campylopus flexuosus* (Hedw.) bird. J Pharm Sci Res 9:292–297

Wood AJ, Oliver MJ, Cove DJ (2000) Bryophytes as model systems. The Bryologist 103:128-133

Wrotniak DW, Gaikwad S, Laskowski D, Dahm H, Niedojadło J (2014) Novel approach towards synthesis of silver nanoparticles from MyxococcusVirescens and their lethality on pathogenic bacterial cells. Austin J Biotechnol Bioeng 1:7 pages

Xiu ZM, Zhang QB, Puppala HL, Colvin VL, Alvarez PJJ (2012) Negligible particle-specific antibacterial activity of silver nanoparticles. Nano Lett 12:4271–4275

You C, Han C, Wang X, Zheng Y, Li Q (2012) The progress of silver nanoparticles in the antibacterial mechanism, clinical application and cytotoxicity. Mol Biol Rep 39:9193–9201

Yu H, Chen M, Rice PM, Wang SX, White RL, Sun S (2005) Dumbbell-like bifunctional Au-Fe$_3$O$_4$ nanoparticles. Nano Lett 5:379–382

Zhang W, Yong C, Kelsey RD, Lena QM (2004) Thiol synthesis and arsenic hyperaccumulation in *Pteris vittata* (Chinese brake fern). Environ Poll 131:337–345

Zhou Y, Lin W, Huang J, Wang W, Lin YGL, Li Q, Lin LDM (2010) Biosynthesis of gold nanoparticles by foliar broths: roles of biocompounds and other attributes of the extracts. Nanoscale Res Lett 5:1351–1359

Chapter 6
Plant and Its Biomolecules on Synthesis of Silver Nanoparticles for the Antibacterial and Antifungal Activity

S. Rajeshkumar, R. V. Santhiyaa, and P. Veena

6.1 Introduction

Silver nanoparticles are used in different biomedical applications such as antifungal, antibacterial, anti-inflammatory, antidiabetic, and anticancer applications. The silver nanoparticles synthesized using different natural resources such as bacteria *Klebsiella planticola* (MTCC 2277), *Vibrio alginolyticus*, *Enterococcus* sp., *Serratia nematodiphila*, *Bacillus* sp., *Phaenerochaete chrysosporium*, *Bacillus licheniformis*, *Escherichia coli*, *Corynebacterium glutamicum*, *Planomicrobium* sp., and *Bacillus subtilis* MTCC 3053 (Rajeshkumar 2016a; Rajeshkumar and Bharath 2017), fungus such as *Trichoderma viride* (5–40 and 2–4 nm), *Aspergillus flavus*, *Aspergillus fumigates*, *Verticillium* sp., *Fusarium oxysporum*, and *Neurospora crassa* (Li et al. 2011). The major application of silver nanoparticles is antimicrobials (Aziz et al. 2014, 2015, 2016). The antimicrobial silver nanoparticles are used in different types of products in detergents, textiles and clothing, the work surfaces and kitchen surface coating, medical implants with Poly(*N*-vinyl pyrrolidone), wastewater treatment and purification of drinking water, food preservation (controlling foodborne pathogens), health supplements, deodorants and washing machines like household items, paints, etc. (Sweet and Singleton 2011).

The plants are the major natural resource used for the environmentally benign silver nanoparticles synthesis, which is shown in Fig. 6.1. Especially the plant parts such as leaves (majorly used for the AgNPs synthesis), flower, fruits, seeds, bark, stem, peels, and sometimes whole plant have been used for the green synthesis of

S. Rajeshkumar (✉)
Department of Pharmacology, Saveetha Dental College and Hospitals, Saveetha Institute of Medical and Technical Sciences, Chennai, Tamil Nadu, India

R. V. Santhiyaa · P. Veena
Nanotherapy Laboratory, School of Bio-Sciences and Technology, Vellore Institute of Technology, Vellore, Tamil Nadu, India

© Springer Nature Switzerland AG 2018
R. Prasad et al. (eds.), *Exploring the Realms of Nature for Nanosynthesis*,
Nanotechnology in the Life Sciences, https://doi.org/10.1007/978-3-319-99570-0_6

Fig. 6.1 Plant-mediated synthesis of silver nanoparticles

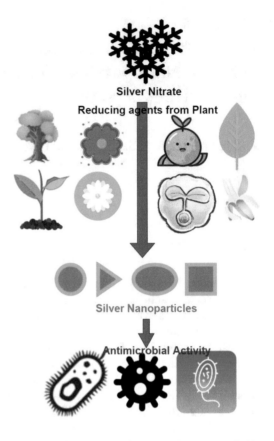

silver nanoparticles described in Tables 6.1, 6.2, 6.3, 6.4, 6.5, 6.6, 6.7, 6.8, and 6.9. The surface plasmon resonance analyzed using UV-vis spectrophotometer, size and shape analyzed using scanning electron microscope and transmission electron microscope, and the phytochemicals present in the plant extract responsible for AgNPs synthesis are depicted in Tables 6.1, 6.2, 6.3, 6.4, 6.5, 6.6, 6.7, 6.8, and 6.9. The different shapes of silver nanoparticles synthesized using plant sources are mentioned in Fig. 6.2. The different functional groups involved in the synthesis of silver nanoparticles are shown in Fig. 6.3.

In recent times, the plant-mediated synthesis of nanoparticles is an emerging out process in the field of nanobiotechnology owing to its several advantages. Various parts of plants such as leaf, stem, root, flower, fruit, and seed were utilized for the synthesis of nanoparticles. Other than the plant-related parts, the waste materials related to the plants were also used for the synthesis of the nanoparticles like hull, peel, and seed of the fruits. The biological compounds in the plant parts were involved in the synthesis of nanoparticles by acting as the reducing as well as capping or stabilizing agent during the synthesis of silver nanoparticles. There is no additional reducing agent added in the green synthesis of nanoparticles. Especially, the existences of phytochemicals like flavonoids, saponins, tannins, terpenoids, and phenols in the plant

Table 6.1 Leaves-mediated synthesis of silver nanoparticles

S. no.	Plant name	UV-Visible (nm)	Size (nm)	Shape	FTIR	References
1	*Phyllanthus amarus*	420	30–42	Flower-like structure	The biomolecules (carbonyl, hydroxyl, alkanes, and amine) present in the leaf extract were responsible for the formation and stabilization of nanoparticles	Ajitha et al. (2018)
2	*Nigella arvensis*	416	5–100	Spherical	The phytochemicals such as flavonoids, alkaloids, and phenols perform a dual as reducing and stabilizing agent	Chahardoli et al. (2018)
3	*Melia azedarach* L.	482	34–48	Spherical	–	Mehmood et al. (2017)
4	*Calliandra haematocephala*	414	70	Spherical	The biomolecules especially gallic acid present in the leaf performed a major role as reducer and stabilizer	Raja et al. (2017)
5	*Cassia roxburghii*	473	15–20	Spherical	The presence of potential bioactive molecules such as alcohols, carboxylates, amines, alkanes, and alkynes act as both reduction and stabilization agent	Moteriya et al. (2017)
6	*Skimmia laureola*	460	46	Spherical	The biomolecules present in the leaf such as skimmidiol, coumarins, and triterpenoids performed as reducing and stabilizing agent	Jamil et al. (2015)

(continued)

Table 6.1 (continued)

S. no.	Plant name	UV-Visible (nm)	Size (nm)	Shape	FTIR	References
7	*Convolvulus arvensis*	430	28	Spherical	The potent bioactive molecules such as organic acids, phenols, and aliphatic amines in the leaf act as a reducing and stabilizing agent for the synthesis of metal ions	Hamedi et al. (2017)
8	*Ficus religiosa*	421	21	Spherical	The biological compounds present in the leaf of Ficus religiosa may be involved in the reduction and capping in the process of synthesis of NPs	Nakkala et al. (2017)
9	*Adhatoda vasica*	450	10–50	Spherical	The functional groups like amides, amino acids, carboxyl and amino groups present in the leaf extract act as reducing and stabilizing agent	Latha et al. (2016)
10	*Physalis angulata*	436	11–96	Spherical	The proteins, amino acids, and phenolic groups were responsible for the synthesis of NPs as stabilizing and reducing agent	Kumar et al. (2017b)
11	*Lantana camara* L.	439	410–450	Spherical	The presence of carboxylic, hydroxyl, phenyl, and carbonyl groups was important for the reduction and stabilization process in the synthesis of NPs	Patil Shriniwas and Kumbhar Subhash (2017)

(continued)

Table 6.1 (continued)

S. no.	Plant name	UV-Visible (nm)	Size (nm)	Shape	FTIR	References
12	*Marsilea quadrifolia*	435	9–42	Spherical	The phenolic compounds, tannins, sugars, alkaloids, and flavonoids present in the leaf of M. quadrifolia might have been used as a reduction and capping agent	Maji et al. (2017)
13	*Olive*	440–458	20–25	Spherical	The presence of functional groups in the olive leaf was responsible for the synthesis of nanoparticles	Khalil (2013)
14	*Tinospora cordifolia*	430	24	Spherical	The biological moieties present in the leaf extract play a dual role in the reducing and stabilizing agent	Selvam et al. (2017)
15	*Andrographis echioides*	426	–	–	The biomolecules like phenolic groups, lipids, and proteins present in the leaf extract of A. echioides act as a reduction and stabilization agent in the synthesis of silver NPs	Elangovan et al. (2015)
16	*Sesuvium portulacastrum*	420	5–20	Spherical	The silver nanoparticles can be synthesized by the extract of callus and leaf of S. portulacastrum due to its excellent reducing and stabilizing property	Sesuvium et al. (2010)

(continued)

Table 6.1 (continued)

S. no.	Plant name	UV-Visible (nm)	Size (nm)	Shape	FTIR	References
17	*Artemisia vulgaris*	420	25	Spherical	The functional groups such as carbonyl, phenols, and aromatic amine present in the phytochemicals of leaf extract might have been acting as a stabilizing and reducing agent for the synthesis of NPs	Rasheed et al. (2017)
18	*Ceropegia thwaitesii*	430	100	Spherical	The presence of methoxy group, triterpenoids, vinyl, carbonyl, and secondary amine in the leaf extract of C. thwaitesii was responsible for the capping and reduction of silver NPs	Muthukrishnan et al. (2014)
19	*Lantana camara*	421	37–29	Spherical	The phytochemicals present in the leaf extract of Lantana camara were responsible for the reduction of silver NPs and stabilization	Ajitha et al. (2015)
20	*Bryophyllum pinnatum*	447	15–40	Spherical	The presence of protein residues in the leaf extract act as a reducing as well as stabilizing agent for the synthesis of silver NPs	Kabir et al. (2017)

(continued)

Table 6.1 (continued)

S. no.	Plant name	UV-Visible (nm)	Size (nm)	Shape	FTIR	References
21	*Alternanthera dentata*	430	50–100	Spherical	The biomolecules present in the extract act as a reducing and capping agent. Thus, the FTIR also confirms that the existence of soluble water fractions in the leaf extract affects the formation of silver NPs	Ashok et al. (2014)
22	*Andrographis echioides*	426	–		The biomolecules like phenolic groups, lipids, and proteins present in the leaf extract of A. echioides act as a reduction and stabilization agent in the synthesis of silver NPs	Elangovan et al. (2015)
23	*Sesuvium portulacastrum*	420	5–20	Spherical	The silver nanoparticles can be synthesized by the extract of callus and leaf of S. portulacastrum due to its excellent reducing and stabilizing property	Sesuvium et al. (2010)
24	*Artemisia vulgaris*	420	25	Spherical	The functional groups such as carbonyl, phenols, and aromatic amine present in the phytochemicals of leaf extract might have been acting as a stabilizing and reducing agent for the synthesis of NPs	Rasheed et al. (2017)

(continued)

Table 6.1 (continued)

S. no.	Plant name	UV-Visible (nm)	Size (nm)	Shape	FTIR	References
25	*Ceropegia thwaitesii*	430	100	Spherical	The presence of methoxy group, triterpenoids, vinyl, carbonyl, and secondary amine in the leaf extract of *C. thwaitesii* was responsible for the capping and reduction of silver NPs	Muthukrishnan et al. (2014)
26	*Lantana camara*	421	37–29	Spherical	The phytochemicals present in the leaf extract of Lantana camara were responsible for the reduction of silver NPs and stabilization	Ajitha et al. (2015)
27	*Bryophyllum pinnatum*	447	15–40	Spherical	The presence of protein residues in the leaf extract act as a reducing as well as stabilizing agent for the synthesis of silver NPs	Kabir et al. (2017)
28	*Alternanthera dentata*	430	50–100	Spherical	The biomolecules present in the extract act as a reducing and capping agent. Thus, the FTIR also confirms that the existence of soluble water fractions in the leaf extract affects the formation of silver NPs	Ashok et al. (2014)

(continued)

Table 6.1 (continued)

S. no.	Plant name	UV-Visible (nm)	Size (nm)	Shape	FTIR	References
29	*Artocarpus altilis*	432	34 and 38	Spherical	The presence of phytoconstituents in the leaf extract act as stabilizing and capping agent for the synthesis of silver nanoparticles	Ravichandran et al. (2016)
30	*Mimusops elengi* L.	434	55–83	Spherical	The presence of biomolecules in the leaf extract of M. elengi has an ability to reduce and stabilize nanoparticles	Prakash et al. (2013)
31	*Euphorbia hirta*	425	15.5	Spherical	The presence of polyphenolic compounds and protein act as both reducing and stabilizing agents	Kumar et al. (2017b)
32	*Ananas comosus*	440–460	12.4	Spherical	The functional groups such as methoxy group and carbonyl group present in the biomolecules of leaf extract act as an excellent reducing and stabilizing agent	Elias et al. (2014)
33	*Acalypha indica*	420	20–30	Spherical	–	Krishnaraj and Jagan (2010)
34	*Sesbania grandiflora*	416	16	Well-dispersed and spherical shape	The FTIR analysis reveals that the leaf extract of S. grandiflora might have been acting as capping and reducing agent	Ajitha et al. (2016)
35	*Hemidesmus indicus*	430	25.24	Spherical	The bioactive molecules in the leaf extract of H. indicus maybe act as a reducing and stabilizing agent	Latha et al. (2015)

(continued)

Table 6.1 (continued)

S. no.	Plant name	UV-Visible (nm)	Size (nm)	Shape	FTIR	References
36	*Rosmarinus officinalis*	450	10–33	Spherical	The functional groups of phenolic compounds in the leaf extract might have been acting as a reducing and stabilizing agent in the synthesis of silver nanoparticles	Ghaedi et al. (2015)
37	*Croton sparsiflorus Morong*	457	22–52	Spherical	The FTIR study exhibits that the functional groups such as carboxyl, amine, hydroxyl, aldehydes, and ketones in the leaf extract exhibit double functional role as reducing and stabilizing agent	Kathiravan et al. (2015)
38	*Atalantia monophylla*	404	35	Spherical	The functional groups of aromatic compounds in the leaf extract act as a reducing and stabilizing agent	Mahadevan et al. (2017)
39	*Ocimum tenuiflorum*	450	25–40	Spherical	Leaf extract of O. tenuiflorum acts as reducing and capping agent for the synthesis of nanoparticles	Patil et al. (2012)
40	*Xanthium strumarium*	436	12–20	Spherical	The phenolic compounds in the leaf extract act as both reducing and stabilizing agent	Kumar et al. (2016b)

(continued)

Table 6.1 (continued)

S. no.	Plant name	UV-Visible (nm)	Size (nm)	Shape	FTIR	References
41	*Mukia maderaspatana*	430	58–458	Spherical	The leaf of M. maderaspatana has a rich source of phenolic compounds which are used as a reducing agent during the synthesis of nanoparticles	Harshiny et al. (2015)
42	*Psidium guajava* L.	435	25–35	Spherical	The presence of polysaccharides in the leaf extract act as reducing and stabilizing agent for the synthesis of nanoparticles	Wang et al. (2017)
43	*Salvinia molesta*	425	12.46	Spherical	The phytoconstituents present in the leaf extract of aquatic fern involved in the bioreduction reaction and stabilizing the nanoparticles to produce stable NPs	Kumar et al. (2016a)
44	*Rauvolfia serpentina Benth*	427	7–10	Spherical	The phytoconstituents such as phenols, flavonoids, tannins, terpenoids, and proteins play a vital role in the reducing as well as capping during the stable synthesis of nanoparticles	Panja et al. (2016)
45	*Abutilon indicum*	455	5–25	Spherical	The phytochemical compounds such as polyphenols, flavonoids, alkaloids, and saponins in the leaf extract of A. indicum were responsible for the reduction reaction	Mata et al. (2015a)

(continued)

Table 6.1 (continued)

S. no.	Plant name	UV-Visible (nm)	Size (nm)	Shape	FTIR	References
46	*Caesalpinia coriaria*	420	78 and 98	Triangle, hexagonal, and spherical	The phytochemical compounds like flavonoids, tannins, terpenoids, and polyphenols in the leaf extract are involved in the reduction reaction as well as stabilizing the metal ions during the synthesis of nanoparticles	Jeeva et al. (2014)
47	*Phoenix dactylifera*	439.5	30 and 85	Spherical	The functional groups of biomolecules in the P. dactylifera act as reducing agent for the formation of silver nanoparticles and capping agent during the synthesis of nanoparticles	Imtiaz et al. (2016)
48	*Santalum album*	423	80–200	Spherical and polydispersity	–	Swamy and Prasad (2012)
49	*Nicotiana tobaccum*	418	8	crystalline nature	Biomolecules and cell-metal ions interaction responsible for formation and stabilization of silver nanoparticles	Prasad et al. (2011)
50	*Solanum lycopersicum*	445	13	Spherical	–	Bhattacharyya et al. (2016)
51	*Trichodesma indicum*	445	20–50	Spherical with face centered cubic	–	Buhroo et al. (2017)

parts were responsible for the conversion of metal ions into nanoparticles (Prasad 2014). The characterization of the nanoparticles was done by UV, FTIR, XRD, SEM, and TEM analysis. The UV-visible spectral analysis was carried out in the synthesis of nanoparticles owing to its optical properties. By performing the UV-visible spectroscopy, the reduction of silver ions to silver nanoparticles was monitored (Ghaedi et al. 2015). The color change was ascribed to the excitation of the surface plasmon

Table 6.2 Fruit-mediated synthesis of silver nanoparticles

S. no	Plant name	UV-Visible (nm)	Size (nm)	Shape	FTIR analysis	References
1	*Cleome viscosa*	410–430	20–50	Spherical and irregular in shape	The phytochemicals (amino acids, alkaloids, phenol, carbohydrates, and tannins) act both reducing and capping agent	Lakshmanan et al. (2017)
2	*Emblica officinalis*	–	15	Spherical	The phytochemicals such as alkaloids, amino acids, tannins, and carbohydrates in the fruit extract were performed as reducing and stabilizing agent	Ramesh et al. (2015)
3	*Crataegus douglasii*	425–475	29.28	Spherical	–	Ghaffari-Moghaddam and Hadi-Dabanlou (2014)
4	*Piper longum*	430	46	Spherical	The phenolic constituents in the leaf extract of *P. longum* were involved in the capping and reduction reaction during the synthesis of nanoparticles	Reddy et al. (2014)
5	*Cassia fistula*	444.5–439.5	69	Spherical	The biomolecules in the fruit extract of *C. fistula* act as both reducing and stabilizing agent	Rashid et al. (2017)
6	*Momordica cymbalaria*	450	15.5	Spherical	The existence of several enzymes and phytochemicals in the fruit extract of *M. cymbalaria* was responsible for the reducing as well as capping agent during the synthesis of nanoparticles	Kumara et al. (2015)
7	*Carissa carandas* (Karonda) berry	420	10–60	Spherical	Carinol (and related resonant compounds) in the berry extract with inductive effect of the proton of methoxy and allyl groups, present at ortho and para positions of the compounds	Joshi et al. (2018)

resonance and the reduction of silver ions by the biomolecules present in the plant extract (Rasheed et al. 2017). The absorbance peak was formed due to the reduction of silver ions and characteristic of the SPR, thus indicating the formation of silver nanoparticles (Daisy and Saipriya 2012). The SEM, TEM, and XRD were used to

Table 6.3 Peel-mediated synthesis of silver nanoparticles

S. no	Plant name	UV-visible (nm)	Size (nm)	Shape	FTIR analysis	References
1	*Carica papaya*	400–435	16–20	Spherical	Phytochemicals and biomolecules such as proteins involved as both reducing and stabilizing agent	Balavijayalakshmi and Ramalakshmi (2017)
2	*Dragon fruit*	430–460	25–26	Spherical	The functional group such as carboxylic acid, phenol, alkanes, alkenes, and carbonyl group in the fruit peel might have been involved as stabilization and reduction agent for the formation of metal ions	Phongtongpasuk et al. (2016)
3	Mango	412–434	7–27	Spherical	The mango peel consists of carboxyl, hydroxyl, aldehydes, and ketones, which were involved in the reduction and stabilization of NPs	Yang and Li (2013)
4	*Citrus sinensis*	445–424	35 ± 2 and 10 ± 1	Spherical	The biomolecules present in the peel of *C. sinensis* act as a reducing and stabilizing agent for the synthesis of NPs	Kaviya et al. (2011)

determine the size, shape, and structure of the nanoparticles. The FTIR analysis was used to determine the functional groups present in the biomolecules. From the above Table 6.9, the highest size of the silver nanoparticles was synthesized by using the leaf of *Lantana camara* which ranges from 410 to 450 nm as well as the lowest size of the AgNPs was synthesized by the seed of *Prosopis farcta* (Patil Shriniwas and Kumbhar Subhash 2017; Miri et al. 2015). The phytochemicals present in the plants are the major sources responsible for the green synthesis of silver nanoparticles shown in Fig. 6.4. The antibacterial and antifungal activities of silver nanoparticles synthesized using different plants have been depicted in Tables 6.10 and 6.11, respectively.

The plant-mediated silver nanoparticles are widely used in various biomedical applications. Predominantly it plays an indispensable role in inhibiting the pathogenic bacteria. The antibacterial activity of silver nanoparticles was determined by performing the antimicrobial assay such as Kirby–Bauer method, minimum inhibitory concentration, and minimum bactericidal concentration. Commonly *Escherichia coli, Staphylococcus aureus, Bacillus subtilis, Pseudomonas aeruginosa, Klebsiella pneumonia, Streptococcus pneumonia*, and some other *Enterobacter* were inhibited by the synthesis of plant-mediated silver nanoparticles. Here are some of the illustrations given below about the AgNPs synthesized by the plants. The AgNPs synthesized by the fruit of *Cleome viscosa* have an ability to inhibit the gram-positive bacteria *S. aureus* and gram-negative bacteria like *E. coli and Klebsiella pneumonia*.

Table 6.4 Root-mediated synthesis of silver nanoparticles

S. no	Plant name	UV-visible (nm)	Size (nm)	Shape	FTIR analysis	References
1	*Dryopteris crassirhizoma*	–	5–60	Spherical	The functional groups (carboxyl, hydroxyl, alkyl, amide) and phytochemicals present in the rhizome may be used as a reducing and stabilizing agent for NPs	Lee et al. (2016)
2	*Coptis chinensis*	428	15–20	Spherical	The polyphenolic compounds present in the leaf extract especially chitosan act as capping and reducing agent	Ahmad et al. (2017)
3	*Bergenia ciliata*	425	35	Spherical	The phenolic compounds present in the rhizome part of plant might have been responsible for the reduction and stabilization of AgNPs	Phull et al. (2016)
4	*Diospyros paniculata*	428	14–28	Spherical	The biomolecules in the root extract act as a reducing agent in the synthesis of NPs	Rao et al. (2016)
5	*Acorus calamus*	421	31.83	Spherical	The presence of alcohol, phenol, carbonyl and aromatic amine in the rhizome extract plays a dual role in the reduction and stabilization of nanoparticles	Nakkala et al. (2014)
6	*Acorus calamus*	420	20–35	Spherical	The functional groups such as carboxylic, carbonyl, amino and hydroxyl groups in the rhizome extract were involved in the conversion of bioreduction of metal ions to nanoparticles	Sudhakar et al. (2015)

It has been determined by the zone of inhibition present in the disc diffusion agar plate. The gram-positive bacteria exhibit the maximum zone of inhibition while comparing to gram-negative bacteria (Lakshmanan et al. 2017). By performing the MIC and MBC assays, it has been found that the AgNPs synthesized by the plant extract of *Aerva lanata* show the excellent antibacterial activity by inhibiting these pathogenic bacteria like *P. aeruginosa, E. coli, B. subtilis, S. aureus, E. faecalis, S. agalactiae,* and *Corynebacterium* (Appapalam and Panchamoorthy 2017). The good diffusion and minimum inhibitory concentration assays were performed to ascertain whether the AgNPs synthesized using the leaf extract of *Nigella arvensis* have an ability to inhibit the pathogenic bacteria. When compared to *P. aeruginosa* and *E. coli, the S. marcescens* shows highest ZOI (Chahardoli et al. 2018). The AgNPs synthesized by the bark extract of *Butea monosperma* exhibit highest antibacterial efficacy against *E. coli* and *B. subtilis* (Pattanayak et al. 2017). The agar diffusion method was performed to determine the antibacterial activity of the AgNPs prepared

Table 6.5 Whole plant-mediated synthesis of silver nanoparticles

S. no	Plant name	UV-visible (nm)	Size (nm)	Shape	FTIR analysis	References
1	*Aerva lanata*	300–700	50	Spherical	The presence of flavonoids, saponins, and polyphenols act as both reducing and stabilizing agent for the synthesis of silver NPs	Appapalam and Panchamoorthy (2017)
2	*Mentha piperita*	450	90	Spherical	The biomolecules present in the plant act as both reducing and stabilizing agent	Mubarakali et al. (2011)
3	*Dioscorea alata*	450	10–20	Spherical	The plant extract of *D. alata* consists of reducing and stabilizing agent property owing to the presence of biomolecules	Pugazhendhi et al. (2016)
4	*Vernonia cinerea* L.	426	40–75	Spherical	The biomolecules present in the plants act as a reducing and capping agent	Ramaswamy et al. (2015)
5	*Boerhavia diffusa*	418	25	Spherical	The presence of potent biomolecules in the plant extract act as reducing and stabilizing agent	Kumar et al. (2014)
6	*Cocos nucifera*	427–428	22	Spherical	The existence of phenolic compounds, terpenoids, and other aromatic compounds plays a vital role in the synthesis of AgNPs as reducing and stabilizing agent	Mariselvam et al. (2014)

(continued)

Table 6.5 (continued)

S. no	Plant name	UV-visible (nm)	Size (nm)	Shape	FTIR analysis	References
7	*Lippia nodiflora*	442	30–60	Spherical	The presence of phenolic compounds in the aerial parts of the plant was responsible for the silver nanoparticles reduction and stabilization.	Sudha et al. (2017)
8	*Artemisia tournefortiana*	420	22.89 ± 14.82	Spherical	The bioactive molecules of hydroxyl and carbonyl groups act as a reducing and stabilizing agent	Baghbani-arani et al. (2017)

Table 6.6 Bark-mediated synthesis of silver nanoparticles

S. no	Plant name	UV-visible (nm)	Size (nm)	Shape	FTIR analysis	References
1	*Butea monosperma*	424	35	Spherical	The phytochemicals present in the bark extract might act as a reducing and stabilizing agent	Pattanayak et al. (2017)
2	*Pongamia pinnata*	420	5–55	Spherical	The presence of phytochemicals such as piperine, phenolic amides, and some other reducing sugars are involved in the reduction and stabilization process	Rajeshkumar (2016a)
3	*Eucommia ulmoides*	413	5–14	Spherical	The potent bioactive molecules present in the bark extract act as a reducing agent as well as capping agent	Lü et al. (2017)
4	*Syzygium cumini*	427	20–60	Polydispersed		Prasad and Swamy (2013)

Table 6.7 Flower-mediated synthesis of silver nanoparticles

S. no	Plant name	UV-visible (nm)	Size (nm)	Shape	FTIR analysis	References
1	*Nyctanthes arbortristis*	–	5–20	Spherical and oval	The phytochemicals in the flower extract of *N. arbortristis* play a vital role in the synthesis of AgNPs as a potent bioreduction and stabilizing agent	Gogoi et al. (2015)
2	*Millettia pinnata*	438	49 ± 0.9	Spherical	The flower extract of *M. pinnata* has the ability to act as reducing and stabilizing agent	Rajakumar et al. (2017)
3	*Tagetes erecta*	430	10–90	Spherical, hexagonal, and irregular shape	The existence of phytochemicals such as flavonoids, saponins, phlobatannins, triterpenes, steroids, and tannins might have been involved in the synthesis of NPs in the form of reducing and capping agent	Padalia et al. (2015)
4	*Plumeria alba*	445	36.19	Spherical	The existence of phytoconstituents groups such as carboxylic, sulfhydryl, and amino act as both reducing and capping agent	Mata et al. (2015b)

Table 6.8 Stem-mediated synthesis of silver nanoparticles

S. no	Plant name	UV-visible (nm)	Size (nm)	Shape	FTIR analysis	References
1	*Garcinia mangostana*	430	30	Spherical in monodispersed nature	The biomolecules present in the stem extract act as a reducing agent	Karthiga (2017)
2	*Hibiscus cannabinus*	444	10	Spherical	The various carboxylic acids present in the *H. cannabinus* were involved in the reduction reaction to reduce the silver ion into silver nanoparticles	Bindhu et al. (2014)

Table 6.9 Other parts of plant-mediated synthesis of silver nanoparticles

S. no	Plant name	UV-visible (nm)	Size (nm)	Shape	FTIR analysis	References
1	*Crocus sativus* L.	450	12–20	Spherical	The biomolecules such as flavonoids, phenols, glycosides, tannins, glycosides, terpenoids with carboxyl, ketone, aldehyde, and other functional groups play a vital role in reduction and stabilization	Bagherzade et al. (2017)
2	*Croton bonplandianum*	436 and 428	4–46	Spherical	The functional groups (hydroxyl, amino, amide, and carboxyl) of phytochemicals such as glucose, tannins, and proteins might have been involved in the reduction and stabilization process	Kumar et al. (2017a)
3	*Azadirachta indica* L.	418	<30	Spherical	The hydroxyl, carboxyl, and amino group present in the neem gum has a property of reducing and stabilizing the silver nanoparticles	Velusamy et al. (2015)
4	*Prosopis farcta*	433	8.5–11	Spherical	The biological molecule present in the extract plays a dual role in reducing as well as capping agent for the formation of NPs	Miri et al. (2015)
5	*Pistacia atlantica*	446	10–50	Spherical	The presence of amide groups in the protein plays an indispensable role in the capping of AgNPs	Sadeghi et al. (2015)

(continued)

Table 6.9 (continued)

S. no	Plant name	UV-visible (nm)	Size (nm)	Shape	FTIR analysis	References
6	*Watermelon*	425	109.97	Spherical	The existence of organic compounds such as flavonoids, lycopene, citrulline, and phenols in the rind of watermelon act as reducing and stabilizing agent for the formation of silver nanoparticles	Patra et al. (2016)
7	*Rosa indica*	441	23.52–60.83	Aggregated and spherical	The hydroxyl and amine groups of *R. indica* act as both reducing and stabilizing agent	(Ramar et al. 2015b)
8	*Coriandrum sativum*	421	13.09	Spherical, non-uniform and polydispersed	*Coriandrum sativum* acts as a capping agent	Nazeruddin et al. (2014)
9	Peanut	450	10–50	Spherical and oval	The phenolic, carboxylic, and amino groups in the peanut hull were responsible for the formation of silver nanoparticles	Velmurugan et al. (2015)

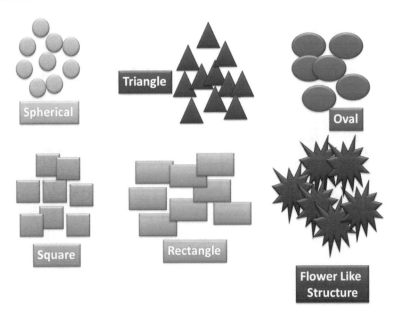

Fig. 6.2 Different shapes of AgNPs synthesized by plant

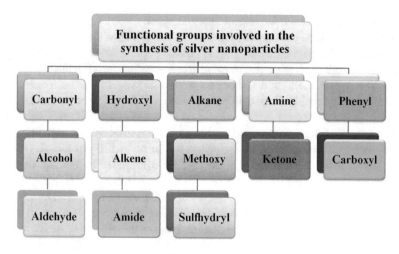

Fig. 6.3 Different functional groups involved in AgNPs synthesis

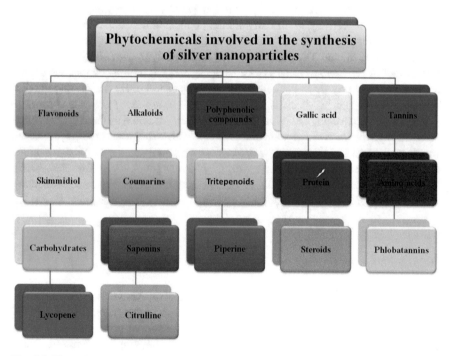

Fig. 6.4 Phytochemicals responsible for plant-mediated AgNPs

by the leaf of the *Skimmia laureola*. It has the capacity to inhibit the pathogenic bacteria like *S. aureus*, *K. pneumonia*, *P. aeruginosa*, and *E. coli* (Jamil et al. 2015). The MIC and MBC assays were carried out to find the antibacterial activity of the AgNPs prepared through the leaf extract of *Artemisia tournefortiana* (Baghbani-arani et al. 2017). The AgNPs callus and leaf extract of *Sesuvium portulacastrum* exhibit antibacterial activity against *S. aureus* and *Micrococcus luteu* (Sesuvium et al. 2010). The disc diffusion method was accomplished to identify the antibacterial activity of the leaf extract of *Artemisia vulgaris*. It has a potential to inhibit the organism like *E. coli*, *S. aureus*, *P. aeruginosa*, *Haemophilus influenza*, and *K. pneumonia* (Rasheed et al. 2017). It has been found that the AgNPs synthesized by the leaf extract of *Lantana camara* comprise antibacterial activity, antioxidant activity, and cytotoxicity (Patil Shriniwas and Kumbhar Subhash 2017). The AgNPs prepared by the flower extract of *Nyctanthes arbortristis* show potential antibacterial activity and cytotoxicity activities (Gogoi et al. 2015). The MIC determines the antibacterial activity of the AgNPs synthesized by the root extract of *Diospyros paniculata* (Rao et al. 2016).

The green synthesis of silver nanoparticles has an antifungal property. It inhibits against various fungal pathogens. Therefore, *Aspergillus* sp., *Penicillium* sp., *Candida* sp., and *Mucor* were most commonly inhibited by the silver nanoparticles. The assay like antimicrobial activity, minimum inhibitory concentration, and minimum bactericidal concentration were carried out to determine whether the fungi are inhibited by the silver nanoparticles. The penicillium was inhibited by the AgNPs synthesis by the leaf extract of *Phyllanthus amarus* (Ajitha et al. 2018). It inhibits by inactivating the sulfhydryl group present in the cell wall and lipids of fungi. By performing the disc diffusion method, it has been identified that the *Aspergillus fumigates*, *Fusarium solani*, *Aspergillus niger*, and *Aspergillus flavus* were inhibited by the AgNPs synthesized through the roots of *Bergenia ciliata* (Phull et al. 2016). The AgNPs synthesized by the root extract of *Diospyros paniculata* have an antibacterial activity. It inhibits against *P. notatum*, *A. flavus*, *S. cerevisiae*, *C. albicans*, and *A. niger* (Rao et al. 2016). The AgNPs produced by the hull of peanut exhibit highest antibacterial activity against *Phytophthora infestans* and *Phytophthora capsici* (Velmurugan et al. 2015).

6.2   Conclusion

This book chapter elucidates about the synthesis of silver nanoparticles using different parts of the plants like leaf, stem, flower, bark, roots, seed, peel, fruit, and aerial parts. It also explicates the phytochemicals and functional groups present in the plants. The plant-mediated synthesis of nanoparticles plays an indispensable role in the field of nanobiotechnology owing to its various applications like antifungal, antibacterial, antioxidant, cytotoxicity activities, and so on. Even though the green synthesis of nanoparticles has some limitations, it has been widely used nowadays when compared to a physical and chemical method of nanoparticle synthesis. From this review, it has been proved that the green synthesis of silver nanoparticles has an ability to inhibit most of the pathogenic bacteria and fungi.

Table 6.10 Antibacterial activity of plant-mediated synthesis of silver nanoparticles

S. no	Plant name	Parts	Bacteria	Remarks	References
1	*Phyllanthus amarus*	Leaf	*Pseudomonas*	Zone of inhibition = 11 mm	Ajitha et al. (2018)
2	*Cleome viscosa*	Fruit	*Staphylococcus aureus, E. coli, Klebsiella pneumonia*	Zone of inhibition against *S. aureus*—17 ± 0.8 mm *E. coli*—16 ± 0.8mm *K. pneumoniae*—14 ± 1.2 mm	Lakshmanan et al. (2017)
3	*Carica papaya*	Peel	*Escherichia coli Staphylococcus aureus*	Zone of inhibition against *E. coli*—75 mm *S. aureus*—65 mm	Balavijayalakshmi and Ramalakshmi (2017)
4	*Lippia nodiflora*	Aerial parts	*Streptococcus pneumonia E. coli*	Zone of inhibition against *S. pneumonia*—24.1 mm *E. coli*—22 mm	Sudha et al. (2017)
5	*Dryopteris crassirhizoma*	Rhizome	*Bacillus cereus Pseudomonas aeruginosa*	Well diffusion method	Lee et al. (2016)
6	*Aerva lanata*	Plant	*P. aeruginosa, E. coli, B. subtilis, S. aureus, E. faecalis, S. agalactiae, and Corynebacterium striatum*	Minimum inhibitory concentration and minimum bacterial concentration method	Appapalam and Panchamoorthy (2017)
7	*Nigella arvensis*	Leaf	*Serratia marcescens, P. aeruginosa, and E. coli*	Well diffusion and MIC method Zone of inhibition against S. marcescens = 20 mm *P. aeruginosa* = 16 mm *E. coli* = 17 mm	Chahardoli et al. (2018)
8	*Melia azedarach* L.	Leaf	*E. coli, K. pneumonia, S. aureus, P. aeruginosa, Proteus* spp.	Paper diffusion method, Minimum inhibitory concentration (MIC)	Mehmood et al. (2017)
9	*Emblica officinalis*	Fruit	*E. coli, K. pneumonia, B. subtilis, and S. aureus*	Disc diffusion method	Ramesh et al. (2015)

Table 6.10 (continued)

S. no	Plant name	Parts	Bacteria	Remarks	References
10	*Crocus sativus* L.	Aqueous extract of saffron	*E. coli, P. aeruginosa, K. pneumonia, S. flexneri, and B. subtilis*	Minimum inhibitory concentration	Bagherzade et al. (2017)
11	*Butea monosperma*	Bark	*E. coli* *Bacillus subtilis*	Agar diffusion method Zone of inhibition against E. coli = 17 mm and B. subtilis = 16 mm	Pattanayak et al. (2017)
12	*Crataegus douglasii*	Fruit	*Staphylococcus aureus* *E. coli*	Disc diffusion method	Ghaffari-Moghaddam and Hadi-Dabanlou (2014)
13	*Calliandra haematocephala*	Leaf	*Escherichia coli*	Well diffusion method	Raja et al. (2017)
14	*Coptis chinensis*	Rhizome	*Escherichia coli* *Bacillus subtilis*	Agar well diffusion method Zone of inhibition against E. coli = 12 ± 1.2 mm B. subtilis = 18 ± 1.6 mm	Ahmad et al. (2017)
15	*Cassia roxburghii*	Leaf	*B. cereus, S. aureus, P. aeruginosa, E. coli*	Antimicrobial assay	Moteriya et al. (2017)
16	*Skimmia laureola*	Leaf	*S. aureus, K. pneumonia, P. aeruginosa, and E. coli*	Agar diffusion method Zone of inhibition against S. aureus = 14.67 mm K. pneumonia and P. aeruginosa = 14.33 mm E. coli = 11.67	Jamil et al. (2015)
17	*Artemisia tournefortiana*	Aerial parts	*Streptococcus pyogenes, P. aeruginosa, and B. subtilis*	Minimum inhibitory concentration and minimum bacterial concentration	Baghbani-arani et al. (2017)
18	*Mentha piperita*	Plant	*S. aureus* *E. coli*	Well diffusion method	Mubarakali et al. (2011)

S. no	Plant name	Parts	Bacteria	Remarks	References
19	*Dragon fruit*	Peel	*Escherichia coli, P. aeruginosa and S. aureus*	Disc diffusion assay Zone of inhibition against E. coli = 15 mm P. aeruginosa = 13 mm S. aureus = 14 mm	Phongtongpasuk et al. (2016)
20	*Convolvulus arvensis*	Leaf	*E. coli*	Disc diffusion method and macro dilution method	Hamedi et al. (2017)
21	*Ficus religiosa*	Leaf	*P. fluorescens, S. typhi, B. subtilis, and E. coli*	Kirby–Bauer disc diffusion method Zone of inhibition against P. fluorescens = 2.9 mm S. typhi = 2.9 mm B. subtilis = 2.8 mm E. coli = 2.7 mm	Nakkala et al. (2017)
22	*Pongamia pinnata*	Bark	*Klebsiella planticola and S. aureus*	Agar well diffusion method Zone of inhibition against K. planticola = 15 mm S. aureus = 13 mm	Rajeshkumar (2016b)
23	*Croton bonplandianum*	Aqueous extract of plant	*E. coli* *S. aureus*	Antimicrobial assay	Kumar et al. (2017a)
24	*Adathoda vasica*	Leaf	*Vibrio parahaemolyticus*	Agar bioassay and well diffusion method	Latha et al. (2016)
25	*Physalis angulata*	Leaf	*E. coli* *S. aureus*	Minimum inhibitory concentration	Kumar et al. (2017b)
26	*Lantana camara*	Leaf	*S. aureus* *P. aeruginosa* *E. coli*	Antibacterial assay Zone of inhibition against S. aureus = 28.1 mm P. aeruginosa = 21.3 mm E. coli = 22.1 mm	Patil Shriniwas and Kumbhar Subhash (2017)

(continued)

Table 6.10 (continued)

S. no	Plant name	Parts	Bacteria	Remarks	References
27	Marsilea quadrifolia	Leaf	Escherichia. coli	Minimum inhibition concentration	Maji et al. (2017)
28	Olive	Leaf	S. aureus, P. aeruginosa, and E. coli	Well diffusion method	Khalil (2013)
29	Eucommia ulmoides	Bark	S. aureus E. coli	Paper disc diffusion method	Lü et al. (2017)
30	Tinospora cordifolia	Leaf	Staphylococcus sp. Klebsiella sp.	Zone of inhibition against Staphylococcus sp. = 13 mm Klebsiella sp. = 12.3 mm	Selvam et al. (2017)
31	Andrographis echioides	Leaf	Escherichia coli Staphylococcus aureus	E. coli = 28 mm S. aureus = 23 mm	Elangovan et al. (2015)
32	Mango	Peel	Escherichia coli Staphylococcus aureus Bacillus subtilis	E. coli = 13 mm S. aureus = 14.5 mm B. subtilis = 11 mm	Yang and Li (2013)
33	Citrus sinensis	Peel	E. coli, P. aeruginosa and S. aureus	Agar well diffusion method	Kaviya et al. (2011)
34	Sesuvium portulacastrum	Callus and leaf	S. aureus and Micrococcus luteus	Zone of inhibition against S. aureus = 23 mm M. luteu = 8 mm	Sesuvium et al. (2010)
35	Azadirachta indica L.	Neem gum	Salmonella enteritidis and Bacillus cereus	Kirby–Bauer disc diffusion method	Velusamy et al. (2015)
36	Prosopis farcta	Seed	S. aureus, B. subtilis, P. aeruginosa, and E. coli	Disc diffusion method	Miri et al. (2015)
37	Dioscorea alata	Plant	Staphylococcus auricularis and Escherichia coli	Agar well diffusion assay	Pugazhendhi et al. (2016)
38	Artemisia vulgaris	Leaf	E. coli, S. aureus, P. aeruginosa, Haemophilus influenza, and K. pneumonia	Disc diffusion method	Rasheed et al. (2017)

S. no	Plant name	Parts	Bacteria	Remarks	References
39	Ceropegia thwaitesii	Leaf	Salmonella typhi and Bacillus subtilis	Disc diffusion method	Muthukrishnan et al. (2014)
40	Solanum trilobatum	Fruit	Streptococcus mutans, enterococcus faecalis, E. coli and K. pneumonia	Antibacterial assay	Ramar et al. (2015a)
41	Lantana camara	Leaf	Pseudomonas spp.. and Bacillus spp.	Kirby–Bauer Disc diffusion method	Ajitha et al. (2015)
42	Bryophyllum pinnatum	Leaf	E. coli and Bacillus megaterium	MIC value for E. coli = 2.65 mm B. megaterium = 6.31 mm MBC value for E. coli = 5.98 mm B. megaterium = 8.34 mm	Kabir et al. (2017)
43	Vernonia cinerea L.	Whole plant	S. aureus E. coli	Disc diffusion method	Ramaswamy et al. (2015)
44	Boerhaavia diffusa	Plant	Flavobacterium branchiophilum, A. hydrophila, and P. fluorescens	MIC against F. branchiophilum-sensitive A. hydrophila-intermediate P. fluorescens-resistant Antibacterial activity F. branchiophilum = 15 mm A. hydrophila = 14 mm P. fluorescens = 12 mm	Kumar et al. (2014)
45	Pistacia atlantica	Seed	S. aureus	Antibacterial assay	Sadeghi et al. (2015)
46	Bergenia ciliata	Rhizome	Micrococcus luteus, S. aureus, Bordetella bronchiseptica, and Enterobactor aerogenes	Disc diffusion method	Phull et al. (2016)

(continued)

Table 6.10 (continued)

S. no	Plant name	Parts	Bacteria	Remarks	References
47	Cocos nucifera	Plant	Klebsiella pneumonia, Plesiomonas shigelloides, Vibrio alginolyticus, and Salmonella paratyphi	Zone of inhibition against K. pneumonia = 24 mm, Plesiomonas shigelloides = 21 mm, V. alginolyticus = 19 mm, and S. paratyphi = 16 mm	Mariselvam et al. (2014)
48	Alternanthera dentata	Leaf	E. coli, P. aeruginosa, K. pneumonia, and E. faecalis	Disc diffusion method	Ashok et al. (2014)
49	Crataegus douglasii	Fruit	E. coli and S. aureus	Disc diffusion method	Ghaffari-Moghaddam and Hadi-Dabanlou (2014)
50	Artocarpus altilis	Leaf	E. coli, P. aeruginosa, and S. aureus	Kirby–Bauer disc diffusion method	Ravichandran et al. (2016)
51	Mimusops elengi L.	Leaf	Klebsiella pneumonia, Micrococcus luteus, and S. aureus	Kirby–Bauer disc diffusion method Zone of inhibition against K. pneumonia = 18 mm M. luteus = 10 mm S. aureus = 11 mm	Prakash et al. (2013)
52	Watermelon	Rind	B. cereus, E. coli, L. monocytogenes, S. aureus, and S. typhimurium	Antibacterial assay Zone of inhibition against S. aureus = 14.54 mm (highest activity) and L. monocytogenes = 9.12 mm (lowest activity) Minimum inhibitory concentration Minimum bactericidal concentration	Patra et al. (2016)
53	Euphorbia hirta	Leaf	E. coli and S. aureus	Antibacterial assay and minimum inhibitory concentration against E. coli and S. aureus (lowest concentration)	Kumar et al. (2017b)
54	Nyctanthes arbortristis	Flower	Escherichia coli	Agar well diffusion method	Gogoi et al. (2015)

S. no	Plant name	Parts	Bacteria	Remarks	References
55	Millettia pinnata	Flower	E. coli, P. aeruginosa, Proteus vulgaris, S. aureus, and K. pneumonia	Agar disc diffusion method E. coli = 20.25 ± 0.91 mm, P. aeruginosa = 17.13 ± 0.80 mm Proteus vulgaris = 0.64 ± 1.09 mm S. aureus = 15.09 ± 0.17 mm and K. pneumonia = 14.81 ± 0.34 mm	Rajakumar et al. (2017)
56	Ananas comosus	Leaf	S. aureus, S. pneumonia, and E. coli	Agar well diffusion method	Elias et al. (2014)
57	Acalypha indica	Leaf	E. coli and V. cholerae	Minimal inhibitory concentration against E. coli and V. cholerae exhibited lowest concentration	Krishnaraj and Jagan (2010)
58	Garcinia mangostana	Stem	E. coli, K. planticola, and B. subtilis	Well diffusion method Contrast to B. subtilis, E. coli, and K. planticola exhibits highest zone of inhibition	Karthiga (2017)
59	Sesbania grandiflora	Leaf	Pseudomonas spp.	Kirby–Bauer disc diffusion method	Ajitha et al. (2016)
60	Tagetes erecta (marigold)	Flower	E. coli and P. aeruginosa	Synergistic antimicrobial activity	Padalia et al. (2015)
61	Hemidesmus indicus	Leaf	S. sonnei	Agar bioassay, well diffusion method and confocal laser scanning microscopic bacterial assay S. sonnei exhibits the highest zone of inhibition = 34 ± 0.2 mm	Latha et al. (2015)
62	Diospyros paniculata	Root	B. subtilis, E. coli, P. aeruginosa K. pneumonia, S. aureus, S. pyogenes, B. pumilus, and P. vulgaris	MIC through microdilution broth method B. subtilis, E. coli, and P. aeruginosa show Maximum activity K. pneumonia, S. aureus, and S. pyogenes exhibit moderate activity Mild activity against B. pumilus and P. vulgaris	Rao et al. (2016)

(continued)

Table 6.10 (continued)

S. no	Plant name	Parts	Bacteria	Remarks	References
63	*Rosmarinus officinalis*	Leaf	*E. coli, P. aeruginosa, S. aureus, and B. subtilis*	Disc diffusion method	Ghaedi et al. (2015)
64	*Croton sparsiflorus morong*	Leaf	*S. aureus, E. coli, and B. subtilis*	Disc diffusion method	Kathiravan et al. (2015)
65	*Rosa indica*	Petal	*Streptococcus mutans, Enterococcus faecalis, E. coli, K. pneumoniae*	Agar well diffusion method	Ramar et al. (2015b)
66	*Prosopis farcta*	Leaf	*S. aureus, B. subtilis, E. coli, and P. aeruginosa*	Disc diffusion method	Miri et al. (2015)
67	*Atalantia monophylla*	Leaf	*S. aureus, B. cereus, K. pneumonia, and E. coli*	Antimicrobial assay by disc diffusion method S. aureus = 37 mm and B. cereus = 36 mm shows maximum ZOI	Mahadevan et al. (2017)
68	*Ocimum tenuiflorum*	Leaf	*E. coli, C. bacterium, and B. subtilis*	Antimicrobial assay by disc diffusion method	Patil et al. (2012)
69	*Xanthium strumarium*	Leaf	*E. coli and S. aureus*	Disc diffusion assay	Kumar et al. (2016b)
70	*Coriandrum sativum*	Seed	*Bacillus subtilis*	Agar diffusion method	Nazeruddin et al. (2014)
71	*Mukia maderaspatana*	Leaf	*B. subtilis, K. pneumonia, S. typhi, and S. aureus*	Disc diffusion assay	Harshiny et al. (2015)
72	*Cassia fistula*	Fruit	*E. coli and K. pneumonia*	Minimum inhibitory concentration	Rashid et al. (2017)
73	*Acorus calamus*	Rhizome	*B. subtilis, B. cereus, and S. aureus*	Kirby–Bauer disc diffusion method	Nakkala et al. (2014)
74	*Plumeria alba*	Flower	*E. coli and B. subtilis*	Agar disc diffusion method Zone of inhibition against E. coli = 28 mm B. subtilis = 20 mm	Mata et al. (2015b)

Table 6.11 Antifungal activity of plant-mediated synthesis of silver nanoparticles

S. no	Plants name	Parts	Fungi	Remarks	References
1	*Phyllanthus amarus*	Leaf	*Penicillium*	The silver NPs have an ability to inactivate the sulfhydryl group present in the lipids and cell wall of fungi	Ajitha et al. (2018)
2	*Vernonia cinerea* L.	Plant	*Candida albicans* and *Penicillium notatum*	Antifungal susceptibility assay	Ramaswamy et al. (2015)
3	*Bergenia ciliata*	Rhizome	*Aspergillus fumigates, Fusarium solani, Aspergillus niger,* and *Aspergillus flavus*	Antifungal assay by disc diffusion method	Phull et al. (2016)
4	*Artocarpus altilis*	Leaf	*Aspergillus vesicolor*	Antimicrobial assay through disc diffusion method *A. vesicolor* shows minimal susceptibility	Ravichandran et al. (2016)
5	*Sesbania grandiflora*	Leaf	*Penicillium* spp.	Kirby–Bauer disc diffusion method	Ajitha et al. (2016)
6	*Tagetes erecta (marigold)*	Flower	*C. albicans, C. neoformans,* and *C. glabrata*	Synergistic antimicrobial activity *Candida albicans* shows maximum zone of inhibition compared to others	Padalia et al. (2015)
7	*Diospyros paniculata*	Root	*P. notatum, A. flavus, S. cerevisiae, C. albicans,* and *A. niger*	MIC through microdilution broth method *P. notatum, A. flavus,* and *S. cerevisiae* show maximum activity *C. albicans* and *A. niger* show moderate activity	Rao et al. (2016)
8	*Rosmarinus officinalis*	Leaf	*Aspergillus oryzae* and *C. albicans*	Disc diffusion method	Ghaedi et al. (2015)
9	*Croton sparsiflorus morong*	Leaf	*Mucor* sp., *Trichoderma* sp., and *Aspergillus niger*	Disc diffusion method	Kathiravan et al. (2015)
10	*Atalantia monophylla*	Leaf	*Candida albicans*	Antimicrobial assay by disc diffusion method *C. albicans* = 34 mm shows a minimum zone of inhibition MIC and MBC were also tested	Mahadevan et al. (2017)

(continued)

Table 6.11 (continued)

S. no	Plants name	Parts	Fungi	Remarks	References
11	*Rauvolfia serpentina Benth*	Leaf	*C. albicans* and *Aspergillus niger*	Agar well diffusion method	Panja et al. (2016)
12	*Peanut*	Shell	*Phytophthora infestans and Phytophthora capsici*	Zone of inhibition 5–6mm	Velmurugan et al. (2015)

References

Ahmad A, Wei Y, Syed F et al (2017) The effects of bacteria-nanoparticles interface on the antibacterial activity of green synthesized silver nanoparticles. Microb Pathog 102:133–142. https://doi.org/10.1016/j.micpath.2016.11.030

Ajitha B, Kumar YA, Shameer S et al (2015) Lantana camara leaf extract mediated silver nanoparticles: antibacterial, green catalyst. J Photochem Photobiol B Biol 149:84–92. https://doi.org/10.1016/j.jphotobiol.2015.05.020

Ajitha B, Kumar YA, Rajesh KM, Reddy PS (2016) Sesbania grandiflora leaf extract assisted green synthesis of silver nanoparticles: antimicrobial activity. Mater Today Proc 3:1977–1984. https://doi.org/10.1016/j.matpr.2016.04.099

Ajitha B, Kumar YA, Jeon H, Won C (2018) Synthesis of silver nanoparticles in an eco-friendly way using Phyllanthus amarus leaf extract: antimicrobial and catalytic activity. Adv Powder Technol 29:86–93. https://doi.org/10.1016/j.apt.2017.10.015

Appapalam ST, Panchamoorthy R (2017) Aerva lanata me diate d phytofabrication of silver nanoparticles and evaluation of their antibacterial activity against wound associated bacteria. J Taiwan Inst Chem Eng 78:539–551. https://doi.org/10.1016/j.jtice.2017.06.035

Ashok D, Palanichamy V, Mohana S (2014) Green synthesis of silver nanoparticles using Alternanthera dentata leaf extract at room temperature and their antimicrobial activity. Spectrochim Acta A Mol Biomol Spectrosc 127:168–171. https://doi.org/10.1016/j.saa.2014.02.058

Aziz N, Fatma T, Varma A, Prasad R (2014) Biogenic synthesis of silver nanoparticles using Scenedesmus abundans and evaluation of their antibacterial activity. J Nanoparticles Article ID 689419. https://doi.org/10.1155/2014/689419

Aziz N, Faraz M, Pandey R, Sakir M, Fatma T, Varma A, Barman I, Prasad R (2015) Facile algae-derived route to biogenic silver nanoparticles: synthesis, antibacterial and photocatalytic properties. Langmuir 31:11605–11612. https://doi.org/10.1021/acs.langmuir.5b03081

Aziz N, Pandey R, Barman I, Prasad R (2016) Leveraging the attributes of Mucor hiemalis-derived silver nanoparticles for a synergistic broad-spectrum antimicrobial platform. Front Microbiol 7:1984. https://doi.org/10.3389/fmicb.2016.01984

Baghbani-arani F, Movagharnia R, Shari A et al (2017) Synthesis of silver nanoparticles from Artemisia tournefortiana Rchb extract. J Photochem Photobiol B Biol 173:640–649. https://doi.org/10.1016/j.jphotobiol.2017.07.003

Bagherzade G, Tavakoli MM, Namaei MH (2017) Green synthesis of silver nanoparticles using aqueous extract of saffron (*Crocus sativus* L.) wastages and its antibacterial activity against six bacteria. Asian Pac J Trop Biomed 7:227–233. https://doi.org/10.1016/j.apjtb.2016.12.014

Balavijayalakshmi J, Ramalakshmi V (2017) Carica papaya peel mediated synthesis of silver nanoparticles and its antibacterial activity against human pathogens. Rev Mex Trastor Aliment 15:413–422. https://doi.org/10.1016/j.jart.2017.03.010

Bhattacharyya A, Prasad R, Buhroo AA, Duraisamy P, Yousuf I, Umadevi M, Bindhu MR, Govindarajan M, Khanday AL (2016) One-pot fabrication and characterization of silver nanoparticles using Solanum lycopersicum: an eco-friendly and potent control tool against Rose Aphid, Macrosiphum rosae. J Nanosci Article ID 4679410, 7 pages. https://doi.org/10.1155/2016/4679410

Bindhu MR, Rekha PV, Umamaheswari T, Umadevi M (2014) Antibacterial activities of Hibiscus cannabinus stem-assisted silver and gold nanoparticles. Mater Lett 131:194–197. https://doi.org/10.1016/j.matlet.2014.05.172

Buhroo AA, Nisa G, Asrafuzzaman S, Prasad R, Rasheed R, Bhattacharyya A (2017) Biogenic silver nanoparticles from Trichodesma indicum aqueous leaf extract against Mythimna separata and evaluation of its larvicidal efficacy. J Plant Protect Res 57(2):194–200. https://doi.org/10.1515/jppr-2017-0026

Chahardoli A, Karimi N, Fattahi A (2018) Nigella arvensis leaf extract mediated green synthesis of silver nanoparticles: their characteristic properties and biological efficacy. Adv Powder Technol 29:202–210. https://doi.org/10.1016/j.apt.2017.11.003

Daisy P, Saipriya K (2012) Biochemical analysis of Cassia fistula aqueous extract and phytochemically synthesized gold nanoparticles as hypoglycemic treatment for diabetes mellitus. Int J Nanomedicine 7:1189–1202. https://doi.org/10.2147/IJN.S26650

Elangovan K, Elumalai D, Anupriya S et al (2015) Phyto mediated biogenic synthesis of silver nanoparticles using leaf extract of Andrographis echioides and its bio-efficacy on anticancer and antibacterial activities. J Photochem Photobiol B Biol 151:118–124. https://doi.org/10.1016/j.jphotobiol.2015.05.015

Elias E, Charles O, Aleruchi C et al (2014) Evaluation of antibacterial activities of silver nanoparticles green-synthesized using pineapple leaf (Ananas comosus). Micron 57:1–5. https://doi.org/10.1016/j.micron.2013.09.003

Ghaedi M, Yousefinejad M, Safarpoor M et al (2015) Rosmarinus officinalis leaf extract mediated green synthesis of silver nanoparticles and investigation of its antimicrobial properties. J Ind Eng Chem 31:167–172. https://doi.org/10.1016/j.jiec.2015.06.020

Ghaffari-Moghaddam M, Hadi-Dabanlou R (2014) Plant mediated green synthesis and antibacterial activity of silver nanoparticles using Crataegus douglasii fruit extract. J Ind Eng Chem:20. https://doi.org/10.1016/j.jiec.2013.09.005

Gogoi N, Jayasekhar P, Mahanta C, Bora U (2015) Green synthesis and characterization of silver nanoparticles using alcoholic fl ower extract of Nyctanthes arbortristis and in vitro investigation of their antibacterial and cytotoxic activities. Mater Sci Eng C 46:463–469. https://doi.org/10.1016/j.msec.2014.10.069

Hamedi S, Abbas S, Mohammadi A (2017) Evaluation of the catalytic, antibacterial and antibio fi lm activities of the Convolvulus arvensis extract functionalized silver nanoparticles. J Photochem Photobiol B Biol 167:36–44. https://doi.org/10.1016/j.jphotobiol.2016.12.025

Harshiny M, Matheswaran M, Arthanareeswaran G (2015) Enhancement of antibacterial properties of silver nanoparticles – ceftriaxone conjugate through Mukia maderaspatana leaf extract mediated synthesis. Ecotoxicol Environ Saf:1–7. https://doi.org/10.1016/j.ecoenv.2015.04.041

Imtiaz M, Hamid L, Qari H et al (2016) One-step synthesis of silver nanoparticles using Phoenix dactylifera leaves extract and their enhanced bactericidal activity. J Mol Liq 223:1114–1122. https://doi.org/10.1016/j.molliq.2016.09.030

Jamil M, Murtaza G, Mehmood A (2015) Green synthesis of silver nanoparticles using leaves extract of Skimmia laureola: characterization and antibacterial activity. Mater Lett 153:10–13. https://doi.org/10.1016/j.matlet.2015.03.143

Jeeva K, Thiyagarajan M, Elangovan V et al (2014) Caesalpinia coriaria leaf extracts mediated biosynthesis of metallic silver nanoparticles and their antibacterial activity against clinically isolated pathogens. Ind Crop Prod 52:714–720. https://doi.org/10.1016/j.indcrop.2013.11.037

Joshi N, Jain N, Pathak A, Singh J, Prasad R, Upadhyaya CP (2018) Biosynthesis of silver nanoparticles using Carissa carandas berries and its potential antibacterial activities. J Sol Gel Sci Technol. https://doi.org/10.1007/s10971-018-4666-2

Kabir F, Fayzunnesa M, Kabir S (2017) Microbial pathogenesis antimicrobial activity of plant-median synthesized silver nanoparticles against food and agricultural pathogens. Microb Pathog 109:228–232. https://doi.org/10.1016/j.micpath.2017.06.002

Karthiga P (2017) Preparation of silver nanoparticles by Garcinia mangostana stem extract and investigation of the antimicrobial properties. Biotechnol Res Innov:1–7. https://doi.org/10.1016/j.biori.2017.11.001

Kathiravan V, Ravi S, Ashokkumar S et al (2015) Green synthesis of silver nanoparticles using Croton sparsiflorus morong leaf extract and their antibacterial and antifungal activities. Spectrochim Acta A Mol Biomol Spectrosc 139:200–205. https://doi.org/10.1016/j.saa.2014.12.022

Kaviya S, Santhanalakshmi J, Viswanathan B et al (2011) Biosynthesis of silver nanoparticles using citrus sinensis peel extract and its antibacterial activity. Spectrochim Acta A Mol Biomol Spectrosc 79:594–598. https://doi.org/10.1016/j.saa.2011.03.040

Khalil MMH (2013) Green synthesis of silver nanoparticles using olive leaf extract and its antibacterial activity. Arab J Chem. https://doi.org/10.1016/j.arabjc.2013.04.007

Krishnaraj C, Jagan EG, Rajasekar S et al (2010) Synthesis of silver nanoparticles using Acalypha indica leaf extracts and its antibacterial activity against water borne pathogens. Colloids Surf B Biointerfaces 76:50–56. https://doi.org/10.1016/j.colsurfb.2009.10.008

Kumar PPNV, Pammi SVN, Kollu P et al (2014) Green synthesis and characterization of silver nanoparticles using Boerhaavia diffusa plant extract and their anti bacterial activity. Ind Crop Prod 52:562–566. https://doi.org/10.1016/j.indcrop.2013.10.050

Kumar D, Hadi S, Mohan R (2016a) Photo-catalyzed and phyto-mediated rapid green synthesis of silver nanoparticles using herbal extract of Salvinia molesta and its antimicrobial efficacy. J Photochem Photobiol B Biol 155:51–59. https://doi.org/10.1016/j.jphotobiol.2015.12.008

Kumar V, Kumar R, Kumar D et al (2016b) Photo-induced rapid biosynthesis of silver nanoparticle using aqueous extract of Xanthium strumarium and its antibacterial and antileishmanial activity. J Ind Eng Chem 37:224–236. https://doi.org/10.1016/j.jiec.2016.03.032

Kumar V, Mohan S, Singh DK et al (2017a) Photo-mediated optimized synthesis of silver nanoparticles for the selective detection of Iron (III), antibacterial and antioxidant activity. Mater Sci Eng C 71:1004–1019. https://doi.org/10.1016/j.msec.2016.11.013

Kumar V, Singh DK, Mohan S et al (2017b) Photoinduced green synthesis of silver nanoparticles using aqueous extract of Physalis angulata and its antibacterial and antioxidant activity. J Environ Chem Eng 5:744–756. https://doi.org/10.1016/j.jece.2016.12.055

Kumara M, Sayeed M, Kumar S, Rani U (2015) Synthesis and characterization of silver nanoparticles using fruit extract of Momordica cymbalaria and assessment of their in vitro antimicrobial, antioxidant and cytotoxicity activitie. Spectrochim Acta A Mol Biomol Spectrosc 151:939–944. https://doi.org/10.1016/j.saa.2015.07.009

Lakshmanan G, Sathiyaseelan A, Kalaichelvan PT, Murugesan K (2017) Plant-mediated synthesis of silver nanoparticles using fruit extract of Cleome viscosa L.: assessment of their antibacterial and anticancer activity. Karbala Int J Mod Sci:1–8. https://doi.org/10.1016/j.kijoms.2017.10.007

Latha M, Sumathi M, Manikandan R et al (2015) Biocatalytic and antibacterial visualization of green synthesized silver nanoparticles using Hemidesmus indicus. Microb Pathog 82:43–49. https://doi.org/10.1016/j.micpath.2015.03.008

Latha M, Priyanka M, Rajasekar P et al (2016) Biocompatibility and antibacterial activity of the Adathoda vasica Linn extract mediated silver nanoparticles. Microb Pathog 93:88–94. https://doi.org/10.1016/j.micpath.2016.01.013

Lee J, Lim J, Velmurugan P et al (2016) Photobiologic-mediated fabrication of silver nanoparticles with antibacterial activity. J Photochem Photobiol B Biol 162:93–99. https://doi.org/10.1016/j.jphotobiol.2016.06.029

Li X, Xu H, Chen ZS, Chen G (2011) Biosynthesis of nanoparticles by microorganisms and their applications. J Nanomater. https://doi.org/10.1155/2011/270974

Lü S, Wu Y, Liu H (2017) Silver nanoparticles synthesized using Eucommia ulmoides bark and their antibacterial efficacy. Mater Lett 196:217–220. https://doi.org/10.1016/j.matlet.2017.03.068

Mahadevan S, Vijayakumar S, Arulmozhi P (2017) Green synthesis of silver nano particles from Atalantia monophylla (L) Correa leaf extract, their antimicrobial activity and sensing capability of H2O2. Microb Pathog 113:445–450. https://doi.org/10.1016/j.micpath.2017.11.029

Maji A, Beg M, Kumar A et al (2017) Spectroscopic interaction study of human serum albumin and human hemoglobin with Mersilea quadrifolia leaves extract mediated silver nanoparticles having antibacterial and anticancer activity. J Mol Struct 1141:584–592. https://doi.org/10.1016/j.molstruc.2017.04.005

Mariselvam R, Ranjitsingh AJA, Raja AU et al (2014) Green synthesis of silver nanoparticles from the extract of the inflorescence of Cocos nucifera (Family : Arecaceae) for enhanced antibacterial activity. Spectrochim Acta A Mol Biomol Spectrosc 129:537–541. https://doi.org/10.1016/j.saa.2014.03.066

Mata R, Nakkala JR, Sadras SR (2015a) Biogenic silver nanoparticles from Abutilon indicum: their antioxidant, antibacterial and cytotoxic effects in vitro. Colloids Surf B Biointerfaces 128:276–286. https://doi.org/10.1016/j.colsurfb.2015.01.052

Mata R, Nakkala JR, Sadras SR (2015b) Catalytic and biological activities of green silver nanoparticles synthesized from Plumeria alba (frangipani) flower extract. Mater Sci Eng C 51:216–225. https://doi.org/10.1016/j.msec.2015.02.053

Mehmood A, Murtaza G, Mahmood T (2017) Phyto-mediated synthesis of silver nanoparticles from Melia azedarach L. leaf extract: characterization and antibacterial activity. Arab J Chem 10:S3048–S3053. https://doi.org/10.1016/j.arabjc.2013.11.046

Miri A, Sarani M, Rezazade M, Darroudi M (2015) Plant-mediated biosynthesis of silver nanoparticles using Prosopis farcta extract and its antibacterial properties. Spectrochim Acta A Mol Biomol Spectrosc 141:287–291. https://doi.org/10.1016/j.saa.2015.01.024

Moteriya P, Padalia H, Chanda S (2017) Characterization, synergistic antibacterial and free radical scavenging efficacy of silver nanoparticles synthesized using Cassia roxburghii leaf extract. J Genet Eng Biotechnol 15:505–513. https://doi.org/10.1016/j.jgeb.2017.06.010

Mubarakali D, Thajuddin N, Jeganathan K, Gunasekaran M (2011) Plant extract mediated synthesis of silver and gold nanoparticles and its antibacterial activity against clinically isolated pathogens. Colloids Surf B Biointerfaces 85:360–365. https://doi.org/10.1016/j.colsurfb.2011.03.009

Muthukrishnan S, Bhakya S, Kumar TS, Rao MV (2014) Biosynthesis, characterization and antibacterial effect of plant-mediated silver nanoparticles using Ceropegia thwaitesii – an endemic species. Ind Crop Prod. https://doi.org/10.1016/j.indcrop.2014.10.022

Nakkala JR, Mata R, Gupta AK, Sadras SR (2014) Biological activities of green silver nanoparticles synthesized with Acorous calamus rhizome extract. Eur J Med Chem 85:784–794. https://doi.org/10.1016/j.ejmech.2014.08.024

Nakkala JR, Mata R, Sadras SR (2017) Green synthesized nano silver: synthesis, physicochemical profiling, antibacterial, anticancer activities and biological in vivo toxicity. J Colloid Interface Sci 499:33–45. https://doi.org/10.1016/j.jcis.2017.03.090

Nazeruddin GM, Prasad NR, Prasad SR et al (2014) Coriandrum sativum seed extract assisted in situ green synthesis of silver nanoparticle and its anti-microbial activity. Ind Crop Prod 60:212–216. https://doi.org/10.1016/j.indcrop.2014.05.040

Padalia H, Moteriya P, Chanda S (2015) Green synthesis of silver nanoparticles from marigold flower and its synergistic antimicrobial potential. Arab J Chem 8:732–741. https://doi.org/10.1016/j.arabjc.2014.11.015

Panja S, Chaudhuri I, Khanra K, Bhattacharyya N (2016) Biological application of green silver nanoparticle synthesized from leaf extract of Rauvolfia serpentina Benth. Asian Pac J Trop Dis 6:549–556. https://doi.org/10.1016/S2222-1808(16)61085-X

Patil Shriniwas P, Kumbhar Subhash T (2017) Antioxidant, antibacterial and cytotoxic potential of silver nanoparticles synthesized using terpenes rich extract of Lantana camara L. leaves. Biochem Biophys Rep 10:76–81. https://doi.org/10.1016/j.bbrep.2017.03.002

Patil RS, Kokate MR, Kolekar SS (2012) Bioinspired synthesis of highly stabilized silver nanoparticles using Ocimum tenuiflorum leaf extract and their antibacterial activity. Spectrochim Acta A Mol Biomol Spectrosc 91:234–238. https://doi.org/10.1016/j.saa.2012.02.009

Patra JK, Das G, Baek K (2016) Phyto-mediated biosynthesis of silver nanoparticles using the rind extract of watermelon (Citrullus lanatus) under photo-catalyzed condition and investigation of its antibacterial, anticandidal and antioxidant efficacy. PT CR. JPB. https://doi.org/10.1016/j.jphotobiol.2016.05.021

Pattanayak S, Rahaman M, Maity D et al (2017) Butea monosperma bark extract mediated green synthesis of silver nanoparticles: characterization and biomedical applications. J Saudi Chem Soc 21:673–684. https://doi.org/10.1016/j.jscs.2015.11.004

Phongtongpasuk S, Poadang S, Yongvanich N (2016) Environmental-friendly method for synthesis of silver nanoparticles from dragon fruit peel extract and their antibacterial activities. Energy Procedia 89:239–247. https://doi.org/10.1016/j.egypro.2016.05.031

Phull A, Abbas Q, Ali A et al (2016) Antioxidant, cytotoxic and antimicrobial activities of green synthesized silver nanoparticles from crude extract of Bergenia ciliata. Future J Pharm Sci 2:31–36. https://doi.org/10.1016/j.fjps.2016.03.001

Prakash P, Gnanaprakasam P, Emmanuel R et al (2013) Green synthesis of silver nanoparticles from leaf extract of Mimusops elengi, Linn. for enhanced antibacterial activity against multi drug resistant clinical isolates. Colloids Surf B Biointerfaces 108:255–259. https://doi.org/10.1016/j.colsurfb.2013.03.017

Prasad R (2014) Synthesis of silver nanoparticles in photosynthetic plants. J Nanoparticles Article ID 963961. https://doi.org/10.1155/2014/963961

Prasad R, Swamy VS (2013) Antibacterial activity of silver nanoparticles synthesized by bark extract of Syzygium cumini. J Nanoparticles. https://doi.org/10.1155/2013/431218

Prasad KS, Pathak D, Patel A, Dalwadi P, Prasad R, Patel P, Kaliaperumal SK (2011) Biogenic synthesis of silver nanoparticles using Nicotiana tobaccum leaf extract and study of their antibacterial effect. Afr J Biotechnol 9(54):8122–8130

Pugazhendhi S, Sathya P, Palanisamy PK, Gopalakrishnan R (2016) Synthesis of silver nanoparticles through green approach using Dioscorea alata and their characterization on antibacterial activities and optical limiting behavior. J Photochem Photobiol B Biol 159:155–160. https://doi.org/10.1016/j.jphotobiol.2016.03.043

Raja S, Ramesh V, Thivaharan V (2017) Green biosynthesis of silver nanoparticles using Calliandra haematocephala leaf extract, their antibacterial activity and hydrogen peroxide sensing capability. Arab J Chem 10:253–261. https://doi.org/10.1016/j.arabjc.2015.06.023

Rajakumar G, Gomathi T, Thiruvengadam M (2017) Evaluation of anti-cholinesterase, antibacterial and cytotoxic activities of green synthesized silver nanoparticles using from Millettia pinnata flower extract. Microb Pathog 103:123–128. https://doi.org/10.1016/j.micpath.2016.12.019

Rajeshkumar S (2016a) Green synthesis of different sized antimicrobial silver nanoparticles using different parts of plants – a review. Int J ChemTech Res 9:197–208

Rajeshkumar S (2016b) Synthesis of silver nanoparticles using fresh bark of Pongamia pinnata and characterization of its antibacterial activity against gram positive and gram negative pathogens. Resour Technol 2:30–35. https://doi.org/10.1016/j.reffit.2016.06.003

Rajeshkumar S, Bharath LV (2017) Mechanism of plant-mediated synthesis of silver nanoparticles – a review on biomolecules involved, characterisation and antibacterial activity. Chem Biol Interact 273. https://doi.org/10.1016/j.cbi.2017.06.019

Ramar M, Manikandan B, Narayanan P et al (2015a) Synthesis of silver nanoparticles using Solanum trilobatum fruits extract and its antibacterial, cytotoxic activity against human breast cancer cell line MCF 7. Spectrochim Acta A Mol Biomol Spectrosc 140:223–228. https://doi.org/10.1016/j.saa.2014.12.060

Ramar M, Manikandan B, Raman T et al (2015b) Biosynthesis of silver nanoparticles using ethanolic petals extract of Rosa indica and characterization of its antibacterial, anticancer and anti-inflammatory activities. Spectrochim Acta A Mol Biomol Spectrosc 138:120–129. https://doi.org/10.1016/j.saa.2014.10.043

Ramaswamy U, Mukundan D, Sreekumar A, Mani V (2015) Green synthesis and characterization of silver nanoparticles using aqueous whole plant extract of Vernonia cinerea L. and its biological activities. Mater Today Proc 2:4600–4608. https://doi.org/10.1016/j.matpr.2015.10.080

Ramesh PS, Kokila T, Geetha D (2015) Plant mediated green synthesis and antibacterial activity of silver nanoparticles using Emblica officinalis fruit extract. Spectrochim Acta A Mol Biomol Spectrosc 142:339–343. https://doi.org/10.1016/j.saa.2015.01.062

Rao NH, Lakshmidevi N, Pammi SVN et al (2016) Green synthesis of silver nanoparticles using methanolic root extracts of Diospyros paniculata and their antimicrobial activities. Mater Sci Eng C 62:553–557. https://doi.org/10.1016/j.msec.2016.01.072

Rasheed T, Bilal M, Iqbal HMN, Li C (2017) Green biosynthesis of silver nanoparticles using leaves extract of Artemisia vulgaris and their potential biomedical applications. Colloids Surf B Biointerfaces 158:408–415. https://doi.org/10.1016/j.colsurfb.2017.07.020

Rashid MI, Hamid L, Ibrahim M et al (2017) Potent bactericidal activity of silver nanoparticles synthesized from Cassia fistula fruit. Microb Pathog 107:354–360. https://doi.org/10.1016/j.micpath.2017.03.048

Ravichandran V, Vasanthi S, Shalini S et al (2016) Mater Lett. https://doi.org/10.1016/j.matlet.2016.05.172

Reddy NJ, Vali DN, Rani M, Rani SS (2014) Evaluation of antioxidant, antibacterial and cytotoxic effects of green synthesized silver nanoparticles by Piper longum fruit. Mater Sci Eng C 34:115–122. https://doi.org/10.1016/j.msec.2013.08.039

Sadeghi B, Rostami A, Momeni SS (2015) Facile green synthesis of silver nanoparticles using seed aqueous extract of Pistacia atlantica and its antibacterial activity. Spectrochim Acta A Mol Biomol Spectrosc 134:326–332. https://doi.org/10.1016/j.saa.2014.05.078

Selvam K, Sudhakar C, Govarthanan M (2017) Eco-friendly biosynthesis and characterization of silver nanoparticles using Tinospora cordifolia (Thunb.) Miers and evaluate its antibacterial, antioxidant potential. J Radiat Res Appl Sci 10:6–12. https://doi.org/10.1016/j.jrras.2016.02.005

Sesuvium L, Nabikhan A, Kandasamy K et al (2010) Synthesis of antimicrobial silver nanoparticles by callus and leaf extracts from. Colloids Surf B Biointerfaces 79:488–493. https://doi.org/10.1016/j.colsurfb.2010.05.018

Sudha A, Jeyakanthan J, Srinivasan P (2017) Green synthesis of silver nanoparticles using Lippia nodiflora aerial extract and evaluation of their antioxidant, antibacterial and cytotoxic effects. Resour Technol 3:506–515. https://doi.org/10.1016/j.reffit.2017.07.002

Sudhakar C, Selvam K, Govarthanan M (2015) Acorus calamus rhizome extract mediated biosynthesis of silver nanoparticles and their bactericidal activity against human pathogens. J Genet Eng Biotechnol 13:93–99. https://doi.org/10.1016/j.jgeb.2015.10.003

Swamy VS, Prasad R (2012) Green synthesis of silver nanoparticles from the leaf extract of Santalum album and its antimicrobial activity. J Optoelectron Biomed Mater 4(3):53–59

Sweet MJ, Singleton I (2011) Silver nanoparticles. A microbial perspective, 1st edn. Elsevier, Amsterdam

Velmurugan P, Sivakumar S, Young-chae S (2015) Synthesis and characterization comparison of peanut shell extract silver nanoparticles with commercial silver nanoparticles and their antifungal activity. J Ind Eng Chem 31:51–54. https://doi.org/10.1016/j.jiec.2015.06.031

Velusamy P, Das J, Pachaiappan R (2015) Greener approach for synthesis of antibacterial silver nanoparticles using aqueous solution of neem gum (Azadirachta indica L.). Ind Crop Prod 66:103–109. https://doi.org/10.1016/j.indcrop.2014.12.042

Wang L, Xie J, Huang T et al (2017) Characterization of silver nanoparticles biosynthesized using crude polysaccharides of Psidium guajava L. leaf and their bioactivities. Mater Lett 208:126–129. https://doi.org/10.1016/j.matlet.2017.05.014

Yang N, Li W (2013) Mango peel extract mediated novel route for synthesis of silver nanoparticles and antibacterial application of silver nanoparticles loaded onto non-woven fabrics. Ind Crop Prod 48:81–88. https://doi.org/10.1016/j.indcrop.2013.04.001

Chapter 7
Plants as Fabricators of Biogenic Platinum Nanoparticles: A Gambit Endeavour

Babita Jha, Anal K. Jha, and Kamal Prasad

7.1 Introduction

Nanotechnology unbolted fenestrations to "think big about small things" and demands daring manoeuvres on precipitous grounds. It not only controls and understands matter at length scales of 1–100 nm but also outright the expression of unique properties of nanomaterials due to small grain size as compared to bulk. The intensification in surface to volume ratio gives birth to active surfaces and discrete energy levels in nano-entities, which increase their probable applications in surface-based applications and electronic usage (Daniel and Astruc 2004). Moreover, advances in development of predefined nanostructures ensure its applicability and plays critical role in many key technologies leading to creation of novel class of nanomaterials. Nanostructured materials are the demand of time and have become the panacea to all present day issues leaving no person novice to it.

One of the blooming applications of nanotechnology is in the field of platinum group metals (PGMs) that include six metals, namely ruthenium (Ru), rhodium (Rh), palladium (Pd), osmium (Os), iridium (Ir), and platinum (Pt). Plethora of metallic nanoparticles have been fabricated and used in various applications but as far as PGMs are concerned the contributions are low and still in its blooming state (Fig. 7.1). These PGMs are in high demand but have low abundance in earth crust, thus making them valuable and expensive (Capeness et al. 2015). Platinum metal, one of the prime members of this group with its name derived from the Spanish word "platina" meaning "little silver", is very scarce in nature around 0.01 ppm (Pedone et al. 2017). It is highly resistant to chemicals, exhibit excellent high-temperature features, has

B. Jha · A. K. Jha
Aryabhatta Centre for Nanoscience and Nanotechnology, Aryabhatta Knowledge University, Patna, India

K. Prasad (✉)
Department of Physics, Tilka Manjhi Bhagalpur University, Bhagalpur, Bihar, India

© Springer Nature Switzerland AG 2018
R. Prasad et al. (eds.), *Exploring the Realms of Nature for Nanosynthesis*,
Nanotechnology in the Life Sciences, https://doi.org/10.1007/978-3-319-99570-0_7

Fig. 7.1 PGMs the field
with blooming applications
of nanotechnology

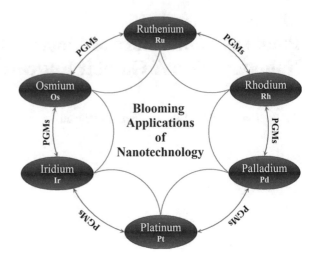

stable electrical properties, and is known for its catalytic properties since ages. With the growing needs of this rare and noble metal in so many sectors, nanotechnology has provided serviceability and platinum nanoparticles are all set to revolutionize the automotive sector (Golunski 2007; Yu et al. 2017), chemical industry (Du et al. 2015; Liu et al. 2015; Oberhauser et al. 2016; Zhang et al. 2016; Tatsumi et al. 2017; Li et al. 2017a, b; Yola and Atar 2017), and biomedical field (Asharani et al. 2010; Porcel et al. 2010; Liu et al. 2011; Manikandan et al. 2013; Shi et al. 2015).

The upraised demand due to widespread uses of platinum nanoparticles is alluring the nanotechnologists to provide possible solutions to its limited availability and look forth to its applicative potentials. Various fabrication routes for platinum nanoparticles are designed in order to give a replacement to bulk platinum metal. Catalysis which is the main essence of its applicability could be increased manifolds when size is reduced to the nanoscale. The biomedical sector would be benefitted in diagnostic and treatment by optimization of its surface structure that determines the catalytic properties. This is a matter of attention and research, where new avenues are being deciphered, biosystems being of prime interest. The transitional hooks between the bulk and nanosized entities are ascertainable fabrication approaches, namely "bottom-up" and "top-down". The "bottom-up" approach accredits that nano-entities "create themselves" by self-assembly of atom by atom, while "top-down" approach postulates splintering of larger entities to form nano-entities (Tiwari et al. 2008; Zhi and Müllen 2008; Merkel et al. 2010; Nugroho et al. 2016). Wherein each approach is brimming with their own advantages and disadvantages, herein, "bottom-up" approach as employed by biosystems with special reference to plants is discussed. This chapter summarizes the recent advances in the field of biogenic Pt NPs fabrication by plants and other biological sources and discusses the various mechanisms proposed behind it. It focuses on the possible benefits of employing plants in comparison to other biological sources.

7.2   Synthesis of Platinum Nanoparticles (Pt NPs)

Platinum nanoparticles owing to its best catalytic action have persuaded the designing of a variety of protocols for their synthesis. The nanoparticles are usually constructed by physical, biological, and chemical methods. The physical method proceeds by the "top-down" approach and the "bottom-up" approach is followed by chemical and biological methods as depicted in Fig. 7.2. These methods are focused to obtain desired physiochemical properties of the Pt NPs as these properties determine the appliance of these nanoparticles in various roles. In case of biomedical applications of Pt NPs besides its physiochemical properties its dispersion state and stability in biotic environment is a matter of contention (Moglianetti et al. 2016; Pedone et al. 2017). Biocompatibility of Pt NPs depends on size, symmetry; surface cladding and so are the synthetic ventures designed accordingly. Herein, we will precisely discuss the physical and chemical methods undertaken for fabrication of Pt NPs with desired characteristics along with its limitation. Then after, the main focus will be on the bioreductive synthesis of Pt NPs and plant in particular.

7.2.1   Physical Methods of Platinum Nanoparticles Fabrication

Physical methods of synthesis pursue the top-down approach and apply mechanism like mechanical pressure as in ball milling, chemical/thermal energy appliance for itching and electro-explosion, kinetic energy in sputtering, thermal energy in laser ablation, etc. to generate nanoparticles (Dhand et al. 2015). In the framework of Pt NPs, laser ablation techniques using laser beams in continuous or pulsed modes

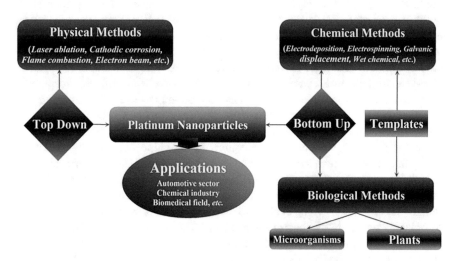

Fig. 7.2 Schematic representation of fabrication protocols of platinum nanoparticles

have been used to prepare platinum and its bimetallics with gold (Chau et al. 2011; Oko et al. 2015) tin (Bommersbach et al. 2008), thin films with cobalt (Rakshit et al. 2008), etc. This method helps to achieve Pt NPs of desired characteristics by varying the pulses, temperature, and other parameters. Aerosol-assisted vapour deposition has been employed to grow highly oriented nanoparticles of platinum (Paschos et al. 2008) and its deposition on tungsten trioxide nanowires (Annanouch et al. 2014). The fabrication of platinum by flame (Strobel and Pratsinis 2009; Choi et al. 2010), electron beam (Ke et al. 2013) and cathodic corrosion (Yanson et al. 2011) are also in practice. Cathodic corrosion method hinges on extreme cathodic polarization of a metal, which in turn produces cation-stabilized metal anions that act as precursors for genesis of nanoparticles. Physical methods bid adieu to the issues of solvent contamination and impurities caused by additives like stabilizer, surfactant as in chemical method. Yet there are limitations to the physical methods as shape and size tunability of nanoparticles is a tedious and difficult affair with resultant lesser yield (Kshirsagar et al. 2011). Besides it, in context of biomedical application their stability and behaviour in biological environment is a matter of research (Correard et al. 2014).

7.2.2 Chemical Methods of Platinum Nanoparticles Fabrication

Chemical synthesis of nanoparticles generally has the involvement of chemical agents like sodium borohydride (Jana et al. 2000), sodium citrate (Ranoszek-Soliwoda et al. 2017), *N,N*-dimethyl formamide (DMF) (Pastoriza-Santos and Liz-Marzán 2009; Zhang et al. 2014), ethyl alcohol (Liz-Marzán and Lado-Touriño 1996; Pal et al. 2009), cetyl trimethylammonium bromide (CTAB) (Moon et al. 2009; Vadivel et al. 2016), poly (*N*-vinyl pyrrolidine) (PVP) (Koczkur et al. 2015), etc. via the chemical reduction method (Bonnemann and Richards 2001). There are various attempts taken to synthesize Pt NPs of controlled shape and sizes via chemical methods (Ahmadi et al. 1996; Petroski et al. 1998; Wang et al. 2008; Leong et al. 2014) electrodeposition (Saminathan et al. 2009; Zhao et al. 2013), electrospinning (Kim et al. 2010), galvanic displacement (Mahmoud et al. 2010; Mahmoud and El-Sayed 2012), and electrochemical reduction (Mahima et al. 2008; Tiwari et al. 2009; Zhou et al. 2010; Li et al. 2012; Gao et al. 2013) are some of the examples. In context to platinum, wet chemical reduction has been reported to be the most powerful fabrication mechanism to generate shaped nanoparticles of required size distribution (Leong et al. 2014). Besides it, dendrimers have been engaged as templates in Pt NPs fabrication, favouring better control over the structure (Crooks et al. 1999; Yamamoto et al. 2009; Wang et al. 2013). These processes overall involve the usage of reducing agents to synthesize the nanoparticles by reduction of platinum salts. The initial platinum seed formation starts with nucleation followed by the growth stage that leads to the Pt NPs genesis. For desired characteristic Pt NPs synthesis, various reducing agents, surfactants, organic acids, solvents, capping agents, etc. are employed; affecting the toxicological profile of the nanoparticles and thus limiting its uses. The large-scale production is also restrained because of the associated environmental issues.

7.2.3 Biological Methods of Platinum Nanoparticles Fabrication

The recent trends of nanoparticles fabrication pioneers the environmental friendly, biologically assisted synthesis processes wherein the biological organisms are employed to produce nontoxic, nano-entities in an easy and inexpensive way. The fabrication of Pt NPs by biological methods manifests to be superior to chemical and physical methods. Nanoscale processes and structures have been integral part of all the biosystems, being in charge of its viability and cognizance. The interaction between inorganic substances and biological entities has led to the focus on their potential applications for synthesis of nanomaterials and thus tempted their employment. These entities have been bestowed with the intrinsic properties to combat high concentration of toxic metal ions through distinct resistance mechanism. The detoxification action advances either by reduction/oxidation of metal ions or by the compilation of insoluble complexes. This very fact is the crux for assuming biological systems as the mantle of nanoparticle production. The organisms hired for the role as nano factories vary from simple prokaryotic cells to complex eukaryotic entities (Korbekandi et al. 2009).

With the advent of nanotechnology and explicitly nano-biotechnology, these biosystems have provided serviceability as factories for fabrication of novel functional nanoparticles and nanomaterials with diverse applications. Nano-biotechnological protocols agreeableness is on account of it being a simple, reliable, worthwhile, eco-friendly venture leading to synthesis of biocompatible nano-entities, unescorted by harmful chemicals. These protocols as a breakthrough in the nano ventures provided an alternative to physical and chemical methods of nanoparticle fabrication. There are ample of communications for bio-assisted syntheses of noble metals like silver and gold (Nadagouda and Varma 2006; Jha et al. 2008, 2009a; Jha and Prasad 2010, 2011a, b, 2013; Prasad et al. 2010; Mallikarjuna et al. 2011; Elia et al. 2014; Prasad 2014; Krithiga et al. 2015) as these are manufactured in a jiffy by even weak reducing agents. Bearing platinum in mind, there are only a few reports in literature describing the fabrication of platinum through the biological entities (Isaac et al. 2013; Siddiqi and Husen 2016; Whiteley et al. 2011). The overall fabrication of platinum nanoparticles can be roundly classified into three categories: Biogenic platinum by microorganisms, by proteins/enzymes/biomolecules as templates, and by plant/plant extracts.

7.2.3.1   Biogenic Platinum by Microorganisms

Microorganisms are endowed with remarkable capability to reduce the heavy metal ions to their respective highly stable nanoparticles with the aid of plethora of enzymes generated via their cellular machinery. It is a kind of defence mechanism to overcome metal toxicity and resultant are novel nanomaterials (Klaus-Joerger et al. 2001; Iravani 2014). The synthesis of nanoparticles by the microorganisms can be assorted into intracellular and extracellular banking upon the location of nanoparticle synthesis (Prasad et al. 2016). The intracellular process relies upon the transportation and enzymatic reduction of metal ions to generate nanoparticles inside the cell. While the extracellular process results in trapping of metal ions outside the cell and their gradual reduction by enzymes.

Bacteria prove to be an excellent source for synthesis of Pt NPs as they are easy to cultivate and manipulate. The first report of platinum fabrication by cyanobacterium, *Plectonema boryanum* UTEX 48 was conveyed by Lengke et al. about a decade ago in 2006. The reduction process was investigated for temperature gradient 25–100 °C. It was observed that by tweaking the incubation temperature and reaction times, the microbes can form nanostructures with distinct morphologies and crystallinity. Thirty to 300 nm nanoparticles where obtained in spherical and dendritic morphology (Lengke et al. 2006). Konishi et al. (2007) reported elemental platinum synthesis by *Shewanella algae* at neutral pH and temperature within 60 min on provision of lactate as the electron donor. Sulphate-reducing bacteria has been employed by researchers (Rashamuse and Whiteley 2007; Riddin et al. 2009) wherein platinum reduction comparative analysis between live cells, heat-killed cells, cell-free extract, and purified hydrogenase enzyme was done. Riddin et al. (2009) also deciphered the bioreductive mechanism of Pt(IV) to Pt(0) via Pt(II) through two different hydrogenase enzymes. The trend set by these initial works, lured many other researchers and diverse number of bacteria's were employed like Cyanobacteria (*Calothrix, Anabaena*) (Brayner et al. 2007), *Pseudomonas aeruginosa* SM1 (Srivastava and Constanti 2012), *Escherichia coli* (Attard et al. 2012; Bennett et al. 2012), *Acinetobacter calcoaceticus* (Gaidhani et al. 2014), *Desulfovibrio alaskensis* G20 (Capeness et al. 2015), *Streptomyces* sp. (Baskaran et al. 2017), and *Desulfovibrio vulgaris* (Martins et al. 2016). *Streptomyces* sp. generated nanoplatinum exhibited cytotoxicity against breast cancer cell lines, catalytic Pt NPs from *Desulfovibrio vulgaris* was used for removal of pharmaceutical compounds.

Fungus-mediated Pt NPs synthesis protocols have gained acceptance on account of it being eco-friendly, inexpensive, easily maintainable, easy to scale-up, quick reduction, and high yield of monodisperse nanoparticles secreted extracellularly (Rai et al. 2011; Whiteley et al. 2011; Kashyap et al. 2013; Soni and Prakash 2012; Honary et al. 2013; Quester, 2013; Yadav et al. 2015; Prasad et al. 2016). The primal report on use of *Fusarium oxysporum f.* sp. *lycopersici* for intercellular and extracellular platinum synthesis was given by Riddin et al. (2006). Varied shapes from hexagons, pentagons, squares, rectangles to circles of 10–100 nm varied size of nanoparticles were observed in micrographs. Extracellular 5–30 nm spherical Pt NPs have also been reported from *Fusarium oxysporum* (Syed and Ahmad 2012). Besides it,

Neurospora crassa and *Penicillium chrysogenum* have also been reported to fabricate Pt NPs (Castro-Longoria et al. 2012; Subramaniyan et al. 2018).

Reports on applicability of algae for platinum nanoparticle production are limited (Shiny et al. 2014, 2016). Seaweed *Padina gymnospora,* a brown alga, rich in sulphated polysaccharides, have been communicated for 5–20 nm sized spherical Pt NPs synthesis. These nanoparticles were assessed cytotoxicity against a 549 lung cancer cell lines. Thus, among the bountiful microbe's only exiguous bacteria, fungi, and algae are reported as bioreactors of platinum synthesis as laid out in Table 7.1.

7.2.3.2 Biogenic Platinum by Proteins/Enzymes/Biomolecules as Templates

The higher cost and immense use of Pt NPs act as an incentive for the researchers to develop unique and green methods of its synthesis. In this framework, several biogenic molecules, proteins, and enzymes were undertaken as templates for platinum synthesis. These macromolecules bioreductive mechanism also proved that removal of spatial constrictions of cells do not affect the potentiality of the cellular enzymes to reduce the respective salts to nanoparticles. Fungi *Fusarium oxysporum* was used for extraction of a cell-free, dimeric hydrogenase enzyme, that passively caused bioreduction of platinum salt (Govender et al. 2009, 2010). Consortium of sulphate-reducing bacteria was the organism of choice to investigate cell-free, cell-soluble protein extract for bioreductive process (Riddin et al. 2010). Apoferritin has also been reported for platinum nanoparticle fabrication (Deng et al. 2009; Fan et al. 2011). Besides it, glucose (Shin et al. 2009; Engelbrekt et al. 2010), antioxidant (gallic acid) (Ko et al. 2015), and vitamin B2 (riboflavin) (Nadaroglu et al. 2017) and bioreduction of Pt NPs have also been communicated. Other biogenic sources like honey (Venu et al. 2011) and egg yolk (Nadaroglu et al. 2017) have also been utilized for genesis of Pt NPs. Fabrication of Pt NPs through templates is guided by bottom-up approach and here lies a narrow articulation in between the chemical and biological methods. There are reports wherein such templates have been categorized under chemical methods. Herein, these templates with biological origins are summarized in Table 7.2.

7.2.3.3 Biogenic Platinum by Plant/Plant Extracts

Metal bioaccumulation in plants convened the ability of plants to reduce the metal ions, either at the site of access or at sites remote from ion penetrations. Further studies interestingly proved that the accumulated deposits were in form of nanoparticles (Makarov et al. 2014). These processes that are means to combat the adverse metal toxicity by plants provided an insight to the fact of using plants as tool for fabricator of various nanoparticles (Akhtar et al. 2013; Kharissova et al. 2013; Mittal et al. 2013; Makarov et al. 2014; Singh et al. 2016). There are reports wherein

Table 7.1 Biogenic platinum synthesized by microorganisms

S. no.	Source	Location	Size (in nm)	Morphology	References
Bacteria					
1	*Plectonema boryanum* UTEX 485	Intracellular and extracellular	30–300	Spherical, dendritic morphology	Lengke et al. (2006)
2.	*Shewanella algae*	Intracellular	5	–	Konishi et al. (2007)
3.	Sulphate-reducing Bacteria	Intracellular	–	–	Rashamuse and Whiteley (2007)
4.	Cyanobacteria (*Calothrix, Anabaena*)	Intracellular	3.2	–	Brayner et al. (2007)
5.	Sulphate-reducing bacteria	Intracellular	–	Amorphous	Riddin et al. (2009)
6.	*Pseudomonas aeruginosa*	Extracellular	450	Circular disks	Srivastava and Constanti (2012)
7.	*Escherichia coli* MC4100	Intracellular	2.3 (1% metal loading) 4.5 (20% metal loading)	Spherical	Attard et al. (2012)
8.	*Escherichia coli*	Intracellular	14	–	Bennett et al. (2012)
9.	*Acinetobacter calcoaceticus* PUCM1011	Intracellular	2–3	–	Gaidhani et al. (2014)
10.	*Desulfovibrio alaskensis* G20	Extracellular	–	–	Capeness et al. (2015)
11.	*Streptomyces* sp.	Extracellular	20–50	Spherical	Baskaran et al. (2017)
12.	*Desulfovibrio vulgaris*	Extracellular	–	–	Martins et al. (2016)
Fungi					
1.	*Fusarium oxysporum f.* Sp. *lycopersici*	Intracellular and extracellular	10–100	Hexagons, pentagons, circles, square, rectangles	Riddin et al. (2006)
2.	*Neurospora crassa*	Intracellular	4–35 20–110 17–76	Spherical single Pt NPs Spherical nanoaggregates Rounded nanoaggregates	Castro-Longoria et al. (2012)
3.	*Fusarium oxysporum*	Extracellular	5–30	Spherical	Syed and Ahmad (2012)

(continued)

Table 7.1 (continued)

S. no.	Source	Location	Size (in nm)	Morphology	References
4.	*Penicillium chrysogenum*	Extracellular	8.5 15	–	Subramaniyan et al. (2018)
Algae					
1.	*Padina gymnospora*	–	5–20	Spherical	Shiny et al. (2014, 2016)

Table 7.2 Biogenic platinum synthesized by proteins/enzymes/biomolecules

S. no.	Source	Size (in nm)	Morphology	References
Protein/enzyme				
1.	Hydrogenase (*Fusarium oxysporum*)	100–140	Spherical	Govender et al. (2009)
2.	Horse spleen Apoferritin	4.7	Spherical	Deng et al. (2009)
3.	Protein extract (Sulphate-reducing Bacteria)	200–1000	Geometric with four straight edges Irregular	Riddin et al. (2010)
4.	Hydrogenase (*Fusarium oxysporum*)	30–40 40–60	Irregular Circular, triangular, pentagonal, hexagonal Nanoplates	Govender et al. (2010)
5.	Apoferritin	1–2	Spherical	Fan et al. (2011)
Biomolecules				
1.	Vitamin B2 (riboflavin)	9.7 8.4	Nanoparticles Nanowires	Nadagouda and Varma (2006)
2.	β-D-glucose	3.8	Nanoparticles Nanowires	Shin et al. (2009)
3.	Glucose	1.6–1.8	–	Engelbrekt et al. (2010)
4.	Antioxidant (Gallic acid)	22	–	Ko et al. (2015)
Other biogenic sources				
1.	Honey	2.2	Nanoparticles, nanowires	Venu et al. (2011)
2.	Egg yolk (quail egg)	7–50	Spherical	Nadaroglu et al. (2017)
3.	Bacterial cellulose matrix	3–4	Spherical	Yang et al. (2009)
4.	Bacterial cellulose matrix	6.3–9.3	Irregular	Aritonang et al. (2014)

whole plants like *Sinapis alba, Lepidium sativum,* and *Medicago sativa* have been studied for accumulation and translocation of Pt NPs within them (Bali et al. 2010; Asztemborska et al. 2015). The whole plant use for nanoparticles fabrication is limited up to detoxification or phytoremediation, as extraction and downstream processing is very cumbersome. The extraction and utilization of phyto-constituents, in the form of plant extracts for nanotransformation is easier, acceptable, and more in practice. With these views in mind, the penetration of biogenic synthesis in the world of PGMs occurred, various plant/plant parts were then after explored for fabrication of nanoparticles, as depicted in Fig. 7.3.

The generalized methodology adopted for Pt NPs fabrication involves the preparation of plant part extract either in aqueous or with other solvents. This extract is then charged with the platinum metal aqueous salt solution like platinum (IV) chloride, platinum (II) chloride, sodium tetrachloroplatinate, etc.; herein, the solubility of the Pt salt in water is an important factor. The processing parameters such as temperature, pH, extract concentration, incubation period, and aeration when properly maintained, the phyto-constituents acting as reducing and capping agents reacts with salt solution to give rise to initial Pt nuclei formation that with subsequent growth leads to stable, homogenous, and capped Pt NPs. General representation of the plant-mediated Pt NPs fabrication is depicted in Fig. 7.4.

The first report of extracellular fabrication of Pt NPs using leaf extracts of *Diospyros kaki* was reported by Song and his coworkers (2010). These researchers besides screening leaf extract as reducing agents for platinum fabrication also investigated the potential effect on the size of the synthesized particles with change in reaction conditions like temperature between 25 and 95 °C and varied concentrations of leaf broth and precursor platinum salt. The rate of Pt NPs synthesis increased with higher leaf broth concentration and increased in reaction temperature maximum being at 95 °C. The success story of this attempt was an initiative for latter works in this area, as tabulated in Table 7.3. Homogenous, highly stable, colloidal suspension of Pt NPs with size 2.4 nm was synthesized in aqueous medium using Gum exudates of Kondagogu tree (*Cochlospermum gossypium*) (Vinod et al. 2011). Soundarrajan et al. (2012) synthesized Pt NPs from *Ocimum sanctum* leaf extracts with conversion carried out at 100 °C, reduced Pt NPs were found as aggregates of irregular shape. The fabrication of palladium and platinum metal nanoparticles employing root extract of *Asparagus racemosus* at room temperature was reported wherein the metal salt platinum tetrachloride was reduced by the root extract and

Fig. 7.3 Plant parts used for platinum nanoparticle fabrication

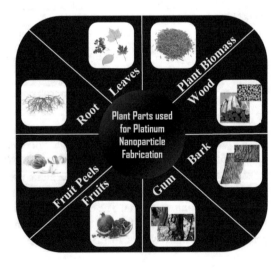

Fig. 7.4 Plant-mediated
fabrication of platinum
nanoparticles

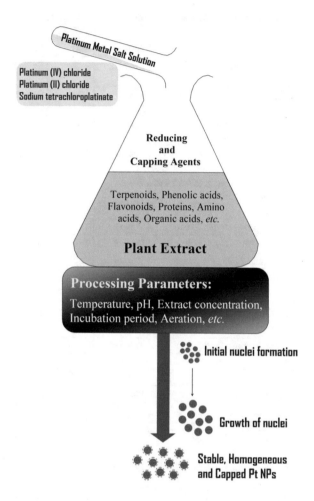

nanoparticles ranging in size between 1 and 6 nm were obtained within 5 min (Raut et al. 2013). *Cacumen platycladi* extract was used for bioreduction of sodium tetra-chloroplatinate (Na_2PtCl_4) to Pt NPs and the variance that occurred in the flavonoid, reducing sugar and protein content of the plant extract was investigated (Zheng et al. 2013). Phytosynthesis of Pt NPs with restrained shape and dimensions has been reported by using plant wood nanomaterials (Lin et al. 2011). Isolated lignin from red pine (*Pinus resinosa*) has also served as a potent source for one-pot synthesis of Pt NPs (Coccia et al. 2012). Bimetallic heterogenous platinum-aurum of 10–40 nm size along with monogenous Pt NPs, 2–4 nm in size were fabricated using green tea (*Camellia sinensis) and* their photo-catalytic degradation of methylene blue was studied by Khalil (2016). With the advancement in this field various plant parts were taken in accord as leaves of *Anacardium occidentale* (Sheny et al. 2013), *Azadirachta indica* (Thirumurugan et al. 2016)*, Mentha piperita* (Yang et al. 2017)*,* Water hyacinth (John Leo and Oluwafemi 2017), *Withania somnifera* (Li

Table 7.3 Biogenic platinum synthesized by plant/plant parts

S. no.	Source	Parts used	Size (in nm)	Morphology	References
1.	*Diospyros kaki*	Leaves	2–12	Spheres Plates	Song et al. (2010)
2.	Hinoki timber	Wood flour	10	Spherical Cubic	Lin et al. (2011)
3.	*Cochlospermum gossypium*	Gum	2.4	–	Vinod et al. (2011)
4.	*Ocimum sanctum*	Leaves	23	Irregular	Soundarrajan et al. (2012)
5.	*Pinus resinosa* (lignin and fulvic acid)	Wood	–	Irregular	Coccia et al. (2012)
6.	Beet	Root	<100	–	Kou and Varma (2012)
7.	*Asparagus racemosus*	Tubers	1–6	Spherical	Raut et al. (2013)
8.	*Anacardium occidentale*	Leaves	–	Irregular Rod shaped	Sheny et al. (2013)
9.	*Cacumen platycladi*	Plant biomass	2.4	–	Zheng et al. (2013)
10.	*Camellia sinensis*	Leaves	2–4	Spherical	Khalil (2016)
11.	*Punica granatum*	Peels	16–23	Spherical	Dauthal and Mukhopadhyay (2014)
12.	*Dioscorea bulbifera*	Tuber	2–5	–	Chopade et al. (2015)
13.	*Citrus sinensis*	Peels	23	Irregular	Castro et al. (2015)
14.	*Quercus glauca*	Leaves	5–15	Spherical	Karthik et al. (2016)
15.	*Fumariae herba*	Plant biomass	10–30	Pentagonal Hexagonal	Dobrucka (2016)
16.	*Alchornea laxiflora*	Bark	5.93	–	Olajire et al. (2017)
17.	*Taraxum laevigatum*	Plant biomass	2–7	Spherical	Tahir et al. (2017)
18.	*Azadirachta indica*	Leaves	5–50	Spherical	Thirumurugan et al. (2016)
19.	*Prunus* × *yedonis*	Gum	10–50	Spherical Oval	Velmurugan et al. (2016)
20.	*Mentha piperita*	Leaves	54.3	Spherical	Yang et al. (2017)
21.	*Punica granatum*	Peels	20.12	Spherical	Şahin et al. (2018)
22.	Water hyacinth	Leaves	3.74	Spherical	John Leo and Oluwafemi (2017)
23.	*Punica granatum*	Fruit	11	–	Jha et al. (2018)
24.	*Withania somnifera*	Leaves	12	Spherical	Li et al. (2017b)

et al. 2017a, b), besides it beet roots (Kou and Varma 2012), and fruit like *Punica granatum* (Jha et al. 2018) have been utilized. Jha et al. (2018) reported in the first hand the use of fruit part, arils of pomegranate for the fabrication of Pt NPs. The arils, extract prepared through it and the colloidal Pt NPs formed along with SEM image, are illustrated in the inset of Fig. 7.5a–d. The broadened XRD peaks clearly conveyed the formation of Pt NPs having face-centred cubic structure which are

Fig. 7.5 The arils (**a**), extract (**b**), and biosynthesized Pt NPs (**c**) with its X-ray diffraction pattern and SEM image (**d**) as fabricated by *Punica granatum*

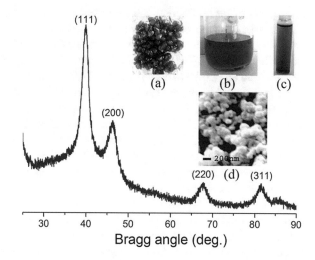

semi-crystalline in nature with very small sized particles (~ nm) as depicted in X-ray diffraction pattern (Fig. 7.5) and SEM image.

The latitude of utilization of plants have been further intensified by using metabolite-rich agro-wastes like peels of pomegranate and orange (Dauthal and Mukhopadhyay 2014; Castro et al. 2015; Şahin et al. 2018) and plant biomass (Zheng et al. 2013; Tahir et al. 2017) for fabrication of this noble metal nanoparticles. The list in the Table 7.3 clearly gives an idea of the multiple numbers of plants utilized and diversification in the plant parts used for Pt NPs synthesis. This also revealed the success rate of fabrication by this approach and led to the need of deciphering the mechanistic aspect of these nanotransformations.

7.3   Mechanistic Aspect of Plant-Mediated Synthesis

Understanding the bioreductive mechanistic aspect underlying in the fabrication of Pt NPs is of core interest. Many researchers are emphasizing in this area although deciphered in chunks it largely remains unexplored. A generalized mechanism for fabrication of Pt NPs from plants is nearly impossible as phyto-constituents varies from plant to plant and the presence of variable phytocomposition among the plant parts makes it more complex. The contents of phytochemicals also fluctuate depending upon season, maturity, and geographical allocation of the plant concerned. The extracellular synthesis using plant extracts (Dauthal and Mukhopadhyay 2016) is more amenable as compared to whole plant usage; as this cocktail of complex structured compounds, with each performing different roles, work out synergistically as green chemical agents yielding the nanoparticles (Alam et al. 2013).

Song and his coworkers were the pioneers in the use of plants for Pt NPs fabrication. In their work using leaf extracts of *Diospyros kaki,* they discovered that for conversion of >90% Pt NPs, 150 min was required at 95 °C as compared to 3 and 11 min when synthesizing silver and gold nanoparticles (Song et al. 2010). This relatively low rate was attributed to the difficulty in the initial nuclei formation that might be further linked to the higher positive electrochemical potential of ag (0.80 V) and au (1.0 V) as compared to Pt (0.74 V) (Haverkamp and Marshall 2009). The results of this study clearly suggested that fabrication of Pt NPs is not an enzyme-mediated process, as the rate of synthesis is greatest at 95 °C and no peaks associated with proteins/enzymes were obtained in FTIR. The functional groups attributed for Pt NPs fabrication included amines, ketones, aldehydes, alcohols, and carboxylic acids present in terpenoids and other reducing sugars. This fact that Pt NPs require higher temperature and longer reaction time was further supported by other researchers in their experiments with *Ocimum sanctum* (Soundarrajan et al. 2012), *Cacumen platycladi* (Zheng et al. 2013) to name a few. Carboxylic acid, phosphine, gallic acid, ascorbic acid, terpenoids, amines, and certain amino acids present in tulsi leaf extract acts as reducing agents for platinum ion reduction. In the fabrication of Pt NPs by tubers of *Asparagus racemosus* (Raut et al. 2013), various bioactive components like alkaloids, polysaccharides, steroids, terpenoids, coumarins, lactones, flavones, linalools, amino acids, and proteins were found to be involved as bioreducing agents in the reduction process at room temperature. Coccia et al. (2012) Reported that the two aromatic naturally occurring derivatives of wood: Lignin and fulvic acid were involved in reduction of platinum metal ions in aqueous solution at 80 °C to the Pt NPs, as indicated by detection of vanillin and other low molecular weight derivatives in the reaction mixture. In another work by Lin et al. (2011), it is documented that wood nanomaterials besides acting as reducing agents also act as an excellent loading and stabilizing media. Zheng and coworkers fabricated Pt NPs with *Cacumen platycladi* extract and tried to explore the bioreductive mechanism underlying behind it. The content of reducing sugar, flavonoids, saccharides, and proteins were assessed before and after the reaction. A significant decrease in reducing sugar and flavonoids was seen as compared to the protein content that was indicative of their probable role in bioreductive process. The variation in these contents before and after the reaction; further varied by extract concentration, reaction temperature, and reaction time (Zheng et al. 2013). *Dioscorea bulbifera* tuber extract were reported to contain phytochemicals like saponins, ascorbic acids, citric acid, flavonoids, reducing sugars, and phenolics that play role in reduction and shape evolution of Pt NPs (Chopade et al. 2015). In neem leaf broth, terpenoids were believed to be the surface active molecules that besides stabilizing the nanoparticles also helped in its bioreduction (Thirumurugan et al. 2016). Castro et al. pointed on the fact that Pt system is more complicated as compared to ag counterpart (Castro et al. 2015). They studied the effect of pH value of the medium on morphology control and valence state of precursor salt in the bioreduction process. The Pt NPs formed in acidic condition where irregular and encapsulated with Sulphur containing organic matter that helped in nanoparticle stabilization. This Sulphur was assumed to be of thiamine present in orange peel. At neutral pH, the shape was small and spherical and particle size and distribution tend towards more uniformity with increasing pH. In the fabrication of Pt NPs using pomegranate peels, Dauthal and

Mukhopadhyay (2014) explored the bioreduction mechanism. Phenolic hydroxyl groups present in polyphenolic compound, ellagic acid, and Pt^{4+} chelated to form an intermediate platinum complex. These phenolic hydroxyls were inductively oxidized to corresponding quinones. The Pt^{4+} to Pt^{2+} reduction occurred in the presence of free electrons or nascent hydrogen produced in the bioreduction process. The Pt^0 atoms further collided and Pt NPs formed, which was stabilized by quinones and ellagic acid. Thus, the bioreduction was attributed to polyphenolic compounds, ellagic acid, gallic acid, and quercetin. In case of Pt NPs fabrication by hot water extract of *Camellia sinensis* (green tea), FTIR analysis indicated polyphenols and caffeine as possible capping and stabilizing agents (Khalil 2016). The reduction and stabilization of Pt NPs by *Quercus glauca* aqueous leaves extract was attributed to functional groups such as flavonoids, carboxyl, tannins, amino, and glycosides or ether groups (Karthik et al. 2016). In the tree bark extract of *Alchornea laxiflora*, quercetin, has been identified as one of the important bioreducing agents along with proteins, polyols, and terpenoids helping in Pt NPs formation (Olajire et al. 2017)

The proposed mechanisms behind the fabrication of Pt NPs by plants are just reasonable hypotheses beyond authentic experimental basis. Further these hypotheses vary from plant to plant. Thus, as far as the specific mechanistic aspect is concerned it has not yet been interpreted well and requires in-depth evaluation. The huge and complex variety of phytochemicals present in various sources makes it challenging to identify specific reducing and stabilizing agents responsible for fabrication and stabilization of nanoparticles. In general, the bioreducing plant secondary metabolites that help in nanotransformation of Pt NPs can be broadly classified into flavonoids, polyphenols, terpenoids, proteins, and organic acids. The hydrogen and electron releasing potency of flavonoids are responsible for its reducing potentials (Zheng et al. 2013). Polyphenols usually form an intermediate complex due to metal chelating ability of its highly nucleophilic aromatic ring; this complex undergoes subsequent reduction to yield Pt NPs (Dauthal and Mukhopadhyay 2014). The hydroxyl group of terpenoids attributes it with reduction capabilities (Song et al. 2010; Raut et al. 2013; Sheny et al. 2013). Proteins and amino acids are complex structured complexes with capability of bioreduction of Pt NPs (Raut et al. 2013). Organic acids like ascorbic acid and others have been reported for their role in nanoparticle fabrication in plant species from mesophytes, xerophytes, and hydrophytes genera (Jha et al. 2009a, b). These organic acids do act as reducing agents for Pt NPs fabrication in many plant species.

7.4 Conclusion

This chapter summarizes the recent research work in the fabrication of biogenic Pt NPs by conferring the various literatures reported so far. In the present scenario, wherein Pt NPs have wide implementation owing to its catalytic and medicinal properties, there is an urge for biological and environmental safety in their production. With increasing demand of precious Pt NPs, there was a need for a simple, reliable, energy efficient, bio-compatible, cost-effective, and eco-friendly fabrication route for its synthesis. Pt NPs fabricated by physical and chemical methods suffered various disadvantages, green chemistry, and biological methods thus came into picture and have become the focal point of research. The fabrication of Pt NPs by plants emerged as advantageous competent to microbes with the presence of broad spectrum of phyto-constituents that act both as reducing and capping agents. Besides eliminating the elaborate process of nurturing cell cultures, the time span for Pt NPs synthesis by plants was much less than that by microorganisms like bacteria and fungi. Plant system can also be aptly scaled up for large-scale synthesis through tissue culture and optimization of downstream processing. In order to exploit and modify the fabrication route for optimized production of biogenic Pt NPs by plants, elucidation of the biochemical and molecular mechanistic aspect of the complete process is necessary. This would help in developing a rational approach and is pre-requisite in making this avenue economically competitive with the conventional methods. One aspect that was clear from this chapter is that mother nature has bestowed us with unique phytodiversity having immense possibilities thus dispensing "a challenging venture with rewarding results" for biogenic platinum nanoparticle fabrication.

References

Ahmadi TS, Wang ZL, Green TC, Henglein A, El-Sayed MA (1996) Shape-controlled synthesis of colloidal platinum nanoparticles. Science 272:1924–1925. https://doi.org/10.1126/science.272.5270.1924

Akhtar MS, Panwar J, Yun YS (2013) Biogenic synthesis of metallic nanoparticles by plant extracts. ACS Sustain Chem Eng 1:591–602. https://doi.org/10.1021/sc300118u

Alam MN, Roy N, Mandal D, Begum NA (2013) Green chemistry for nanochemistry: exploring medicinal plants for the biogenic synthesis of metal NPs with fine-tuned properties. RSC Adv 3:11935–11956. https://doi.org/10.1039/c3ra23133j

Annanouch FE, Haddi Z, Llobet E (2014) Aerosol assisted chemical vapor deposition for $C_6 H_6$ and NO_2 detection. IEEE Trans Sensors 6–9. https://doi.org/10.1109/ICSENS.2014.6984932

Aritonang HF, Onggo D, Ciptati C, Radiman CL (2014) Synthesis of platinum nanoparticles from $K_2 PtCl_4$ solution using bacterial cellulose matrix. J Nanoparticles 2014:1–6 Article ID 285954. https://doi.org/10.1155/2014/285954

Asharani PV, Xinyi N, Hande MP, Valiyaveettil S (2010) DNA damage and p53-mediated growth arrest in human cells treated with platinum nanoparticles. Nanomedicine (Lond) 5:51–64. https://doi.org/10.2217/nnm.09.85

Asztemborska M, Steborowski R, Kowalska J, Bystrzejewska-Piotrowska G (2015) Accumulation of platinum nanoparticles by *Sinapis alba* and *Lepidium sativum* plants. Water Air Soil Pollut 226:126–127. https://doi.org/10.1007/s11270-015-2381-y

Attard G, Casadesús M, MacAskie LE, Deplanche K (2012) Biosynthesis of platinum nanoparticles by *Escherichia coli* MC4100: can such nanoparticles exhibit intrinsic surface enantioselectivity? Langmuir 28:5267–5274. https://doi.org/10.1021/la204495z

Bali R, Siegele R, Harris AT (2010) Biogenic Pt uptake and nanoparticle formation in Medicago sativa and Brassica juncea. J Nanopart Res 12:3087–3095. https://doi.org/10.1007/s11051-010-9904-7

Baskaran B, Muthukumarasamy A, Chidambaram S, Sugumaran A, Ramachandran K, Rasu Manimuthu T (2017) Cytotoxic potentials of biologically fabricated platinum nanoparticles from Streptomyces sp. on MCF-7 breast cancer cells. IET Nanobiotechnol 11:241–246. https://doi.org/10.1049/iet-nbt.2016.0040

Bennett JA, Attard GA, Deplanche K, Casadesus M, Huxter SE, Macaskie LE, Wood J (2012) Improving selectivity in 2-butyne-1,4-diol hydrogenation using biogenic Pt catalysts. ACS Catal 2:504–511. https://doi.org/10.1021/cs200572z

Bommersbach P, Chaker M, Mohamedi M, Guay D (2008) Physico-chemical and electrochemical properties of platinum - tin nanoparticles synthesized by pulsed laser ablation for ethanol oxidation. J Phys Chem C 112:14672–14681. https://doi.org/10.1021/jp801143a

Bonnemann H, Richards RM (2001) Nanoscopic metal particles - synthetic methods and potential applications. Eur J Inorg Chem 2001:2455–2480. https://doi.org/10.1002/1099-0682(200109)2001:10<2455::AID-EJIC2455>3.0.CO;2-Z

Brayner R, Barberousse H, Hemadi M, Djedjat C, Yéprémian C, Coradin T, Livage J, Fiévet F, Couté A (2007) Cyanobacteria as bioreactors for the synthesis of au, ag, Pd, and Pt nanoparticles via an enzyme-mediated route. J Nanosci Nanotechnol 7:2696–2708. https://doi.org/10.1166/jnn.2007.600

Capeness MJ, Edmundson MC, Horsfall LE (2015) Nickel and platinum group metal nanoparticle production by *Desulfovibrio alaskensis* G20. New Biotechnol 32:727–731. https://doi.org/10.1016/j.nbt.2015.02.002

Castro L, Blázquez ML, González F, Muñoz JÁ, Ballester A (2015) Biosynthesis of silver and platinum nanoparticles using orange peel extract: characterisation and applications. IET Nanobiotechnol 9:252–258. https://doi.org/10.1049/iet-nbt.2014.0063

Castro-Longoria E, Moreno-Velásquez SD, Vilchis-Nestor AR, Arenas-Berumen E, Avalos-Borja M (2012) Production of platinum nanoparticles and nanoaggregates using *Neurospora crassa*. J Microbiol Biotechnol 22:1000–1004. https://doi.org/10.4014/jmb.1110.10085

Chau JLH, Chen CY, Yang MC, Lin KL, Sato S, Nakamura T, Yang CC, Cheng CW (2011) Femtosecond laser synthesis of bimetallic Pt-au nanoparticles. Mater Lett 65:804–807. https://doi.org/10.1016/j.matlet.2010.10.088

Choi ID, Lee H, Shim YB, Lee D (2010) A one-step continuous synthesis of carbon-supported Pt catalysts using a flame for the preparation of the fuel electrode. Langmuir 26:11212–11216. https://doi.org/10.1021/la1005264

Chopade B, Ghosh S, Nitnavare R, Dewle A, Tomar GB, Chippalkatti R, More P, Kitture R, Kale S, Bellare J (2015) Novel platinum–palladium bimetallic nanoparticles synthesized by Dioscorea bulbifera: anticancer and antioxidant activities. Int J Nanomedicine 10:7477–7490. https://doi.org/10.2147/IJN.S91579

Coccia F, Tonucci L, Bosco D, Bressan M, d'Alessandro N (2012) One-pot synthesis of lignin-stabilised platinum and palladium nanoparticles and their catalytic behaviour in oxidation and reduction reactions. Green Chem 14:1073–1078. https://doi.org/10.1039/c2gc16524d

Correard F, Maximova K, Estève MA, Villard C, Roy M, Al-Kattan A, Sentis M, Gingras M, Kabashin AV, Braguer D (2014) Gold nanoparticles prepared by laser ablation in aqueous biocompatible solutions: assessment of safety and biological identity for nanomedicine applications. Int J Nanomedicine 9:5415–5430. https://doi.org/10.2147/IJN.S65817

Crooks RM, Zhao M, Sun L, Chechik V, Yeung LK (1999) Dendrimer-encapsulated metal nanoparticles: synthesis, characterization, and applications to catalysis. Adv Mater 11:217–220. https://doi.org/10.1002/(SICI)1521-4095(199903)11:3<217::AID-ADMA217>3.0.CO;2-7

Daniel MCM, Astruc D (2004) Gold nanoparticles: assembly, supramolecular chemistry, quantum-size related properties and applications toward biology, catalysis and nanotechnology. Chem Rev 104:293–346. https://doi.org/10.1021/cr030698

Dauthal P, Mukhopadhyay M (2014) Biofabrication, characterization, and possible bio-reduction mechanism of platinum nanoparticles mediated by agro-industrial waste and their catalytic activity. J Ind Eng Chem 22:185–191. https://doi.org/10.1016/j.jiec.2014.07.009

Dauthal P, Mukhopadhyay M (2016) Noble metal nanoparticles: plant-mediated synthesis, mechanistic aspects of synthesis, and applications. Ind Eng Chem Res 55:9557–9577. https://doi.org/10.1021/acs.iecr.6b00861

Deng QY, Yang B, Wang JF, Whiteley CG, Wang XN (2009) Biological synthesis of platinum nanoparticles with apoferritin. Biotechnol Lett 31:1505–1509. https://doi.org/10.1007/s10529-009-0040-3

Dhand C, Dwivedi N, Loh XJ, Jie Ying AN, Verma NK, Beuerman RW, Lakshminarayanan R, Ramakrishna S (2015) Methods and strategies for the synthesis of diverse nanoparticles and their applications: a comprehensive overview. RSC Adv 5:105003–105037. https://doi.org/10.1039/C5RA19388E

Dobrucka R (2016) Biofabrication of platinum nanoparticles using *Fumariae herba* extract and their catalytic properties. Saudi J Biol Sci, in press. https://doi.org/10.1016/j.sjbs.2016.11.012

Du Y, Su J, Luo W, Cheng G (2015) Graphene-supported nickel-platinum nanoparticles as efficient catalyst for hydrogen generation from hydrous hydrazine at room temperature. ACS Appl Mater Interfaces 7:1031–1034. https://doi.org/10.1021/am5068436

Elia P, Zach R, Hazan S (2014) Green synthesis of gold nanoparticles using plant extracts as reducing agents. Int J Nanomedicine 9:4007–4021. https://doi.org/10.2147/IJN.S57343

Engelbrekt C, Sørensen KH, Lübcke T, Zhang J, Li Q, Pan C, Bjerrum NJ, Ulstrup J (2010) 1.7 nm platinum nanoparticles: synthesis with glucose starch, characterization and catalysis. Chem Phys Chem 11:2844–2853. https://doi.org/10.1002/cphc.201000380

Fan J, Yin JJ, Ning B, Wu X, Hu Y, Ferrari M, Anderson GJ, Wei J, Zhao Y, Nie G (2011) Direct evidence for catalase and peroxidase activities of ferritin-platinum nanoparticles. Biomaterials 32:1611–1618. https://doi.org/10.1016/j.biomaterials.2010.11.004

Gaidhani SV, Yeshvekar RK, Shedbalkar UU, Bellare JH, Chopade BA (2014) Bio-reduction of hexachloroplatinic acid to platinum nanoparticles employing *Acinetobacter calcoaceticus*. Process Biochem 49:2313–2319. https://doi.org/10.1016/j.procbio.2014.10.002

Gao F, Yang N, Obloh H, Nebel CE (2013) Shape-controlled platinum nanocrystals on boron-doped diamond. Electrochem Commun 30:55–58. https://doi.org/10.1016/j.elecom.2013.02.004

Golunski SE (2007) Why use platinum in catalytic converters? Platin Met Rev 51:162. https://doi.org/10.1595/147106707X205857

Govender Y, Riddin T, Gericke M, Whiteley CG (2009) Bioreduction of platinum salts into nanoparticles: a mechanistic perspective. Biotechnol Lett 31:95–100. https://doi.org/10.1007/s10529-008-9825-z

Govender Y, Riddin TL, Gericke M, Whiteley CG (2010) On the enzymatic formation of platinum nanoparticles. J Nanopart Res 12:261–271. https://doi.org/10.1007/s11051-009-9604-3

Haverkamp RG, Marshall AT (2009) The mechanism of metal nanoparticle formation in plants: limits on accumulation. J Nanopart Res 11:1453–1463. https://doi.org/10.1007/s11051-008-9533-6

Honary S, Gharaei-fathabad E, Barabadi H, Naghibi F (2013) Fungus-mediated synthesis of gold nanoparticles: a novel biological approach to nanoparticle synthesis. J Nanosci Nanotechnol 13:1427–1430. https://doi.org/10.1166/jnn.2013.5989

Iravani S (2014) Bacteria in nanoparticle synthesis: current status and future prospects. Int Sch Res Not 2014:1–18 Article ID 359316. https://doi.org/10.1155/2014/359316

Isaac R, Gobalakrishnan S, Rajan G, Wu R-J, Pamanji SR, Khagga M, Baskaralingam V, Chavali M (2013) An overview of facile green biogenic synthetic routes and applications of platinum nanoparticles. Adv Sci Eng Med 5:763–770. https://doi.org/10.1166/asem.2013.1377

Jana NR, Wang ZL, Sau TK, Pal T (2000) Seed-mediated growth method to prepare cubic copper nanoparticles. Curr Sci 79:1367–1370

Jha AK, Prasad K (2010) Green synthesis of silver nanoparticles using *Cycas* leaf. Int J Green Nanotechnol Phys Chem 1:110–117. https://doi.org/10.1080/19430871003684572

Jha AK, Prasad K (2011a) Green fruit of chili (*Capsicum annum* L.) synthesizes nano silver! Dig J Nanomater Biostruct 6:1717–1723

Jha AK, Prasad K (2011b) Biosynthesis of gold nanoparticles using bael (*Aegle marmelos*) leaf: mythology meets technology. Int J Green Nanotechnol Biomed 3:92–97. https://doi.org/10.10 80/19430892.2011.574560

Jha AK, Prasad K (2013) Rose (Rosa sp.) petals assisted green synthesis of gold nanoparticles. J Bionanosci 7:245–250. https://doi.org/10.1166/jbns.2013.1139

Jha AK, Prasad K, Kulkarni AR (2008) Yeast mediated synthesis of silver nanoparticles. Int J Nanosci Nanotechnol 4:17–22

Jha AK, Prasad K, Kumar V, Prasad K (2009a) Biosynthesis of silver nanoparticles using eclipta leaf. Biotechnol Prog 25:1476–1479. https://doi.org/10.1002/btpr.233

Jha AK, Prasad K, Prasad K, Kulkarni AR (2009b) Plant system: Nature's nanofactory. Colloids Surf B Biointerfaces 73:219–223. https://doi.org/10.1016/j.colsurfb.2009.05.018

Jha B, Rao M, Chattopadhyay A, Bandyopadhyay A, Prasad K, Jha AK (2018) *Punica granatum* fabricated platinum nanoparticles: a therapeutic pill for breast cancer. AIP Conf Proc 30087:2–5. https://doi.org/10.1063/1.5032422

John Leo A, Oluwafemi OS (2017) Plant-mediated synthesis of platinum nanoparticles using water hyacinth as an efficient biomatrix source – an eco-friendly development. Mater Lett 196:141–144. https://doi.org/10.1016/j.matlet.2017.03.047

Karthik R, Sasikumar R, Chen SM, Govindasamy M, Vinoth Kumar J, Muthuraj V (2016) Green synthesis of platinum nanoparticles using *Quercus glauca* extract and its electrochemical oxidation of hydrazine in water samples. Int J Electrochem Sci 11:8245–8255. https://doi.org/10.20964/2016.10.62

Kashyap PL, Kumar S, Srivastava AK, Sharma AK (2013) Myconanotechnology in agriculture: a perspective. World J Microbiol Biotechnol 29:191–207. https://doi.org/10.1007/s11274-012-1171-6

Ke X, Bittencourt C, Bals S, Van Tendeloo G (2013) Low-dose patterning of platinum nanoclusters on carbon nanotubes by focused-electron-beam induced deposition as studied by TEM. Beilstein J Nanotechnol 4:77–86. https://doi.org/10.3762/bjnano.4.9

Khalil M (2016) Biosynthesis and characterization of Pt and au- Pt nanoparticles and their photo catalytic degradation of methylene blue. Int J Adv Res 2:694–703

Kharissova OV, Dias HVR, Kharisov BI, Pérez BO, Pérez VMJ (2013) The greener synthesis of nanoparticles. Trends Biotechnol 31:240–248. https://doi.org/10.1016/j.tibtech.2013.01.003

Kim JM, Joh HI, Jo SM, Ahn DJ, Ha HY, Hong SA, Kim SK (2010) Preparation and characterization of Pt nanowire by electrospinning method for methanol oxidation. Electrochim Acta 55:4827–4835. https://doi.org/10.1016/j.electacta.2010.03.036

Klaus-Joerger T, Joerger R, Olsson E, Granqvist CG (2001) Bacteria as workers in the living factory: metal-accumulating bacteria and their potential for materials science. Trends Biotechnol 19:15–20. https://doi.org/10.1016/S0167-7799(00)01514-6

Ko Y-L, Krishnamurthy S, Yun Y-S (2015) Facile synthesis of monodisperse pt and pd nanoparticles using antioxidants. J Nanosci Nanotechnol 15:412–417. https://doi.org/10.1166/jnn.2015.8375

Koczkur KM, Mourdikoudis S, Polavarapu L, Skrabalak SE, Koczkur KM, Mourdikoudis S, Polavarapu L, Skrabalak SE (2015) Polyvinylpyrrolidone (PVP) in nanoparticle synthesis. Dalton Transac R Soc Chem 44:17883–17905. https://doi.org/10.1039/C5DT02964C HAL Id: hal-01217114

Konishi Y, Ohno K, Saitoh N, Nomura T, Nagamine S, Hishida H, Takahashi Y, Uruga T (2007) Bioreductive deposition of platinum nanoparticles on the bacterium Shewanella algae. J Biotechnol 128:648–653. https://doi.org/10.1016/j.jbiotec.2006.11.014

Korbekandi H, Iravani S, Abbasi S (2009) Production of nanoparticles using organisms. Crit Rev Biotechnol 29:279–306. https://doi.org/10.3109/07388550903062462

Kou J, Varma RS (2012) Beet juice utilization: expeditious green synthesis of noble metal nanoparticles (ag, au, Pt, and Pd) using microwaves. RSC Adv 2:10283–10290. https://doi.org/10.1039/c2ra21908e

Krithiga N, Rajalakshmi A, Jayachitra A (2015) Green synthesis of silver nanoparticles using leaf extracts of *Clitoria ternatea* and *Solanum nigrum* and study of its antibacterial effect against common nosocomial pathogens. J Nanosci 2015:1–8. https://doi.org/10.1155/2015/928204

Kshirsagar P, Sangaru SS, Malvindi MA, Martiradonna L, Cingolani R, Pompa PP (2011) Synthesis of highly stable silver nanoparticles by photoreduction and their size fractionation by phase transfer method. Colloids Surf A Physicochem Eng Asp 392:264–270. https://doi.org/10.1016/j.colsurfa.2011.10.003

Lengke MF, Fleet ME, Southam G (2006) Synthesis of platinum nanoparticles by reaction of filamentous cyanobacteria with platinum (IV) - chloride complex. Langmuir 22:7318–7323. https://doi.org/10.1021/la060873s

Leong GJ, Schulze MC, Strand MB, Maloney D, Frisco SL, Dinh HN, Pivovar B, Richards RM (2014) Shape-directed platinum nanoparticle synthesis: nanoscale design of novel catalysts. Appl Organomet Chem 28:1–17. https://doi.org/10.1002/aoc.3048

Li Y, Jiang Y, Chen M, Liao H, Huang R, Zhou Z, Tian N, Chen S, Sun S (2012) Electrochemically shape-controlled synthesis of trapezohedral platinum nanocrystals with high electrocatalytic activity. Chem Commun 48:9531. https://doi.org/10.1039/c2cc34322c

Li X, Wang Y, Li L, Huang W, Xiao Z, Wu P, Wenbo Z, Guo W, Jiang P, Liang M (2017a) Deficient copper decorated platinum nanoparticles for selective hydrogenation of chloronitrobenzene. J Mater Chem A 5:11294–11300. https://doi.org/10.1039/C7TA01587A

Li Y, Zhang J, Gu J, Chen S, Wang C, Jia W (2017b) Biosynthesis of polyphenol-stabilised nanoparticles and assessment of anti-diabetic activity. J Photochem Photobiol B Biol 169:96–100. https://doi.org/10.1016/j.jphotobiol.2017.02.017

Lin X, Wu M, Wu D, Kuga S, Endo T, Huang Y (2011) Platinum nanoparticles using wood nano-materials: eco-friendly synthesis, shape control and catalytic activity for p-nitrophenol reduction. Green Chem 13:283–287. https://doi.org/10.1039/C0GC00513D

Liu Y, Li D, Sun S (2011) Pt-based composite nanoparticles for magnetic, catalytic, and biomedical applications. J Mater Chem 21:12579–12587. https://doi.org/10.1039/c1jm11605c

Liu Y, Wu H, Chong Y, Wamer WG, Xia Q, Cai L, Nie Z, Fu PP, Yin JJ (2015) Platinum nanopar-ticles: efficient and stable catechol oxidase mimetics. ACS Appl Mater Interfaces 7:19709–19717. https://doi.org/10.1021/acsami.5b05180

Liz-Marzán LM, Lado-Touriño I (1996) Reduction and stabilization of silver nanoparticles in etha-nol by nonionic surfactants. Langmuir 12:3585–3589. https://doi.org/10.1021/la951501e

Mahima S, Kannan R, Komath I, Aslam M, Pillai VK (2008) Synthesis of platinum Y-junction nanostructures using hierarchically designed alumina templates and their enhanced electro-catalytic activity for fuel-cell applications. Chem Mater 20:601–603. https://doi.org/10.1021/cm702102b

Mahmoud MA, El-Sayed MA (2012) Metallic double shell hollow nanocages: the challenges of their synthetic techniques. Langmuir 28:4051–4059. https://doi.org/10.1021/la203982h

Mahmoud MA, Saira F, El-Sayed MA (2010) Experimental evidence for the nanocage effect in catalysis with hollow nanoparticles. Nano Lett 10:3764–3769. https://doi.org/10.1021/nl102497u

Makarov VV, Love AJ, Sinitsyna OV, Makarova SS, Yaminsky IV, Taliansky ME, Kalinina NO (2014) "Green" nanotechnologies: synthesis of metal nanoparticles using plants. Acta Nat 6:35–44. https://doi.org/10.1039/c1gc15386b

Mallikarjuna K, Narasimha G, Dillip GR, Praveen B, Shreedhar B, Lakshmi CS, Reddy BVS, Raju BDP (2011) Green synthesis of silver nanoparticles using ocimum leaf extract and their characterization. Dig J Nanomater Biostruct 6:181–186

Manikandan M, Hasan N, Wu HF (2013) Platinum nanoparticles for the photothermal treat-ment of neuro 2A cancer cells. Biomaterials 34:5833–5842. https://doi.org/10.1016/j.biomaterials.2013.03.077

Martins M, Mourato C, Sanches S, Noronha JP, Crespo MTB, Pereira IAC (2016) Biogenic plati-num and palladium nanoparticles as new catalysts for the removal of pharmaceutical com-pounds. Water Res 108:160–168. https://doi.org/10.1016/j.watres.2016.10.071

Merkel TJ, Herlihy KP, Nunes J, Orgel RM, Rolland JP, Desimone JM (2010) Scalable, shape-specific, top-down fabrication methods for the synthesis of engineered colloidal particles. Langmuir 26:13086–13096. https://doi.org/10.1021/la903890h

Mittal AK, Chisti Y, Banerjee UC (2013) Synthesis of metallic nanoparticles using plant extracts. Biotechnol Adv 31:36–356. https://doi.org/10.1016/j.biotechadv.2013.01.003

Moglianetti M, De Luca E, Pedone D, Marotta R, Catelani T, Sartori B, Amenitsch H, Retta SF, Pompa PP (2016) Platinum nanozymes recover cellular ROS homeostasis in an oxidative stress-mediated disease model. Nanoscale 8:3739–3752. https://doi.org/10.1039/C5NR08358C

Moon SY, Kusunose T, Sekino T (2009) CTAB-assisted synthesis of size- and shape-controlled gold nanoparticles in SDS aqueous solution. Mater Lett 63:2038–2040. https://doi.org/10.1016/j.matlet.2009.06.047

Nadagouda MN, Varma RS (2006) Green and controlled synthesis of gold and platinum nanomaterials using vitamin B2: density-assisted self-assembly of nanospheres, wires and rods. Green Chem 8:516–518. https://doi.org/10.1039/b601271j

Nadaroglu H, Alayli A, Ince S, Babagil A (2017) Green synthesis and characterisation of platinum nanoparticles using quail egg yolk. Spectrochim Acta A Mol Biomol Spectrosc 172:43–47. https://doi.org/10.1016/j.saa.2016.05.023

Nugroho FAA, Iandolo B, Wagner JB, Langhammer C (2016) Bottom-up nanofabrication of supported noble metal alloy nanoparticle arrays for plasmonics. ACS Nano 10:2871–2879. https://doi.org/10.1021/acsnano.5b08057

Oberhauser W, Evangelisti C, Tiozzo C, Vizza F, Psaro R (2016) Lactic acid from glycerol by ethylene-stabilized platinum-nanoparticles. ACS Catal 6:1671–1674. https://doi.org/10.1021/acscatal.5b02914

Oko DN, Garbarino S, Zhang J, Xu Z, Chaker M, Ma D, Guay D, Tavares AC (2015) Dopamine and ascorbic acid electro-oxidation on au, AuPt and Pt nanoparticles prepared by pulse laser ablation in water. Electrochim Acta 159:174–183. https://doi.org/10.1016/j.electacta.2015.01.192

Olajire AA, Adeyeye GO, Yusuf RA (2017) Alchornea laxiflora bark extract assisted green synthesis of platinum nanoparticles for oxidative desulphurization of model oil. J Clust Sci 28:1565–1578. https://doi.org/10.1007/s10876-017-1167-3

Pal A, Shah S, Devi S (2009) Microwave-assisted synthesis of silver nanoparticles using ethanol as a reducing agent. Mater Chem Phys 114:530–532. https://doi.org/10.1016/j.matchemphys.2008.11.056

Paschos O, Choi P, Efstathiadis H, Haldar P (2008) Synthesis of platinum nanoparticles by aerosol assisted deposition method. Thin Solid Films 516:3796–3801. https://doi.org/10.1016/j.tsf.2007.06.123

Pastoriza-Santos I, Liz-Marzán LM (2009) N, N-Dimethylformamide as a reaction medium for metal nanoparticle synthesis. Adv Funct Mater 19:679–688. https://doi.org/10.1002/adfm.200801566

Pedone D, Moglianetti M, De Luca E, Bardi G, Pompa PP (2017) Platinum nanoparticles in nanobiomedicine. Chem Soc Rev 46:4951–4975. https://doi.org/10.1039/C7CS00152E

Petroski JM, Wang ZL, Green TC, El-sayed MA (1998) Kinetically controlled growth and shape formation mechanism of platinum nanoparticles. J Phys Chem B 102:3316–3320. https://doi.org/10.1021/jp981030f

Porcel E, Liehn S, Remita H, Usami N, Kobayashi K, Furusawa Y, Le Sech C, Lacombe S (2010) Platinum nanoparticles: a promising material for future cancer therapy? Nanotechnology 21:85103–85107. https://doi.org/10.1088/0957-4484/21/8/085103

Prasad R (2014) Synthesis of silver nanoparticles in photosynthetic plants. J Nanoparticles Article ID 963961. https://doi.org/10.1155/2014/963961

Prasad K, Jha AK, Prasad K, Kulkarni AR (2010) Can microbes mediate nano-transformation? Indian J Phys 84:1355–1360. https://doi.org/10.1007/s12648-010-0126-8

Prasad R, Pandey R, Barman I (2016) Engineering tailored nanoparticles with microbes: quo vadis. WIREs Nanomed Nanobiotechnol 8:316–330. https://doi.org/10.1002/wnan.1363

Quester K (2013) Biosynthesis and microscopic study of metallic nanoparticles. Micron 54–55:1–27. https://doi.org/10.1016/j.micron.2013.07.003

Rai M, Gade A, Yadav A (2011) Biogenic nanoparticles: an introduction to what they are, how they are synthesized and their applications. In: Rai M, Duran N (eds) Metal nanoparticles in microbiology. Springer, Berlin, pp 1–14. https://doi.org/10.1007/978-3-642-18312-6

Rakshit RK, Bose SK, Sharma R, Budhani RC, Vijaykumar T, Neena SJ, Kulkarni GU (2008) Correlations between morphology, crystal structure, and magnetization of epitaxial cobalt-platinum films grown with pulsed laser ablation. J Appl Phys 103:023915–023915. https://doi.org/10.1063/1.2832763

Ranoszek-Soliwoda K, Tomaszewska E, Socha E, Krzyczmonik P, Ignaczak A, Orlowski P, Krzyzowska M, Celichowski G, Grobelny J (2017) The role of tannic acid and sodium citrate in the synthesis of silver nanoparticles. J Nanopart Res 19:273–287. https://doi.org/10.1007/s11051-017-3973-9

Rashamuse KJ, Whiteley CG (2007) Bioreduction of Pt (IV) from aqueous solution using sulphate-reducing bacteria. Appl Microbiol Biotechnol 75:1429–1435. https://doi.org/10.1007/s00253-007-0963-3

Raut RW, Haroon ASM, Malghe YS, Nikam BT, Kashid SB (2013) Rapid biosynthesis of platinum and palladium metal nanoparticles using root extract of asparagus racemosus Linn. Adv Mater Let 4:650–654. https://doi.org/10.5185/amlett.2012.11470

Riddin TL, Gericke M, Whiteley CG (2006) Analysis of the inter- and extracellular formation of platinum nanoparticles by *Fusarium oxysporum* F. Sp. lycopersici using response surface methodology. Nanotechnology 17:3482–3489. https://doi.org/10.1088/0957-4484/17/14/021

Riddin TL, Govender Y, Gericke M, Whiteley CG (2009) Two different hydrogenase enzymes from sulphate-reducing bacteria are responsible for the bioreductive mechanism of platinum into nanoparticles. Enzym Microb Technol 45:267–273. https://doi.org/10.1016/j.enzmictec.2009.06.006

Riddin T, Gericke M, Whiteley CG (2010) Biological synthesis of platinum nanoparticles: effect of initial metal concentration. Enzym Microb Technol 46:501–505. https://doi.org/10.1016/j.enzmictec.2010.02.006

Şahin B, Aygün A, Gündüz H, Şahin K, Demir E, Akocak S, Şen F (2018) Cytotoxic effects of platinum nanoparticles obtained from pomegranate extract by the green synthesis method on the MCF-7 cell line. Colloids Surf B Biointerfaces 163:119–124. https://doi.org/10.1016/j.colsurfb.2017.12.042

Saminathan K, Kamavaram V, Veedu V, Kannan AM (2009) Preparation and evaluation of electro-deposited platinum nanoparticles on in situ carbon nanotubes grown carbon paper for proton exchange membrane fuel cells. Int J Hydrog Energy 34:3838–3844. https://doi.org/10.1016/j.ijhydene.2009.03.009

Sheny DS, Philip D, Mathew J (2013) Synthesis of platinum nanoparticles using dried *Anacardium occidentale* leaf and its catalytic and thermal applications. Spectrochim Acta A Mol Biomol Spectrosc 114:267–271. https://doi.org/10.1016/j.saa.2013.05.028

Shi Y, Lin M, Jiang X, Liang S (2015) Recent advances in FePt nanoparticles for biomedicine. J Nanomater 2015:1–13. Article ID 467873. https://doi.org/10.1155/2015/467873

Shin Y, Bae IT, Exarhos GJ (2009) "Green" approach for self-assembly of platinum nanoparticles into nanowires in aqueous glucose solutions. Colloids Surf A Physicochem Eng Asp 348:191–195. https://doi.org/10.1016/j.colsurfa.2009.07.013

Shiny PJ, Mukherjee A, Chandrasekaran N (2014) Haemocompatibility assessment of synthesised platinum nanoparticles and its implication in biology. Bioprocess Biosyst Eng 37:991–997. https://doi.org/10.1007/s00449-013-1069-1

Shiny PJ, Mukherjee A, Chandrasekaran N (2016) Biosynthesised silver and platinum nanoparticles. RSC Adv 6:27775–27787. https://doi.org/10.1039/C5RA27185A

Siddiqi KS, Hu sen A (2016) Green synthesis, characterization and uses of palladium/platinum nanoparticles. Nanoscale Res Lett 11:482–494. https://doi.org/10.1186/s11671-016-1695-z

Singh P, Kim YJ, Zhang D, Yang DC (2016) Biological synthesis of nanoparticles from plants and microorganisms. Trends Biotechnol 34:588–599. https://doi.org/10.1016/j.tibtech.2016.02.006

Song JY, Kwon EY, Kim BS (2010) Biological synthesis of platinum nanoparticles using Diospyros kaki leaf extract. Bioprocess Biosyst Eng 33:159–164. https://doi.org/10.1007/s00449-009-0373-2

Soni N, Prakash S (2012) Efficacy of fungus mediated silver and gold nanoparticles against Aedes aegypti larvae. Parasitol Res 110:175–184. https://doi.org/10.1007/s00436-011-2467-4

Soundarrajan C, Sankari A, Dhandapani P, Maruthamuthu S, Ravichandran S, Sozhan G, Palaniswamy N (2012) Rapid biological synthesis of platinum nanoparticles using Ocimum sanctum for water electrolysis applications. Bioprocess Biosyst Eng 35:827–833. https://doi.org/10.1007/s00449-011-0666-0

Srivastava SK, Constanti M (2012) Room temperature biogenic synthesis of multiple nanoparticles (ag, Pd, Fe, Rh, Ni, Ru, Pt, co, and li) by Pseudomonas aeruginosa SM1. J Nanopart Res 14:831–840. https://doi.org/10.1007/s11051-012-0831-7

Strobel R, Pratsinis SE (2009) Flame synthesis of supported platinum group metals for catalysis and sensors. Platinum Metals Rev 53:11–20. https://doi.org/10.1595/147106709X392993

Subramaniyan SA, Sheet S, Vinothkannan M, Yoo DJ, Lee YS, Belal SA, Shim KS (2018) One-pot facile synthesis of Pt nanoparticles using cultural filtrate of microgravity simulated grown P. chrysogenum and their activity on bacteria and cancer cells. J Nanosci Nanotechnol 18:3110–3125. https://doi.org/10.1166/jnn.2018.14661

Syed A, Ahmad A (2012) Extracellular biosynthesis of platinum nanoparticles using the fungus Fusarium oxysporum. Colloids Surf B Biointerfaces 97:27–31. https://doi.org/10.1016/j.colsurfb.2012.03.026

Tahir K, Nazir S, Ahmad A, Li B, Khan AU, Khan ZUH, Khan FU, Khan QU, Khan A, Rahman AU (2017) Facile and green synthesis of phytochemicals capped platinum nanoparticles and in vitro their superior antibacterial activity. J Photochem Photobiol B Biol 166:246–251. https://doi.org/10.1016/j.jphotobiol.2016.12.016

Tatsumi H, Liu F, Han HL, Carl LM, Sapi A, Somorjai GA (2017) Alcohol oxidation at platinum-gas and platinum-liquid interfaces: the effect of platinum nanoparticle size, water coadsorption, and alcohol concentration. J Phys Chem C 121:7365–7371. https://doi.org/10.1021/acs.jpcc.7b01432

Thirumurugan A, Aswitha P, Kiruthika C, Nagarajan S, Christy AN (2016) Green synthesis of platinum nanoparticles using Azadirachta indica - an eco-friendly approach. Mater Lett 170:175–178. https://doi.org/10.1016/j.matlet.2016.02.026

Tiwari DK, Behari J, Sen P (2008) Time and dose-dependent antimicrobial potential of ag nanoparticles synthesized by top-down approach. Curr Sci 95:647–655

Tiwari JN, Pan F-M, Lin K-L (2009) Facile approach to the synthesis of 3D platinum nanoflowers and their electrochemical characteristics. New J Chem 33:1482–1485. https://doi.org/10.1039/b901534p

Vadivel M, Babu RR, Ramamurthi K, Arivanandhan M (2016) CTAB cationic surfactant assisted synthesis of CoFe$_2$O$_4$ magnetic nanoparticles. Ceram Int 42:19320–19328. https://doi.org/10.1016/j.ceramint.2016.09.101

Velmurugan P, Shim J, Oh B (2016) Prunus x yedoenis tree gum mediated synthesis of platinum nanoparticles with antifungal activity against phytopathogens. Mater Lett 174:61–65. https://doi.org/10.1016/j.matlet.2016.03.069

Venu R, Ramulu TS, Anandakumar S, Rani VS, Kim CG (2011) Bio-directed synthesis of platinum nanoparticles using aqueous honey solutions and their catalytic applications. Colloids Surf A Physicochem Eng Asp 384:733–738. https://doi.org/10.1016/j.colsurfa.2011.05.045

Vinod VTP, Saravanan P, Sreedhar B, Devi DK, Sashidhar RB (2011) A facile synthesis and characterization of ag, au and Pt nanoparticles using a natural hydrocolloid gum kondagogu. Colloids Surf B Biointerfaces 83:291–298. https://doi.org/10.1016/j.colsurfb.2010.11.035

Wang C, Daimon H, Onodera T, Koda T, Sun S (2008) A general approach to the size- and shape-controlled synthesis of platinum nanoparticles and their catalytic reduction of oxygen. Angew Chem Int Ed 120:3644–3647. https://doi.org/10.1002/ange.200800073

Wang X, Zhang Y, Li T, Tian W, Zhang Q, Cheng Y (2013) Generation 9 polyamidoamine dendrimer encapsulated platinum nanoparticle mimics catalase size, shape, and catalytic activity. Langmuir 29:5262–5270. https://doi.org/10.1021/la3046077

Whiteley C, Govender Y, Riddin T, Rai M (2011) Enzymatic synthesis of platinum nanoparticles: prokaryote and eukaryote systems. In: Rai M, Duran N (eds) Metal nanoparticles in microbiology. Springer, Berlin, pp 103–134. https://doi.org/10.1007/978-3-642-18312-6_5

Yadav A, Kon K, Kratosova G, Duran N, Ingle AP, Rai M (2015) Fungi as an efficient mycosystem for the synthesis of metal nanoparticles: progress and key aspects of research. Biotechnol Lett 37:2099–2120. https://doi.org/10.1007/s10529-015-1901-6

Yamamoto K, Imaoka T, Chun WJ, Enoki O, Katoh H, Takenaga M, Sonoi A (2009) Size-specific catalytic activity of platinum clusters enhances oxygen reduction reactions. Nat Chem 1:397–402. https://doi.org/10.1038/nchem.288

Yang J, Sun D, Li J, Yang X, Yu J, Hao Q, Liu W, Liu J, Zou Z, Gu J (2009) In situ deposition of platinum nanoparticles on bacterial cellulose membranes and evaluation of PEM fuel cell performance. Electrochim Acta 54:6300–6305. https://doi.org/10.1016/j.electacta.2009.05.073

Yang C, Wang M, Zhou J, Chi Q (2017) Bio-synthesis of peppermint leaf extract polyphenols capped nano-platinum and their in-vitro cytotoxicity towards colon cancer cell lines (HCT 116). Mater Sci Eng C 77:1012–1016. https://doi.org/10.1016/j.msec.2017.04.020

Yanson AI, Rodriguez P, Garcia-Araez N, Mom RV, Tichelaar FD, Koper MTM (2011) Cathodic corrosion: a quick, clean, and versatile method for the synthesis of metallic nanoparticles. Angew Chem Int Ed 50:6346–6350. https://doi.org/10.1002/anie.201100471

Yola ML, Atar N (2017) Electrochemical detection of atrazine by platinum nanoparticles/carbon nitride nanotubes with molecularly imprinted polymer. Ind Eng Chem Res 56:7631–7639. https://doi.org/10.1021/acs.iecr.7b01379

Yu S, Li F, Yang H, Li G, Zhu G, Li J, Zhang L, Li Y (2017) Pt-nanoflower as high performance electrocatalyst for fuel cell vehicle. Int J Hydrog Energy 42:29971–29976. https://doi.org/10.1016/j.ijhydene.2017.06.228

Zhang Y, Shi R, Yang P (2014) Synthesis of ag nanoparticles with tunable sizes using N,N-dimethyl formamide. J Nanosci Nanotechnol 14:3011–3016. https://doi.org/10.1166/jnn.2014.8558

Zhang N, Bu L, Guo S, Guo J, Huang X (2016) Screw thread-like platinum-copper nanowires bounded with high-index facets for efficient electrocatalysis. Nano Lett 16:5037–5043. https://doi.org/10.1021/acs.nanolett.6b01825

Zhao W, Zhou X, Xue Z, Wu B, Liu X, Lu X (2013) Electrodeposition of platinum nanoparticles on polypyrrole-functionalized graphene. J Mater Sci 48:2566–2573. https://doi.org/10.1007/s10853-012-7047-1

Zheng B, Kong T, Jing X, Odoom-Wubah T, Li X, Sun D, Lu F, Zheng Y, Huang J, Li Q (2013) Plant-mediated synthesis of platinum nanoparticles and its bioreductive mechanism. J Colloid Interface Sci 396:138–145. https://doi.org/10.1016/j.jcis.2013.01.021

Zhi L, Müllen K (2008) A bottom-up approach from molecular nanographenes to unconventional carbon materials. J Mater Chem 18:1472–1484. https://doi.org/10.1039/b717585j

Zhou ZY, Huang ZZ, Chen DJ, Wang Q, Tian N, Sun SG (2010) High-index faceted platinum nanocrystals supported on carbon black as highly efficient catalysts for ethanol electrooxidation. Angew Chem Int Ed 49:411–414. https://doi.org/10.1002/anie.200905413

Chapter 8
Hidden Treasures for Nanomaterials Synthesis!

Niraj Kumari, Priti Kumari, Anal K. Jha, and Kamal Prasad

8.1 Introduction

Right from settlement of the mother earth, plant systems have played a crucial role in phylogenic diversity. The systematic origin of plant life has been evolved from aquatic system to top soil leading to formation of an adorable phytodiversity. The chemical constituents and metabolic promises decide the phylogenetic and functional fate of plants present in nature. Morphological organization, molecular distribution, and interaction of metabolites during metabolic fluxes (both primary as well as secondary) in plants contribute them with nature's blessing for adaptability and survival against different environmental harshness ranging from extremely cold alpine regions to deep ocean beds through deserts. Chemical constituents (or metabolic status) of plants provide strength for development of resistance against harsh environment and these chemical constituents also provide tremendous implications for surviving the humanity in nature. The interaction of inorganic nanoparticles with biological structures is one of the most promising areas of research in modern nanoscience and technology. The term nanotechnology is defined as the fabrications, manipulation, and utilization of materials at a scale smaller than 1 mm. The "nano" is a word which means dwarf or extremely small derived from a Greek word and used as prefix to any parameter. Physicist Professor Richard Feynman had given the concept of nanotechnology in his historic lecture "there's plenty of room at the bottom" in 1959 and the term was introduced by Professor Norio Taniguchi in Tokyo Science University in 1974. Nanoscience and technology is the area of research and

N. Kumari · P. Kumari · A. K. Jha
Aryabhatta Centre for Nanoscience and Nanotechnology, Aryabhatta Knowledge University, Patna, India

K. Prasad (✉)
Department of Physics, Tilka Manjhi Bhagalpur University, Bhagalpur, Bihar, India
e-mail: prasad_k@tmbuniv.ac.in

© Springer Nature Switzerland AG 2018
R. Prasad et al. (eds.), *Exploring the Realms of Nature for Nanosynthesis*,
Nanotechnology in the Life Sciences, https://doi.org/10.1007/978-3-319-99570-0_8

development which deals with the development of synthesis methods and surface analytical tools for building structures and materials, as well as to understand the change in chemical and physical properties due to miniaturization, and also provides an opportunity of using such properties for the development of novel and functional materials and devices. Nanoparticles are described as particulate distribution of solid particles with at least one dimension of size range 10–1000 nm. The most important significance of nanoparticles is surface area to volume facet ratio which affirms to interact with other particles easier. There is alteration in the properties of conventional materials at nano level due to the quantum effect (spatial confinements); large surface area, large surface energy, and the behavior of surfaces start to lead the behavior of bulk materials. By amenably manipulating the structure, size, shape, and composition of the nanoscale materials, it is quite possible to tune their physical, chemical, electrical, optical, magnetic, mechanical, catalytic, and biological properties (Jha and Prasad 2011, 2014, 2016a, b, c; Vivek et al. 2012; Kaviya et al. 2011; Punuri et al. 2012; Venu et al. 2011; Daniel and Astruc 2004; Schmid 1992; Rastogi and Arunachalam 2013; Sen et al. 2013; Krishnaraj et al. 2012; Krolikowska et al. 2003; Valtchev and Tosheva 2013; Gurunathan et al. 2009; Fayaz et al. 2011; Yilmaz et al. 2011; Castro et al. 2010; Pingali et al. 2005; Flores et al. 2013; Shin et al. 2004; Wang et al. 2007; Song and Kim 2009a, b; Kumar et al. 2012a, b, c, d, e, f; Jha et al. 2009a, b, c, d; Prasad et al. 2016). So the interface of nanotechnology in combination with biotechnology is arising to produce nanobiotechnology and its tremendous applications have been found in the field of medical science to engineering science and daily life. It is observed that there are several synthesis procedures for the preparation of ultrafine nanoparticles and few of them have been emerged with time. Generally there are two approaches involved in the syntheses of nanoparticles, a bottom-up approach (self-assembly) and a top-down approach. In bottom-up approaches, the nanoparticles can be synthesized by self-assembly of atom by atom, molecule by molecule and this is mostly depending on Gibbs free energy and/or follows the laws of thermodynamics, so that synthesized nanoparticles are in a state closure to a thermodynamic equilibrium state. In top-down approaches, nanoparticles can be synthesized by the use of conventional methods. A conventional method of nanosynthesis contains physical and chemical methods. So that nanosynthesis can be accomplished through several physical, chemical, biological, and amalgamated techniques (methods) such as extensive ball milling, mechanical grinding condensation or co-precipitation, sol-gel method, drawing glassy materials, biological fabrication and template based methods like materials around/within templates, etc. (Fig. 8.1). Synthesis of nanoparticles is one of the most important aims of nanotechnology with well-defined sizes, shapes, and controlled monodispersity. The most common method of nanosynthesis is physical and chemical methods; each of these methods synthesizes a mixture of nanoparticles with poor morphology which are either capital and/or labor, energy intensive, time consuming, employ hazardous chemicals such as strong reducing agents, polymer capping agents and surfactants, and also crave high temperature for reaction. The use of such toxic and perilous chemicals is responsible for various biological risks (Jia et al. 2009, Jha and Prasad 2010a, b, c; Chandran et al. 2006; Prasad 2014;

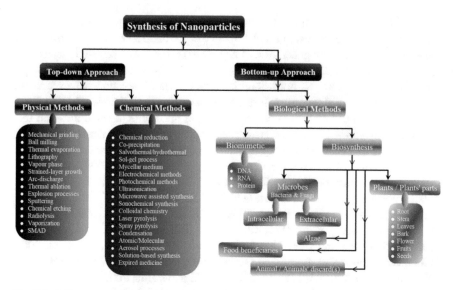

Fig. 8.1 Different synthetic routes for nanomaterials

Song et al. 2003, 2010; Xie et al. 2007; Sinha et al. 2009; Jha and Prasad 2011; Li et al. 2007; Kumar et al. 2011a, b, c; Prasad 2014; Prasad et al. 2016). Therefore, it is a demand of recent research approaches to develop a reliable, high yield, nontoxic, cost-effective, clean, less time consuming, eco-friendly method, which occurs at ambient conditions for the nanosynthesis. The biologically inspired nanosynthesis is evolving into an important branch of nanotechnology. Thus the biologically inspired nanosynthesis has been termed as biogenic synthesis/green synthesis/biological synthesis/biochemical synthesis/biomineralization in which a wide range of biological resources has been used for nanosynthesis like microorganisms (bacteria, yeast, fungi, algae, and viruses) (Prasad et al. 2016), plants, plant parts, animals, animal debris, expired medicine, etc. which studied under nanobiotechnology. The microbe-based protocols have been developed by several authors but the major drawback of microbe-based nanosynthesis is the obligatory constraint of aseptic conditions which requires longer time, trained staff, and the scaling-up cost. The plant-mediated nanosynthesis reduces the metal ions of corresponding nanoparticles in short time at low cost with easy availability of plant in nature which make them more preferred biological resources as compared to microbes. So the use of plant extracts for nanosynthesis is potentially advantageous which controls morphology and provides natural capping agents for the stabilization of nanoparticles. A lot of literature survey has been reported during the last two decades on plant-mediated nanosynthesis like silver, gold, platinum, iron, copper, copper oxide, zinc, zinc oxide, iron oxide, palladium, etc. because plant extracts have antioxidant or reducing properties which are responsible for them. In recent years, nanobiotechnology has originated as an upcoming field for the nanosynthesis. The growth of ecologically benign, "green" synthesis cues is in consonance with the recent RoHS and WEEE legislation stipulated by the EU. It has been well known that living cells are the best examples of machines that operate at the nano level and perform a number of jobs ranging from generation of energy to extraction of targeted materials at very high efficiency. The ribosome, histone protein, DNA, chromatin, Golgi apparatus, interior structure of mitochondria, photosynthetic reaction center, and the fabulous ATPase are all nanostructures, which work quite efficiently. The cell factories have proved to be encouraging tools for the upcoming technologies and biomedical applications. Plant extracts have remarkable advantage in terms of metabolic flux (primary as well as secondary metabolites) which is good source of oxidizing/reducing agents for nanosynthesis (Leela and Vivekanandan 2008; Singh et al. 2011a, b; Jha and Prasad 2010b, 2012a, b, 2013; Agnihotri et al. 2009; Ahmad et al. 2002; Niraimathi et al. 2013; Kumar et al. 2008, 2012a, b, c, d, e, f; Song et al. 2009; Dubey et al. 2009; Dameron et al. 1989; Dameron and Winge 1990; Dean et al. 1997; Krishnamurthy et al. 2011; Prathna et al. 2011; Ahmad et al. 2003; Ankamwar et al. 2005; Fortin and Beveridge 2000; Ghosh et al. 2012a, b; Jha et al. 2008, 2009a; Krolikowska et al. 2003; Li et al. 1999; Philip 2009; Wang et al. 2009; Prasad and Jha 2009, 2010; Prasad 2014; Sanghi and Verma 2009; Saha et al. 2010; Zhu et al. 2011; Tripathy et al. 2010; Pandey et al. 2012; Jobitha et al. 2012; Karuppiah and Rajmohan 2013; Ali 2011; Babu and Gunasekaran 2009; Marchiol 2012). This chapter will focus on how nature and material science can work together to create a "green" way of synthesizing metal/oxide/chalcogenide nanoparticles for a wide ambit of its application in different industries.

8.2 Bionanotechnology

Bionanotechnology is considered as one of the upcoming branches of nanotechnology in which biological sources can effectively be used for the synthesis of a plethora of nanoparticles of desired shape, size, and morphology. The nanosynthesis involving fabrication of different metal, oxide, or chalcogenide nanoparticles apparently this is the cumulative response of the biological system which being taken in to use and its immediate chemical ambience along with basic metabolic fluxes, metabolite content and signal transduction of the organism being engaged for the purpose. Plants are generally multicellular, eukaryotic, autotrophic organisms having cellulose containing cell walls and reproducing mostly sexually but asexual reproduction can also be common. Plants are characterized by the presence of green photosynthesizing unit chloroplasts. They have got naturally bestowed property to synthesize the complex molecules like carbohydrates, proteins, lipids as primary metabolites alkaloids, benzenoids, coumarins, flavonoids, lignans, lipid, sterol, tannins, triterpene, saponins, glycosides storage as secondary metabolites and these compounds are known to exhibit the physiological activities and medicinal properties (Tripathy et al. 2010; Bawankar et al. 2014; Ahmed and Hussain 2013; Salem et al. 2015; Premanathan et al.

2011; Prasad 2014; Azam et al. 2012; Gordon et al. 2011; Koli 2015; Namratha and Monica 2013). The mechanism of reduction of metals could be due to the conjugated oxidation-reduction reactions in which release of electrons are due to NADPH-dependent oxidoreductase. Most of the plant extract contain oxidoreductases, quinone, and anthraquinone which have played an important role in nanofabrication (Prasad et al. 2016). Presence of these metabolites in mostly all plant extract triggers a redox reaction due to tautomerization (keto to enol) leading to nanoparticles synthesis. The reduction of metal ions can also occur by means of the electrons carrier from NADPH to NADPH-dependent oxidoreductase. The oxidoreductases are pH sensitive and work in substitute manner. Oxidases get activated during lower value of pH, whereas reductases activated during higher value of pH. The expected mechanism for the synthesis of metal and metal oxide nanoparticles is illustrated in Fig. 8.2. The reduction potential of the plant broth seems to decide the fate of metal ions exposed. A high value of reduction potential at higher pH (>7) values is required for the formation of metal NPs while oxide NPs are resulted due to lower reduction potential at lower pH (< 7). The synthesis of plant-mediated metal/oxide nanoparticles and its conformation by different instruments are represented in form of flowchart in Fig. 8.3. It is well noted that pH value along with other parameters like kinetics, agitation rate, incubation time, temperature, reductant concentration, mixing ratio, and solution chemistry are deciding the moirai of metal ions to bio-reductant of plant extracts. So that protein has played an important role in the reduction of metal ions into its corresponding metal/oxide/chalcogenide nanoparticles and its stabilization. The formation of metal/oxide nanoparticles could be conformed using UV-Spectroscopy technique, afterwards these nanoparticles may be taken into use for further study (Shankar et al. 2004a, b; Kora et al. 2010; Jha and Prasad 2011, 2014, 2015, 2016a, b, c; Rao and Paria 2013).

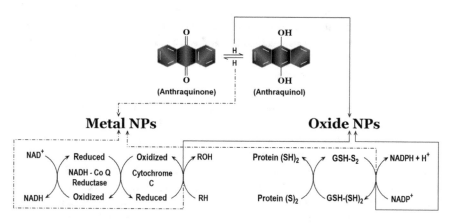

Fig. 8.2 Probable mechanism of plant-mediated nanosynthesis

Fig. 8.3 Flow diagram for the conformation of plant-mediated nanosynthesis

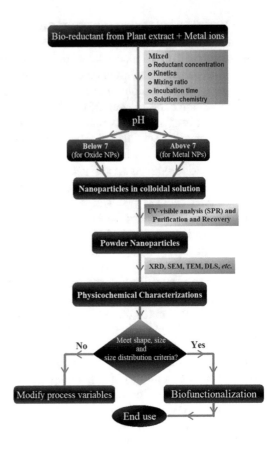

8.3 Methodology

A large number of plants have been described to facilitate nanosynthesis, which are discussed briefly in Table 8.1. The protocol of nanosyntheses involves: the collection of plant/plant parts from the available sites and washed thoroughly twice/thrice with tap water followed by distilled water to remove associated debris. Prior to the start of experiment, plant/plant parts like root, stem, bark, leaf, flower, fruit, seed, etc. was rinsed thoroughly by de-ionized water. Plant extract was prepared by taking defined amount (15 gm) of plant part thoroughly washed, dried, cut into fine pieces and mixed with 100 mL of 50% ethanol in a 250 mL conical flask and mixture was boiled for 20 min till the color changes from clear transparent to light green. The plant mass has been pressed by wrapping in serene cloth and 50 mL extract was collected under laminar flow which treated as source extract. A finite small portion of it (25 mL) was filtered and two times diluted by mixing distilled water. Now, 20 mL of 0.025(M) metal ion solution was mixed to the plant extracts solution. The mixed solution was heated on steam bath up to 60 °C for 10–20 min until appearance of deposition at the bottom of the flask, indicative of the inception of nano-transformation. Value of pH

Table 8.1 List of plant/plant parts used for biosynthesis of different nanoparticles

Plant	Plant parts	Nanoparticles	References
Aegle marmelos	Leaf	Ag	Jha and Prasad (2011)
Annona squamosal	Leaf	Ag	Song and Kim (2009a, b), Kumar et al. (2012a, b, c, d, e, f)
Argemone mexicana	Leaf	Ag	Singh et al. (2010a, b), Jha and Prasad (2014)
Acalypha indica	Leaf, seed, bark	Ag	Krishnaraj et al. (2010), Rao and Paria (2013)
Alternanthera sessilis	Leaf	Ag	Niraimathi et al. (2013)
Andrographis paniculata	Leaf	Ag	Suriyakalaa et al. (2013)
Artemisia nilagirica	Leaf	Ag	Song and Kim (2009a, b)
Artemisia nilagirica	Whole plant	Ag	Kumar et al. (2013a, b)
Azolla sp.	Leaf	Ag	Jha and Prasad (2016a, b, c)
Azadirachta indica	Leaf, seed, bark	Ag, Au, ZnO	Shankar et al. (2004a, b), Tripathy et al. (2010), Babu and Gunasekaran (2009), Ramezani et al. (2008), Thirumurugan et al. (2010), Bhuyan et al. (2015)
Aegle marmelos	Leaf	Au	Rao and Paria (2013)
Alfalfa	Leaf	Au	Montes et al. (2011)
Ananas comosus	Leaf	Au	Basavegowda et al. (2013)
Allium sativum	Bulb	Au	Rafiee et al. (2012)
Allium cepa	Bulb	Ag	Saxena et al. (2010)
Achillea biebersteinii	Leaf	Ag	Baharara et al. (2014, 2015)
Arbutus Unedo	Leaf	Ag	Kouvaris et al. (2012)
Aloe vera	Leaf	Ag, Au, Zn/ZnO, Copper Oxide, In_2O_3, TiO_2, Fe, Mg doped ZnO	Chandran et al. (2006), Kathirvelu et al. (2009), Bhuyan et al. (2015), Maensiri et al. (2008), Medda et al. (2015), Muralikrishna et al. (2014), Kumar et al. (2015), Nithya et al. (2013), Rao et al. (2015), Yadav et al. (2016)
Aloe barbadensis miller	Leaf	Zn/ZnO	Sangeetha et al. (2011)
Banana	Peel	Au	Bankar et al. (2010a, b)
Barbated Skullcup	Leaf	Au	Wang et al. (2009)
Benincasa hispida	Seed	Au	Aromal and Philip (2012a, b)
Beta vulgaris	Sugar beet pulp	Au	Castro et al. (2011)
Bryophyllum sp.	Leaf	Ag	Jha et al. (2009a, b, c, d)
Boswellia serrata	Leaf	Ag	Kora et al. (2012)
Brevibacterium casei	Leaf	Ag	Kalishwaralal et al. (2010)

(continued)

Table 8.1 (continued)

Plant	Plant parts	Nanoparticles	References
Boswellia ovalifoliolata	Stem, bark	Ag	Savithramma et al. (2011)
Basella alba	Leaf	Ag	Leela and Vivekanandan (2008)
Brevibacterium casei	Leaf	Au	Yeary et al. (2005)
Calotropis procera	Leaf	Zn/ZnO	Singh et al. (2011a, b)
Catharanthus roseus	Leaf	TiO$_2$	Velayutham et al. (2012)
Cycas circinalis	Leaf	Ag	Johnson and Prabu (2015), Ali (2011)
Commelina benghalensis	Leaf	Ag	Johnson and Prabu (2015)
Citrus sinensis	Peel	Ag	Konwarh et al. (2011), Zahir and Rahuman (2012)
Citrus limon (lemon)	Juice	Ag	Prathna et al. (2011)
Cynodon dactylon	Leaf	Ag	Sahu (2013)
Cardiospermum halicacabum	Leaf	Ag	Mitra (2012)
Cinnamon zeylanicum	Bark	Ag	Kumar et al. (2009a, b, c)
Cycas	Leaf	Ag	Jha and Prasad (2010a, b, c)
Capsicum annum L.	Fruit	Ag	Jha and Prasad (2011), Li et al. (2007)
Carica papaya	Fruit	Ag	Jha and Prasad (2015), Jain et al. (2009)
Curcuma longa	Tuber power	Ag	Kumar et al. (2011a, b, c)
Cyperus sp.	Leaf	Ag	Jha et al. (2009a, b, c, d)
Coleus aromaticus	Leaf	Ag	Vanaja and Annadurai (2013)
Calotropis procera	Leaf	Ag	Mohamed et al. (2014)
Citrullus colocynthis	Calli	Ag	Satyavani et al. (2011a, b)
Clerodendron infortunatum	Leaf	Ag	Jha and Prasad (2016a, b, c)
Citrullus colocynthis	Leaf	Ag	Satyavani et al. (2011a, b)
Cassia auriculata	Leaf	Ag	Udayasoorian et al. (2011)
Cissus quadrangularis	Leaf	Ag	Kumar et al. (2012a, b, c, d, e, f), Valli and Vaseeharan (2012)
Camellia sinensis	Leaf	Ag, ZnO	Nestor et al. (2008)
Cocos nucifera	Coir	Ag	Roopan et al. (2013)
Cinnamomum camphora	Bark, leaf	Ag	Huang et al. (2007)
Catharanthus roseus	Leaf	Ag	Kotakadi et al. (2013)
Cochlospermum gossypium	Leaf	Ag	Kora et al. (2010), Vinod et al. (2011)
Coccinia grandis	Leaf	Ag	Arunachalam et al. (2012)
Chenopodium album	Leaf	Ag	Dwivedi and Gopal (2010)
Crossandra infundibuliformis	Leaf	Ag	Kaviya et al. (2012)

(continued)

Table 8.1 (continued)

Plant	Plant parts	Nanoparticles	References
Cymbopogon flexuosus	Leaf	Au	Shankar et al. (2004a, b)
Cicer arietinum	Bean	Au	Ghule et al. (2006)
Camellia sinensis	Leaf	Au	Sharma et al. (2007)
Coriandrum sativum	Leaf	Au	Narayanan and Sakthivel (2008), Boruah et al. (2012)
Cinnamomum zeylanicum	Bark	Au	Smitha et al. (2009)
Cuminum cyminum	Seed	Au	Krishnamurthy et al. (2011)
Cassia fistula	Stem	Au	Daisy and Saipriya (2012)
Cinnamomum camphora	Whole plant	Au	Ahmad et al. (2003)
Cochlospermum gossypium	Leaf	Au, Pt	Agnihotri et al. (2009)
Cypress	Leaf	Au	Noruzi et al. (2012)
Crocus sativus (saffron)	Leaf	Au	Kumar et al. (2011a, b, c)
Chenopodium album	Leaf	Au	Satyavani et al. (2011a, b)
Cassia auriculata	Leaf	Au	Kumar et al. (2011a, b, c)
Cacumen Platycladi	Leaf	Au	Zhan et al. (2011)
Carissa carandas	Berries	Ag	Joshi et al. (2018)
Centella asiatica	Leaf	Au	Das et al. (2010)
Calotropis procera L.	Latex	Cu	Harne et al. (2012)
Cinnamomum zeylanicum	Bark	Pd	Kumar et al. (2009a, b, c)
Curcuma longa	Tuber	Pd	Kumar et al. (2009a, b, c)
Cinnamomum camphora	Leaf	Pd	Yang et al. (2010)
Diospyros kaki	Leaf	Pt	Song et al. (2010)
Diospyros kaki	Leaf	Ag	Song and Kim (2009a, b)
Dillenia indica	Fruit	Ag	Singh et al. (2013)
Dioscorea bulbifera	Tuber	Ag	Ghosh et al. (2012a, b)
Dalbergia sissoo	Leaf	Ag	Singh et al. (2012)
Desmodium triflorum	Leaf	Ag	Ahmad et al. (2011)
Diospyrus kaki	Leaf	Au	Song et al. (2009)
Dalbergia sissoo	Leaf	Au	Patil et al. (2012a, b)
Eucalyptus camaldulensis	Leaf	Au	Ramezani et al. (2008)
Eclipta	Leaf	Ag	Jha et al. (2009a, b, c, d)
Eclipta prostrata	Leaf	TiO_2	Kumar et al. (2012a, b, c, d, e, f)

(continued)

Table 8.1 (continued)

Plant	Plant parts	Nanoparticles	References
Euphorbia hirta	Leaf	Ag	Elumalai et al. (2010), Priyadarshini et al. (2012)
Eucalyptus sp.	Leaf	Ag	Jha and Prasad (2012a, b)
Euphorbia prostrata	Leaf	Ag	Zahir and Rahuman (2012)
Emblica officinalis	Fruit	Ag, Au	Ankamwar et al. (2005), Jha et al. (2009a, b, c, d)
Eucalyptus hybrida (safeda)	Leaf	Ag	Dubey et al. (2009)
Eclipta prostrate	Leaf	Ag	Kumar and Rahuman (2011)
Elettaria Cardamomum	Leaf	Ag	Jobitha et al. (2012)
Ficus amplissima	Fruit, Flower	Ag	Johnson and Prabu (2015)
Ficus benghalensis	Fruit, Flower	Ag	Saxena et al. (2012)
Ginkgo biloba	Leaf	Ag	Song and Kim (2009a, b)
Gelsemium sempervirens	Leaf	Ag	Das et al. (2013a, b)
Gracilaria corticata	Leaf	Ag	Kumar et al. (2013a, b)
Guar gum	Latex	Au	Pandey et al. (2013)
Gluconacetobacter xylinum	Leaf	Iron/iron oxide	Zhu et al. (2011)
Gardenia jasminoides	Leaf	Pd	Jia et al. (2009)
Glycine max	Leaf	Pd	Petla et al. (2012)
Goat	Slaughter waste	ZnO	Jha and Prasad (2016a, b, c)
Helianthus annuus	Leaf	Ag	Leela and Vivekanandan (2008)
Hydrilla sp.	Whole plant	Ag	Jha et al. (2009a, b, c, d)
Hydrastis canadensis	Whole plant	Ag	Das et al. (2013a, b)
Hevea brasiliensis	Whole plant	Ag	Guidelli et al. (2011)
Henna	Whole plant	Ag, Au	Kasthuri et al. (2009), Banerjee and Narendhirakannan (2011)
Hibiscus rosa sinensis	Flower and leaf	Ag, Au	Singh et al. (2013)
Hibiscus sabdariffa	Leaf	ZnO	Bala et al. (2015)
Honey		Ag, Au, Pt	Kumar et al. (2009a, b, c), Philip (2009), Soundarrajan et al. (2012)
Ixora coccinea	Whole plant	Ag	Karuppiah and Rajmohan (2013)
Iresine herbstii	Whole plant	Ag	Dipankar and Murugan (2012)
Jatropha curcas	Latex	Ag	Bar et al. (2009a)
Jatropha curcas	Seed	Ag	Bar et al. (2009b)

(continued)

Table 8.1 (continued)

Plant	Plant parts	Nanoparticles	References
Justicia gendarussa burm	Leaf	Au	Fazaludeen et al. (2012)
Jatropha curcas	Whole plant	TiO_2	Hudlikar et al. (2012)
Lippia nodiflora	Whole plant	Ag	Johnson and Prabu (2015)
Lippia citriodora	Whole plant	Ag	Cruz et al. (2010)
Loquat	Whole plant	Ag	Awwad et al. (2013)
Lonicera japonica L.	Whole plant	Ag	Kumar and Yadav (2011)
Lonicera japonica L.	Whole plant	Au	Ali (2011)
Lemon grass (*Cymbopogon flexuosus*)	Juice	Au	Shankar et al. (2005), Singh et al. (2011a, b)
Musa paradisiac	Peeled banana	Pd	Bankar et al. (2010a, b)
Mangosteen	Whole plant	Ag	Veerasamy (2011)
Mentha piperita	Leaf	Au	Prasad and Jha (2009)
Momordica charantia	Whole plant	Au	Pandey et al. (2012)
Moringa oleifera	Whole plant	Ag	Prasad and Elumalai (2011), Mubayi et al. (2012)
Magnolia Kobus	Whole plant	Ag, Au	Song and Kim (2009a, b), Song et al. (2009)
Mentha sp.	Leaf	Ag	Jha and Prasad (2012a, b)
Melia azedarach	Whole plant	Ag	Sukirtha et al. (2012)
Mukia scabrella	Whole plant	Ag	Prabakar et al. (2013)
Memecylon edule	Whole plant	Ag, Au	Elavazhagan and Arunachalam (2011), Jha et al. (2007)
Mentha piperita	Whole plant	Ag	Ali et al. (2011)
Murraya koenigii	Leaf	Ag, Au, ZnO	Philip et al. (2011), Jha et al. (2009a, b, c, d), Divyapriya et al. (2014)
Mangifera indica	Leaf, fruit	Ag, Au	Phillip (2011), Phillip (2010)
Mirabilis jalapa	Flowers	Au	Vankar and Bajpai (2010)
Manilkara zapota	Leaf	Ag	Kumar and Rahuman (2012)
Medicago sativa	Seed	Ag	Lukman et al. (2011)
Macrotyloma uniflorum	Leaf	Ag, Au	Vidhu et al. (2011), Aromal et al. (2012)
Malva parviflora	Leaf	Ag	Zayed et al. (2012)
Madhuca longifolia	Leaf	Au	Sharma et al. (2007)
Nicotiana tabacum	Leaf	Ag	Prasad et al. (2011)
Nyctanthes arbortristis	Flower	Au	Das et al. (2011)
Ocimum tenuiflorum	Leaf	Ag	Patil et al. (2012a, b)
Oryza sativa		Ag	Leela and Vivekanandan (2008)
Ocimum sanctum	Root, stem	Ag	Ahmad et al. (2010a, b)

(continued)

Table 8.1 (continued)

Plant	Plant parts	Nanoparticles	References
Ocimum basilicum	Leaf	Au	Singhal et al. (2012)
Ocimum sp.	Leaf	Ag	Jha and Prasad (2012a, b)
Orange juice	Fruit	Ag, Zn/ZnO	Jha et al. (2011), Jha and Prasad (2015)
Ocimum sanctum	Leaf	Ag, Au, Pt	Singhal et al. (2011), Philip and Unni (2011), Rao et al. (2013), Mohamed et al. (2014), Soundarrajan et al. (2012)
Pelargonium, roseum	Leaf	Au	Ramezani et al. (2008)
Psidium guajava	Leaf	Au	Raghunandan et al. (2010)
Piper betle	Leaf	Au	Punuri et al. (2012), Krishnamurthy et al. (2010)
Punica granatum	Peels	Au	Sharma and Singh (2013)
Phyla nodiflora Linn.	Whole plant		Sharma and Singh (2013)
Panicum virgatum	Leaf		Mason et al. (2012)
Pelargonium graveolens	Leaf	Ag, Au	Leela and Vivekanandan (2008), Shankar et al. (2003a, b)
Pinus densiflora	Leaf	Ag	Song and Kim (2009a, b)
Platanus orientalis	Leaf	Ag	Song and Kim (2009a, b)
Piper betle	Leaf	Ag	Mallikarjuna et al. (2012), Rani and Reddy (2011)
Phytolacca decandra	Leaf	Ag	Das et al. (2013a, b)
Pergularia daemia	Plant latex	Ag	Patil et al. (2012a, b)
Papaver somniferum	Leaf	Ag	Raghavan et al. (2012)
Pithecellobium dulce	Leaf	Ag	Raman et al. (2012)
Punica granatum	Peels	Ag, ZnO	Ahmad et al. (2012), Mishra and Sharma (2015)
Piper betle (Paan)	Leaf	Ag	Khan et al. (2012)
Pandanus odorifer (Forssk.)	Leaf	Ag	Panda et al. (2011)
Pinus resinosa	Bark	Pd, Pt	Coccia et al. (2012), Ingle et al. (2009)
Parthenium hysterophorus	Leaf	Zn/ZnO	Rajiv et al. (2013), Sindhura et al. (2014)
Polygala tenuifolia	Leaf	Zn/ZnO	Nagajyothi et al. (2015)
Rice husk	Dry powder	SiO_2	Rafiee et al. (2012)
Rosa sp.	Leaf, flower	Ag	Jha and Prasad (2013)
Rosa damascene	Leaf, flower	Ag, Au	Ghoreishi et al. (2011), Shankar et al. (2003a, b)
Rhizophora mucronata	Leaf	Ag	Umashankari et al. (2012)
Rosa hybrid	Petal	Au	Noruzi et al. (2011)
Syzygium aromaticum	Whole plant	Au	Deshpande et al. (2010)
Santalum album	Leaf	Ag	Swamy and Prasad (2012)

(continued)

Table 8.1 (continued)

Plant	Plant parts	Nanoparticles	References
Stevia rebaudiana	Whole plant	Au	Mishra et al. (2010)
Sphearanthus amaranthoids	Whole plant	Au	Nellore et al. (2012)
Sorbus aucuparia	Whole plant	Au	Phillip (2011)
S. torvum (Solanaceae)	Whole plant	Au	Govindaraju et al. (2011)
Saccharum officinarum	Whole plant	Ag	Leela and Vivekanandan (2008)
Sorghum bicolor Zea mays	Seed	Ag	Leela and Vivekanandan (2008)
Syzygium cumini	Whole plant	Ag	Kumar et al. (2010a, b), Kumar and Yadav (2012)
Syzygium cumini	Seed	Ag	Kumar et al. (2010a, b)
Syzygium cumini	Seed	Ag	Banerjee and Narendhirakannan (2011)
Syzygium cumini	Leaf	Ag	Prasad et al. (2012)
Syzygium cumini	Bark	Ag	Prasad and Swamy (2013)
Svensonia hyderabadensis	Whole plant	Ag	Rao and Savithramma (2011)
Shorea tumbuggaia	Stem bark	Ag	Savithramma et al. (2011)
Solanum lycopersicum	Fruit	Ag	Bhattacharyya et al. (2016)
Sorghum spp.	Bran powder	Ag	Njagi et al. (2010)
Spirulina platensis	Whole plant	Ag	Mahdieh et al. (2012)
Sesbania grandiflora	Whole plant	Ag	Das et al. (2010)
Sesuvium portulacastrum L.	Whole plant	Ag	Nabikhan et al. (2010)
Sorbus aucuparia	Whole plant	Ag	Dubey et al. (2010a, b)
Solanum nigrum	Whole plant	Zn/ZnO	Ramesh et al. (2015)
Sorghum spp.	Bran powder	Iron/iron oxide	Njagi et al. (2011)
Tanacetum vulgare	Fruit	Au	Jha et al. (2009a, b, c, d)
Terminalia catappa	Leaf	Au	Ankamwar (2010)
Trigonella foneugraecum	Leaf	Au	Aromal and Philip (2012a, b)
Terminalia chebula	Leaf	Au	Kumar et al. (2012a, b, c, d, e, f)
Tea polyphenols	Leaf	Ag	Moulton et al. (2010)
Tinospora cordifolia	Leaf	Ag	Jayaseelan et al. (2011)
Thuja occidentalis	Leaf	Ag	Das et al. (2013a, b)
Tanacetum vulgare	Fruit	Ag	Kumar et al. (2012a, b, c, d, e, f)
Terminalia chebula	Leaf	Ag	Dubey et al. (2010a, b)
Trachyspermum ammi	Leaf	Ag	Raghavan et al. (2012)

(continued)

Table 8.1 (continued)

Plant	Plant parts	Nanoparticles	References
Trichodesma indicum	Leaf	Ag	Buhroo et al. (2017)
Trianthema decandra	Leaf	Ag	Geethalakshmi and Sarada (2010)
Tribulus terrestris L.	Fruit	Ag	Gopinath et al. (2012)
Withania somnifera	Leaf	Ag	Nagati et al. (2012)
Wood	Stem	Pt	Lin et al. (2011)
Zingiber officinale rhizome	Leaf	Ag	Kumar et al. (2012a, b, c, d, e, f)
Zingiber officinale	Leaf	Ag	Kumar et al. (2012a, b, c, d, e, f), Singh et al. (2010a, b)
Zingiber officinale	Leaf	Au	Jha and Prasad (2010a, b, c), Kumar et al. (2011a, b, c)

of the mixed plant extract solution is suitably adjusted at this stage depending upon targeted task synthesis of a metal or an oxide or chalcogenide. It is allowed to cool further in the laboratory ambience. After overnight, the mixed plant extract solution is noticed to have distinctly marked deposits. It is filtered through a Whatman filter paper and dried for subsequent characterization. The syntheses of nanoparticles were checked by UV-visible spectroscopy.

8.4   Conclusion

This chapter encompasses the various methods for nanoparticle synthesis and with changing pace of time the demand of nanoparticles in different fields is bound to rise so this can very well be met by going for green options like plant based synthesis. These natural plants are wonderful amenable resource for nanofabrication with less time consuming, eco-friendly, nontoxic, and easy protocols for scaling up. Hence, this green synthesis of nanoparticles will result in a significant payoff for the field of medical science.

References

Agnihotri M, Joshi S, Kumar AR, Zinjarde S, Kulkarni S (2009) Biosynthesis of gold nanoparticles by the tropical marine yeast *Yarrowia lipolytica* NCIM 3589. Mater Lett 63:1231–1234

Ahmed M, Hussain F (2013) Chemical composition and biochemical activity of *Aloe vera* (*Aloe barbadensis* Miller) leaves. Int J Chem Biochem Sci 3:29–33

Ahmad A, Mukherjee P, Mandal D, Senapati S, Khan MI, Kumar R, Sastry M (2002) Enzyme mediated extracellular synthesis of CdS nanoparticles by the fungus, *Fusarium oxysporum*. J Am Chem Soc 124:12108–12109

Ahmad A, Mukherjee P, Senapati S, Mandal D, Khan MI, Kumar R, Sastry M (2003) Extracellular biosynthesis of silver nanoparticles using the fungus *Fusarium oxysporum*. Colloids Surf B Biointerfaces 28:313–318

Ahmad N, Sharma S, Singh VN, Shamsi SF, Fatma A, Mehta BR (2010a) Biosynthesis of silver nanoparticles from *Desmodium triflorum*: A novel approach towards weed utilization. Biotechnol Res Int 2011:1–8

Ahmad N, Sharma S, Alam MK, Singh VN, Shamsi SF, Mehta BR, Fatma A (2010b) Rapid synthesis of silver nanoparticles using dried medicinal plant of basil. Colloids Surf B: Biointerfaces 81:81–86

Ahmad N, Sharma S, Rai R (2012) Rapid green synthesis of silver and gold nanoparticles using peels of *Punica granatum*. Adv Mater Lett 3:376–380

Ahmad MK, Ansari BA (2011) Toxicity of neem based pesticide azacel to the embryo and fingerlings of zebra fish *Danio rerio (Cyprinidae)*. World J Zool 6:47–51

Ali DM, Thajuddin N, Jeganathan K, Gunasekhran M (2011) Plant extract mediated synthesis of silver and gold nanoparticles and its antibacterial activity against clinically isolated pathogens. Colloids Surf B: Biointerfaces 85:360–365

Ali SA (2011) Antimicrobial studies of aqueous extract of the leaves of *Lophira lanceolata*. Res J Pharm Biol Chem Sci 2:637–643

Ankamwar B, Damle C, Ahmad A, Satry M (2005) Biosynthesis of gold and silver nanoparticles using *Emblica Officinalis* fruit extract, their phase transfer and transmetallation in an organic solution. J Nanosci Nanotechnol 5:1665–1671

Ankamwar B (2010) Biosynthesis of gold nanoparticles (green-gold) using leaf extract of *Terminalia Catappa*. E-J Chem 7:1334–1339

Aromal SA, Philip D (2012a) *Benincasa hispida* seed mediated green synthesis of gold nanoparticles and its optical nonlinearity. Phys E 44:1329–1334

Aromal SA, Vidhu VK, Philip D (2012) Green synthesis of well-dispersed gold nanoparticles using *Macrotyloma uniflorum*. Spectrochim Acta A Mol Biomol Spectrosc 85:99–104

Aromal SA, Philip D (2012b) Green synthesis of gold nanoparticles using *Trigonella foenum-graecum* and its size-dependent catalytic activity. Spectrochim Acta A Mol Biomol Spectrosc 97:1–5

Arunachalam R, Singh SD, Kalimuthu B, Uthirappan M, Rose C, Mandal AB (2012) Phytosynthesis of silver nanoparticles using *Coccinia grandis* leaf extract and its application in the photocatalytic degradation. Colloids Surf B: Biointerfaces 94:226–230

Awwad AM, Salem NM, Abdeen AO (2013) Biosynthesis of silver nanoparticles using Loquat leaf extract and its antibacterial activity. Adv Mater Lett 4:338–342

Azam A, Arham SA, Oves M, Khan MS, Habib SS, Memic A (2012) Antimicrobial activity of metal oxide nanoparticles against Gram-positive and Gram-negative bacteria: a comparative study. Int J Nanomed 7:6003–6009

Babu MMG, Gunasekaran P (2009) Production and structural characterization of crystalline silver nanoparticles from *Bacillus cereus* isolate. Colloids Surf A 74:191–195

Baharara J, Namvar F, Ramezani T, Mousavi M, Mohamad R (2014) Green synthesis of silver nanoparticles using *Achillea biebersteinii* flower extract and its anti-angiogenic properties in the rat aortic ring model. Molecules 19:4624–4634

Baharara J, Namvar F, Ramezani T, Mousavi M, Mohamad R (2015) Silver nanoparticles biosynthesized using *Achillea biebersteinii* flower extract: apoptosis induction in MCF-7 cells via caspase activation and regulation of Bax and Bcl-2 gene expression. Molecules 20:2693–2706

Bala N, Saha S, Chakraborty M, Maiti M, Das S, Basub R, Nandy P (2015) Green synthesis of zinc oxide nanoparticles using *Hibiscus subdariffa* leaf extract: effect of temperature on synthesis, anti-bacterial activity and anti-diabetic activity. RSC Adv 5:4993–5003

Banerjee J, Narendhirakannan RT (2011) Biosynthesis of silver nanoparticles from *Syzygium cumini* (L.) seed extract and evaluation of their in vitro antioxidant activities. Dig J Nanomater Biostruct 6:961–968

Bankar A, Joshi B, Kumar AR, Zinjarde S (2010a) Banana peel extract mediated synthesis of gold nanoparticles. Colloids Surf B: Biointerfaces 80:45–50

Bankar A, Joshi B, Kumar AR, Zinjarde S (2010b) Banana peel extract mediated novel route for the synthesis of palladium nanoparticles. Mater Lett 64:1951–1953

Bar H, Bhui DK, Sahoo GP, Sarkar P, De SP, Misra A (2009a) Green synthesis of silver nanoparticles using latex of *Jatropha curcas*. Colloids Surf A Physicochem Eng Asp 339:134–139

Bar H, Bhui DK, Sahoo GP, Sarkar P, Pyne S, Misra A (2009b) Green synthesis of silver nanoparticles using seed extract of *Jatropha curcas*. Colloids Surf A Physicochem Eng Asp 348:212–216

Basavegowda N, Kupiec AS, Malina D, Yathirajan HS, Keerthi VR, Chandrashekar N, Salman D, Liny P (2013) Plant mediated synthesis of gold nanoparticles using fruit extracts of *Ananas comosus* (L.) (Pine apple) and evaluation of biological activities. Adv Mater Lett 4:332–337

Bawankar R, Singh P, Babu S (2014) Bioactive compounds and medicinal properties of *Aloe vera* L.: an update. J Plant Sci 2:102–107

Bhattacharyya A, Prasad R, Buhroo AA, Duraisamy P, Yousuf I, Umadevi M, Bindhu MR, Govindarajan M, Khanday AL (2016) One-pot fabrication and characterization of silver nanoparticles using Solanum lycopersicum: an eco-friendly and potent control tool against Rose Aphid, Macrosiphum rosae. J Nanosci 2016:4679410. https://doi.org/10.1155/2016/4679410

Bhuyan T, Mishra K, Khanuja M, Prasad R, Varma A (2015) Biosynthesis of zinc oxide nanoparticles from *Azadirachta indica* for antibacterial and photocatalytic applications. Mater Sci Semicond Process 32:55–61

Boruah SK, Boruah PK, Sharma P, Medhi C, Medhi OK (2012) Green synthesis of god nanoparticles using *Camellia sinensis* and kinetics of the reaction. Adv Mater Lett 3:481–486

Buhroo AA, Nisa G, Asrafuzzaman S, Prasad R, Rasheed R, Bhattacharyya A (2017) Biogenic silver nanoparticles from *Trichodesma indicum* aqueous leaf extract against Mythimna separata and evaluation of its larvicidal efficacy. J Plant Protect Res 57(2):194–200. https://doi.org/10.1515/jppr-2017-0026

Castro L, Blazquez ML, Munoz JA, Gonzalez F, Balboa CG, Ballester A (2011) Biosynthesis of gold nanowires using sugar-beet pulp. Process Biochem 46:1076–1082

Castro L, Blázquez ML, González F, Munoz JA, Ballester A (2010) Extracellular biosynthesis of gold nanoparticles using sugar beet pulp. Chem Eng J 164:92–97

Chandran SP, Chaudhary M, Pasricha R, Ahmad A, Sastry M (2006) Synthesis of gold nanotriangles and silver nanoparticles using *Aloe vera* plant extract. Biotechnol Prog 22:577–583

Coccia F, Tonucci L, Bosco D, Bressan M, Alessandro N (2012) One-pot synthesis of lignin-stabilised platinum and palladium nanoparticles and their catalytic behaviour in oxidation and reduction reactions. Green Chem 14:1073–1078

Cruz D, Fale PL, Mourato A, Vaz PD, Serralheiro ML, Lino ARL (2010) Preparation and physicochemical characterization of Ag nanoparticles biosynthesized by *Lippia citriodora* (Lemon Verbena). Colloids Surf B Biointerfaces 81:67–73

Daisy P, Saipriya K (2012) Biochemical analysis of *Cassia fistula* aqueous extract and phytochemically synthesized gold nanoparticles as hypoglycemic treatment for diabetes mellitus. Int J Nanomed 7:1189–1202

Dameron CT, Reese RN, Mehra RK, Kortan AR, Carroll PJ, Steigerwald ML, Brus LE, Winge DR (1989) Biosynthesis of cadmium sulphide quantum semiconductor crystallites. Nature 338:596–597

Dameron CT, Winge DR (1990) Peptide mediated formation of quantum semiconductors. Trends Biotechnol 8:3–6

Daniel MC, Astruc D (2004) Gold nanoparticles: assembly, supramolecular chemistry, quantumsizerelated properties, and applications toward biology, catalysis, and nanotechnology. Chem Rev 104:293–346

Das J, Das MP, Velusamy P (2013a) *Sesbania grandiflora* leaf extract mediated green synthesis of antibacterial silver nanoparticles against selected human pathogens. Spectrochim Acta A Mol Biomol Spectrosc 104:265–270

Das RK, Borthakur BB, Bora U (2010) Green synthesis of gold nanoparticles using ethanolic leaf extract of *Centella asiatica*. Mater Lett 64:1445–1447

Das RK, Gogoi N, Bora U (2011) Green synthesis of gold nanoparticles using *Nyctanthes arbortristis* flower extract. Bioprocess Biosyst Eng 34:615–619

Das S, Das J, Samadder A, Bhattacharyya SS, Das D, Khudabukhsh AR (2013b) Biosynthesized silver nanoparticles by ethanolic extracts of *Phytolacca decandra*, *Gelsemium sempervirens*, *Hydrastis canadensis* and *Thuja occidentalis* induce differential cytotoxicity through G2/M arrest in A375 cells. Colloids Surf B: Biointerfaces 101:325–336

Dean RT, Fu S, Stocker R, Davies MJ (1997) Biochemistry and pathology of radical mediated protein oxidation. Biochem J 324:1–18

Deshpande R, Bedre MD, Basavaraja S, Sawle B, Manjunath SY, Venkataraman A (2010) Rapid biosynthesis of irregular shaped gold nanoparticles from macerated aqueous extracellular dried clove buds (*Syzygium aromaticum*) solution. Colloids Surf B: Biointerfaces 79:235–240

Dipankar C, Murugan S (2012) The green synthesis, characterization and evaluation of the biological activities of silver nanoparticles synthesized from *Iresine herbstii* leaf aqueous extracts. Colloids Surf B: Biointerfaces 98:112–119

Divyapriya S, Sowmia C, Sasikala S (2014) Synthesis of zinc oxide nanoparticles and antimicrobial activity of *Murraya koenigii*. World J Pharm Pharm Sci 3:1635–1645

Dubey M, Bhadauria S, Kushwah BS (2009) Green synthesis of nanosilver particles from extract of *eucalyptus hybrida* (safeda) leaf. Dig J Nanomater Biostruct 4:537–543

Dubey SP, Lahtinen M, Sillanpää M (2010a) Tansy fruit mediated greener synthesis of silver and gold nanoparticles. Process Biochem 45:1065–1071

Dubey SP, Lahtinen M, Särkkä H, Sillanpää M (2010b) Bioprospective of Sorbus aucuparia leaf extract in development of silver and gold nanocolloids. Colloids Surf B: Biointerfaces 80:26–33

Dwivedi AD, Gopal K (2010) Biosynthesis of silver and gold nanoparticles using *Chenopodium album* leaf extract. Colloids Surf A Physicochem Eng Asp 369:27–33

Elavazhagan T, Arunachalam KD (2011) *Memecylon edule* leaf extract mediated green synthesis of silver and gold nanoparticles. Int J Nanomed 6:1265–1278

Elumalai EK, Prasad TNVKV, Hemachandran J, Therasa SV, Thirumalai T, David E (2010) Extracellular synthesis of silver nanoparticles using leaves of *Euphorbia hirta* and their antibacterial activities. J Pharm Sci Res 2:549–554

Fayaz AM, Girilal M, Venkatesan R, Kalaichelvan PT (2011) Biosynthesis of anisotropic gold nanoparticles using *Maduca longifolia* extract and their potential in infrared absorption. Colloids Surf B Biointerfaces 88:287–291

Fazaludeen MF, Manickam C, Ibraheem MAA, Ahmed MQ, Beg QZ (2012) Synthesis and characterizations of gold nanoparticles by *Justicia gendarussa burm* F leaf extract. J Microbiol Biotech Res 2:23–34

Flores CY, Miñán AG, Grillo CA, Salvarezza RC, Vericat C, Schilardi PL (2013) Citrate-capped silver nanoparticles showing good bactericidal effect against both planktonic and sessile bacteria and a low cytotoxicity to osteoblastic cells. ACS Appl Mater Interfaces 5:3149–3159

Fortin D, Beveridge TJ (2000) From biology to biotechnology and medical applications. In: Aeuerien E (ed) Biomineralization. Wiley VCH, Weinheim

Geethalakshmi R, Sarada DVL (2010) Synthesis of plant-mediated silver nanoparticles using *Trianthema decandra* extract and evaluation of their anti microbial activities. Int J Eng Sci Technol 2:970–975

Ghoreishi SM, Behpour M, Khayatkashani M (2011) Green synthesis of silver and gold nanoparticles using *Rosa damascena* and its primary application in electrochemistry. Phys E 44:97–104

Ghosh S, Patil S, Ahire M, Kitture R, Kale S, Pardesi K, Cameotra SS, Bellare J, Dhavale DD, Jabgunde A, Chopad BA (2012a) Synthesis of silver nanoparticles using *Dioscorea bulbifera* tuber extract and evaluation of its synergistic potential in combination with antimicrobial agents. Int J Nanomedicine 7:483–496

Ghosh S, Patil S, Ahire M, Kitture R, Gurav DD, Jabgunde AM, Kale S, Pardesi K, Shinde V, Bellare J, Dhavale DD, Chopade BA (2012b) *Gnidia glauca* flower extract mediated synthesis of gold nanoparticles and evaluation of its chemocatalytic potential. J Nanobiotechnol 1:10–17

Ghule K, Ghule AV, Liu JY, Ling YC (2006) Microscale size triangular gold prisms synthesized using Bengal gram beans (*Cicer arietinum* L.) extract and HAuCl$_4$· 3H$_2$O: a green biogenic approach. J Nanosci Nanotechnol 6:3746–3751

Gopinath V, Ali DM, Priyadarshini S, Priyadharsshini NM, Thajuddin N, Velusamy P (2012) Biosynthesis of silver nanoparticles from *Tribulus terrestris* and its antimicrobial activity: a novel biological approach. Colloids Surf B: Biointerfaces 96:69–74

Gordon T, Perlstein B, Houbara O, Felner I, Banin E, Margel S (2011) Synthesis and characterization of zinc/iron oxide composite nanoparticles and their antibacterial properties. Colloids Surf A Physicochem Eng Asp 374:1–8

Govindaraju K, Kiruthiga V, Manikandan R, Ashokkumar T, Singaravelu G (2011) β-glucosidase assisted biosynthesis of gold nanoparticle: a green chemistry approach. Mater Lett 65:256–259

Guidelli EJ, Ramos AP, Zaniquelli ME, Baffa O (2011) Green synthesis of colloidal silver nanoparticles using natural rubber latex extracted from *Hevea brasiliensis*. Spectrochim Acta A Mol Biomol Spectrosc 82:140–145

Gurunathan S, Kalishwaralal K, Vaidyanathan R, Deepak V, Pandian SRK, Muniyandi J, Hariharan N, Eom SH (2009) Biosynthesis, purification and characterization of silver nanoparticles using *Escherichia coli*. Colloids Surf B: Biointerfaces 74:328–335

Harne S, Sharma A, Dhaygude M, Joglekar S, Kodam K, Hudlikar M (2012) Novel route for rapid biosynthesis of copper nanoparticles using aqueous extract of *Calotropis procera* L. latex and their cytotoxicity on tumor cells. Colloids Surf B: Biointerfaces 95:284–288

Huang J, Li Q, Sun D, Lu Y, Su Y, Yang X, Wang H, Wang Y, Shao W, He N, Hong J, Chen C (2007) Biosynthesis of silver and gold nanoparticles by novel sundried *Cinnamomum camphora* leaf. Nanotechnology 18:105104–105115

Hudlikar M, Joglekar S, Dhaygude M, Kodam K (2012) Green synthesis of TiO$_2$ nanoparticles by using aqueous extract of *Jatropha curcas* L. latex. Mater Lett 75:196–199

Ingle A, Rai M, Gade A, Bawaskar M (2009) *Fusarium solani*: a novel biological agent for the extracellular synthesis of silver nanoparticles. J Nanopart Res 11:2079–2085

Joshi N, Jain N, Pathak A, Singh J, Prasad R, Upadhyaya CP (2018) Biosynthesis of silver nanoparticles using *Carissa carandas* berries and its potential antibacterial activities. J Sol-Gel Sci Technol 86(3):682–689. https://doi.org/10.1007/s10971-018-4666-2

Jain D, Daima HK, Kachhwala S, Kothari SL (2009) Synthesis of plant-mediated silver nanoparticles using papaya fruit extract and evaluation of their anti microbial activities. Dig J Nanomater Biostruct 4:557–563

Jha AK, Prasad K (2011) Biosynthesis of gold nanoparticles using bael (*Aegle marmelos*) leaf: mythology meets technology. Int J Green Nanotechnol Phys Chem 3:92–97

Jha AK, Prasad K (2014) Green synthesis of silver nanoparticles and its activity on SiHa cervical cancer cell line. Adv Mater Lett 5:501–505

Jha AK, Prasad K (2016c) Aquatic fern (*Azolla Sp.*) assisted synthesis of gold nanoparticles. Int J Nanosci 15:1650008–1650005

Jha AK, Prasad K, Prasad K, Kulkarni AR (2009a) Plant system: nature's nanofactory. Colloids Surf B Biointerfaces 73:219–223

Jha AK, Prasad K (2010a) Green synthesis of silver nanoparticles using *Cycas* leaf. Int J Green Nanotechnol Phys Chem 1:110–117

Jha AK, Prasad K (2015) Facile green synthesis of metal and oxide nanoparticles using papaya juice. J Bionanosci 9:311–314

Jha AK, Prasad K (2016a) Green synthesis and antimicrobial activity of silver nanoparticles onto cotton fabric: an amenable option for textile industries. Adv Mater Lett 7:42–46

Jha AK, Prasad K, Kumar V, Prasad K (2009d) Biosynthesis of silver nanoparticles using *Eclipta* leaf. Biotechnol Prog 25:1476–1479

Jha AK, Prasad K (2012a) Biosynthesis of gold nanoparticles using common aromatic plants. Int J Green Nanotechnol Phy Chem 4:219–224

Jha AK, Kumar V, Prasad K (2011) Biosynthesis of metal and oxide nanoparticles using orange juice. J Bionanosci 5:162–165

Jha AK, Prasad K (2013) Rose (*Rosa* sp.) petals assisted green synthesis of gold nanoparticles. J Bionanosci 7:245–250

Jha AK, Prasad K (2012b) PbS nanoparticles: biosynthesis and characterization. Int J Nanopart 5:369–379

Jha AK, Prasad K, Kulkarni AR (2008) Synthesis of nickel nanoparticles: bioreduction method. Nanosci Nanotechnol Ind J 2:26–29

Jha AK, Prasad K, Kulkarni AR (2007) Microbe-mediated nanotransformation: cadmium. Nano 2:239–242

Jha AK, Prasad K, Prasad K (2009b) Biosynthesis of Sb_2O_3 nanoparticles: a low-cost green approach. Biotechnol J 4:1582–1585

Jha AK, Prasad K, Kulkarni AR (2009c) Synthesis of TiO_2 nanoparticles using microorganisms. Colloids Surf B: Biointerfaces 71:226–229

Jha AK, Prasad K (2010b) Biosynthesis of metal and oxide nanoparticles using *Lactobacilli* from yoghurt and probiotic spore tablets. Biotechnol J 5:285–291

Jha AK, Prasad K (2010c) Synthesis of $BaTiO_3$ nanoparticles: a new sustainable green approach. Integr Ferroelectr 117:49–54

Jha AK, Prasad K (2016b) Synthesis of ZnO nanoparticles from goat slaughter waste for environmental protection. Int J Curr Eng Technol 6:147–151

Jia L, Zhang Q, Li Q, Song H (2009) The biosynthesis of palladium nanoparticles by antioxidants in *Gardenia jasminoides Ellis*: long lifetime nanocatalysts for p-nitrotoluene hydrogenation. Nanotechnology 20:385601–385611

Jobitha GG, Annadurai G, Kannan C (2012) Green synthesis of silver nanoparticle using *Elettaria cardamomom* and assessment of its antimicrobial activity. Int J Pharm Sci Res 3:323–330

Johnson I, Prabu HJ (2015) Green synthesis and characterization of silver nanoparticles by leaf extracts of *Cycas circinalis*, *Ficus amplissima*, *Commelina benghalensis* and *Lippia nodiflora*. Int Nano Lett 5:43–51

Jayaseelan C, Rahuman AA, Kumar GR, Kirthi AV, Kumar TS, Marimuthu S (2011) Synthesis of pediculocidal and larvicidal silver nanoparticles by leaf extract from heart leaf moonseed plant, *Tinospora cordifolia* Miers. Parasitol Res 109:185–194

Kalishwaralal K, Venkataraman D, Pandian SRK, Kottaisamy M, Selvaraj B, Bose K, Sangiliyandi G (2010) Biosynthesis of silver and gold nanoparticles using *Brevibacterium casei*. Colloids Surf B: Biointerfaces 77:257–262

Karuppiah M, Rajmohan R (2013) Green synthesis of silver nanoparticles using *Ixora coccinea* leaves extract. Mater Lett 97:141–143

Kasthuri J, Veerapandian S, Rajendiran N (2009) Biological synthesis of silver and gold nanoparticles using apiin as reducing agent. Colloids Surf B: Biointerfaces 68:55–60

Kathirvelu S, D'Souza L, Dhurai B (2009) UV protection finishing of textiles using ZnO nanoparticles. Indian J Fibre Text Res 34:267–273

Kaviya S, Santhanalakshmi J, Viswanathan B, Muthumary J, Srinivasan K (2011) Biosynthesis of silver nanoparticles using *Citrus sinensis* peel extract and its antibacterial activity. Spectrochim Acta A 79:594–598

Kaviya S, Santhanalakshmi J, Viswanathan B (2012) Biosynthesis of silver nano-flakes by *Crossandra infundibuliformis* leaf extract. Mater Lett 67:64–66

Khan Z, Bashir O, Hussain JI, Kumar S, Ahmad R (2012) Effects of ionic surfactants on the morphology of silver nanoparticles using Paan (*Piper betle*) leaf petiole extract. Colloids Surf B: Biointerfaces 98:85–90

Koli A (2015) Biological synthesis of stable Zinc oxide nanoparticles and its role as anti-diabetic and anti-microbial agents. Int J Acad Res 2:139–143

Kora AJ, Sashidhar RB, Arunachalam J (2012) Aqueous extract of gum oblibanum (*Boswellia serrata*): a reductant and stabilizer for the biosynthesis of antibacterial silver nanoparticles. Process Biochem 47:1516–1520

Kora AJ, Sashidhar RB, Arunachalam J (2010) Gum kondagogu (*Cochlospermum gossypium*): a template for the green synthesis and stabilization of silver nanoparticles with antibacterial application. Carbohydr Polym 82:670–679

Konwarh R, Gogoi B, Philip R, Laskar MA, Karak N (2011) Biomimetic preparation of polymer-supported free radical scavenging, cytocompatible and antimicrobial "green" silver nanoparticles using aqueous extract of *Citrus sinensis* peel. Colloids Surf B: Biointerfaces 84:338–345

Kotakadi VS, Rao YS, Gaddam SA, Prasad TNVKV, Reddy AV, Gopal DVRS (2013) Simple and rapid biosynthesis of stable silver nanoparticles using dried leaves of *Catharanthus roseus*. Linn. *G. Donn* and its anti microbial activity. Colloids Surf B: Biointerfaces 105:194–198

Kouvaris P, Delimitis A, Zaspalis V, Papadopoulos D, Tsipas SA, Michailidis N (2012) Green synthesis and characterization of silver nanoparticles produced using *Arbutus unedo* leaf. Extract Mater Lett 76:18–20

Krishnamurthy S, Kumar MS, Lee SY, Bae MA, Yun YS (2011) Biosynthesis of Au nanoparticles using cumin seed powder extract. J Nanosci Nanotechnol 11:1811–1814

Krishnamurthy S, Sathishkumar M, Kim S, Yun YS (2010) Counter ions and temperature incorporated tailoring of biogenic gold nanoparticles. Process Biochem 45:1450–1458

Krishnaraj C, Jagan EG, Rajasekar S, Selvakumar P, Kalaichelvan PT, Mohan N (2010) Synthesis of silver nanoparticles using *Acalypha indica* leaf extracts and its antibacterial activity against water borne pathogens. Colloids Surf B: Biointerfaces 76:50–56

Krishnaraj C, Ramachandran R, Mohan K, Kalaichelvan PT (2012) Optimization for rapid synthesis of silver nanoparticles and its effect on phytopathogenic fungi. Spectrochim Acta A Mol Biomol Spectrosc 93:95–99

Krolikowska A, Kudelski A, Michota A, Bukowska J (2003) SERS studies on the structure of thioglycolic acid monolayers on silver and gold. Surf Sci 532:227–232

Kumar MV, Priya K, Nancy FT, Noorlidah A, Ahmed ABA (2013a) Biosynthesis, characterisation and anti-bacterial effect of plant-mediated silver nanoparticles using *Artemisia nilagirica*. Ind Crop Prod 41:235–240

Kumar R, Roopan SM, Prabhakarn A, Khanna VG, Chakroborty S (2012a) Agricultural waste *Annona squamosa* peel extract: Biosynthesis of silver nanoparticles. Spectrochim Acta A Mol Biomol Spectrosc 90:173–176

Kumar MS, Sneha K, Won SW, Cho CW, Kim S, Yun YS (2009a) *Cinnamon Zeylanicum* bark extract and powder mediated green synthesis of nano-crystalline silver particles and its bactericidal activity. Colloids Surf B: Biointerfaces 73:332–338

Kumar MS, Krishnamurthy S, Yun YS (2010a) Immobilization of silver nanoparticles synthesized using *Curcuma longa* tuber powder and extract on cotton cloth for bactericidal activity. Bioresour Technol 101:7958–7965

Kumar TS, Rahuman AA, Bagavan A, Marimuthu S, Jayaseelan C, Kirthi AV, Kamaraj C, Kumar GR, Zahir AA, Elango G, Velayutham K, Iyappan M, Siva C, Karthik L, Rao KVB (2012b) Evaluation of stem aqueous extract and synthesized silver nanoparticles using *Cissus quadrangularis* against *Hippobosca maculate* and *Rhipicephalus (Boophilus) microplus*. Exp Parasitol 132:156–165

Kumar GR, Rahuman AA (2011) Larvicidal activity of synthesized silver nanoparticles using *Eclipta prostrata* leaf extract against filariasis and malaria vectors. Acta Trop 118:196–203

Kumar P, Selvi SS, Govindaraju M (2013b) Seaweed-mediated biosynthesis of silver nanoparticles using *Gracilaria corticata* for its antifungal activity against Candida spp. Appl Nanosci 3:495–500

Kumar MS, Sneha K, Kwak IS, Mao J, Tripathy SJ, Yun YS (2009b) Phyto-crystallization of palladium through reduction process using *Cinnamom zeylanicum* bark extract. J Hazard Mater 171:400–404

Kumar V, Yadav SK (2011) Synthesis of stable, polyshaped silver and gold nanoparticles using leaf extract of *Lonicera japonica* L. Int J Green Nanotechnol 3:281–291

Kumar GR, Rahuman AA (2012) Acaricidal activity of aqueous extract and synthesized silver nanoparticles from *Manilkara zapota* against Rhipicephalus (Boophilus) microplus. Res Vet Sci 93:303–309

Kumar V, Yadav SC, Yadav SK (2010b) *Syzygium cumini* leaf and seed extract mediated biosynthesis of silver nanoparticles and their characterization. J Chem Technol Biotechnol 85:1301–1309

Kumar V, Yadav SK (2012) Characterization of gold nanoparticles synthesized by leaf and seed extract of *Syzygium cumini* L. J Exp Nanosci 7:440–451

Kumar KM, Sinha M, Mandal BK, Ghosh AR, Kumar KS, Reddy PS (2012c) Green synthesis of silver nanoparticles using *Terminalia chebula* extract at room temperature and their antimicrobial studies. Spectrochim Acta A Mol Biomol Spectrosc 91:228–233

Kumar KP, Paul W, Sharma CP (2012d) Green synthesis of silver nanoparticles with *Zingiber officinale* extract and study of its blood compatibility. BioNanoScience 2:144–152

Kumar RV, Devi V, Adavallan K, Saranya D (2011a) Green synthesis and characterization of gold nanoparticles using extract of anti-tumor potent *Crocus sativus*. Phys E 44:665–671

Kumar VG, Gokavarapu SD, Rajeswari A, Dhas TS, Karthick V, Kapadia Z, Shrestha T, Barathy IA, Roy A, Sinha S (2011b) Facile green synthesis of gold nanoparticles using leaf extract of antidiabetic potent *Cassia auriculata*. Colloids Surf B: Biointerfaces 87:159–163

Kumar KM, Mandal BK, Sinha M, Kumar VK (2012e) *Terminalia chebula* mediated green and rapid synthesis of gold nanoparticles. Spectrochim Acta A Mol Biomol Spectrosc 86:490–494

Kumar KP, Paul W, Sharma CP (2011c) Green synthesis of gold nanoparticles with *Zingiber officinale* extract: characterization and blood compatibility. Process Biochem 46:2007–2013

Kumar GR, Rahuman AA, Priyamvada B, Khanna VG, Kumar DK, Sujin PJ (2012f) *Eclipta prostrata* leaf aqueous extract mediated synthesis of titanium dioxide nanoparticles. Mater Lett 68:115–117

Kumar MS, Sneha K, Yun YS (2009c) Palladium nanocrystals synthesis using *Curcuma longa* tuber extract. Int J Mater Sci 4:11–17

Kumar SA, Peter YA, Nadeau JL (2008) Facile biosynthesis, separation and conjugation of gold nanoparticles to doxorubicin. Nanotechnology 19:495101

Kumar PPNV, Shameem U, Kollu P, Kalyani RL, Pammi SVN (2015) Green synthesis of copper oxide nanoparticles using *Aloe vera* leaf extract and its antibacterial activity against fish bacterial pathogens. BioNanoScience 5:135–139

Leela A, Vivekanandan M (2008) Tapping the unexploited plant resources for the synthesis of silver nanoparticles. Afr J Biotechnol 7:3162–3165

Li S, Qui L, Shen Y, Xie A, Yu X, Zhang L, Zhang Q (2007) Green synthesis of silver nanoparticles using *Capsicum annuum* L. extract. Green Chem 9:852–858

Li Y, Duan X, Qian Y, Liao H (1999) Nanocrystalline silver particles: synthesis, agglomeration, and sputtering induced by electron beam. J Colloid Interface Sci 209:347–349

Lin X, Wu M, Wu D, Kuga S, Endo T, Huang Y (2011) Platinum nanoparticles using wood nanomaterials: eco-friendly synthesis, shape control and catalytic activity for *p*-nitrophenol reduction. Green Chem 13:283–287

Lukman AI, Gong B, Marjo CE, Roessner U, Harris AT (2011) Facile synthesis, stabilization, and anti-bacterial performance of discrete Ag nanoparticles using *Medicago sativa* seed exudates. J Colloid Interface Sci 353:433–444

Maensiri S, Laokul P, Klinkaewnarong J, Phokha S, Promarak V, Seraphin S (2008) Indium oxide (In$_2$O$_3$) nanoparticles using *Aloe vera* plant extract: synthesis and optical properties. J Optoelectron Adv Mater 10:161–165

Mahdieh M, Zolanvari A, Azimee AS, Mahdieh M (2012) Green biosynthesis of silver nanoparticles by *Spirulina platensis*. Sci Iran 19:926–929

Mallikarjuna K, Dillip GR, Narasimha G, Sushma NJ, Raju BDP (2012) Phytofabrication and characterization of silver nanoparticles from *Piper betle* broth. Res J Nanosci Nanotechnol 2:17–23

Marchiol L (2012) Synthesis of metal nanoparticles in living plants. Ital J Agron 3:37–41

Mason C, Vivekanandhan S, Misra M, Mohanty AK (2012) Switchgrass (*Panicum virgatum*) extract mediated green synthesis of silver nanoparticles. World J Nano Sci Eng 2:47–52

Medda S, Hajra A, Dey U, Bose P, Mondal NK (2015) Biosynthesis of silver nanoparticles from *Aloe vera* leaf extract and antifungal activity against *Rhizopus* sp. and *Aspergillus* sp. Appl Nanosci 5:875–880

Mishra AN, Bhadauria S, Gaur MS, Pasricha R, Kushwah BS (2010) Synthesis of gold nanoparticles by leaves of zero-calorie sweetener herb (*Stevia rebaudiana*) and their nanoscopic characterization by spectroscopy and microscopy. Int J Green Nanotechnol Phys Chem 1:118–124

Mishra V, Sharma R (2015) Green synthesis of zinc oxide nanoparticles using fresh peels extract of *Punica granatum* and its antimicrobial activities. Int J Pharma Res Health Sci 3:694–699

Mitra B (2012) Green-synthesis and characterization of silver nanoparticles by aqueous leaf extracts of *Cardiospermum helicacabum* L. Drug Invent Today 4:340–344

Mohamed NH, Ismail MA, Abdel-Mageed WM, Shoreit AAM (2014) Antimicrobial activity of latex silver nanoparticles using *Calotropic procera*. Asian Pac J Trop Biomed 4:876–883

Montes MO, Mayoral A, Deepak FL, Parsons JG, Yacaman MJ, Videa JRP, Torresdey JLG (2011) Anisotropic gold nanoparticles and gold plates biosynthesis using *alfalfa* extracts. J Nanopart Res 13:3113–3120

Moulton MC, Stolle LKB, Nadagouda MN, Kunzelman S, Hussaina SM, Varma RS (2010) Synthesis, characterization and biocompatibility of "green" synthesized silver nanoparticles using tea polyphenols. Nanoscale 2:763–770

Mubayi A, Chatterji S, Rai PM, Watal G (2012) Evidence based green synthesis of nanoparticles. Adv Mater Lett 3:519–525

Muralikrishna T, Pattanayak M, Nayak PL (2014) Green synthesis of gold nanoparticles using (*Aloe vera*) aqueous extract. World J Nano Sci Technol 3:45–51

Nabikhan A, Kandasamy K, Raj A, Alikunhi NM (2010) Synthesis of antimicrobial silver nanoparticles by callus and leaf extracts from saltmarsh plant, *Sesuvium portulacastrum* L. Colloids Surf B: Biointerfaces 79:488–493

Nagajyothi PC, Cha SJ, Yang IJ, Shin HM (2015) Antioxidant and anti-inflammatory activities of zinc oxide nanoparticles synthesized using *Polygala tenuifolia* root extract. J Photochem Photobiol B Biol 146:10–17

Nagati VB, Alwala J, Koyyati R, Donda MR, Banala R, Padigya PRM (2012) Green synthesis of plant-mediated silver nanoparticles using *Withania somnifera* leaf extract and evaluation of their anti microbial activity. Asian Pac J Trop Biomed 2:1–5

Namratha N, Monica PV (2013) Green synthesis of silver nanoparticles using *Mentha asiatica* (Mint) extract and evaluation of their antimicrobial potential. J Pharm Tech 3:170–174

Narayanan KB, Sakthivel N (2008) Coriander leaf mediated biosynthesis of gold nanoparticles. Mater Lett 62:4588–4590

Nellore J, Pauline PC, Amarnath K (2012) Biogenic synthesis of *Sphearanthus amaranthoids* towards the efficient production of the biocompatible gold nanoparticles. Dig J Nanomater Biostruct 7:123–133

Nestor ARV, Mendieta VS, Lopez MAC, Espinosa RMG, Lopez MAC, Alatorre JAA (2008) Solvent less synthesis and optical properties of Au and Ag nanoparticles using *Camellia sinensis*. Mat Lett 62:3103–3105

Niraimathi KL, Sudha V, Lavanya R, Brindha P (2013) Biosynthesis of silver nanoparticles using *Alternanthera sessilis* (Linn.) extract and their antimicrobial, antioxidant activities. Colloids Surf B: Biointerfaces 102:288–291

Nithya A, Rokesh K, Katachalam KJ (2013) Biosynthesis, characterization and application of titanium dioxide nanoparticles. Nano Vision 3:169–174

Njagi EC, Huang H, Stafford L, Genuino H, Galindo HM, Collins JB, Hoag GE, Suib SL (2011) Biosynthesis of iron and silver nanoparticles at room temperature using aqueous sorghum bran extracts. Langmuir 27:264–271

Njagi LW, Nyaga PN, Mbuthia PG, Bebora LC, Michieka JN, Kibe JK, Minga UM (2010) Prevalence of Newcastle disease virus in village indigenous chickens in varied agro-ecological zones in Kenya. Livestock Res Rural Development 22:1–8

Noruzi M, Zare D, Davoodi D (2012) A rapid biosynthesis route for the preparation of gold nanoparticles by aqueous extract of cypress leaves at room temperature. Spectrochim Acta A Mol Biomol Spectrosc 94:84–88

Noruzi M, Zare D, Khoshnevisan K, Davoodi D (2011) Rapid green synthesis of gold nanoparticles using *Rosa hybrida* petal extract at room temperature. Spectrochim Acta A Mol Biomol Spectrosc 79:1461–1465

Panda KK, Achary VM, Krishnaveni R, Padhi BK, Sarangi SN, Sahu SN, Panda BB (2011) In vitro biosynthesis and genotoxicity bioassay of silver nanoparticles using plants. Toxicol Vitro 25:1097–1105

Pandey S, Oza G, Mewada A, Sharon M (2012) Green synthesis of highly stable gold nanoparticles using *Momordica charantia* as nano fabricator. Arch Appl Sci Res 4:1135–1141

Pandey S, Goswami GK, Nanda KK (2013) Green synthesis of polysaccharide/gold nanoparticle nanocomposite: an efficient ammonia sensor. Carbohyd Polym 94:229–234

Patil CD, Borase HP, Patil SV, Salunkhe RB, Salunke BK (2012a) Larvicidal activity of silver nanoparticles synthesized using *Pergularia daemia* plant latex against *Aedes aegypti* and *Anopheles stephensi* and nontarget fish *Poecillia reticulata*. Parasitol Res 111:555–562

Patil RS, Kokate MR, Kolekar SS (2012b) Bioinspired synthesis of highly stabilized silver nanoparticles using *Ocimum tenuiflorum* leaf extract and their antibacterial activity. Spectrochem Acta A 91:234–238

Petla RK, Vivekanandhan S, Misra M, Mohanty AK, Satyanarayana N (2012) Soybean (*Glycine Max*) leaf extract based green synthesis of palladium nanoparticles. J Biomater Nanobiotechnol 3:14–19

Philip D (2009) Honey mediated green synthesis of gold nanoparticles. Spectrochim Acta A Mol Biomol Spectrosc 73:650–653

Phillip D (2010) Rapid green synthesis of spherical gold nanoparticles using *Mangifera indica* leaf. Spectrochim Acta A Mol Biomol Spectrosc 77:807–810

Phillip D (2011) *Mangifera indica* leaf-assisted biosynthesis of well-dispersed silver nanoparticles. Spectrochim Acta A Mol Biomol Spectrosc 78:327–331

Philip D, Unni C (2011) Extracellular biosynthesis of gold and silver nanoparticles using Krishna tulsi (*Ocimum sanctum*) leaf. Phys E 43:1318–1322

Philip D, Unni C, Aromal SA, Vidhu VK (2011) *Murraya Koenigii* leaf-assisted rapid green synthesis of silver and gold nanoparticles. Spectrochim Acta A Mol Biomol Spectrosc 78:899–904

Pingali KC, Rockstraw DA, Deng S (2005) Silver nanoparticles from ultrasonic spray pyrolysis of aqueous silver nitrate. Aerosol Sci Technol 39:1010–1014

Prabakar K, Sivalingam P, Rabeek SIM, Muthuselvam M, Devarajan N, Arjunan A, Karthick R, Suresh MM, Wembonyama JP (2013) Evaluation of antibacterial efficacy of phyto fabricated silver nanoparticles using *Mukia scabrella* (Musumusukkai) against drug resistance nosocomial gram negative bacterial pathogens. Colloids Surf B: Biointerfaces 104:282–288

Prasad K, Jha AK (2009) ZnO nanoparticles: synthesis and adsorption study. Nat Sci 1:129–135

Prasad K, Jha AK (2010) Biosynthesis of CdS nanoparticles: an improved green and rapid procedure. J Colloid Interface Sci 342:68–72

Prasad KS, Pathak D, Patel A, Dalwadi P, Prasad R, Patel P, Selvaraj K (2011) Biogenic synthesis of silver nanoparticles using *Nicotiana tobaccum* leaf extract and study of their antibacterial effect. Afr J Biotechnol 10:8122–8130

Prasad R (2014) Synthesis of silver nanoparticles in photosynthetic plants. J Nanopart 11:963955–963961

Prasad R, Pandey R, Barman I (2016) Engineering tailored nanoparticles with microbes: quo vadis. WIREs Nanomed Nanobiotechnol 8:316–330. https://doi.org/10.1002/wnan.1363

Prasad R, Swamy VS (2013) Antibacterial activity of silver nanoparticles synthesized by bark extract of *Syzygium cumini*. J Nanopart 2013:431218. https://doi.org/10.1155/2013/431218

Prasad R, Swamy VS, Varma A (2012) Biogenic synthesis of silver nanoparticles from the leaf extract of *Syzygium cumini* (L.) and its antibacterial activity. Int J Pharm Bio Sci 3(4):745–752

Prasad TN, Elumalai E (2011) Biofabrication of Ag nanoparticles using *Moringa oleifera* leaf extract and their antimicrobial activity. Asian Pac J Trop Biomed 1:439–442

Prathna TC, Chandrasekaran N, Raichur AM, Mukhergee A (2011) Biomimetic synthesis of silver nanoparticles by *Citrus limon* (lemon) aqueous extract and theoretical prediction of particle size. Colloids Surf B: Biointerfaces 82:152–159

Premanathan M, Karthikeyan K, Jeyasubramanian K, Manivannan G (2011) Selective toxicity of ZnO nanoparticles toward Gram-positive bacteria and cancer cells by apoptosis through lipid peroxidation. Nanomedicine: NBM 7:184–192

Priyadarshini A, Murugan K, Panneerselvam K, Ponarulselvam C, Hwang S, Nicoletti JS (2012) Biolarvicidal and pupicidal potential of silver nanoparticles synthesized using *Euphorbia hirta* against *Anopheles stephensi* Liston (*Diptera: Culicidae*). Parasitol Res 111:997–1006

Punuri JB, Sharma P, Sibyala S, Tamuli R, Bora U (2012) *Piper betle*-mediated green synthesis of biocompatible gold nanoparticles. Int Nano Lett 2:18–27

Rafiee E, Shahebrahimi S, Feyzi M, Shaterzadeh M (2012) Optimization of synthesis and characterization of nanosilica produced from rice husk (a common waste material). Int Nano Lett 2:29–37

Raghavan KV, Nalini SP, Prakash NU, Kumar DM (2012) One step green synthesis of silver nano/microparticles using extracts of *Trachyspermum ammi* and *Papaver somniferum*. Colloids Surf B: Biointerfaces 94:114–117

Raghunandan D, Basavaraja S, Mahesh B, Balaji S, Manjunath SY, Venkataraman A (2010) Microwave-assisted rapid extracellular synthesis of stable bio-functionalized silver nanoparticles from guava (*Psidium guajava*) leaf extract. J Nanopart Res 5:34–41

Rajiv P, Rajeshwari S, Venckatesh R (2013) Bio-fabrication of zinc oxide nanoparticles using leaf extract of *Parthenium hysterophorus* L. and its size-dependent antifungal activity against plant fungal pathogens. Spectrochim Acta A Mol Biomol Spectrosc 112:384–387

Raman N, Sudharsan S, Kumar VV, Pravin N, Vithiya K (2012) *Pithecellobium dulce* mediated extra-cellular green synthesis of larvicidal silver nanoparticles. Spectrochim Acta A Mol Biomol Spectrosc 96:1031–1037

Ramesh M, Anbuvannan M, Viruthagiri G (2015) Green synthesis of ZnO nanoparticles using Solanum nigrum leaf extract and their antibacterial activity. Spectrochim Acta A Mol Biomol Spectrosc 136:864–870

Ramezani N, Ehsanfar Z, Shamsa F, Amin G, Shahverdi HR, Esfahani HRM, Shamsaie A, Bazaz RD, Shahverdi AR (2008) Screening of medicinal plant methanol extracts for the synthesis of gold nanoparticles by their reducing potential. Z Naturforsch 63:903–908

Rani PU, Reddy PR (2011) Green synthesis of silver-protein (core-shell) nanoparticles using *Piper betle* L. leaf extract and its ecotoxicological studies on *Daphnia manga*. Colloids Surf A Physicochem Eng Asp 389:188–194

Rao KG, Ashok CH, Rao KV, Chakra CHS, Tambur P (2015) Green synthesis of TiO$_2$ nanoparticles using *Aloe vera* extract. Int J Adv Res Phys Sci 2:28–34

Rao KJ, Paria S (2013) Green synthesis of silver nanoparticles from aqueous *Aegle marmelos* leaf extract. Mater Res Bull 48:628–634

Rao ML, Savithramma N (2011) *Biological synthesis of silver nanoparticles using Svensonia hyderabadensis* leaf extract and evaluation of their antimicrobial efficacy. J Pharm Sci Res 3:1117–1121

Rao YS, Kotakadi VS, Prasad TNVKV, Reddy AV, Gopal DVRS (2013) Green synthesis and spectral characterization of silver nanoparticles from Lakshmi tulasi (*Ocimum sanctum*) leaf extract. Spectrochim Acta A Mol Biomol Spectrosc 103:156–159

Rastogi L, Arunachalam J (2013) Green synthesis route for the size controlled synthesis of biocompatible gold nanoparticles using aqueous extract of garlic (*Allium sativum*). Adv Mater Lett 4:548–555

Roopan SM, Madhumitha RG, Rahuman AA, Kamaraj C, Bharathi A, Surendra TV (2013) Low-cost and eco-friendly phyto-synthesis of silver nanoparticles using *Coos nucifera* coir extract and its larvicidal activity. Ind Crop Prod 43:631–635

Sahu N (2013) Synthesis and characterization of silver nanoparticles using *Cynodon dactylon* leaves and assessment of their antibacterial activity. Bioprocess Biosyst Eng 36:999–1004

Saha S, Pal A, Kundu S, Basu S, Pal T (2010) Photochemical green synthesis of calcium-alginatestabilized Ag and Au nanoparticles and their catalytic application to 4-nitrophenol reduction. Langmuir 26:2885–2893

Salem W, Leitner DR, Zingl FG, Schratter G, Prassl R, Goessler W, Reidl J, Schild S (2015) Antibacterial activity of silver and zinc nanoparticles against *Vibrio cholerae* and enterotoxic *Escherichia coli*. Int J Med Microbiol 305:85–95

Sangeetha G, Rajeshwari S, Venckatesh R (2011) Green synthesis of zinc oxide nanoparticles by aloe barbadensis miller leaf extract: structure and optical properties. Mater Res Bull 46:2560–2566

Sanghi R, Verma P (2009) Biomimetic synthesis and characterisation of protein capped silver nanoparticles. Bioresour Technol 100:501–504

Satyavani K, Gurudeeban S, Ramanathan T, Balasubramanian T (2011a) Biomedical potential of silver nanoparticles synthesized from calli cells of *Citrullus colocynthis* (L.) Schrad. J Nanobiotechnol 9:43–51

Satyavani K, Ramanathan T, Gurudeekan S (2011b) Green synthesis of silver nanoparticles using stem dried callus extract of bitter apple (*Citrullus colocynthis*). Dig J Nanomater Biostruct 6:1019–1024

Savithramma N, Rao ML, Devi PS (2011) Evaluation of antibacterial efficacy of biologically synthesized silver nanoparticles using stem barks of *Boswellia ovalifoliolata* Bal. and Henry and *Shorea tumbuggai*. J Biol Sci 11:39–45

Saxena A, Tripathi RM, Singh RP (2010) Biological synthesis of silver nanoparticles by using onion (*Allium cepa*) extract and their antibacterial activity. Dig J Nanomater Biostruct 5:427–432

Saxena A, Tripathi RM, Zafar F, Singh P (2012) Green synthesis of silver nanoparticles using aqueous solution of *Ficus benghalensis* leaf extract and characterization of their antibacterial activity. Mater Lett 67:91–94

Schmid G (1992) Large clusters and colloids, metals in the embryonic state. Chem Rev 92:1709–1727

Sen IK, Maity K, Islam SS (2013) Green synthesis of gold nanoparticles using a glucan of an edible mushroom and study of catalytic activity. Carbohydr Polym 91:518–528

Shankar SS, Rai A, Ahmad A, Sastry M (2004a) Rapid synthesis of Au, Ag, and bimetallic Au core-Ag shell nanoparticles using Neem (*Azadirachta indica*) leaf broth. J Colloid Interface Sci 275:496–502

Shankar SS, Rai A, Ankamwar B, Singh A, Ahmad A, Sastry M (2004b) Biological synthesis of triangular gold nanoprisms. Nat Mater 3:482–488

Shankar SS, Rai A, Ahmad A, Sastry M (2005) Controlling the optical properties of lemongrass extract synthesized gold nanotriangles and potential application in infrared-absorbing optical coatings. Chem Mater 17:566–572

Shankar SS, Ahmad A, Pasricha R, Sastry M (2003a) Bioreduction of chloroaurate ions by geranium leaves and its endophytic fungus yields gold nanoparticles of different shapes. J Mater Chem 13:1822–1826

Shankar SS, Ahmad A, Sastry M (2003b) Geranium leaf assisted biosynthesis of silver nanoparticles. Biotechnol Prog 19:1627–1631

Sharma NC, Sahi SV, Nath S, Parsons JG, Torresdey JLG, Pal T (2007) Synthesis of plant-mediated gold nanoparticles and catalytic role of biomatrix-embedded nanomaterials. Environ Sci Technol 41:5137–5242

Sharma RA, Singh R (2013) A review on *Phyla nodiflora* L.: a wild wetland medicinal herb. Int J Pharm Sci Rev Res 20:57–63

Shin HS, Yang HJ, Kim SB, Lee SS (2004) Mechanism of growth of colloidal silver nanoparticles stabilized by polyvinyl pyrrolidone in γ-irradiated silver nitrate solution. J Colloid Interface Sci 274:89–94

Sindhura KS, Prasad TNVKV, Selvam PP, Hussain OM (2014) Synthesis, characterization and evaluation of effect of phytogenic zinc nanoparticles on soil exo-enzymes. Appl Nanosci 4:819–827

Singh A, Jain D, Upadhyay MK, Khandelwal N, Verma HN (2010a) Green synthesis of silver nanoparticles using *Argemone mexicana* leaf extract and evaluation of their antimicrobial activities. Dig J Nanomater Biostruct 5:483–489

Singh A, Shukla R, Hassan S, Bhonde RR, Sastry M (2011a) Cytotoxicity and cellular internalization studies of biogenic gold nanotriangles in animal cell lines. Int J Green Nanotechnol 3:251–263

Singh C, Baboota RK, Naik PK, Singh H (2012) Biocompatible synthesis of silver and gold nanoparticles using leaf extract of *Dalbergia sissoo*. Adv Mater Lett 3:279–285

Singh C, Sharma V, Naik PK, Khandelwal V, Singh H (2010b) A green biogenic approach for synthesis of gold and silver nanoparticles using *zingiber officinale*. Dig J Nanomater Biostruct 6:535–542

Singh RP, Shukla VK, Yadav RS, Sharma PK, Singh PK, Pandey AC (2011b) Biological approach of zinc oxide nanoparticles formation and its characterization. Adv Mater Lett 2:313–317

Singh S, Saikia JP, Buragohain AK (2013) A novel "green" synthesis of colloidal silver nanoparticles (SNP) using *Dillenia indica* fruit extract. Colloids Surf B Biointerfaces 102:83–85

Singhal G, Bhavesh R, Kasariya K, Sharma AR, Singh RP (2011) Biosynthesis of silver nanoparticles using *Ocimum sanctum* (Tulsi) leaf extract and screening its antimicrobial activity. J Nanopart Res 13:2981–2988

Singhal GS, Riju B, Ranjan A, Rajendra SP (2012) Ecofriendly biosynthesis of gold nanoparticles using medicianally important *Ocimum basilicum* leaf extract. Adv Sci Eng Med 4:62–66

Sinha S, Pan I, Chanda P, Sen SK (2009) Nanoparticles fabrication using ambient biological resources. J Appl Biosci 19:1113–1130

Smitha SL, Philip D, Gopchandran KG (2009) Green synthesis of gold nanoparticles using *Cinnamomum zeylanicum* leaf broth. Spectrochim Acta A Mol Biomol Spectrosc 74:735–739

Song JY, Kim BS (2009a) Rapid biological synthesis of silver nanoparticles using plant leaf extracts. Bioprocess Biosyst Eng 32:79–84

Song JY, Kim BS (2009b) Biological synthesis of bimetallic Au/Ag nanoparticles using persimmom (*Diospyros kaki*) leaf extract. Korean J Chem Eng 25:808–811

Song JY, Jang HK, Kim BS (2009) Biological synthesis of gold nanoparticles using *Magnolia kobus* and *Diopyros kaki* leaf extracts. Process Biochem 44:1133–1138

Song JY, Kwon EY, Kim BS (2010) Biological synthesis of platinum nanoparticles using *Diopyros kaki* leaf extract. Bioprocess Biosyst Eng 33:159–164

Song WY, Sohn EJ, Martinoia E, Lee YJ, Yang YY, Jasinski M, Forestier C, Hwang I, Lee Y (2003) Engineering tolerance and accumulation of lead and cadmium in transgenic plants. Nat Biotechnol 21:914–919

Soundarrajan C, Sankari A, Dhandapani P, Maruthamuthu S, Ravichandran S, Sozhan G, Palaniswamy N (2012) Rapid biological synthesis of platinum nanoparticles using *Ocimum sanctum* for water electrolysis applications. Bioprocess Biosyst Eng 35:827–833

Sukirtha R, Priyanka KM, Antony JJ, Kamalakkannan S, Thangam R, Gunasekaran P, Krishnan M, Achiraman S (2012) Cytotoxic effect of Green synthesized silver nanoparticles using *Melia azedarach* against in vitro HeLa cell lines and lymphoma mice mode. Process Biochem 47:273–279

Suriyakalaa U, Antony JJ, Suganya S, Siva D, Sukirtha R, Kamalakkannan S, Pichiah PBT, Achiraman S (2013) Hepatocurative activity of biosynthesized silver nanoparticles fabricated using *Andrographis paniculata*. Colloids Surf B: Biointerfaces 102:189–194

Swamy VS, Prasad R (2012) Green synthesis of silver nanoparticles from the leaf extract of *Santalum album* and its antimicrobial activity. J Optoelectron Biomed Mater 4(3):53–59

Thirumurugan A, Jiflin GJ, Rajagomathi G, Tomy NA, Ramachandran S, Jaiganesh R (2010) Biotechnological synthesis of gold nanoparticles of *Azardirachta indica* leaf extract. Int J Biol Technol 1:75–77

Tripathy A, Raichur AM, Chandrasekaran N, Prathna TC, Mukherjee A (2010) Process variables in biomimetic synthesis of silver nanoparticles by aqueous extract of *Azadirachta indica* (Neem) leaves. J Nanopart Res 12:237–241

Udayasoorian C, Kumar KV, Jayabalakrishnan RM (2011) Extracellular synthesis of silver nanoparticles using leaf extract of *Cassia auriculata*. J Nanomater Biostruct 6:279–283

Umashankari J, Inbakandan D, Kumar TTA, Balasubramanian T (2012) Mangrove plant, *Rhizophora mucronata* (Lamk, 1804) mediated one pot green synthesis of silver nanoparticles and its antibacterial activity against aquatic pathogens. Saline Syst 8:11–19

Valli JS, Vaseeharan B (2012) Biosynthesis of silver nanoparticles by *Cissus quadrangularis* extracts. Mater Lett 82:171–173

Valtchev V, Tosheva L (2013) Porous nanosized particles: preparation, properties, and applications. Chem Rev 113:6734–6760

Vanaja M, Annadurai G (2013) *Coleus aromaticus* leaf extract mediated synthesis of silver nanoparticles and its bactericidal activity. Appl Nanosci 3:217–223

Vankar PS, Bajpai D (2010) Preparation of gold nanoparticles from *Mirabilis jalapa* flowers. Indian J Biochem Biophys 47:157–160

Veerasamy R (2011) Biosynthesis of silver nanoparticles using *mangosteen* leaf extract and evaluation of their antimicrobial activities. J Saudi Chem Soc 15:113–120

Velayutham K, Rahuman AA, Rajakumar G, Kumar TS, Marimuthu T, Jayaseelan C, Bagavan A, Kirthi AV, Kamaraj C, Zahir AA, Elango G (2012) Evaluation of Catharanthus roseus leaf extract-mediated biosynthesis of titanium dioxide nanoparticles against Hippobosca maculata and Bovicola ovis. Parasitol Res 111:2329–2337

Venu R, Ramulu TS, Anandakumar S, Rani VS, Kim CG (2011) Bio-directed synthesis of platinum nanoparticles using aqueous honey solutions and their catalytic applications. Colloids Surf A Physicochem Eng Asp 384:733–738

Vidhu VK, Aromal SA, Philip D (2011) Green synthesis of silver nanoparticles using *Macrotyloma uniflorum*. Spectrochim Acta A 83:392–397

Vinod VTP, Saravanan P, Sreedhar B, Devi DK, Sashidhar RB (2011) A facile synthesis and characterization of Ag, Au and Pt nanoparticles using a natural hydrocolloid gum kondagogu (*Cochlospermum gossypium*). Colloids Surf B: Biointerfaces 83:291–298

Vivek R, Thangam R, Muthuchelian K, Gunasekaran P, Kaveri K, Kannan S (2012) Green biosynthesis of silver nanoparticles from *Annona squamosa* leaf extract and its in vitro cytotoxic effect on MCF-7 cells. Process Biochem 47:2405–2410

Wang Y, He X, Wang K, Zhang X, Tan W (2009) *Barbated skullcup* herb extract-mediated biosynthesis of gold nanoparticles and its primary application in electrochemistry. Colloids Surf B: Biointerfaces 73:75–79

Wang X, Zuo J, Keil P, Grundmeier G (2007) Comparing the growth of PVD silver nanoparticles on ultra-thin fluorocarbon plasma polymer films and self-assembled fluoroalkyl silane monolayers. Nanotechnology 18:265303

Xie H, Lee JY, Wang DIC, Ting YP (2007) Silver nanoplates: from biological to biomimetic synthesis. ACS Nano 1:429–439

Yadav JP, Kumar S, Budhwar L, Yadav A, Yadav M (2016) Characterization and antibacterial activity of synthesized silver and iron nanoparticles using *Aloe vera*. J Nanomed Nanotechnol 7:1000384–1000387

Yang X, Li Q, Wang H, Huang J, Lin L, Wang W, Sun D, Su Y, Opiyo JB, Hong L, Wang Y, He N, Jia L (2010) Green synthesis of palladium nanoparticles using broth of *Cinnamomum camphora* leaf. J Nanopart Res 12:1589–1598

Yeary LW, Moon J, Love LJ, Thompson JR, Rawn CJ, Phelps TJ (2005) Magnetic properties of biosynthesized magnetite nanoparticles. IEEE Trans Magn 41:4384–4389

Yilmaz M, Turkdemir H, Kilic MA, Bayram E, Cicek A, Mete A, Ulug B (2011) Biosynthesis of silver nanoparticles using leaves of *Stevia rebaudiana*. Mater Chem Phys 130:1195–1202

Zahir AA, Rahuman AA (2012) Evaluation of different extracts and synthesized silver nanoparticles from leaves of *Euphorbia prostrate* against the plant *Haemaphysalis bispinosa* and *Hippobosca maculate*. Vet Parasitol 187:511–520

Zayed MF, Eisa WH, Shabaka AA (2012) Malvaparviflora extract assisted green synthesis of silver nanoparticles. Spectrochim Acta A Mol Biomol Spectrosc 98:423–428

Zhan G, Huang J, Du M, Rauf IA, Ma Y, Li Q (2011) Green synthesis of Au–Pd bimetallic nanoparticles: Single-step bioreduction method with plant extract. Mater Lett 65:2989–2991

Zhu H, Jia S, Wan T, Jia Y, Yang H, Li J, Yan L, Zhong C (2011) Biosynthesis of spherical Fe_3O_4 bacterial cellulose nanocomposites as adsorbents for heavy metal ions. Carbohydr Polym 86:1558–1564

Chapter 9
Synthesis of Functionalized Nanoparticles for Biomedical Applications

Priti Kumari, Niraj Kumari, Anal K. Jha, K. P. Singh, and Kamal Prasad

9.1 Introduction

History of nanoscience and technology can be traced back to the ancient Indian and Greek literatures. Ever inquisitive nature of the human beings might have led to the exploration and establishment of a plant as a source of medicine. The time changed its face and observations got transcribed in meticulous inscriptions and consequently translated into applications. Charak and Sushruta Samhita are the two such examples from our ancient literatures. Especially, Charak Samhita, Ras ratnakar, and Ark Prakash written by Maharishi Charak, Nagarjuna, and Lankapati Ravan, respectively, are the most authentic source of ancient knowledge as far as interaction of plant extracts and metal ions is concerned. Indeed, our timeworn and treasured mythology has got a new face of technology!

Nanotechnology is considered as a truly multidisciplinary field of research and development which is witnessing an exponential progress during recent times. It rises from the convergence of different disciplines of science and technology like physics, chemistry, biology, materials science, engineering science, medical science, and other sciences at nanometer (nm) scale (<100 nm), which provides ways of synthesis, characterization and manipulation of plethora of nanoparticles at nanoscale. The concept of reduction in particle size of materials is clearly provided in age-old knowledge of Ayurveda too, which utilizes the medicinal values of

P. Kumari · N. Kumari · A. K. Jha
Aryabhatta Centre for Nanoscience and Nanotechnology, Aryabhatta Knowledge University, Patna, India

K. P. Singh
University Department of Zoology, Vinoba Bhave University, Hazaribag, India

K. Prasad (✉)
Department of Physics, Tilka Manjhi Bhagalpur University, Bhagalpur, Bihar, India
e-mail: prasad_k@tmbuniv.ac.in

© Springer Nature Switzerland AG 2018
R. Prasad et al. (eds.), *Exploring the Realms of Nature for Nanosynthesis*,
Nanotechnology in the Life Sciences, https://doi.org/10.1007/978-3-319-99570-0_9

various plant and/or plant parts. So the first relation between nanoscale and human life was developed naturally in Ayurveda about 5000 year ago as an Indian system of medicine. Recently, modern science has been started exploring nanoscience and nanotechnology in twenty-first century. Ayurvedic medicine has metal nanoparticles like silver, gold, platinum, tin, zinc, iron, copper, etc. Nanoparticle has multifunctional properties based on their specific characteristics such as size, shape, distribution, and morphology. It has very broad spectrum of applications in different fields such as medicine, diagnosis, biosensors, bioimaging, targeted drug delivery, nutrition, catalysis, ultrasensitive chemical sensors, cosmetics, nanodevices fabrication, nanomedicine, agriculture, waste water treatment, cosmetics, and food industry. Apart from medical application of nanoparticles, it has been also used in electronics, magnetic, optoelectronics, energy storage, mechanics, instrumentations, and storage devices (Sondi and Salopek-Sondi 2004; Chandran et al. 2006; Song and Kim 2009; Monda et al. 2011; Bar et al. 2009; Kim et al. 2007; Joerger et al. 2000; Panigrahi et al. 2004; Oliveira et al. 2005; Sathishkumar et al. 2009; Dubey et al. 2010; Khalil et al. 2013; Ahmad et al. 2003; Popescu et al. 2010; Baruwati et al. 2009; Daniel and Astruc 2004; Dhuper et al. 2012; Kumar et al. 2014; Vivek et al. 2012; Jha and Prasad 2011a; Prasad et al. 2014, 2016, 2017). Recently, extensive work has been done on development of new drugs from natural products for the last two decades due to the reported resistance of microorganisms from the existing drugs. Nanoparticles have been synthesized through different physical and chemical processes which have their own merits and demerits. Physical and chemical methods of nanoparticle synthesis has required high energy, temperature, pressure, radiation and highly concentrated reducing and stabilizing agents that are harmful to the ecosystem. Therefore, biological synthesis/green synthesis of nanoparticles has been using less energy, low temperature and pressure and occurs in single steps that produce functionalized nanoparticles. Monodispersed and functionalized nanoparticles are challenged for biomaterials science which is synthesized by green technology with specific size, shape, and morphology. So, the green synthesis of nanoparticle has been evolving over other classical (physical and chemical) methods due to the availability of more biological entities and eco-friendly protocols which are non-toxic, non-hazardous, clean, high yield, and cost effective (Prasad 2014). Plant-mediated green synthesis of nanoparticles has been highly explored due to availability of more biodiversity, easily found in nature, and economically feasible. Indian traditional medical system has utilized approx. 600 medicinal plants for development of medicine to cure the chronic diseases during seventeenth century. These plant/plants parts contain novel phytochemicals in form of primary and secondary metabolites such as proteins, amino acids, enzymes, carbohydrates, polysaccharides, alkaloids, flavonoid, tannins, terpenoids, saponins, vitamins, and phenolic acid. A large number of medicinal plants and its various metabolites have been reported in Table 9.1. These phytochemicals (primary and secondary metabolites) have ability to reduce the metal ions into corresponding metal nanoparticles. Hence, the metabolites of different medicinal plant possess different properties and act as bioreductants, stabilizing and capping agents of synthesized nanoparticles. Medicinal plant/plant parts have been utilized to fabricate

Table 9.1 Images of plants/plants' parts with their botanical names along with their phytochemicals

Plants/parts with their botanical names and images	Phytochemicals	References
Aegle marmelos (Leaf)	Akimmianine, aegelin, lupeol, cineole, citral, citronellal, cuminaldehyde, eugenol, and marmesinin	Jha and Prasad (2011a)
Mentha arvensis (Leaf)	d-carvone, careen, d-sylvesterene, and citronellol	Jha and Prasad (2012)
Ocimum sanctum (Leaf)	Eugenol, carvacrol, caryophyllene, limonene, and mucilage	Jha and Prasad (2012)
Eucalyptus globules (Leaf)	Cineole, pinenes, sesquiterpene alcohols, aromadendrene, cuminaldehyde eugenol, and terpenoids	Jha and Prasad (2012)
Azolla sp. (Leaf)	Phenolics, tannins, anthraquinone, glycosides, and sugars	Jha and Prasad (2016a)
Clerodendrum infortunatum (Leaf)	Alkaloids, steroids, terpenoids, phenolics, flavonoids, tannins, and saponins	Jha and Prasad (2016b)
Eclipta alba (Leaf)	Flavonoids, polyphenols, flavones, quinones, terpenoids, and organic acids	Jha et al. (2009)

(continued)

Table 9.1 (continued)

Plants/parts with their botanical names and images	Phytochemicals	References
Cycas (Leaf)	Flavonoids, biflavanone, tetrahydrohinokiflavone, amentoflavone, metallothionein, phytochelatins, superoxide dismutase, catalase, glutathione, malate, citrate, oxalate, succinate, aconitate, α-ketoglutarate, peroxidase	Jha and Prasad (2010)
Azadirachta indica (leaf)	Nimbin, nimbanene, nimban-diol, nimbolide, n-hexacosanol, nimbiol, azadirachtin, tartaric acid, different alkaloids, flavonoids, steroids, terpenoids, organic acids	Kumari et al. (2018a, b), Bhuyan et al. (2015)
Nyctanthes arbortristis (leaf)	Carbohydrate, tannins, saponin, flavonoids, glycosides, phenols, anthraquinones, terpenoids, alkaloids, phlobotamins, and organic acids	Kumari et al. (2018a, b)
Withania somnifera (Leaf)	Methyl 7-oxooctadecanoate, flavonoids, tannins, steroids, amino acids, glycosides, glucose	Nagati et al. (2012)
Acalypha indica (Leaf)	Quercetin, flavonoids, phenolic compounds	Krishnaraj et al. (2010)
Aloe vera (Leaf)	Biomolecules, vitamins, enzymes, lignin, saponins, salicylic acid, amino acids, and sugars	Maensiri et al. (2008)
Andrographis paniculata (Leaf)	Alkaloids, flavonoids amino acids, saponins, tannins, and terpenoids	Suriyakalaa et al. (2013)
Artemisia nilagirica (Leaf)	9-Octadecenoic acid (Z)-tetradecylester, Agaric acid, Tripalmitin, Ergosta-5, 7, 22-trien-3-ol, acetate, Cycloheptasiloxane, tetradecamethyl	Song and Kim (2009)

(continued)

Table 9.1 (continued)

Plants/parts with their botanical names and images	Phytochemicals	References
Tinospora cordifolia (Leaf)	Alkaloids, diterpenoid lactones, glycosides, steroids, sesquiterpenoid, phenolics, aliphatic compounds, polysaccharides	Jayaseelan et al. (2011)
Melia azedarach (Leaf)	Tannic acid, polyphenols, fatty acid, di-terpenoids, sterols, phenol, flavonoid, glycosides, lactones, azadirachtin, nimbin, nimboslin, quercetin, escopolnenin, azadiyeron, azadiyeradion, -14 opeksi, azadyeridon, gedonin, nimosinulnimosinolid, nimbandion, salanol, nimbinen, -6 dastilnimbinen, margozonolid, isomargozonolid, meyanetriol, salannin (its 14 derivatives), feraksinoloz, sinamat, meliasin, A and B nimbolins and limonoid	Sukirtha et al. (2012)
Cinnamomum zeylanicum (Leaf)	Water-soluble organics, ascorbic acid, etc.	Sathishkumar et al. (2009)
Iresine herbstii (Leaf)	Phytoalexin, isothiocyanates, allicins, anthocyanins, essentials oils, tannins, polyphenols, and terpenoids	Dipankar and Murugan (2012)
Euphorbia prostrate (Leaf)	Anthraquinones, protein, polyphenols, flavonoids, phenols, phlobatannins, polysaccharides, saponins, tannins, and terpenoids	Zahir and Rahuman (2012)
Mentha piperita (Leaf)	Menthol, terpenoid, some essential oil	MubarakAli et al. (2011)
Piper betle (Leaf)	Isoverbascoside, terpenoids, volatile oil, and tannins	Cruz et al. (2010)

(continued)

Table 9.1 (continued)

Plants/parts with their botanical names and images	Phytochemicals	References
Piper betle (Leaf)	Alkaloid, tannins, chavicol, betelphenol, eugenol, allyl pyrocatechin, terpene, cineol, caryophyllene, cadinene, menthone phenyl, propane, sesquiterpene, cyneole, sugar, and some essential oil	Mallikarjuna et al. (2012)
Gelsemium sempervirens (Whole)	Protein, amide, saponins, flavonoids, and carbohydrates	Das et al. (2011)
Alternanthera sessilis (Whole)	Amine and carboxyl group such as proteins, amino acids, enzymes, polysaccharides, alkaloids, tannins, phenolics, saponins, terpenoids, and vitamins	Niraimathi et al. (2012)
Mirabilis jalapa (Flower)	Polysaccharides, flavones, terpenoids, and proteins	Vankar and Bajpai (2010)
Rosa sp. (Flower)	Monoterpene hydrocarbons, sesquiterpenes hydrocarbons, oxygenated sesquiterpenes, alkanes/alkenes, alcohols, furan derivatives (O-heterocyclic), kaempferol and quercetin, glycosides	Jha and Prasad (2013)
Citrus aurantium (Fruit)	Flavonoids, vitamin C, flavanones, hesperetin, naringenin, eriodictyol, etc.	Jha et al. (2011)

(continued)

Table 9.1 (continued)

Plants/parts with their botanical names and images	Phytochemicals	References
 Capsicum annum (Fruit)	*trans-p*-feruloyl-β-D-glucopyranoside, *trans-p*-sinapoyl-β-D-glucopyranoside, quercetin 3-*O*-α-l-rhamnopyranoside-7-*O*-β-D-glucopyranoside, *trans-p*-ferulyl alcohol-4-*O*-[6-(2-methyl-3-hydroxypropionyl] glucopyranoside, luteolin 6-*C*-β-D-glucopyranoside-8-*C*-α-larabinopyranoside, apigenin 6-*C*-β-D-glucopyranoside-8-*C*-α-l-arabinopyranoside, luteolin 7-*O*-[2-(β-D-apiofuranosyl)-β-D-glucopyranoside], quercetin 3-*O*-α-l-rhamnopyranoside, and luteolin 7-*O*-[2-(β-D-apiofuranosyl)-4-(β-D-glucopyranosyl)-6-malonyl]-β-D-glucopyranoside	Jha and Prasad (2011b)
 Caria papaya (Fruit)	Hydroxyl flavones, catechins, vitamins, phenols, proteolytic enzyme	Jain et al. (2009), Jha and Prasad (2015)
 Dillenia indica (Fruit)	Biomolecules, lupeol, betulinaldehyde, stigmasterol, betulinic acid, beta-sitosterol, dillentin, betulin, and triterpenoids	Singh et al. (2013)
 Cucumis sativus (Fruit)	Glycosides, alkaloids, tannins, proteins and amino acids, phytosterols, steroids, terpenoids, saponins	Roy et al. (2015)
 Allium cepa	Flavonoids, alkaloids, anthocyanins, polyphenols, diallyl disulfide	Saxena et al. (2010)

(continued)

Table 9.1 (continued)

Plants/parts with their botanical names and images	Phytochemicals	References
Dioscorea bulbifera (Tuber)	Diosgenin, ascorbic acid, flavonol glycosides and proteins	Ghosh et al. (2012)
Citrus sinensis (Peel)	Water-soluble compounds, citric acid, ascorbic acid	Kaviya et al. (2011)
Punica granatum (Peel)	Saponins, quinones, flavonoids, alkaloids, glycosides, coumarins, anthocyanin, and betacyanin	Jayaprakash and Sangeetha (2015)
Trigonella-foenum graecum (Seed)	Flavonoids, proteins	Aromal and Philip (2012)
Cassia fistula (Stem)	Phenol, flavonoid, proanthocyanidin, lupeol, β-sitosterol, hexacosanol	Daisy and Saipriya (2012)
Citrullus colocynthis (Calli)	Polyphenols, catechin gallate, epicatechin gallate, and theaflavin	Satyavani et al. (2011)
Boswellia serrata (Gum)	Protein, enzyme, α-pinene, p-cymene, d-limonene, monoterpenes, diterpenes, and sesquiterpenes	Kora et al. (2012)

different structure, shape, and size like wire, tube, and flower type of nanoparticles which works as functionalized nanoparticles. Functionalized nanoparticles have been used potentially in different areas of medical science and pharmaceutical industries such as diagnosis, therapeutics, drug delivery systems and development of advanced instruments, surgical bandages/nanodevices, and commercial products. Hence, plant-mediated nanoparticle has been considered as building blocks of the forthcoming generations to control various diseases (Jha et al. 2009; Nagati et al. 2012; Krishnaraj et al. 2010; Maensiri et al. 2008; Jha and Prasad 2016a; Sukirtha et al. 2012; MubarakAli et al. 2011; Prasad et al. 2011, 2012; Valtchev and Tosheva 2013; Weiss et al. 2006; Wang et al. 2007; Vanaja et al. 2013; Shameli et al. 2012; Gopinath et al. 2012; Amarnath et al. 2012; Prasad and Swamy 2013, Swamy and Prasad 2012; Mittal et al. 2013; Jacob et al. 2012; Kuppusamy et al. 2016; Ahmed et al. 2016; Makarov et al. 2014; Jayaprakash and Sangeetha 2015; Haverkamp and Marshall 2009; Cruz et al. 2010; Roy and Das 2015; Kumari et al. 2017; Joshi et al. 2018). This chapter will focus on synthesis of functionalized nanoparticles and its application in medical science.

9.2 Fabrication of Functionalized Nanoparticles

Nanoparticles are delineated as particulate dispersions of solid particles with at least one dimension at a size range of 10–1000 nm. When the inorganic nanoparticles are surrounded by the organic molecules as an interfacial layer to stabilize and cap them or are attached as a functional group is defined as functionalized nanoparticles. Most of the plants in nature have been possessing organically active biomolecules which are already established, and these active biomolecules (primary and secondary metabolites) act as reducing, stabilizing, and capping agents for fabrication of inorganic nanoparticles. This fabricated nanoparticle has been occurring in a single step which is non-pathogenic, economically feasible, and ecofriendly. The protocol for the fabrication of different functionalized nanoparticle involves the collection of plants/plant parts from available place, then it has to be washed thoroughly twice/thrice with tap water followed by sterile distilled water to remove associated debris. Approximately 10 gm clean and fresh source (plant/plants' part) has to be dried, cut into fine pieces, and mixed with 100 mL of 50% ethanol in a flask, and the mixture is allowed to be heated and boiled for 15 min before decanting till the color changes from clear transparent to light green (for example). The source mass is then pressed by wrapping in serene cloth and 50 mL extract (say) has to be collected under laminar flow. It has to be doubled in volume and treat it as a source extract. For the reduction of metal ions, this 50 mL of source extract is to be allowed to mix with 5 mL of 0.025 M aqueous solution of metal salt constant stirring at 60 °C on orbital-sacker. As soon as, it starts to change the color of the mixture from deep straw to different colors due to excitation of surface plasmon resonance which indicates the formation of metal nanoparticles. The synthesized nanoparticles through source extract have to be centrifuged at 5500 rpm for 15 min and subsequently re-dispersed in de-ionized water to get rid of any uncoordinated biological molecules. Plant extracts are been used to prepare the nanoparticles, and different techniques have been employed to characterize these nanoparticles. The synthesized nanoparticles through source extract were centrifuged at 5500 rpm for 15 min and subsequently re-dispersed in de-ionized water to get rid of any uncoordinated biological molecules. Plant extracts have been used to prepare the nanoparticles, and different techniques were employed to characterize these nanoparticles. Transmission electron microscopy (TEM) and scanning electron microscope (SEM) analysis provide the size and shape of nanoparticles. The FTIR spectroscopic technique is being used to characterize the nanoparticles which define the biomolecules that have a primary amine group, carbonyl group, and hydroxyl groups, and other stabilizing functional groups bind with inorganic (metal) nanoparticles (Jha et al. 2009, 2011; Das et al. 2011; Jha and Prasad 2010, 2012; Ghosh et al. 2012; Umoren et al. 2014; Tripathy et al. 2010; Bankar et al. 2010; Ahmed and Hussain 2013; Salem et al. 2015; Premanathan et al. 2011; Azam et al. 2012; Gordon et al. 2011; Namratha and Monica 2013; Rao and Paria 2013). Figure 9.1 shows *Nyctanthes arbortristis* leaves (as an example) mediated synthesis of functionalized silver nanoparticles (AgNPs) which may be used for various biomedical applications.

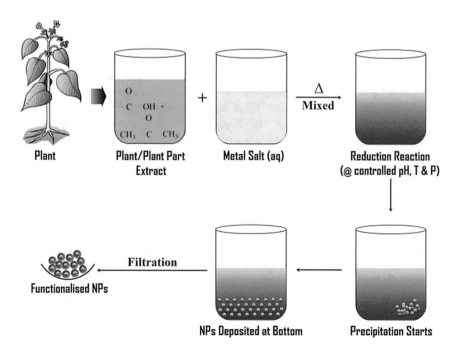

Fig. 9.1 Green synthesis of silver nanoparticles using *Nyctanthes arbortristis* leaves

9.3 Application of Functionalized Silver Nanoparticles

Functionalized nanoparticle has been attached with different types of ligands or capping agents which make chemically more flexible surfaces. This can be turned into biological tools and technique like imaging agents, fluorescent tags, biosensors, molecular scale fluorescent tags, and targeted molecular delivery vehicles which easily distinguish diseased and healthy tissue. This can provide therapy at a molecular level with the help of tools and technique, thus treating the disease and assisting in study of the pathogenesis. This has been designing the drugs with greater degree of cell specificity, more efficacy, and less adverse effects. Here, the application of functionalized silver nanoparticles (AgNPs) has been discussed in various biomedical to industrial applications such as antibacterial, antifungal, and anti-cancer (Mokhtari et al. 2009; Barkalina et al. 2014; Surendiran et al. 2009; Kumari et al. 2017; Nikalje 2015; Cormode et al. 2009; Rai et al. 2013; Han et al. 2007; Fakruddin et al. 2012; Daniel and Astruc 2004; Du et al. 2007; Fayaz et al. 2009; Fortin and Beveridge 2000; Krolikowska et al. 2003; Mallikarjuna et al. 2012; Ninganagouda et al. 2014; Valtchev and Tosheva 2013; Flores et al. 2013; Aziz et al. 2014, 2015, 2016; Ghodake et al. 2013; Liu et al. 2013; Prabhu and Poulose 2012; Ray et al. 2011; Zhang et al. 2012; Kumari et al. 2018a, b). A schematic diagram has been representing various applications of AgNPs which is provided in Fig. 9.2.

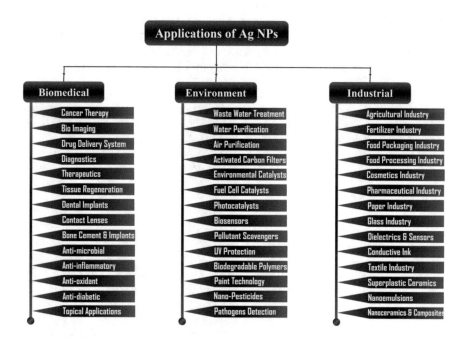

Fig. 9.2 Applications of AgNPs in different areas of science, technology, and industry

9.4   Antimicrobial Activity of AgNPs

Nanoparticle interacts with biological molecules at the nanoscale in both intracellular and extracellular manner. Nanoparticles are obtaining greater importance due to the presence of larger surface area that enables them to have a better contact with microbes as compared to their massive counterparts which shows high antimicrobial activity at the nanoscale. Hence, the various nanoparticles are being used as the potent antimicrobial agents due to the nanosize which changes the physio-chemical properties of metal and oxide. Nanoparticle is an effective and fast acting antimicrobial agent against a broad spectrum of pathogenic bacteria and fungi which have been utilized in various processes in the field of medical science. Silver nanoparticle is one of the nanoparticles which showed effective antimicrobial activity against both bacteria and fungi. The AgNPs interact with thiol group compound present in bacterial and fungal membranes which exhibit a very strong association with them to destroy the activity of bacteria (Gram-positive and Gram-negative bacteria) and fungi cells. The antibacterial activity of AgNPs was studied against *Escherichia coli* (*E. coli*) using agar well diffusion method. The zones of inhibition (clear zones) were observed after the incubation time against the test organism with dose-dependent AgNPs (Fig. 9.3). When *E. coli* cells treated with AgNPs it becomes the formation of "pits" in the bacterial cell walls showed the accumulation of AgNPs which leads to cell death. Another way to get the enhanced efficiency, the functionalized AgNPs have to be mixed with different antibiotics such as erythromycin, penicillin G, clindamycin, amoxicillin, and vancomycin and were evaluated against different microbes such as *E. coli* and *S. aureus* which showed better results. The mechanisms of the action of AgNPs in *E. coli* have been observed through the leakage of reducing sugars and proteins which induce cell death. In addition, the AgNPs not only interact with cell wall and cell membranes but it also perforates into the

Fig. 9.3 Antimicrobial activity of silver nanoparticles (AgNPs) on *E. coli*

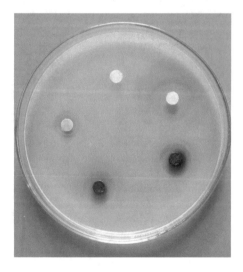

bacterial and fungal cells. After perforation of the AgNPs into the cells, it binds to the cell wall, cell membrane, genomes and inhibits the respiratory process. Therefore the presence of AgNPs inside the cell has inhibited the uptake of phosphate and the release of mannitol, succinate, proline, and glutamine molecules in *E.* coli cells. This study represents the action of AgNPs attached to the sulfur-containing proteins on the bacterial cell membrane, and it can also combine with phosphorus moieties in DNA, which inactivate the replication of DNA and enhance the membrane permeability, inhibition of enzyme functions finally leading to cell death. It shows that the active surfaces of AgNPs release the Ag-ions to induce the generation of intracellular reactive oxygen species (ROS) in bacterial cells and uncouple the electron transport system. The antimicrobial activities of these nanoparticles depend on chemisorbed Ag^+, formed on the surface of AgNPs by dint of extreme sensitivity to oxygen. This nanoparticle have been studied to synthesize composites for use as coating materials for several mechanisms, disinfecting filters and as a medium for antibiotic delivery which have been offered to explain the inhibitory effect of AgNPs on bacteria. These have also been used to treat skin infections like burns, wounds, and healing. AgNPs appear to be alternative antibacterial agents to antibiotics and have the ability to overcome the bacterial resistance against antibiotics. Therefore, silver nanoparticles can be used as effective growth inhibitor in various microbes and they are applicable to different antibacterial and antifungal control system (Amin et al. 2012; Singhal et al. 2011; Singh et al. 2008; Sinha et al. 2009; El-Rafie et al. 2010; Geethalakshmi and Sarada 2010; Fatima et al. 2015; Verma et al. 2010; Sathishkumar et al. 2009; Durán et al. 2007; Aziz et al. 2014, 2015, 2016; Birla et al. 2009; Sharma et al. 2014; Gade et al. 2010; Bankura et al. 2012; Flores et al. 2013; Fortin and Beveridge 2000; Ghodake et al. 2013; Lokina et al. 2015; Krolikowska et al. 2003; Liu et al. 2013; Manikandan et al. 2014; Prabhu and Poulose 2012; Ray et al. 2011; Tripathi et al. 2007; Zhang et al. 2012; Barkalina et al. 2014; Mallikarjuna et al. 2012; Surendiran et al. 2009; Nikalje 2015; Cormode et al. 2009; Ninganagouda et al. 2014; Roy et al. 2015; Rai et al. 2013; Fakruddin et al. 2012; Bindhu and Umadevi 2013; Ali et al. 2011; Amarnath et al. 2012; Zhang et al. 2016).

9.5 Anticancer Activity of AgNPs

Cancer is a disorder causing death worldwide which is characterized by an abnormal growth of cells or tissues developed through various cellular functions such as cell signaling, proliferations, and apoptosis which changes the molecular interaction within the cell. It is a second major cause of death in the human. It is treated by different methods like surgery, chemotherapy, radiotherapy, and others. These standard methods of treatments are expensive and have side effects; therefore alternatives of such treatments are urgently needed which should be inexpensive, non-toxic, and effective with minimal side effects for wider acceptability. So there are increasing demands for anticancer therapy which is a challenge to find drugs and therapies for the treatments of various cancers. The battle against cancer is most difficult particularly in the development of therapies for multiplying several tumors. Chemotherapy has been used for the treatment of cancer but still it reveals low specificity and restricted dose to limit toxicity. Therefore the conventional methods have been developed to require the combination of controlled released technology and targeted drug delivery that is more effective and less harmful. Nanoparticles are expected to revolutionize the cancer treatment as diagnosis and therapy. It can provide unique interactions with biomolecules that may be useful in diagnosis and treatment of cancer. Recently, various nanoparticles (Ag, Au, Pt, Cu, ZnO, CuO) have been used in anticancer therapy for several types of cancer like HT-29 cell lines, Hep2 cell line, THP-1 human leukemia cell lines, Vero cell line and breast cancer lines, liver cancer cell lines (HepG2) and lung cancer cell lines (A549). In this chapter, investigation of functionalized silver nanoparticles (AgNPs) and their anticancer activity against THP-1 human leukemia cell lines has been presented. Recently, experiments observed that AgNPs induce apoptosis and sensitize cancerous cells. It can also induce alterations in cell morphology, metabolic activity, DNA damage, decreased cell viability and increased production of reactive oxygen species (ROS), which might enhance the oxidative stress leading to mitochondrial damage. The cellular uptake of AgNPs has been occurred through endocytosis. When such cells are treated with nanoparticles, they exhibit various abnormalities like upregulation of metallothionein, downregulation of major actin, myosin, and filamin binding proteins which arrest mitotic division. The morphology analysis of cancer cells observed that functionalized AgNPs could induce cell death very significantly. Figure 9.4 shows the morphological analysis of THP-1 leukemia cell line treated with different concentrations of functionalized AgNPs synthesized form *Nyctanthes arbortristis* leaf and compared with untreated cell line. The cellular and molecular mechanisms of AgNPs-treated cancerous and normal cell lines could be understood as the AgNPs are capable of adsorbing cytosolic proteins on the surface that may influence the intracellular function of the cancerous cells, which can also regulate gene expression and production of inflammatory cytokines. The AgNPs could be altering the regulation of about more than 1000 genes in a cell but among several genes, heat shock protein, metallothionein, and histone proteins are most

Fig. 9.4 Comparison of morphological changes in untreated and treated (with different concentrations of functionalized AgNPs synthesized using *Nyctanthes arbortristis* leaves) THP-1 leukemia cell line

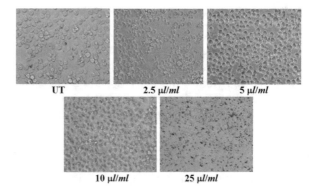

significant. Recently, AgNPs are also inducing autophagy through the accumulation of autophagolysosomes inside cancerous cells. Autophagy is a concentration-dependent process which has a dual function, at lower level of concentration it can enhance the cell survival and at higher level of concentration it can cause cell death. Functionalized AgNPs manifested pronounced toxic effect against carcinoma cells than non-carcinoma cells which indicates that AgNPs could target cell-specific toxicity and could lower level of pH in the cancer cells. So the targeted delivery is an essential requirement for the treatment of cancer (Flores et al. 2013; Ramar et al. 2014; Banerjee and Narendhirakannan 2011; Sathishkumar et al. 2009; Lokina et al. 2015; Azzazy et al. 2012; Daisy and Saipriya 2012; Fortina et al. 2007; Fatima et al. 2015; Gan and Li 2012; Jacob et al. 2011; Jacob et al. 2012; Subramanian 2012; Safaepour et al. 2009; Asharani et al. 2009; Sanpui et al. 2011; Guo et al. 2013; Rajeshkumar 2016; Phull et al. 2016; Majeed et al. 2018; Krishnan et al. 2016; Nagajyothi et al. 2017; Majeed et al. 2016).

9.6 Conclusion

Nature has indeed been elegant and resourceful which enables us to create the most efficient miniaturized functional nanoparticles. The use of green technology for metal nanoparticles synthesis leads a desire to develop eco-friendly techniques. These functionalized nanoparticles have been used in various fields such as pharmaceuticals, diagnosis, therapeutics, sustainable energy, agriculture, environments, and other commercial products. The plant-mediated nanoparticles have projected impact on treatment and diagnosis of different diseases with controlled side effects. In future, the plant metabolites have wide perspective for the synthesis of functionalized metal nanoparticles in biomedical applications and commercial products.

References

Ahmad A, Mukherjee P, Senapati S, Mandal D, Khan MI, Kumar R, Sastry M (2003) Extracellular biosynthesis of silver nanoparticles using the fungus *Fusarium oxysporum*. Colloids Surf B Biointerfaces 28:313–318

Ahmed M, Hussain F (2013) Chemical composition and biochemical activity of *Aloe vera* (*Aloe barbadensis* miller) leaves. Int J Chem Biochem Sci 3:29–33

Ahmed S, Ahmad M, Swami BL, Ikram S (2016) A review on plants extract mediated synthesis of silver nanoparticles for antimicrobial applications: a green expertise. J Adv Res 7:17–28

Ali DM, Thajuddin N, Jeganathan K, Gunasekaran M (2011) Plant extract mediated synthesis of silver and gold nanoparticles and its antibacterial activity against clinically isolated pathogens. Colloids Surf B Biointerfaces 85:360–365

Amarnath K, Kumar J, Reddy T, Mahesh V, Ayyappan SR, Nellore J (2012) Synthesis and characterization of chitosan and grape polyphenols stabilized palladium nanoparticles and their antibacterial activity. Colloids Surf B Biointerfaces 92:254–261

Amin M, Anwar F, Janjua MRSA, Iqbal MA, Rashid U (2012) Green synthesis of silver nanoparticles through reduction with *Solanum xanthocarpum* L. berry extract: characterization, antimicrobial and urease inhibitory activities against *Helicobacter pylori*. Int J Mol Sci 13:9923–9941

Aromal SA, Philip D (2012) Green synthesis of gold nanoparticles using *Trigonella foenumgraecum* and its size dependent catalytic activity. Spectrochim Acta A 97:1–5

Asharani PV, Mun GLK, Hande MP, Valiyaveettil S (2009) Cytotoxicity and genotoxicity of silver nanoparticles in human cells. ACS Nano 3:279–290

Azam A, Ahmed AS, Oves M, Khan MS, Habib SS, Memic A (2012) Antimicrobial activity of metal oxide nanoparticles against Gram-positive and Gram-negative bacteria: a comparative study. Int J Nanomed 2012:6003–6009

Aziz N, Fatma T, Varma A, Prasad R (2014) Biogenic synthesis of silver nanoparticles using *Scenedesmus abundans* and evaluation of their antibacterial activity. J Nanopart 2014:689419. https://doi.org/10.1155/2014/689419

Aziz N, Faraz M, Pandey R, Sakir M, Fatma T, Varma A, Barman I, Prasad R (2015) Facile algae-derived route to biogenic silver nanoparticles: synthesis, antibacterial and photocatalytic properties. Langmuir 31:11605–11612. https://doi.org/10.1021/acs.langmuir.5b03081

Aziz N, Pandey R, Barman I, Prasad R (2016) Leveraging the attributes of *Mucor hiemalis*-derived silver nanoparticles for a synergistic broad-spectrum antimicrobial platform. Front Microbiol 7:1984. https://doi.org/10.3389/fmicb.2016.01984

Azzazy HME, Mansour MMH, Samir TM, Franco R (2012) Gold nanoparticles in the clinical laboratory: principles of preparation and applications. Clin Chem Lab Med 50:193–209

Banerjee J, Narendhirakannan RT (2011) Biosynthesis of silver nanoparticles from *Syzygium cumini* (L.) seed extract and evaluation of their in vitro antioxidant activities. Dig J Nanomater Biostruct 6:961–968

Bankar A, Joshi B, Kumar AR, Zinjarde S (2010) Banana peel extract mediated novel route for the synthesis of silver nanoparticles. Colloids Surf A Physicochem Eng Asp 368:58–63

Bankura P, Maity D, Mollick MM, Mondal D, Bhowmick B, Bain MK, Chakraborty A, Sarkar J, Acharya K, Chattopadhyay D (2012) Synthesis, characterization and antimicrobial activity of dextran stabilized silver nanoparticles in aqueous medium. Carbohydr Polym 1:1159–1165

Bar H, Bhui DK, Sahoo GP, Sarkar P, De SP (2009) Green synthesis of silver nanoparticles using latex of *Jatropha Curcas*. Colloids Surf A Physicochem Eng Aspects 339:134–139

Barkalina N, Charalambous C, Jones C, Coward K (2014) Nanotechnology in reproductive medicine: emerging applications of nanomaterials. Nanomedicine 10(5):921–938

Baruwati B, Polshettiwar V, Varma RS (2009) Glutathione promoted expeditious green synthesis of silver nanoparticles in water using microwaves. Green Chem 11:926–930

Bhuyan T, Mishra K, Khanuja M, Prasad R, Varma A (2015) Biosynthesis of zinc oxide nanoparticles from Azadirachta indica for antibacterial and photocatalytic applications. Mater Sci Semicond Process 32:55–61

Bindhu MR, Umadevi M (2013) Synthesis of monodispersed silver nanoparticles using *Hibiscus cannabinus* leaf extract and its antimicrobial activity. Spectrochim Acta A Mol Biomol Spectrosc 101:184–190

Birla SS, Tiwari VV, Gade AK, Ingle AP, Yadav AP, Rai MK (2009) Fabrication of silver nanoparticles by *Phoma glomerata* and its combined effect against *Escherichia coli, Pseudomonas aeruginosa* and *Staphylococcus aureus*. Lett Appl Microbiol 1:2173–2179

Chandran SP, Chaudhary M, Pasricha R, Ahmad A, Sastry M (2006) Synthesis of gold nanotriangles and silver nanoparticles using *Aloe vera* plant extract. Biotechnol Prog 22:577–583

Cruz D, Fale PL, Mourato A, Vaz PD, Serralheiro ML, Lino ARL (2010) Preparation and physicochemical characterization of Ag nanoparticles biosynthesized by *Lippia citriodora* (*Lemon Verbena*). Colloids Surf B Biointerfaces 81:67–73

Daisy P, Saipriya K (2012) Biochemical analysis of *Cassia fistula* aqueous extracts and phytochemically synthesized gold nanoparticles as hypoglycemic treatment for *diabetes mellitus*. Int J Nanomed 7:1189–1202

Daniel MC, Astruc D (2004) Gold nanoparticles: assembly, supramolecular chemistry, quantum size related properties, and applications toward biology, catalysis, and nanotechnology. Chem Rev 104:293–346

Das RK, Gogoi N, Bora U (2011) Green synthesis of gold nanoparticles using *Nyctanthes arbortristis* flower extract. Bioprocess Biosyst Eng 34:615–619

Cormode DP, Skajaa T, Fayad ZA, Willem JMM (2009) Nanotechnology in medical imaging: probe design and applications. Arterioscler Thromb Vasc Biol 29:992–1000

Dipankar C, Murugan S (2012) The green synthesis, characterization and evaluation of the biological activities of silver nanoparticles synthesized from *Iresine herbstii* leaf aqueous extracts. Colloids Surf B Biointerfaces 98:112–119

Dubey SP, Lahtinen M, Sarkka H, Sillanpaa M (2010) Bioprospective of *Sorbus aucuparia* leaf extract in development of silver and gold nanocolloids. Colloids Surf B Biointerfaces 80:26–33

Dhuper S, Panda D, Nayak PL (2012) Green synthesis and characterization of zero valent iron nanoparticles from the leaf extract of *Mangifera indica*. Nano Trends: J Nanotech App 13:16–22

Du L, Jiang H, Liu X, Wang E (2007) Biosynthesis of gold nanoparticles assisted by *Escherichia coli* DH5α and its application on direct electrochemistry of hemoglobin. Electrochem Commun 9:1165–1170

Durán N, Marcato PD, De Souza GIH, Alves OL, Esposito E (2007) Antibacterial effect of silver nanoparticles produced by fungal process on textile fabrics and their effluent treatment. J Biomed Nanotechnol 3:203–208

El-Rafie MH, Mohamed AA, Shaheen TI, Hebeish A (2010) Antimicrobial effect of silver nanoparticles produced by fungal process on cotton fabrics. Carbohydr Polym 80:779–782

Fakruddin M, Hossain Z, Afroz H (2012) Prospects and applications of nanobiotechnology: a medical perspective. J Nanobiotechnol 10:1–8

Fatima F, Bajpai P, Pathak N, Singh S, Priya S, Verma SR (2015) Antimicrobial and immunomodulatory efficacy of extracellularly synthesized silver and gold nanoparticles by a novel phosphate solubilizing fungus *Bipolaris tetramera*. BMC Microbiol 15:52–61

Fayaz AM, Balaji K, Girilal M, Kalaichelvan PT, Venkatesan R (2009) Mycobased synthesis of silver nanoparticles and their incorporation into sodium alginate films for vegetable and fruit preservation. J Agric Food Chem 57:6246–6252

Flores CY, Miñán AG, Grillo CA, Salvarezza RC, Vericat C, Schilardi PL (2013) Citrate capped silver nanoparticles showing good bactericidal effect against both planktonic and sessile bacteria and a low cytotoxicity to osteoblastic cells. ACS Appl Mater Interfaces 5:3149–3159

Fortin D, Beveridge TJ (2000) From biology to biotechnology and medical applications. In: Aeuerien E (ed) Biomineralization. Weinheim, Wiley VCH

Fortina P, Kricka LJ, Graves DJ, Park J, Hyslop T, Tam F (2007) Applications of nanoparticles to diagnostics and therapeutics in colorectal cancer. Trends Biotechnol 25:145–152

Gade A, Gaikwad S, Tiwari V, Yadav A, Ingle A, Rai M (2010) Biofabrication of silver nanoparticles by *Opuntiaficus-indica*: in vitro antibacterial activity and study of the mechanism involved in the synthesis. Curr Nanosci 6:370–375

Gan PP, Li SFY (2012) Potential of plant as a biological factory to synthesize gold and silver nanoparticles and their applications. Rev Environ Sci Biotechnol 11:169–206

Geethalakshmi R, Sarada DVL (2010) Synthesis of plant mediated silver nanoparticles using Trianthema decandra extract and evaluation of their antimicrobial activities. Int J Eng Sci Technol 2:970–975

Ghodake G, Lim SR, Lee DS (2013) Casein hydrolytic peptides mediated green synthesis of antibacterial silver nanoparticles. Colloids Surf B Biointerfaces 108:147–151

Ghosh S, Patil S, Ahire M, Kitture R, Kale S, Pardesi K, Cameotra SS, Bellare J, Dhavale DD, Jabgunde A, Chopad BA (2012) Synthesis of silver nanoparticles using *Dioscorea bulbifera* tuber extract and evaluation of its synergistic potential in combination with antimicrobial agents. Int J Nanomed 7:483–496

Gopinath V, MubarakAli D, Priyadarshini S, Priyadharsshini NM, Thajuddin N, Velusamy P (2012) Biosynthesis of silver nanoparticles from *Tribulus terrestris* and its antimicrobial activity a novel biological approach. Colloids Surf B Biointerfaces 96:69–74

Gordon T, Perlstein B, Houbara O, Felner I, Banin E, Margel S (2011) Synthesis and characterization of zinc/iron oxide composite nanoparticles and their antibacterial properties. Colloids Surf A Physicochem Eng Aspects 374:1–8

Guo D, Zhu L, Huang Z, Zhou H, Ge Y, Ma W, Wu J, Zhang X, Zhou X, Zhang Y (2013) Anti-leukemia activity of PVP-coated silver nanoparticles via generation of reactive oxygen species and release of silver ions. Biomaterials 34:7884–7894

Han G, Ghosh P, Rotello VM (2007) Functionalized gold nanoparticles for drug delivery. Nanomedicine 2:113–123

Haverkamp RG, Marshall AT (2009) The mechanism of metal nanoparticle formation in plants: limits on accumulation. J Nanopart Res 11:1453–1463

Jha AK, Prasad K (2010) Green synthesis of silver nanoparticles using *Cycas* leaf. Int J Green Nanotechnol Phy Chem 1:110–117

Jha AK, Prasad K (2011a) Biosynthesis of gold nanoparticles using bael (*Aegle marmelos*) leaf: mythology meets technology. Int J Green Nanotechnol Phys Chem 3:92–97

Jha AK, Prasad K (2011b) Green fruit of chili (*Capsicum annum* L.) synthesizes nano silver. Dig J Nanomater Biostruct 6:1717–1723

Jha AK, Prasad K (2012) Biosynthesis of gold nanoparticles using common aromatic plants. Int J Green Nanotechnol Phy Chem 4:219–224

Jha AK, Prasad K (2013) Rose (Rosa sp.) petals assisted green synthesis of gold nanoparticles. J Bionanosci 7:1–6

Jha AK, Prasad K (2015) Facile green synthesis of metal and oxide nanoparticles using Papaya juice. J Bionanosci 9:311–314

Jha AK, Prasad K (2016a) Aquatic fern (*Azolla Sp.*) assisted synthesis of gold nanoparticles. Int J Nanosci 15:1650008–1650005

Jha AK, Prasad K (2016b) Green synthesis and antimicrobial activity of silver nanoparticles onto cotton fabric: an amenable option for textile industries. Adv Mater Lett 7:42–46

Jha AK, Prasad K, Kumar V, Prasad K (2009) Biosynthesis of silver nanoparticles using *Eclipta* leaf. Biotechnol Prog 25:1476–1479

Jha AK, Kumar V, Prasad K (2011) Biosynthesis of metal and oxide nanoparticles using orange juice. J Bionanosci 5:162–166

Jayaprakash A, Sangeetha R (2015) Phytochemical screening of *Punica granatum* Linn. peel extracts. J Acad Indust Res 4:160–162

Jain D, Daima HK, Kachhwaha S, Kothari SL (2009) Synthesis of plant-mediated silver nanoparticles using papaya fruit extract and evaluation of their anti-microbial activities. Dig J Nanomat Biostruct 4:557–563

Jayaseelan C, Rahuman AA, Rajakumar G, Kirthi AV, Santhoshkumar T, Marimuthu S (2011) Synthesis of pediculocidal and larvicidal silver nanoparticles by leaf extract from heart leaf moon seed plant *Tinospora cordifolia* Miers. Parasitol Res 109:185–194

Joerger R, Klaus T, Granqvist CG (2000) Biologically produced silver–carbon composite materials for optically functional thin-film coatings. Adv Mater 12:407–409

Joshi N, Jain N, Pathak A, Singh J, Prasad R, Upadhyaya CP (2018) Biosynthesis of silver nanoparticles using *Carissa carandas* berries and its potential antibacterial activities. J Sol-Gel Sci Techn 86(3):682–689. https://doi.org/10.1007/s10971-018-4666-2

Jacob SJP, Finub JS, Narayanan A (2012) Synthesis of silver nanoparticles using *Piper longum* leaf extracts and its cytotoxic activity against Hep-2 cell line. Colloids Surf B Biointerfaces 91:212–214

Jacob S, Finub J, Narayanan A (2011) Synthesis of silver nanoparticles using Piper longum leaf extracts and its cytotoxic activity against Hep-2 cell line. Colloids Surf B Biointerfaces 91:212–214

Kaviya S, Santhanalakshmi J, Viswanathan B, Muthumary J, Srinivasan K (2011) Biosynthesis of silver nanoparticles using *Citrus sinensis* peel extract and its antibacterial activity. Spectrochim Acta A 79:594–598

Khalil KA, Fouad H, Elsarnagawy T, Almajhdi FN (2013) Plant mediated green synthesis of silver nanoparticles—a review. Int J Electrochem Sci 8:3483–3493

Kim JS, Kuk E, Yu KN, Jong-Ho K, Park SJ, Lee HJ, Kim SH (2007) Antimicrobial effects of silver nanoparticles. Nanomedicine 3:95–101

Kora AJ, Sashidhar RB, Arunachalam J (2012) Aqueous extract of gum olibanum (*Boswellia serrata*): a reductant and stabilizer for the biosynthesis of antibacterial silver nanoparticles. Process Biochem 47:1516–1520

Krishnan V, Bupesh G, Manikandan E, Thanigai AK, Magesh S (2016) Green synthesis of silver nanoparticles using *Piper nigrum* concoction and its anticancer activity against MCF-7 and Hep-2 cell lines. J Antimicrob Agents 2:123–128

Krishnaraj C, Jagan EG, Rajasekar S, Selvakumar P, Kalaichelvan PT, Mohan N (2010) Synthesis of silver nanoparticles using *Acalypha indica* leaf extracts and its antibacterial activity against water borne pathogens. Colloids Surf B Biointerfaces 76:50–56

Krolikowska A, Kudelski A, Michota A, Bukowska J (2003) SERS studies on the structure of thioglycolic acid monolayers on silver and gold. Surf Sci 532:227–232

Kumar DA, Palanichamy V, Roopan SM (2014) Green synthesis of silver nanoparticles using *Alternanthera dentata* leaf extract at room temperature and their antimicrobial activity. Spectrochim Acta A Mol Biomol Spectrosc 127:168–171

Kumari N, Kumari P, Jha AK, Prasad K (2018a) Enhanced antimicrobial activity in biosynthesized ZnO nanoparticles. AIP Conf Proc 1953:030054

Kumari P, Kumari N, Jha AK, Singh KP, Prasad K (2018b) *Nyctanthes arbortristis* mediated synthesis of silver nanoparticles: cytotoxicity assay against THP-1 human leukemia cell lines. AIP Conf Proc 1953:030071

Kumari N, Jha AK, Prasad K (2017) Fungal nanotechnology and biomedicine. In: Prasad R (ed) Fungal nanotechnology: applications in agriculture, industry, and medicine, Fungal biology. Springer. ISBN: 978-3-319-68423-9

Kuppusamy P, Yusoff MM, Maniam GP, Govindan N (2016) Biosynthesis of metallic nanoparticles using plant derivatives and their new avenues in pharmacological applications—an updated report. Saudi Pharm J 24:473–484

Liu L, Yang J, Xie J, Luo Z, Jiang J, Yang Y (2013) The potent antimicrobial properties of cell penetrating peptide conjugated silver nanoparticles with excellent selectivity for Gram positive bacteria over erythrocytes. Nanoscale 5:3834–3840

Lokina S, Stephen A, Kaviyarasan V, Arulvasu C, Narayanan V (2015) Cytotoxicity and antimicrobial studies of silver nanoparticles synthesized using *Psidium guajava* L. extract. Synth React Inorg Met-Org Nano-Metal Chem 45:426–432

Maensiri S, Laokul P, Klinkaewnarong J, Phokha S, Promarak V, Seraphin S (2008) Indium oxide (In_2O_3) nanoparticles using *Aloe vera* plant extract: synthesis and optical properties. J Optoelectron Adv Mater 10:161–165

Majeed S, Danish M, Zahrudin AHB, Dash GK (2018) Biosynthesis and characterization of silver nanoparticles from fungal species and its antibacterial and anticancer effect. Karbala Int J Modern Sci 4:86–92

Majeed S, Abdullah MS, Nanda A, Ansari MT (2016) In vitro study of the antibacterial and anticancer activities of silver nanoparticles synthesized from *Penicillium brevicompactum* (MTCC-1999). J Taibah Univ Sci 10:614–620

Makarov VV, Love AJ, Sinitsyna OV, Makarova SS, Yaminsky IV, Taliansky ME, Kalinina NO (2014) "Green" nanotechnologies: synthesis of metal nanoparticles using plants. Acta Nat 6:35–44

Mallikarjuna K, Dillip GR, Narashima G, Sushma NJ, Raju BDP (2012) Phytofabrication and characterization of silver nanoparticles from *Piper betel* broth. Res J Nanosci Nanotech 2:17–23

Manikandan D, Prakash DG, Gandhi NN (2014) A rapid and green route to synthesis of silver nanoparticles from *Plectranthus barbatus* (coleus forskohlii) root extract for antimicrobial activity. Int J Chem Tech Res 6:4391–4396

Mittal AK, Chisti Y, Banerjee UC (2013) Synthesis of metallic nanoparticles using plant extracts. Biotechnol Adv 31:346–356

Mokhtari N, Daneshpajouh S, Seyedbagheri S, Atashdehghan R, Abdi K, Sarkar S, Minaian S, Shahverdi HR, Shahverdi AR (2009) Biological synthesis of very small silver nanoparticles by culture supernatant of *Klebsiella pneumonia*: the effects of visible light irradiation and the liquid mixing process. Mater Res Bull 44:1415–1421

Monda S, Roy N, Laskar RA, Sk I, Basu S, Mandal D, Begum NA (2011) Biogenic synthesis of Ag, Au and bimetallic Au/Ag alloy nanoparticles using aqueous extract of mahogany (*Swietenia mahogani* JACQ.) leaves. Colloids Surf B Biointerfaces 82:497–504

MubarakAli D, Thajuddin N, Jeganathan K, Gunasekaran M (2011) Plant extract mediated synthesis of silver and gold nanoparticles and its antibacterial activity against clinically isolated pathogens. Colloids Surf B Biointerfaces 85:360–365

Nagajyothi PC, Muthuraman P, Sreekanth TVM, Kim DH, Shim J (2017) Green synthesis: in-vitro anticancer activity of copper oxide nanoparticles against human cervical carcinoma cells. Arab J Chem 10:215–225

Nagati VB, Alwala J, Koyyati R, Donda MR, Banala R, Padigya PRM (2012) Green synthesis of plant-mediated silver nanoparticles using *Withania somnifera* leaf extract and evaluation of their anti-microbial activity. Asian Pac J Trop Biomed 2:1–5

Namratha N, Monica PV (2013) Green synthesis of silver nanoparticles using *Mentha asiatica* (Mint) extract and evaluation of their antimicrobial potential. J Pharm Tech 3:170–174

Nikalje AP (2015) Nanotechnology and its applications in medicine. Med Chem 5:081–089

Ninganagouda S, Rathod V, Singh D (2014) Characterization and biosynthesis of silver nanoparticles using a fungus *Aspergillus niger*. Int Lett Nat Sci 10:49–57

Niraimathi KL, Sudha V, Lavanya R, Brindha P (2012) Biosynthesis of silver nanoparticles using *Alternanthera sessilis* (Linn.) extract and their antimicrobial, antioxidant activities. Colloids Surf B Biointerfaces 88:34–39

Oliveira MM, Ugarte D, Zanchet D, Zarbin AJG (2005) Influence of synthetic parameters on the size, structure, and stability of dodecanethiol-stabilized silver nanoparticles. J Colloid Interface Sci 292:429–435

Panigrahi S, Kundu S, Ghosh S, Nath S, Pal T (2004) General method of synthesis for metal nanoparticles. J Nanopart Res 6:411–414

Phull AR, Abbas Q, Ali A, Raza H, Kim SJ, Zia M, Haq I (2016) Antioxidant, cytotoxic and anti-microbial activities of green synthesized silver nanoparticles from crude extract of *Bergenia ciliate*. Future J Pharm Sci 2:31–36

Popescu M, Velea A, Lorinczi A (2010) Biogenic production of nanoparticles. Dig J Nanomater Biostruct 5:1035–1040

Prabhu S, Poulose EK (2012) Silver nanoparticles: mechanism of antimicrobial action, synthesis, medical applications, and toxicity effects. Int Nano Lett 2:32–41

Prasad KS, Pathak D, Patel A, Dalwadi P, Prasad R, Patel P, Kaliaperumal SK (2011) Biogenic synthesis of silver nanoparticles using Nicotiana tobaccum leaf extract and study of their anti-bacterial effect. Afr J Biotechnol 9(54):8122–8130

Prasad R (2014) Synthesis of silver nanoparticles in photosynthetic plants. J Nanopart 2014:963961. https://doi.org/10.1155/2014/963961

Prasad R, Bhattacharyya A, Nguyen QD (2017) Nanotechnology in sustainable agriculture: recent developments, challenges, and perspectives. Front Microbiol 8:1014. https://doi.org/10.3389/fmicb.2017.01014

Prasad R, Kumar V, Prasad KS (2014) Nanotechnology in sustainable agriculture: present concerns and future aspects. Afr J Biotechnol 13(6):705–713

Prasad R, Pandey R, Barman I (2016) Engineering tailored nanoparticles with microbes: quo vadis. WIREs Nanomed Nanobiotechnol 8:316–330. https://doi.org/10.1002/wnan.1363

Prasad R, Swamy VS (2013) Antibacterial activity of silver nanoparticles synthesized by bark extract of *Syzygium cumini*. J Nanopart 2013:431218. https://doi.org/10.1155/2013/431218

Prasad R, Swamy VS, Varma A (2012) Biogenic synthesis of silver nanoparticles from the leaf extract of *Syzygium cumini* (L.) and its antibacterial activity. Int J Pharm Bio Sci 3(4):745–752

Premanathan M, Karthikeyan K, Jeyasubramanian K, Manivannan G (2011) Selective toxicity of ZnO nanoparticles toward Gram-positive bacteria and cancer cells by apoptosis through lipid peroxidation. Nanomedicine 7:184–192

Rai M, Ingle AP, Gupta IR, Birla SS, Yadav AP, Abd-Elsalam KA (2013) Potential role of biological systems in formation of nanoparticles: mechanism of synthesis and biomedical applications. Curr Nanosci 9:576–587

Rajeshkumar S (2016) Anticancer activity of eco-friendly gold nanoparticles against lung and liver cancer cells. J Genet Eng Biotechnol 14:195–202

Ramar M, Manikandan B, Marimuthu PN, Raman T, Mahalingam A, Subramanian P (2014) Synthesis of silver nanoparticles using *Solanum trilobatum* fruits extract and its antibacterial, cytotoxic activity against human breast cancer cell line MCF 7. Spectrochim Acta A 140:223–228

Rao KJ, Paria S (2013) Green synthesis of silver nanoparticles from aqueous *Aegle marmelos* leaf extract. Mater Res Bull 48:628–638

Ray S, Sarkar S, Kundu S (2011) Extracellular biosynthesis of silver nanoparticles using the mycorrhhizal mushroom *Tricholoma crassum* (BERK.) SACC: its antimicrobial activity against pathogenic bacteria and fungus, including multidrug resistant plant and human bacteria. Dig J Nanomater Biostruct 6:1289–1299

Roy S, Das TK (2015) Plant mediated green synthesis of silver nanoparticles—a review. Int J Plant Biol Res 3:1044–1055

Roy K, Sarkar CK, Ghosh CK (2015) Single-step novel biosynthesis of silver nanoparticles using *Cucumis sativus* fruit extract and study of its photcatalytic and antibacterial activity. Dig J Nanomater Biostruct 10:107–115

Safaepour M, Shahverdi AR, Shahverdi HR, Khorramizadeh MR, Gohari AR (2009) Green synthesis of small silver nanoparticles using geraniol and its cytotoxicity against *Fibrosarcoma–Wehi* 164. Avicenna J Med Biotechnol 1:111–115

Salem W, Leitner DR, Zingl FG, Schratter G, Prassl R, Goessler W, Reidl J, Schild S (2015) Antibacterial activity of silver and zinc nanoparticles against *Vibrio cholerae* and enterotoxic *Escherichia coli*. Int J Med Microbiol 305:85–95

Sanpui P, Chattopadhyay A, Ghosh SS (2011) Induction of apoptosis in cancer cells at low silver nanoparticle concentrations using chitosan nanocarrier. ACS Appl Mater Interfaces 3:218–228

Sathishkumar M, Sneha K, Won SW, Cho CW, Kim S, Yun YS (2009) *Cinnamomum zeylanicum* bark extract and powder mediated green synthesis of nano-crystalline silver particles and its bactericidal activity. Colloids Surf B Biointerfaces 73:332–338

Satyavani K, Gurudeeban S, Ramanathan T, Balasubramanian T (2011) Biomedical potential of silver nanoparticles synthesized from calli cells of *Citrullus colocynthis* (L.) Schrad. J Nanobiotechnol 9:43–51

Saxena A, Tripathi RM, Singh RP (2010) Biological synthesis of silver nanoparticles by using onion (*Allium cepa*) extract and their antibacterial activity. Dig J Nanomater Biostruct 5:427–432

Shameli K, Ahmad M, Al-Mulla EAJ, Ibrahim NA, Shabanzadeh P, Rustaiyan A, Abdollahi Y (2012) Green biosynthesis of silver nanoparticles using *Callicarpa maingayi* stem bark extraction. Molecules 17:8506–8517

Sharma G, Sharma AR, Kurian M, Bhavesh R, Nam JS, Lee SS (2014) Green synthesis of silver nanoparticle using *Myristica fragrans* (nutmeg) seed extract and its biological activity. Dig J Nanomater Biostruct 9:325–332

Singh M, Singh S, Prasad S, Gambhir IS (2008) Nanotechnology in medicine and antibacterial effect of silver nanoparticles. Dig J Nanomater Biostruct 3:115–122

Singh S, Saikia JP, Buragohain AK (2013) A novel "Green" synthesis of colloidal silver nanoparticles (SNP) using *Dillenia indica* fruit extract. Colloids Surf B Biointerfaces 102:83–85

Singhal G, Bhavesh R, Kasariya K, Sharma AR, Singh RP (2011) Biosynthesis of silver nanoparticles using *Ocimum sanctum* (Tulsi) leaf extract and screening its antimicrobial activity. J Nanopart Res 13:2981–2988

Sinha S, Pan I, Chanda P, Sen SK (2009) Nanoparticles fabrication using ambient biological resources. J Appl Biosci 19:1113–1130

Sondi I, Salopek-Sondi B (2004) Silver nanoparticles as antimicrobial agent: a case study on *E. coli* as a model for Gram-negative bacteria. J Colloid Interface Sci 275:177–182

Song JY, Kim BS (2009) Biological synthesis of bimetallic Au/Ag nanoparticles using Persimmon (*Diospyros kaki*) leaf extract. Korean J Chem Eng 25:808–811

Subramanian V (2012) Green synthesis of silver nanoparticles using *Coleus amboinicus* lour, antioxitant activity and invitro cytotoxicity against Ehrlich's Ascite carcinoma. J Pharm Res 5:1268–1272

Sukirtha R, Priyanka KM, Antony JJ, Kamalakkannan S, Thangam R, Gunasekaran P, Krishnan M, Achiraman S (2012) Cytotoxic effect of Green synthesized silver nanoparticles using *Melia azedarach* against in vitro HeLa cell lines and lymphoma mice mode. Process Biochem 47:273–279

Surendiran A, Sandhiya S, Pradhan SC, Adithan C (2009) Novel applications of nanotechnology in medicine. Indian J Med Res 130:689–701

Suriyakalaa U, Antony JJ, Suganya S, Siva D, Sukirtha R, Kamalakkannan S, Pichiah PBT, Achiraman S (2013) Hepatocurative activity of biosynthesized silver nanoparticles fabricated using *Andrographis paniculata*. Colloids Surf B Biointerfaces 102:189–194

Swamy VS, Prasad R (2012) Green synthesis of silver nanoparticles from the leaf extract of *Santalum album* and its antimicrobial activity. J Optoelectron Adv M 4(3):53–59

Tripathy A, Raichur AM, Chandrasekaran N, Prathna TC, Mukherjee A (2010) Process variables in biomimetic synthesis of silver nanoparticles by aqueous extract of *Azadirachta indica* (Neem) leaves. J Nanopart Res 12:237–241

Tripathi AK, Harsh NSK, Gupta N (2007) Fungal treatment of industrial effluents: a mini review. Life Sci J 4:78–81

Umoren SA, Obot IB, Gasem ZM (2014) Green synthesis and characterization of silver nanoparticles using red apple (*Malus domestica*) fruit extract at room temperature. J Mater Environ Sci 5:907–914

Valtchev V, Tosheva L (2013) Porous nanosized particles: preparation, properties, and applications. Chem Rev 113:6734–6760

Vanaja M, Rajeshkumar S, Paulkumar K, Gnanajobitha G, Malarkodi C, Annadurai G (2013) Phytosynthesis and characterization of silver nanoparticles using stem extract of *Coleus aromaticus*. Int J Mater Biomat Appl 3:1–4

Vankar PS, Bajpai D (2010) Preparation of gold nanoparticles from *Mirabilis jalapa* flowers. Indian J Biochem Biophys 47:157–160

Verma VC, Kharwar RN, Gange AC (2010) Biosynthesis of antimicrobial silver nanoparticles by the endophytic fungus *Aspergillus clavatus*. Nanomedicine 5:33–40

Vivek R, Thangam R, Muthuchelian K, Gunasekaran P, Kaveri K, Kannan S (2012) Green biosynthesis of silver nanoparticles from *Annona squamosa* leaf extract and its in vitro cytotoxic effect on MCF-7 cells. Process Biochem 47:2405–2410

Wang X, Zuo J, Keil P, Grundmeier G (2007) Comparing the growth of PVD silver nanoparticles on ultra-thin fluorocarbon plasma polymer films and self-assembled fluoroalkyl silane monolayers. Nanotechnology 18:265303

Weiss J, Takhistov P, Julianmcclements D (2006) Functional materials in food nanotechnology. J Food Sci 71:107–116

Zahir AA, Rahuman AA (2012) Evaluation of different extracts and synthesized silver nanoparticles from leaves of *Euphorbia prostrate* against the plant *Haemaphysalis bispinosa* and *Hippobosca maculate*. Vet Parasitol 187:511–520

Zhang H, Smith JA, Oyanedel-Craver V (2012) The effect of natural water conditions on the antibacterial performance and stability of silver nanoparticles capped with different polymers. Water Resour 46:691–699

Zhang S, Tang Y, Vlahovic B (2016) A review on preparation and applications of silver-containing nanofibers. Nanoscale Res Lett 11:1–8

Chapter 10
Degradation Dye Using Gold and Silver Nanoparticles Synthesized by Using Green Route and Its Characteristics

S. Rajeshkumar and R. V. Santhiyaa

10.1 Introduction

Currently, Nanobiotechnology becomes very apparent in the field of research owing to its multidisciplinary role (Vanaja et al. 2013). The size of the nanomaterial extended between 1 and 100 nm (Rajeshkumar et al. 2016). Diverse procedures were carried out to synthesize nanoparticles such as physical, chemical, and biological (Menon et al. 2017). Earlier, they are used for the synthesis of nanoparticles. However, these techniques have been faced several consequences like high cost, toxicity, and production of intense heat and pressure. To overcome these issues, living organisms such as bacteria (Shahverdi et al. 2007; Rajeshkumar et al. 2013), algae (Singaravelu et al. 2007; González et al. 2013; Aziz et al. 2014, 2015), fungi (Shaligram et al. 2009; Hemath et al. 2010; Aziz et al. 2016), and plants (Salam et al. 2012; Kiruba Daniel et al. 2013; Prasad 2014) plays a key role in the eco-friendly nanoparticles. Among the other nanoparticles, gold and silver nanoparticles have numerous applications in the field of biomedical like diagnosis, therapeutic, sensors, and bio imaging (Salata 2004; Prasad et al. 2016). However, green synthesis by plants possess numerous advantages than the microorganisms owing to its easy availability, safe to handle, and also has a great potential to reduce and stabilize the nanoparticles by the presence of phytochemicals (Rajeshkumar et al. 2013; Vanaja et al. 2013, 2014; Prasad et al. 2016; Asha et al. 2017; Menon et al. 2017).

The silver nanoparticles synthesized by plants such as *Catharanthus roseus* (Ks et al. 2011), *Allium sativum* (Ahamed et al. 2011), *Parthenium* (Parashar et al.

S. Rajeshkumar (✉)
Department of Pharmacology, Saveetha Dental College and Hospitals, Saveetha Institute of Medical and Technical Sciences, Chennai, TN, India

R. V. Santhiyaa
Nanotherapy Laboratory, School of Bio-Sciences and Technology, Vellore Institute of Technology, Vellore, TN, India

© Springer Nature Switzerland AG 2018
R. Prasad et al. (eds.), *Exploring the Realms of Nature for Nanosynthesis*,
Nanotechnology in the Life Sciences, https://doi.org/10.1007/978-3-319-99570-0_10

2009), *Mirabilis jalapa* (Vankar and Bajpai 2010), *Euphorbia hirta* (Elumalai et al. 2010), *Nicotiana tobaccum* (Prasad et al. 2011), *Syzygium cumini* (Prasad et al. 2012; Prasad and Swamy 2013), *Santalum album* (Swamy and Prasad 2012), *Vitis vinifera* (Hassan et al. 2013), *Trichodesma indicum* (Buhroo et al. 2017)*, Solanum lycopersicum* (Bhattacharyya et al. 2016), and *Carissa carandas* (Joshi et al. 2018). The synthesis of gold nanoparticles using microbes such as *Penicillium denitrificans* (Bennur et al. 2016), *Bacillus stearothermophilus* (Luo et al. 2014), *Shewanella neidensis* (Suresh et al. 2011), *Staphylococcus epidermidis* (Ogi et al. 2010), *Penicillium chrysogenum* (Sawle et al. 2008), *Penicillium crustosum* (Roy and Das 2016), and *Rhizopus oryzae* (Ahmad et al. 2003; Sanghi et al. 2011). This review explains about the green synthesis of silver and gold nanoparticles and their ability in dye removal. The synthesis of different parts of plant and fruit mediated gold nanoparticles such as *Bacopa monnieri* (Babu et al. 2013), *Mangifera indica* (Philip 2010), *Mimosa pudica* (Suganya et al. 2016), *Citrus maxima* (Yu et al. 2016), *Punica granatum* (Ganeshkumar et al. 2013), *Cassia auriculata* (Venkatachalam et al. 2013), *Moringa oleifera* (Anand et al. 2015), *Cucurbita pepo* (Gonnelli et al. 2015), *Abelmoschus esculentus* (Jayaseelan et al. 2013), *Cassia fistula* (Daisy and Saipriya 2012), and *Ficus religiosa* (Wani et al. 2013) (Fig. 10.1).

10.2   Silver Nanoparticles and Silver Nanoparticles Based Nanocomposites

Table 10.1 shows the silver nanoparticles and silver nanoparticles based nanocomposites used in dye degradation (Fig. 10.2), and characters of the nanoparticles are also described. The procyanidin present in the grape seeds plays a major role in the conversion of Ag + into Ag NPs by forming the intermediate compound Ag + −procyanidin. Once the silver nitrate added into the seed extract solution it turns into dark brown. The direct orange dye was degraded by the process of catalytic reaction by the addition of reducing agent $NaBH_4$ (Ping et al. 2017). The leaf extract of *Zanthoxylum armatum* mediated AgNPs has the ability to degrade the aromatic and aliphatic compounds such as Safranine, methyl red, methyl orange, and methylene blue. From the analysis of FTIR, the biological role of *Z. armatum* can be identified as amino group and hydroxyl group indicates the presence of phenols, the nitro compound, and amine group aids to form proteins. This protein binds to the AgNPs and acts as a stabilizing agent (Jyoti and Singh 2016). Due to the scavenging activity of Ag, it trapped the electrons acquired on the CU_2O catalyst and Ag + ions turned into metallic Ag once it gets reacted with the electrons. The CU_2O system helps to degrade the methyl orange dye (Jiana et al. 2016). The exopolysaccharide surface is capable of degrading the azo dyes. The AgNPs weakens the azo-double bond by binding to the nitrogen bond, sulfur, and oxygen atoms of the azo dye through conjugation process. The hetero atoms present in the EPS exhibit the hydrophilic interaction between the azo dyes; this may bring the dye in close contact with the

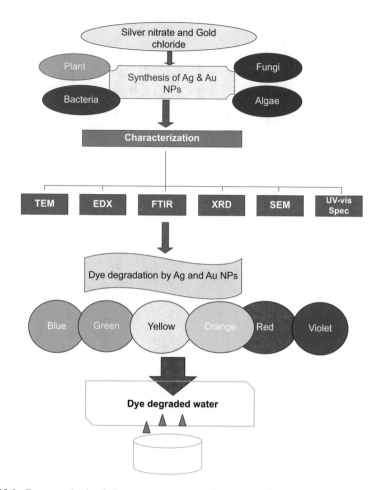

Fig. 10.1 Green synthesis of silver and gold nanoparticles in dye degradation

catalytic sites (Saravanan et al. 2017). The egg shell of Anas platyrhynchos mediated AgNPs plays a vital role in the degradation of harmful dyes. The FTIR analysis was done to find out the functional groups and biomolecules present in the egg shell that are responsible for reducing and stabilizing the nanoparticles. The presence of functional groups such as O-H, C=C, C=O, and N-H interprets the presence of collagen in the egg shell. It consists of glycine, aminoacids, hydroxyproline, and hydroxylysine components which help to stabilize the nanoparticles (Sinha and Ahmaruzzaman 2015a). *Achillea millefolium* and peach kernel shell are used for the synthesis of AgNPs due to its biological activities. The phenolic compounds include flavonoids and phenol carbonic acids are present in the *Achillea millefolium L.* which has the ability to reduce metallic ions and synthesizing nanoparticles. It also plays an efficient role in the degradation of hazardous dye by the process of catalytic reaction due to its low-commercial value (Khodadadi et al. 2017a). The silica

Table 10.1 Silver nanoparticles and its based nanocomposites characteristics and its dye removal

S. no.	Nanoparticles or nanocomposites	Reducing or stabilizing agent	Characteristics	Applications	References
1	Silver nanoparticles	Grape seed extract	TEM, EDX SIZE = 54.8 nm	Direct Orange dye 26	Ping et al. (2017)
2	Silver nanoparticles	Zanthoxylum armatum	SEM, TEM, SAED, EDS, XRD SIZE = 15–50 nm	Methylene blue, Methyl red, Safranine, and Methyl orange	Jyoti and Singh (2016)
3	Silver doped Cu₂O catalyst	–	–	Methyl orange	Jiana et al. (2016)
4	Silver nanoparticle	Exopolysaccharide	TEM, SEM, AFM SIZE = 35 nm SHAPE = spherical	Congo red and Methyl orange	Saravanan et al. (2017)
5	Silver and gold-silver core shell NPs	Anas platyrhynchos (egg shell)	TEM, STEM, XRD, SAED For Ag, SIZE = 6–26 nm SHAPE = spherical For Au-Ag NPs, TEM, STEM, XRD, SAED = 9–18 nm SHAPE = spherical and oval	Methylene blue, Methyl violet 6B, and Rose Bengal	Sinha and Ahmaruzzaman (2015a)
6	Silver nanoparticles	Achillea millefolium L. and peach kernel shell	–	Methylene blue, 4-Nitrophenol and Methyl orange	Khodadadi et al. (2017a)
7	Silicon dioxide NPs, silver and gold coated with SiO₂ NPs (core shell)	Nanocore and Nanoshell	TEM SIZE = SiO₂ = 30–10 nm, Ag NPs = 20–5 nm, and Au NPs = 15–3 nm	Methyl orange	Badr and Mahmoud (2007)
8	Silver nanoparticle	Gmelina arborea (fruit extract)	TEM, HRTEM, SAED SIZE = 8–32 nm SHAPE = spherical and crystalline	Methylene blue	Saha et al. (2017)
9	Silver and gold nanoparticles	Punica granatum (pomegranate juice)	TEM SIZE = for AgNPs = 36 nm, for AuNPs = 18 nm SHAPE = spherical	Cationic phenothiazine dye, anionic mono azo dye, and cationic fluorescent dye	Meenakumari and Philip (2015)

S. no.	Nanoparticles or nanocomposites	Reducing or stabilizing agent	Characteristics	Applications	References
10	Silver nanoparticle	*Terminalia cuneata* (bark extract)	XRD, DLS, HRTEM, EDS SIZE = 25–50 nm SHAPE = spherical	Direct Yellow 12 dye	Edison et al. (2016b)
11	Silver nanoparticle	*Polygonum hydropiper* (plant)	FTIR, FESEM SIZE = 60 nm SHAPE = spherical and aggregated into irregular structure	Methylene blue	Bonnia et al. (2016)
12	Silver nanoparticle-titanium dioxide (nanocomposite)	*Carpobrotus acinaciformis* (leaf and flower extract)	FTIR, FESEM, XRD, EDS SIZE = 20–50 nm	Congo red and Methyl orange	Rostami-Vartooni et al. (2016)
13	Silver nanoparticles	*Biophytum sensitivum* (leaf extract)	FTIR, XRD, and HR-TEM SIZE = 19.06 nm SHAPE = spherical	Methylene blue and Methyl orange	Joseph and Mathew (2015a)
14	Silver nanoparticles	(Palm shell extract)	UV, IR, XRD, TEM, EDX and Raman analysis SIZE = 50 nm SHAPE = clusters	Reactive blue-21 dye, Copper phthalocyanine-based dye, Reactive blue-21, Rhodamine-6G, and Xanthene dye	Vanaamudan et al. (2016)
15	Silver nanoparticles	*Mussaenda erythrophylla* (leaf extract)	SEM, XRD, EDAX, FTIR, PCS SIZE = 0.1–10,000 SHAPE = crystallite	Methyl orange dye and toxic azo dyes	Varadavenkatesan et al. (2016)
16	Silver nanoparticles	Cationic cetyl trimethyl ammonium bromide (surfactant)	TEM, FTIR, SEAD SIZE = 5–20 nm SHAPE = spherical, truncated triangular highly Poly-dispersed	Methyl orange	Zaheer and Aazam (2017)
17	Silver nanoparticles	Montmorillonite (clay)	XRD, FTIR, SEM, BET, EDX SIZE = 12 + −2 nm	Methylene blue	Wang et al. (2017)

(continued)

Table 10.1 (continued)

S. no.	Nanoparticles or nanocomposites	Reducing or stabilizing agent	Characteristics	Applications	References
18	Graphene oxide-silver nanoparticles (GO-AgNPs)	*Picrasma quassioides* (bark extract)	SEM-EDX, TEM-SAED SIZE = for AgNPs = 17.5–66.5 nm For GO-AgNPs = 10–49.5 nm SHAPE = spherical and crystallite	Methylene blue	Sreekanth et al. (2016)
19	Silver nanoparticles	Aqueous extract of tea	FTIR, XRD, TEM SIZE = 45 nm SHAPE = torispherical	Ethyl violet and Methylene blue	Qing et al. (2017)
20	Silver nanoparticles	*Sterculia acuminata* (fruit extract)	XRD, TEM, FTIR, PSA SIZE = 10.3 nm SHAPE = crystallite and spherical	Methyl orange, Direct blue 24, Methylene blue, 4-Nitrophenol and Phenol red	Bogireddy et al. (2016)
21	Silver nanoparticles	*Lagerstroemia speciosa*	FTIR, XRD, TEM SIZE = 12 nm SHAPE = spherical	Methyl orange and methylene blue	Sai Saraswathi et al. (2017)
22	Silver nanoparticles	*Dimocarpus longan*	XRD, TEM, FTIR SIZE = 40 nm SHAPE = face-centered cubic structures	Methylene blue, 4-Nitrophenol, and 4-Aminophenol	Khan et al. (2016a)
23	Silver nanoparticles (AgNPs)	*Cicer arietinum* (chick pea)	DLS, TEM, EDAX, and FTIR SIZE = 88.8 nm SHAPE = spherical	4-Nitrophenol, Methylene blue and Congo red	Arya et al. (2017)
24	Silver nanoparticles (AgNPs)	*Ulva lactuca* (seaweed)	FTIR, HRSEM, XRD SIZE = 48.59 nm SHAPE = spherical	Methyl orange	Kumar et al. (2013)

25	Silver nanoparticles AgNPs	*Trigonella foenum-graecum* seeds	TEM, HRTEM, SAED, FTIR, XRD SIZE = 22 nm, 28 nm, and 32 nm SHAPE = polycrystalline	Methylene blue, Methyl orange, and eosin	Vidhu and Philip (2014)
26	Au and AgNPs	*Costus pictus* (leaf extract)	FTIR, SEM, DLS, EDX SIZE = for CPAgNPs = 46.7 nm CPAuNPs = 37.2 nm SHAPE = spherical	Methylene blue and Methyl red	Nakkala et al. (2015)
27	AgNPs	*Mesoporous titania*	XRD, EDX, HRTEM, DRS, FESEM SIZE = 34 and 18 nm SHAPE = crystallite	Methylene blue and indigo carmine dyes	Abdel Messih et al. (2017)
28	AgNPs	*Anacardium Occidentale Testa*	FTIR, XRD, HRTEM, EDS SIZE = 25 nm SHAPE = distorted spherical	Congo red and Methyl orange	Edison et al. (2016a)
29	AgNPs	Aqueous medium using starch	FTIR, XRD and HR-TEM SIZE = 18.2 ± 0.97 nm SHAPE = spherical	Methyl orange and Rhodamine	Joseph and Mathew (2015b)
30	AgNPs	*Cirsium japonicum*	HRTEM, XRD, FTIR, EDX SIZE = 2-8 nm SHAPE = spherical and no aggregation	Bromo phenyl blue	Khan et al. (2016b)

(continued)

Table 10.1 (continued)

S. no.	Nanoparticles or nanocomposites	Reducing or stabilizing agent	Characteristics	Applications	References
31	Silver-graphene oxide (Ag/GO) Nanocomposite	–	TEM, HRTEM, FTIR XRD SIZE = 9.22 ± 0.18 nm SHAPE = crystallite	Congo red and Bismarck brown	Borthakur et al. (2017)
32	Silver nanoparticles (AgNPs)	Ag/clinoptilolite—Vaccinium macrocarpon (fruit extract)	TEM, FTIR, EDS SIZE = 15-30 nm SHAPE = spherical	Methyl orange, Methylene blue, Rhodamine B, and Congo red	Khodadadi et al. (2017b)
33	Silver phosphate/graphene nanocomposite (Ag$_3$PO$_4$/G)	–	–	Heterocyclic dye, indigo dye, anthraquinone dye, and azo dye	Xu et al. (2017)
34	Silver nanoparticles AgNPs	Caralluma edulis	HRTEM, FTIR, XRD, EDX SIZE = 2–10 nm SHAPE = spherical and crystallite	Bromothymol blue	Khan et al. (2016c)
35	Silver capped on commercial diamond NPs	–	–	Methylene blue, Orange II, Acid Red 1 and Rhodamine B	Manickam-Periyaraman et al. (2016)
36	Silver nanoparticles AgNPs	Polygonum hydropiper	FESEM SIZE = 60 nm SHAPE = spherical and aggregated in irregular form	Methylene blue	Bonnia et al. (2016)
37	Silver nanoparticles AgNPs	Mussaenda erythrophylla (leaf extract)	SEM, XRD, EDAX, FTIR, PCS SIZE = 0.1–10,000 nm SHAPE = crystallite	Methyl orange	Varadavenkatesan et al. (2016)
38	Silver nanoparticle and colemanite ore waste Nanocomposites	–	TEM, XRD, XPS, BET SIZE = 5-30 nm SHAPE = spherical	Reactive Yellow 86 and Reactive Red 2	Yola et al. (2013)

#	Nanoparticle	Source	Characterization	Dye	Reference
39	Silver nanoparticle (AgNPs)	Thymbra spicata (leaf extract)	FESEM, TEM, XRD, EDS, FTIR SIZE = 7 nm SHAPE = crystallite	4-Nitrophenol, Rhodamine B, and Methylene blue	Veisi et al. (2018)
40	Silver nanoparticle-Solanum nigrum (Sn-AgNPs)	–	FTIR, TEM, XRD SIZE = 20 and 30 nm SHAPE = spherical	Degradation of dye effluent	Malaikozhundan et al. (2017)
41	Silver nanoparticle (AgNPs)	Morinda tinctoria (leaf extract)	XRD, SEM, EDX, FTIR SIZE = 79–96 nm SHAPE = spherical and rod	Methylene blue	Vanaja et al. (2014)
42	Silver nanoparticle (AgNPs)	Tangerine peel (orange peel)	SIZE = 30.29 ± 5.1 nm, 16.68 ± 5.7 nm, and 25.85 ± 8.4 nm SHAPE = spherical	Methyl orange	Alzahrani (2015)
43	Silver nanoparticles (AgNPs)	Lagerstroemia Speciosa	XRD, HRTEM SIZE = 12 nm SHAPE = spherical	Methyl orange	Saraswathi and Santhakumar (2017)
44	Silver nanoparticles (AgNPs)	Gracilaria corticata (red algae)	HRSEM SIZE = <100 nm SHAPE = spherical	Malachite green	Poornima and Valivittan (2017)
45	Silver nanoparticles (AgNPs)	Hypnea musciformis	SEM, XRD, AFM, FTIR SIZE = for 2D view = 2–55.8 nm. SHAPE = spherical SIZE = for 3D view = 12 nm SHAPE = crystalline	Methyl orange	Ganapathy Selvam and Sivakumar (2015)
46	Silver nanoparticle (AgNPs)	Amaranthus gangeticus Linn (leaf extract)	HR-TEM, SAED, FTIR SIZE = 11–15 nm SHAPE = globular	Congo red	Kolya et al. (2015)

Fig. 10.2 Role of silver nanoparticles in dye degradation

(core shell) coated AgNPs and AuNPs core showed an excellent activity in reduction of methyl orange dye by photocatalytic degradation under xenon lamp (Badr and Mahmoud 2007). The availability of biomolecules (polyphenols, caffeine, proteins, polysaccharides, aminoacids) and phytochemicals in the fruit extract of *Gmelina arborea* plant has the properties of reducing the metal ions and stabilizing the nanoparticles by the eco-friendly green synthesis method (Saha et al. 2017). By doing the FTIR analysis, the biomolecules that are attached to the surface of the metallic ion can be identified. The presence of biomolecules such as phenolic hydroxyl, peptide bond, and carboxyl group in the *Punica granatum* acts as both stabilizing and reducing agent. The phenolic hydroxyl compound is involved during the reduction process of the metallic ion. The Ag mediated *Punica granatum* nanoparticles are very potential to degrade the anionic mono azo dye cationic phenothiazine and fluorescent types of dyes (Meenakumari and Philip 2015). The bark extract of *Terminalia cuneata* are rich in various phytochemical constituents such as polyphenols, tannic acids, gallic acid, flavanoids, phytosterol, ellagic acid, and triterpenoid saponins with high numbers. These phytoconstituents act as a reducing agent for the conversion of silver ions to silver atoms and for stabilizing the nanoparticles. These biological activities in *Terminalia cuneata* were analyzed by the performing FTIR analysis (Edison et al. 2016b). The FTIR analysis indicates the presence of phenolic compound, hydroxyl group, free amines, carbonyl, and proteins in the plant of *Polygonum hydropiper*. These biomolecules help in the formation AgNPs by reducing the ions and stabilizing the NPs with appropriate capping agent (Bonnia et al. 2016). The extract of *Carpobrotus acinaciformis* is used as a stabilizing and reducing agent for production the immobilized AgNPs on the surface of titanium due to its phenolic and flavanoids constituents (Rostami-Vartooni et al. 2016). The FTIR analysis determined the presence of phenolic and flavonoid in the leaf extract of *Biophytum sensitivum* which plays a major role in both stabilizing and reducing agent (Joseph and Mathew 2015a). The polyphenols present in the palm tree extract were used as a capping agent for the synthesis of silver nanoparticles. The FTIR reveals that the presence of phenols, polysaccharides, and proteins in the palm shell extract act as a reducing and stabilizing agent. The silver nanoparticles have shown extraordinary performance in degrading the cationic and anionic dye using catalytic reaction (Vanaamudan et al. 2016).

10.3   Gold Nanoparticles and Gold Nanoparticles Based Nanocomposites

Table 10.2 shows gold nanoparticles and gold nanoparticles based nanocomposites, characteristics and dye degradation were elucidates (Fig. 10.3). The nanocomposites of gold-sodium niobate were synthesized by the hydrothermal method. The Au doped NaNbo$_3$ nanocomposite using dye degradation was carried out by photocatalytic degradation under visible light. While comparing to gold nanoparticle, the gold doped nanocomposites shows a greater performance in degrading the malachite green dye (Baeissa 2016). The leaves extract of *Dalbergia coromandeliana* expressed an excellent activity towards the dye degradation due to the presence of phytoconstituents such as terpenoids, neoflavonoids, steroids, and isoflavonoids. These play a vital role in reducing the Au ions into Au atoms by donating the electron and hydrogen atom as well as stabilizing the nanoparticles with the binding agent (Umamaheswari et al. 2018). From the study of FTIR, it was identified that the presence of these functional groups (O-H, C=O, and N-H) denotes the availability of the protein in the egg shell extract. These functional groups get involved in the production of NPs and act as both reducing agent and capping agent. The dye degradation was carried out by catalytic reaction (Sinha and Ahmaruzzaman 2015b). Carboxylic acids, nitro compounds, aromatic and aliphatic amines, and alkanes are the functional groups present in the flower extract of *Plumeria alba*. From the FTIR analysis, it has been detected that the presence of these functional groups makes the efficient binding between the *Plumeria alba* and gold nanoparticles for the production and capping of PAGNPs. Due to the catalytic reaction, the 4-nitrophenol reduced into 4-aminophenol (Mata et al. 2016). For the first time, the gold nanoparticles were developed extracellularly from the cell-free extract of *Bacillus*

Table 10.2 Characterization of gold nanoparticles and its dye removal

S. no	Nanoparticles or nanocomposites	Reducing or stabilizing agent	Characteristics	Applications	References
1	AgNPs, Au-AgNPs core shell NPs	Anas platyrhynchos (egg shell)	FTIR, TEM, SAED, STEM, XRD, EDAX = for AgNPs, 6–26 nm. SHAPE = spherical For Ag-AuNPs = 9–18 nm. SHAPE = spherical and oval	Rose Bengal, Methyl Violet 6B, and Methylene blue	Sinha and Ahmaruzzaman (2015a)
2	Au/NaNbO$_3$ Nanocomposite	–	TEM, XRD SIZE = 50 nm SHAPE = nanocube structure	Malachite green	Baeissa (2016)
3	Gold nanoparticles (AuNPs)	*Dalbergia coromandeliana*	HRTEM, XRD, SAED, FTIR = 532 nm SHAPE = crystalline	Congo red and methyl orange	Umamaheswari et al. (2018)
4	AuNPs	Egg shells of Anas platyrhynchos	TEM, SAED, FTIR = 540–880 nm	Eosin Y dye	Sinha and Ahmaruzzaman (2015b)
5	AuNPs	*Plumeria alba* (flower extract)	TEM, EDAX, XRD, FTIR = 28 ± 5.6 and 15.6 ± 3.4 nm SHAPE = spherical	Methylene blue, eosin Y, 4-nitrophenol, methyl red, Congo red, and ethidium bromide	Mata et al. (2016)
6	AuNPs	*Bacillus marisflavi*	XRD, FESEM, TEM and DLS = 14 nm SHAPE = spherical, face-centered cubic structures	Congo red into a naphthylamine and phthalic acid and methylene blue to 2-methyl benzothiazole	Nadaf and Kanase (2016)
7	AuNPs	Sodium rhodizonate	TEM, SEM, XRD = 11 nm–7 nm	4-Nitrophenol, Methyl orange and Methylene blue	Islam et al. (2017)
8	AuNPs	*Aspergillus* sp.	TEM = 12–48 nm SHAPE = spherical and pseudo-spherical	2-Nitrophenol, 3-nitrophenol, 4-nitrophenol, o-nitroaniline and m-nitroaniline	Qu et al. (2017)

(continued)

Table 10.2 (continued)

S. no	Nanoparticles or nanocomposites	Reducing or stabilizing agent	Characteristics	Applications	References
9	AuNPs	*Suaeda Fruticosa* (plant extract)	XRD, HRTEM, SEM, and FTIR SIZE = 6–8 nm SHAPE = spheroid	Phenolic azo dyes	Khan et al. (2017)
10	Au/polydimethy-L aminoethyl acrylate brushes (PDMAEMA)/RGO Nano hybrids	–	FTIR, XRD, TEM = 200 to 800 nm SHAPE = spherical and worm	Rhodamine B, Methyl orange and Eosine Y	Mogha et al. (2017)
11	AuNPs	*Sterculia acuminata* (fruit extract)	XRD, TEM, FTIR = 9.37–38.12 nm SHAPE = spherical	4-nitrophenol, Methylene blue, methyl orange, and direct blue 24	Bogireddy et al. (2015)
12	AuNPs	*Pogostemon benghalensis* (leaf extract)	XRD, TEM, FTIR = 10–50 nm SHAPE = spherical and triangular	Methylene blue	Paul et al. (2015)
13	AuNPs	*Mimosa pudica*	XRD, TEM, HRTEM, FTIR = 16 nm SHAPE = spherical with narrow	Rhodamine B and o-nitrophenol	Devi et al. (2015)
14	Gold nanoparticles-titanium dioxide (Au/TiO$_2$)	*Cinnamomum tamala* (leaf extract)	FTIR, TEM, XRD SIZE = 8–20 nm SHAPE = crystalline	Methyl orange	Naik et al. (2013)
15	AuNPs	Kashayam, Guggulutiktham	TEM, XRD, FTIR SIZE = 15–50 nm SHAPE = crystallite	Methylene blue	Suvith and Philip (2014)

marisflavi isolated from the estuaries. It has been proved that this organism has the ability to degrade the organic dyes such as methylene blue into 2-methyl benzothiazole and congo red into alpha naphthylamine. From the FTIR studies, the functional groups which are responsible for the reduction and capping process were identified such as amine, aldehyde, and the functional groups of protein. The cell-free extract mediated gold nanoparticles were bound to the proteins and the aldehydes via conjugation process (Nadaf and Kanase 2016). The gold nanoparticles carried on the cellulose fibers have high catalytic activity. These catalysts that have been employed in the catalytic reaction of organic dyes include methyl orange, 4-nitrophenol, and methylene blue. Sodium rhodizonate acts as both reducing and stabilizing agent (Islam et al. 2017).

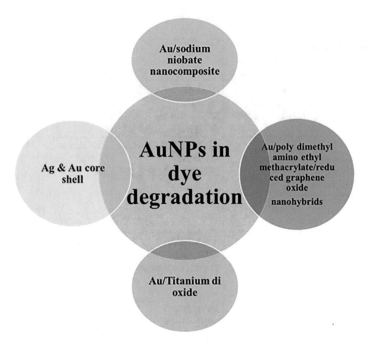

Fig. 10.3 Role of gold nanoparticles in dye degradation

10.4   Conclusion

Previously, several studies have been proved that the silver and gold nanoparticles possess various beneficial activities. Hence, this review proves that the green synthesis of silver and gold nanoparticles plays a prominent role in the degradation of hazardous dyes such as direct orange 26, methylene blue, congo red, rose Bengal, Safranine, 4-nitrophenol, 4-aminophenol, methyl red, methyl orange, rhodamine B, indigo, cationic phenothiazine dye, eosin, acid red, reactive red 2, direct yellow, m-nitroaniline, o-nitroaniline, bromophenyl blue, malachite green, rhodamine-6G, ethidium bromide, 2-nitrophenol, 3-nitrophenol, Bismarck brown, reactive yellow 86, xanthene dye, anionic mono azo dye, and ethyl violet.

References

Abdel Messih MF, Ahmed MA, Soltan A, Anis SS (2017) Facile approach for homogeneous dispersion of metallic silver nanoparticles on the surface of mesoporous titania for photocatalytic degradation of methylene blue and indigo carmine dyes. J Photochem Photobiol A Chem 335:40–51. https://doi.org/10.1016/j.jphotochem.2016.11.001

Ahamed M, Majeed Khan MA, Siddiqui MKJ et al (2011) Green synthesis, characterization and evaluation of biocompatibility of silver nanoparticles. Phys E Low Dimen Syst Nanostruct 43:1266–1271. https://doi.org/10.1016/j.physe.2011.02.014

Ahmad A, Senapati S, Khan MI et al (2003) Extracellular biosynthesis of monodisperse gold nanoparticles by a novel extremophilic actinomycete, thermomonospora sp. Langmuir 19:3550–3553. https://doi.org/10.1021/la0267721

Alzahrani E (2015) Eco-friendly production of silver nanoparticles from peel of tangerine for degradation of dye. World J Nano Sci Eng 5:10–16

Anand K, Gengan RM, Phulukdaree A, Chuturgoon A (2015) Agroforestry waste moringa oleifera petals mediated green synthesis of gold nanoparticles and their anti-cancer and catalytic activity. J Ind Eng Chem 21:1105–1111. https://doi.org/10.1016/j.jiec.2014.05.021

Arya G, Sharma N, Ahmed J et al (2017) Degradation of anthropogenic pollutant and organic dyes by biosynthesized silver nano-catalyst from *Cicer arietinum* leaves. J Photochem Photobiol B Biol 174:90–96. https://doi.org/10.1016/j.jphotobiol.2017.07.019

Asha S, Thirunavukkarasu P, Rajeshkumar S (2017) Eco-friendly synthesis of silver nanoparticles using aqueous leaves extract of clome gynandra and their antibacterial activity. Int J Pharmaceut Res 9:32–37

Aziz N, Fatma T, Varma A, Prasad R (2014) Biogenic synthesis of silver nanoparticles using *Scenedesmus abundans* and evaluation of their antibacterial activity. J Nanopart 2014:689419 https://doi.org/10.1155/2014/689419

Aziz N, Faraz M, Pandey R, Sakir M, Fatma T, Varma A, Barman I, Prasad R (2015) Facile algae-derived route to biogenic silver nanoparticles: synthesis, antibacterial and photocatalytic properties. Langmuir 31:11605–11612. https://doi.org/10.1021/acs.langmuir.5b03081

Aziz N, Pandey R, Barman I, Prasad R (2016) Leveraging the attributes of *Mucor hiemalis*-derived silver nanoparticles for a synergistic broad-spectrum antimicrobial platform. Front Microbiol 7:1984. https://doi.org/10.3389/fmicb.2016.01984

Babu PJ, Sharma P, Saranya S, Bora U (2013) Synthesis of gold nanoparticles using ethonolic leaf extract of *Bacopa monnieri* and UV irradiation. Mater Lett 93:431–434. https://doi.org/10.1016/j.matlet.2012.11.034

Badr Y, Mahmoud MA (2007) Photocatalytic degradation of methyl orange by gold silver nano-core/silica nano-shell. J Phys Chem Solids 68:413–419. https://doi.org/10.1016/j.jpcs.2006.12.009

Baeissa ES (2016) Photocatalytic degradation of malachite green dye using Au/NaNbO3 nanoparticles. J Alloys Compd 672:564–570. https://doi.org/10.1016/j.jallcom.2016.02.024

Bennur T, Khan Z, Kshirsagar R et al (2016) Biogenic gold nanoparticles from the Actinomycete *Gordonia amarae*: application in rapid sensing of copper ions. Sensors Actuators B Chem 233:684–690. https://doi.org/10.1016/j.snb.2016.04.022

Bhattacharyya A, Prasad R, Buhroo AA, Duraisamy P, Yousuf I, Umadevi M, Bindhu MR, Govindarajan M, Khanday AL (2016) One-pot fabrication and characterization of silver nanoparticles using *Solanum lycopersicum*: an eco-friendly and potent control tool against rose aphid, *Macrosiphum rosae*. J Nanosci 2016:4679410, 7 pages. https://doi.org/10.1155/2016/4679410

Bogireddy NKR, Hoskote Anand KK, Mandal BK (2015) Gold nanoparticles- synthesis by *Sterculia acuminata* extract and its catalytic efficiency in alleviating different organic dyes. J Mol Liq 211:868–875. https://doi.org/10.1016/j.molliq.2015.07.027

Bogireddy NKR, Kiran Kumar HA, Mandal BK (2016) Biofabricated silver nanoparticles as green catalyst in the degradation of different textile dyes. J Environ Chem Eng 4:56–64. https://doi.org/10.1016/j.jece.2015.11.004

Bonnia NN, Kamaruddin MS, Nawawi MH et al (2016) Green biosynthesis of silver nanoparticles using "Polygonum Hydropiper" and study its catalytic degradation of methylene blue. Procedia Chem 19:594–602. https://doi.org/10.1016/j.proche.2016.03.058

Borthakur P, Boruah PK, Hussain N et al (2017) Specific ion effect on the surface properties of Ag/reduced graphene oxide nanocomposite and its influence on photocatalytic efficiency towards azo dye degradation. Appl Surf Sci 423:752–761. https://doi.org/10.1016/j.apsusc.2017.06.230

Buhroo AA, Nisa G, Asrafuzzaman S, Prasad R, Rasheed R, Bhattacharyya A (2017) Biogenic silver nanoparticles from *Trichodesma indicum* aqueous leaf extract against *Mythimna sepa-*

rata and evaluation of its larvicidal efficacy. J Plant Protect Res 57(2):194–200. https://doi.org/10.1515/jppr-2017-0026

Daisy P, Saipriya K (2012) Biochemical analysis of *Cassia fistula* aqueous extract and phytochemically synthesized gold nanoparticles as hypoglycemic treatment for *diabetes mellitus*. Int J Nanomed 7:1189–1202. https://doi.org/10.2147/IJN.S26650

Devi HS, Rajmuhon Singh N, Premananda Singh H, David Singh T (2015) Facile synthesis of biogenic gold nanocatalyst for efficient degradation of organic pollutants. J Environ Chem Eng 3:2042–2049. https://doi.org/10.1016/j.jece.2015.07.014

Edison TNJI, Atchudan R, Sethuraman MG, Lee YR (2016a) Reductive-degradation of carcinogenic azo dyes using *Anacardium occidentale* testa derived silver nanoparticles. J Photochem Photobiol B Biol 162:604–610. https://doi.org/10.1016/j.jphotobiol.2016.07.040

Edison TNJI, Lee YR, Sethuraman MG (2016b) Green synthesis of silver nanoparticles using *Terminalia cuneata* and its catalytic action in reduction of direct yellow-12 dye. Spectrochim Acta Part A Mol Biomol Spectrosc 161:122–129. https://doi.org/10.1016/j.saa.2016.02.044

Elumalai EK, Prasad TNVKV, Hemachandran J, Therasa SV (2010) Extracellular synthesis of silver nanoparticles using leaves of Euphorbia hirta and their antibacterial activities. J Pharm Sci Res 2:549–554

Ganapathy Selvam G, Sivakumar K (2015) Phycosynthesis of silver nanoparticles and photocatalytic degradation of methyl orange dye using silver (Ag) nanoparticles synthesized from *Hypnea musciformis* (Wulfen) J.V. Lamouroux. Appl Nanosci 5:617–622. https://doi.org/10.1007/s13204-014-0356-8

Ganeshkumar M, Sathishkumar M, Ponrasu T et al (2013) Spontaneous ultra fast synthesis of gold nanoparticles using *Punica granatum* for cancer targeted drug delivery. Colloids Surf B Biointerfaces 106:208–216. https://doi.org/10.1016/j.colsurfb.2013.01.035

Gonnelli C, Cacioppo F, Giordano C et al (2015) Cucurbita pepo l. extracts as a versatile hydrotropic source for the synthesis of gold nanoparticles with different shapes. Green Chem Lett Rev 8:39–47. https://doi.org/10.1080/17518253.2015.1027288

González F, Blázquez ML, Muñoz JA et al (2013) Biological synthesis of metallic nanoparticles using algae. IET Nanobiotechnol 7:109–116. https://doi.org/10.1049/iet-nbt.2012.0041

Hassan K et al (2013) Optimization of biological synthesis of silver nanoparticles using *Fusarium oxysporum*. Iran J Pharm Res 12:289–298

Hemath NKS, Kumar G, Karthik L, Bhaskara Rao KV (2010) Extracellular biosynthesis of silver nanoparticles using the filamentous fungus Penicillium sp. Arch Appl Sci Res 2:161–167

Islam MT, Dominguez N, Ahsan MA et al (2017) Sodium rhodizonate induced formation of gold nanoparticles supported on cellulose fibers for catalytic reduction of 4-nitrophenol and organic dyes. J Environ Chem Eng 5:4185–4193. https://doi.org/10.1016/j.jece.2017.08.017

Jayaseelan C, Ramkumar R, Rahuman AA, Perumal P (2013) Green synthesis of gold nanoparticles using seed aqueous extract of *Abelmoschus esculentus* and its antifungal activity. Ind Crop Prod 45:423–429. https://doi.org/10.1016/j.indcrop.2012.12.019

Jiana L, Caia D, Sua G, Lina D, Lina M, Jiayong Lia JL, Wan X, Tie S, Lan S (2016) The accelerating effect of silver ion on the degradation of methyl orange in Cu2O system. Appl Catal A Gen 512:74–84. https://doi.org/10.1016/j.apcata.2007.12.033

Joseph S, Mathew B (2015a) Microwave-assisted green synthesis of silver nanoparticles and the study on catalytic activity in the degradation of dyes. J Mol Liq 204:184–191. https://doi.org/10.1016/j.molliq.2015.01.027

Joseph S, Mathew B (2015b) Facile synthesis of silver nanoparticles and their application in dye degradation. Mater Sci Eng B 195:90–97. https://doi.org/10.1016/j.mseb.2015.02.007

Joshi N, Jain N, Pathak A, Singh J, Prasad R, Upadhyaya CP (2018) Biosynthesis of silver nanoparticles using Carissa carandas berries and its potential antibacterial activities. J Sol Gel Sci Technol 86:682. https://doi.org/10.1007/s10971-018-4666-2

Jyoti K, Singh A (2016) Green synthesis of nanostructured silver particles and their catalytic application in dye degradation. J Genet Eng Biotechnol 14:311–317. https://doi.org/10.1016/j.jgeb.2016.09.005

Khan FU, Chen Y, Khan NU et al (2016a) Antioxidant and catalytic applications of silver nanoparticles using *Dimocarpus longan* seed extract as a reducing and stabilizing agent. J Photochem Photobiol B Biol 164:344–351. https://doi.org/10.1016/j.jphotobiol.2016.09.042

Khan ZUH, Khan A, Shah A et al (2016b) Enhanced photocatalytic and electrocatalytic applications of green synthesized silver nanoparticles. J Mol Liq 220:248–257. https://doi.org/10.1016/j.molliq.2016.04.082

Khan ZUH, Khan A, Shah A et al (2016c) Photocatalytic, antimicrobial activities of biogenic silver nanoparticles and electrochemical degradation of water soluble dyes at glassy carbon/silver modified past electrode using buffer solution. J Photochem Photobiol B Biol 156:100–107. https://doi.org/10.1016/j.jphotobiol.2016.01.016

Khan ZUH, Khan A, Chen Y et al (2017) Photo catalytic applications of gold nanoparticles synthesized by green route and electrochemical degradation of phenolic Azo dyes using AuNPs/GC as modified paste electrode. J Alloys Compd 725:869–876. https://doi.org/10.1016/j.jallcom.2017.07.222

Khodadadi B, Bordbar M, Nasrollahzadeh M (2017a) *Achillea millefolium* L. extract mediated green synthesis of waste peach kernel shell supported silver nanoparticles: application of the nanoparticles for catalytic reduction of a variety of dyes in water. J Colloid Interface Sci 493:85–93. https://doi.org/10.1016/j.jcis.2017.01.012

Khodadadi B, Bordbar M, Yeganeh-Faal A, Nasrollahzadeh M (2017b) Green synthesis of Ag nanoparticles/clinoptilolite using *Vaccinium macrocarpon* fruit extract and its excellent catalytic activity for reduction of organic dyes. J Alloys Compd 719:82–88. https://doi.org/10.1016/j.jallcom.2017.05.135

Kiruba Daniel SCG, Mahalakshmi N, Sandhiya J et al (2013) Rapid synthesis of Ag nanoparticles using henna extract for the fabrication of photoabsorption enhanced dye sensitized solar cell (PE-DSSC). Adv Mater Res 678:349–360. https://doi.org/10.4028/www.scientific.net/AMR.678.349

Kolya H, Maiti P, Pandey A, Tripathy T (2015) Green synthesis of silver nanoparticles with antimicrobial and azo dye (Congo red) degradation properties using *Amaranthus gangeticus* Linn leaf extract. J Anal Sci Technol 6:33. https://doi.org/10.1186/s40543-015-0074-1

Ks M, Ek E, Patel TN, Murty VR (2011) Catharanthus roseus: a natural source for the synthesis of silver nanoparticles. Asian Pac J Trop Biomed 1:270–274. https://doi.org/10.1016/S2221-1691(11)60041-5

Kumar P, Govindaraju M, Senthamilselvi S, Premkumar K (2013) Photocatalytic degradation of methyl orange dye using silver (Ag) nanoparticles synthesized from *Ulva lactuca*. Colloids Surfaces B Biointerfaces 103:658–661. https://doi.org/10.1016/j.colsurfb.2012.11.022

Luo P, Liu Y, Xia Y et al (2014) Aptamer biosensor for sensitive detection of toxin a of Clostridium difficile using gold nanoparticles synthesized by Bacillus stearothermophilus. Biosens Bioelectron 54:217–221. https://doi.org/10.1016/j.bios.2013.11.013

Malaikozhundan B, Vijayakumar S, Vaseeharan B et al (2017) Two potential uses for silver nanoparticles coated with *Solanum nigrum* unripe fruit extract: biofilm inhibition and photodegradation of dye effluent. Microb Pathog 111:316–324. https://doi.org/10.1016/j.micpath.2017.08.039

Manickam-Periyaraman P, Espinosa SM, Espinosa JC et al (2016) Dyes decolorization using silver nanoparticles supported on nanometric diamond as highly efficient photocatalyst under natural sunlight irradiation. J Environ Chem Eng 4:4485–4493. https://doi.org/10.1016/j.jece.2016.10.011

Mata R, Bhaskaran A, Sadras SR (2016) Green-synthesized gold nanoparticles from Plumeria alba flower extract to augment catalytic degradation of organic dyes and inhibit bacterial growth. Particuology 24:78–86. https://doi.org/10.1016/j.partic.2014.12.014

Meenakumari M, Philip D (2015) Degradation of environment pollutant dyes using phytosynthesized metal nanocatalysts. Spectrochim Acta Part A Mol Biomol Spectrosc 135:632–638. https://doi.org/10.1016/j.saa.2014.07.037

Menon S, Rajeshkumar S, Venkat Kumar S (2017) Resource-efficient technologies and its applications. Resour Technol 3:1–12. https://doi.org/10.1016/j.reffit.2017.08.002

Mogha NK, Gosain S, Masram DT (2017) Gold nanoworms immobilized graphene oxide polymer brush nanohybrid for catalytic degradation studies of organic dyes. Appl Surf Sci 396:1427–1434. https://doi.org/10.1016/j.apsusc.2016.11.182

Nadaf NY, Kanase SS (2016) Biosynthesis of gold nanoparticles by Bacillus marisflavi and its potential in catalytic dye degradation. Arab J Chem. https://doi.org/10.1016/j.arabjc.2016.09.020

Naik GK, Mishra PM, Parida K (2013) Green synthesis of Au/TiO2 for effective dye degradation in aqueous system. Chem Eng J 229:492–497. https://doi.org/10.1016/j.cej.2013.06.053

Nakkala JR, Bhagat E, Suchiang K, Sadras SR (2015) Comparative study of antioxidant and catalytic activity of silver and gold nanoparticles synthesized from *Costus pictus* leaf extract. J Mater Sci Technol 31:986–994. https://doi.org/10.1016/j.jmst.2015.07.002

Ogi T, Saitoh N, Nomura T, Konishi Y (2010) Room-temperature synthesis of gold nanoparticles and nanoplates using *Shewanella* algae cell extract. J Nanopart Res 12:2531–2539. https://doi.org/10.1007/s11051-009-9822-8

Parashar V, Parashar R, Sharma B, Pandey AC (2009) Parthenium leaf extract mediated synthesis of silver nanoparticles: a novel approach towards weed utilization. Dig J Nanomater Biostruct 4:45–50

Paul B, Bhuyan B, Dhar Purkayastha D et al (2015) Green synthesis of gold nanoparticles using *Pogostemon benghalensis* (B) O. Ktz. leaf extract and studies of their photocatalytic activity in degradation of methylene blue. Mater Lett 148:37–40. https://doi.org/10.1016/j.matlet.2015.02.054

Philip D (2010) Rapid green synthesis of spherical gold nanoparticles using *Mangifera indica* leaf. Spectrochim Acta Part A Mol Biomol Spectrosc 77:807–810. https://doi.org/10.1016/j.saa.2010.08.008

Ping Y, Zhang J, Xing T et al (2017) Green synthesis of silver nanoparticles using grape seed extract and their application for reductive catalysis of Direct Orange 26. J Ind Eng Chem 58:74. https://doi.org/10.1016/j.jiec.2017.09.009

Poornima S, Valivittan K (2017) Degradation of malachite green (dye) by using photo-catalytic biogenic silver nanoparticles synthesized using red algae (Gracilaria corticata) aqueous extract. Int J Curr Microbiol Appl Sci 6:62–70

Prasad R (2014) Synthesis of silver nanoparticles in photosynthetic plants. J Nanopart 2014:963961, 8 pages. https://doi.org/10.1155/2014/963961

Prasad R, Swamy VS (2013) Antibacterial activity of silver nanoparticles synthesized by bark extract of *Syzygium cumini*. J Nanopart 2013:1. https://doi.org/10.1155/2013/431218

Prasad KS, Pathak D, Patel A, Dalwadi P, Prasad R, Patel P, Kaliaperumal SK (2011) Biogenic synthesis of silver nanoparticles using *Nicotiana tobaccum* leaf extract and study of their antibacterial effect. Afr J Biotechnol 9(54):8122–8130

Prasad R, Swamy VS, Varma A (2012) Biogenic synthesis of silver nanoparticles from the leaf extract of *Syzygium cumini* (L.) and its antibacterial activity. Int J Pharma Bio Sci 3(4):745–752

Prasad R, Pandey R, Barman I (2016) Engineering tailored nanoparticles with microbes: quo vadis. WIREs Nanomed Nanobiotechnol 8:316–330. https://doi.org/10.1002/wnan.1363

Qing W, Chen K, Wang Y et al (2017) Green synthesis of silver nanoparticles by waste tea extract and degradation of organic dye in the absence and presence of H2O2. Appl Surf Sci 423:1019–1024. https://doi.org/10.1016/j.apsusc.2017.07.007

Qu Y, Pei X, Shen W et al (2017) Biosynthesis of gold nanoparticles by Aspergillum sp. WL-au for degradation of aromatic pollutants. Phys E Low Dimen Syst Nanostruct 88:133–141. https://doi.org/10.1016/j.physe.2017.01.010

Rajeshkumar S, Malarkodi C, Vanaja M et al (2013) Antibacterial activity of algae mediated synthesis of gold nanoparticles from turbinaria conoides. Der Pharma Chem 5:224–229

Rajeshkumar S, Malarkodi C, Vanaja M, Annadurai G (2016) Anticancer and enhanced antimicrobial activity of biosynthesized silver nanoparticles against clinical pathogens. J Mol Struct 1116:165–173. https://doi.org/10.1016/j.molstruc.2016.03.044

Rostami-Vartooni A, Nasrollahzadeh M, Salavati-Niasari M, Atarod M (2016) Photocatalytic degradation of azo dyes by titanium dioxide supported silver nanoparticles prepared by a green method using *Carpobrotus acinaciformis* extract. J Alloys Compd 689:15–20. https://doi.org/10.1016/j.jallcom.2016.07.253

Roy S, Das TK (2016) Effect of biosynthesized silver nanoparticles on the growth and some biochemical parameters of @@ *Aspergillus foetidus*. J Environ Chem Eng 4:4. https://doi.org/10.1016/j.jece.2016.02.010

Saha J, Begum A, Mukherjee A, Kumar S (2017) A novel green synthesis of silver nanoparticles and their catalytic action in reduction of methylene blue dye. Sustain Environ Res 27:245–250. https://doi.org/10.1016/j.serj.2017.04.003

Sai Saraswathi V, Kamarudheen N, BhaskaraRao KV, Santhakumar K (2017) Phytoremediation of dyes using *Lagerstroemia speciosa* mediated silver nanoparticles and its biofilm activity against clinical strains *Pseudomonas aeruginosa*. J Photochem Photobiol B Biol 168:107–116. https://doi.org/10.1016/j.jphotobiol.2017.02.004

Salam HA, Rajiv P, Kamaraj M et al (2012) Plants: green route for nanoparticle synthesis. Int Res J Biol Sci 1:85–90. https://doi.org/10.1007/s11051-009-9621-2

Salata OV (2004) Nanoparticles—known and unknown health risks. J Nanobiotechnol 6:1–6. https://doi.org/10.1186/1477-3155-2-12

Sanghi R, Verma P, Puri S (2011) Enzymatic formation of gold nanoparticles using Phanerochaete Chrysosporium. Adv Chem Eng Sci 1:154–162. https://doi.org/10.4236/aces.2011.13023

Saraswathi VS, Santhakumar K (2017) Green synthesis of silver nanoparticles mediated using lagerstroemia speciosa and photocatalytic activity against azo dye. Mech Mater Sci Eng J. https://doi.org/10.2412/mmse.72.63.602

Saravanan C, Rajesh R, Kaviarasan T et al (2017) Synthesis of silver nanoparticles using bacterial exopolysaccharide and its application for degradation of azo-dyes. Biotechnol Rep 15:33–40. https://doi.org/10.1016/j.btre.2017.02.006

Sawle BD, Salimath B, Deshpande R et al (2008) Biosynthesis and stabilization of Au and Au-Ag alloy nanoparticles by fungus, *Fusarium semitectum*. Sci Technol Adv Mater 9:035012. https://doi.org/10.1088/1468-6996/9/3/035012

Shahverdi AR, Fakhimi A, Shahverdi HR, Minaian S (2007) Synthesis and effect of silver nanoparticles on the antibacterial activity of different antibiotics against *Staphylococcus aureus* and *Escherichia coli*. Nanomed Nanotechnol Biol Med 3:168–171. https://doi.org/10.1016/j.nano.2007.02.001

Shaligram NS, Bule M, Bhambure R et al (2009) Biosynthesis of silver nanoparticles using aqueous extract from the compactin producing fungal strain. Process Biochem 44:939–943. https://doi.org/10.1016/j.procbio.2009.04.009

Singaravelu G, Arockiamary JS, Kumar VG, Govindaraju K (2007) A novel extracellular synthesis of monodisperse gold nanoparticles using marine alga, Sargassum wightii Greville. Colloids Surf B Biointerfaces 57:97–101. https://doi.org/10.1016/j.colsurfb.2007.01.010

Sinha T, Ahmaruzzaman M (2015a) High-value utilization of egg shell to synthesize silver and gold-silver core shell nanoparticles and their application for the degradation of hazardous dyes from aqueous phase-A green approach. J Colloid Interface Sci 453:115–131. https://doi.org/10.1016/j.jcis.2015.04.053

Sinha T, Ahmaruzzaman M (2015b) A novel green and template free approach for the synthesis of gold nanorice and its utilization as a catalyst for the degradation of hazardous dye. Spectrochim Acta Part A Mol Biomol Spectrosc 142:266–270. https://doi.org/10.1016/j.saa.2015.02.020

Sreekanth TVM, Jung M-J, Eom I-Y (2016) Green synthesis of silver nanoparticles, decorated on graphene oxide nanosheets and their catalytic activity. Appl Surf Sci 361:102–106. https://doi.org/10.1016/j.apsusc.2015.11.146

Suganya USU, Govindaraju K, Kumar GG et al (2016) Anti-proliferative effect of biogenic gold nanoparticles against breast cancer cell lines (MDA-MB-231 & MCF-7). Appl Surf Sci 371:415–424. https://doi.org/10.1016/j.apsusc.2016.03.004

Suresh AK, Pelletier DA, Wang W et al (2011) Biofabrication of discrete spherical gold nanoparticles using the metal-reducing bacterium *Shewanella oneidensis*. Acta Biomater 7:2148–2152. https://doi.org/10.1016/j.actbio.2011.01.023

Suvith VS, Philip D (2014) Catalytic degradation of methylene blue using biosynthesized gold and silver nanoparticles. Spectrochim Acta Part A Mol Biomol Spectrosc 118:526–532. https://doi.org/10.1016/j.saa.2013.09.016

Swamy VS, Prasad R (2012) Green synthesis of silver nanoparticles from the leaf extract of *Santalum album* and its antimicrobial activity. J Optoelectron Biomed Mater 4(3):53–59

Umamaheswari C, Lakshmanan A, Nagarajan NS (2018) Green synthesis, characterization and catalytic degradation studies of gold nanoparticles against congo red and methyl orange. J Photochem Photobiol B Biol 178:33–39. https://doi.org/10.1016/j.jphotobiol.2017.10.017

Vanaamudan A, Soni H, Padmaja Sudhakar P (2016) Palm shell extract capped silver nanoparticles—as efficient catalysts for degradation of dyes and as SERS substrates. J Mol Liq 215:787–794. https://doi.org/10.1016/j.molliq.2016.01.027

Vanaja M, Gnanajobitha G, Paulkumar K et al (2013) Phytosynthesis of silver nanoparticles by *Cissus quadrangularis*: influence of physicochemical factors. J Nanostruct Chem 3:17. https://doi.org/10.1186/2193-8865-3-17

Vanaja M, Paulkumar K, Baburaja M et al (2014) Degradation of methylene blue using biologically. Bioinorg Chem Appl 2014:1–8. https://doi.org/10.1155/2014/742346

Vankar PS, Bajpai D (2010) Preparation of gold nanoparticles from *Mirabilis jalapa* flowers. Indian J Biochem Biophys 47:157–160

Varadavenkatesan T, Selvaraj R, Vinayagam R (2016) Phyto-synthesis of silver nanoparticles from *Mussaenda erythrophylla* leaf extract and their application in catalytic degradation of methyl orange dye. J Mol Liq 221:1063–1070. https://doi.org/10.1016/j.molliq.2016.06.064

Veisi H, Azizi S, Mohammadi P (2018) Green synthesis of the silver nanoparticles mediated by *Thymbra spicata* extract and its application as a heterogeneous and recyclable nanocatalyst for catalytic reduction of a variety of dyes in water. J Clean Prod 170:1536–1543. https://doi.org/10.1016/j.jclepro.2017.09.265

Venkatachalam M, Govindaraju K, Mohamed Sadiq A et al (2013) Functionalization of gold nanoparticles as antidiabetic nanomaterial. Spectrochim Acta Part A Mol Biomol Spectrosc 116:331–338. https://doi.org/10.1016/j.saa.2013.07.038

Vidhu VK, Philip D (2014) Catalytic degradation of organic dyes using biosynthesized silver nanoparticles. Micron 56:54–62. https://doi.org/10.1016/j.micron.2013.10.006

Wang N, Hu Y, Zhang Z (2017) Sustainable catalytic properties of silver nanoparticles supported montmorillonite for highly efficient recyclable reduction of methylene blue. Appl Clay Sci 150:47–55. https://doi.org/10.1016/j.clay.2017.08.024

Wani K, Choudhari A, Chikate R et al (2013) Synthesis and characterization of gold nanoparticle using Ficus religiosa extract. Carbon Sci Technol 1:203–210

Xu L, Wang Y, Liu J et al (2017) High-efficient visible-light photocatalyst based on graphene incorporated Ag_3PO_4 nanocomposite applicable for the degradation of a wide variety of dyes. J Photochem Photobiol A Chem 340:70–79. https://doi.org/10.1016/j.jphotochem.2017.02.022

Yola ML, Eren T, Atar N, Wang S (2013) Adsorptive and photocatalytic removal of reactive dyes by silver nanoparticle-colemanite ore waste. Chem Eng J 242:333–340. https://doi.org/10.1016/j.cej.2013.12.086

Yu J, Xu D, Guan HN et al (2016) Facile one-step green synthesis of gold nanoparticles using *Citrus maxima* aqueous extracts and its catalytic activity. Mater Lett 166:110–112. https://doi.org/10.1016/j.matlet.2015.12.031

Zaheer Z, Aazam ES (2017) Cetyltrimethylammonium bromide assisted synthesis of silver nanoparticles and their catalytic activity. J Mol Liq 242:1035–1041. https://doi.org/10.1016/j.molliq.2017.07.123

Chapter 11
Nanomaterials: An Upcoming Fortune to Waste Recycling

Mugdha Rao, Anal K. Jha, and Kamal Prasad

11.1   Introduction

The role of nanotechnology in waste minimization is an oozing trend as it deals with the dual concept of "from waste to treat waste" and "from waste to valuable products." Nanotechnology is definitely a province that encircles subject matters like biology, materials science, and chemistry. It handles maneuvering materials at nanoscale called nanoparticles (NPs). Nanomaterials are cobbled up through either of the two approaches: (a) bottom up and (b) top down. Bottom-up way is the preferred one as the nanofabrication occurs through atom by atom, molecules by molecules stacked to obtain thermodynamically controlled composition of nanomaterial. Gibbs free energy found to be whimsical in such stacking process (Rao et al. 2017a). Nanostructures developed through biological process follows bottom-up approach. Top-down approach requires a progenitor material that undergoes reduction either through chemical or physical method (Gade et al. 2010). Waste mediated NPs (WMNPs) are fabricated mostly through biological (Bankar et al. 2010; Kaviya et al. 2011; Ahmad et al. 2012; Edison and Sethuraman 2013; Shankar et al. 2014) and chemical route. Chemical methods for synthesizing WMNPs involves irradiation-assisted chemical method (Park et al. 2014), pyrolysis (Rajarao et al. 2014; Wu et al. 2016; Kumar et al. 2014), and CVD (chemical vapor deposition) (Hintsho et al. 2014; Hintsho et al. 2016; Suriani et al. 2016; Wu et al. 2017).

The caption "waste" is not at all neoteric to us rather it has spawned facilely than anything else could have. An unforeseen jack up in residents and their comforts in accretion to their hustled lifestyle has crowned waste at the top. Nowadays,

M. Rao · A. K. Jha
Aryabhatta Centre for Nanoscience and Nanotechnology, Aryabhatta Knowledge University, Patna, Bihar, India

K. Prasad (✉)
Department of Physics, Tilka Manjhi Bhagalpur University, Bhagalpur, Bihar, India

© Springer Nature Switzerland AG 2018
R. Prasad et al. (eds.), *Exploring the Realms of Nature for Nanosynthesis*,
Nanotechnology in the Life Sciences, https://doi.org/10.1007/978-3-319-99570-0_11

researchers are dynamically involved to combat waste to prevent ecological pollution in turn lives on the globe. Merely working on pollution control at the level of academics is not solely to be advantageous rather public awareness is of prime importance. Nanotechnology in the context of environmental remediation is blooming rapidly (Fig. 11.1). The reason could be the effective pertinence of nanomaterial that helps in simplifying the treatment steps, which in turn effects the time, cost, and energy requirement for executed process (Dermatas et al. 2018).

The chapter details about the waste classification in brief and then followed up by general waste management concept. Further role of nanotechnology at omnifarious scaffolds of waste management like prevention, reduction, reuse, recycle, energy recovery, and disposal were elaborated. The chapter aimed at conveying that nanotechnology might be an option other than conventional one to waste utilization. Nanomaterial fabrications from variegated sources like garbage, combustible materials, ash, pharmaceuticals, agricultural and microbial waste are accompanied in this chapter. The application of WMNPs is also brought into the limelight. In the immediate future, nanotechnology will roll on through the speculation of scale up in utilization of waste to prevent biosphere pollution.

11.2   Classification of Wastes

Before getting into details about types of waste, a question to be answered, "What is the purpose of classifying waste? Is it really important?" Definitely a big Yes! According to Lymer (2016), the storage, treatment, and its disposal are very dependent on the category to which waste belongs. Further, it aids in deciding the error-free landfill sites.

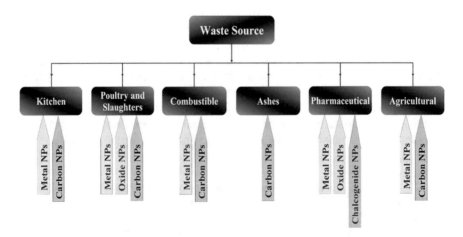

Fig. 11.1 Flow diagram of different groups of NPs fabricated from various waste source as an indication of nanotechnology role in environmental remediation

The expression "waste" hails from the Latin sound vastus that expresses unoccupied or uncultivated. Although waste can be ascertained in multiple ways but in agreement to Basel Convention of the Control of Transboundary Movements of Hazardous Wastes and their Disposal, 1989 "wastes" are the substances, which are spaced out or are intended to be spaced out of or required to be spaced out by provisions of rational law. Although classifying refuse is a tedious task, there is no unique concept of categorizing all types of waste. In depth, it is indeed the role of chemist to categorize waste as hazardous or non-hazardous.

As per our intellection, systematizing waste depends on the context in which we are working into like various fountainhead of waste in the locality, industrial scale, or the fractional elements of the offal. In coalition with execution of nanotechnology in waste conversion to valuable products, classification of waste is mentioned in Table 11.1. We have classified waste into six categories as per their origination source and the burden they create. Primary source of approximately all the offal are either households or the industries.

11.3   Waste Management (General Concept)

The escalating population with time is also raising the quantity of waste globally. There are numerous waste source concisely named to be industrial and municipal solid waste. But the uppermost concern is waste management. Not only the management methods are required to be updated rather folk educational awareness towards minimization of waste generation and its management, that too at personal level must be polished. We have seen common folks in developing countries like ours receiving immense pleasure after burning the waste as such in the context of cleaning up the area overcrowded with garbage. Waste management in developed homeland is quite contrastive to developed countries. Having clean and hygienic environment is the most favorable one because we are alive due to air that we breathe, the water that we drink, and the surroundings where we live. Waste

Table 11.1 Classification of waste as per their source of origin

Classification	Origin point	Burden
1. Garbage	Domiciliary/merchandise	Kitchen waste (food), poultry/slaughter house
2. Settlings	Domiciliary/retailer	Dead leaves, wood wool, shredders waste, waste tyre rubber, polyethelene, plastic bottles, etc.
3. Ashes	Households/industries	Rice husk ashes, coal fly ashes
4. Pharmaceuticals	Industries/households	Expired medicines, scraps etc.
5. Waste microbes	Industries/laboratories	Metal ion-treated cells, fungi refusals, bacterial cellulose discard, etc.
6. Agricultural waste	Beverage industries/food processing industries	Palm oil mill effluent, defatted seeds, orange peels, etc.

management is the sequential interconnected system that comprises of 6 gears and is depicted in Fig. 11.2.

In correspondence to the waste hierarchy, deterrence of waste generation is of utmost importance. If the waste prevention step is surpassed, then the hierarchical steps will be followed. Apart from the above steps to be followed, after disposal monitoring of the area is also a concern. According to Demirbas (2011), pyrolysis-incineration, landfill, and biogas production are the reassuring waste treatment processes. In another Handbook (2014), he has mentioned treatment methods for municipal waste like sanitary landfill, incineration, gasification, biodegradable process, composting, and anaerobic digestion. These followed-up treatment protocol that has its own gratification and hindrance is depicted in Table 11.2.

Although there are enormous methods of waste disposal, many more has to be seized into account as nanotechnology-based waste conversion to unburden the load on conventional protocols. Above all if we concentrate much on reduction of waste generation that will actually benefit in controlling environmental pollution. It is well exemplified by Rahman (2000) in his guide that reducing dependency on plastic bags, plastic film containers, tin can, rubber boot sole, plastic cups, aluminum can, cigarette butt, disposable diaper, plastic bottle, glass bottle, etc. will save the environment to a great extent.

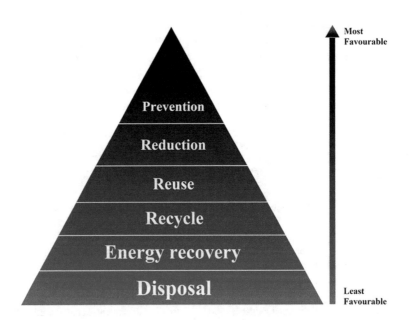

Fig. 11.2 Waste management steps detailed as pyramid format in context of favorability

Table 11.2 Waste treatment method protocols pros and cons

Process	Advantages	Disadvantages
Sanitary landfill	– No emission of methane and toxic gases – Limited land/water table contamination – No bad odor – No problem created by birds/rodents	– Designing cost – Land requirement
Incineration	– Size of waste reduce – Reduction in dumping – No disease causing issues	– Air pollution – Toxic gases emission – Ash treatment requirement – High maintenance cost
Gastrification	– Gas handling is easier – No hazardous gas emission – Less ash production	– Costly maintenance – Tar and volatile poison release
Biodegradation process		
(a) Composting	– Recycling of waste – Generate manure – Cost-effective	– Time consuming – Odor problem – Attract birds and rodents
(b) Anaerobic digestion	– Less energy requirement – Simple construction	– Pre-processing of slurry – Sludge disposal – Odor problem – Maintenance of plants
(c) Waste stabilization pond system	– Simple control, operation, and maintenance – No electro-mechanical required	– Huge land requirement – Odor problem – Groundwater contamination
(d) Duckweed pond system	– Proper covering prevents mosquito breeding and odor – Yield of proteinaceous material as animal feed	– Low pathogen removal
(e) Facultative aerated lagoon	– Simple operation – Simple installation issue – Lower energy cost	– Possibility of groundwater contamination
(f) Activated sludge treatment	– Less land requirement	– Costly – Foam formation affects the process
(g) Biological, filtration, and oxygenated reactor technology	– Space saving – Odor-free – Less operational supervision	– Sludge formation

11.4   Nanotechnology in Waste Management

Nowadays, nanotechnology is not only limited to sectors like energy (Luther 2008), food safety (Joseph and Morrison 2006; Rao et al. 2017a, b), electronics (Bhat 2007), agriculture (Joseph and Morrison 2006; Rao et al. 2017a, b; Sangeetha et al. 2017a, c), medicine (Nikalje 2015) rather has occupied a space in environmental protection and waste management is delineated in Fig. 11.3. Nanotechnology is compassionately involved with materials of dimensions ranging from 1–100 nm. The inducement for nanomaterial versatility and lessen volume required in application is mainly because of tiny size that frills enhanced reactivity for adsorption, catalysis of several contaminant groups such as organics, for example, polycyclic aromatic hydrocarbons (Yang et al. 2006), heavy metals like barium (Çelebi et al. 2007) and inorganics like nitrate (Joo and Cheng 2006).

Nanotechnology is definitely proven as a screen raiser at each stage of waste management that is PR3ED (prevention-reduction-reuse-recycle-energy recovery-disposal) and is briefly described as follows.

11.4.1   PR3ED

Nanomaterials credibly is a good replacer as in making the process greener by substituting materials that undergo high-energy consumption, curtailment of toxic chemicals, solvent thus enhancing manufacturing efficiency and so reduction of waste and pollution (Shan et al. 2009). According to Kumar et al. (2016), nanomaterials like fullerene and carbon nanotubes have superior mechanical and electronics properties that frame it as meritorious in various applications like field emission display, battery storage media, and nanoelectronic devices. They explained that use

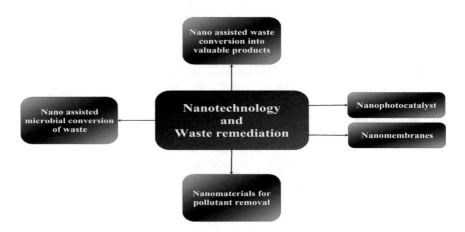

Fig. 11.3 Panoramic view of nanotechnology in waste remediation

of carbon nanotubes (p-type semiconductor) in solar cells enhances its absorption efficiency. Further, they concluded that multi-walled carbon nanotube absorbs additional power than single walled nanotube in the solar cells. In a research performed by Li et al. (2013), graphene and carbon nanotubes endowed as fetching material for photovoltaic system. These materials possess optical and electronic properties that propel its direct use for improving the energy conversion in thin film solar cells. Alturaif et al. (2014) in their review explained that silicon-based solar cells are used at large scale but these cells possess major setback of not being cheaper. They considered organic material occupied solar cells as the potential candidate. They explained organic material to be the best alternative, as it possesses conducting as well as semi-conductive properties, low cost, simple manufacturing protocol, and high throughput. Above all goodness, organic solar cells power conversion efficiency elevates with increment in temperature in comparison to the inorganic solar cells, which decreases upon elevated temperature. Thus, nanomaterials are rather to be called as magic bullets at the pollution prevention stage, as being explained in the above examples.

Shan et al. (2009) in their review mentioned that nanocatalyst supplemented to diesel fuel helps in high efficient burning thus eventuate in release of fewer flue gases. Also explained that nanofabricated sorbents provide better adsorption of pollutants. To expunge CO_2 from smokestack nanocrystal proved to be effective. Nanocrystal works by conferring an electron to CO_2 to react with other molecules and making it non-toxic. They cited another example wherein functionalized sorbents are very effective for diminution of mercury, a pollutant transported globally from coal combustion. A couple of mercury removal sorbents like titanium oxide nanocrystal and iron oxide NPs were explained, where former one has to undergo maintenance periodically.

Reuse of materials after its implication to waste remediation subsist the third most favorable step in waste hierarchy. According to Lunge et al. (2014), magnetic NPs as such or bolstered adsorbents can be effortlessly reused after their separation through outwardly applied magnetic field. These NPs have become center of attraction for environmental remediation, as it owes high surface area and super paramagnetic properties. They reported that magnetic NPs could remove organic pollutants like phenols and dyes and ions like arsenic, mercury, nickel, copper, fluoride, chromium, etc. Group of scientist Tran et al. (2010) fabricated oxide magnetic NPs like magnetite, manganite, and spinel ferrite. Among them magnetite magnetic NPs were utilized for confiscation of heavy toxic metals like lead and arsenic from desecrated water. Recycling and energy recovery are the interlinked stages in waste hierarchy. Recycling is a protocol wherein waste materials either are converted to fresh usable material or object. Therefore, recycling contribute as in reduction of crude raw materials and appositeness of waste with potential of being useful. The role of a few nanomaterial-assisted waste conversion is mentioned in Table 11.3.

The least favorable step in the waste pyramid is disposal. According to Shan et al. (2009), incineration of waste for its disposal possess many advantages like complete destruction, reduction in usage of land, and aids in energy recovery too. However, incineration can impose environmental issues like emission of dioxin

Table 11.3 Catalogue of nanomaterial-assisted waste conversion

Waste source	Waste-treated Nanomaterial	Other	References
Waste cooking oil	Mesoporous silica/ superparamagnetic iron oxide core shell NPs	Lipase enzyme	Karimi et al. (2013)
Olive mill wastewater	Titanium(IV) oxide anatase, iron (III) oxide nanorods NPs	Fungi	Nogueira et al. (2015)
Waste cooking oil	Magnetic NPs	Lipase enzyme	Yu et al. (2013)
Rice husk	Calcium hydroxyapetite NPs	Cellulose and xylanase	Dutta et al. (2014)
Water with high phosphate content	Iron NPs	Spondias purpurea seed waste	Arshadi et al. (2015)
Waste cooking oil	Molybdenum oxide/Zirconia NPs	Ferric manganese	Alhassan et al. (2015)
Organic pollutant	Nitrogen-doped carbon nanotube	Iron	Yao et al. (2017)

from incineration process. They reported in their review about the coalition disposal by employing incineration and nanophotocatalyst. The list of few nanophotocatalyst for environmental pollutants removal mentioned in the research article of Shan and co-researcher is mentioned in Fig. 11.4.

The fact is plastics are not biodegradable due to inertness of material it is being made of. Fa et al. (2011) came up with an approach to make plastics degradable by embedding functionalized TiO_2 NPs into commercial poly(vinyl) chloride (PVC). The TiO_2 NPs were modified with perchlorinated phthalocyanine iron (II) (TP) to embellish their photocatalytic task. It was illustrated that the TP composite degradation efficiency was incomparably elevated under UV and sunlight luminosity than the primeval PVC and the composite $PVCTiO_2$. Table 11.4 comprises of list of nanomaterials knotted in environmental pollutants remediation (Shan et al. 2009; Gangadhar et al. 2012; Wang et al. 2012a, b; Amin et al. 2014).

11.5   Trash: Repository of Nanomaterial

11.5.1   Nanomaterial from Garbage

Garbage is the unpretentious materials discarded by human beings excluding toxic waste products. It includes kitchen or food items like poultry/slaughterhouse waste. The following tract deals with the repository of nanomaterial in garbage.

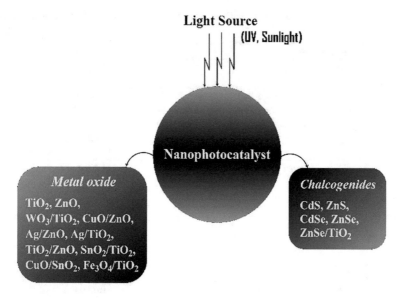

Fig. 11.4 Nanophotocatalyst for environmental pollutants removal

11.5.1.1 Kitchen Waste

Everyday our kitchen generates lots of biodegradable green waste. It primarily comprises of fruit and vegetable peels. These waste peels are abounded in metabolites that have the potential to achieve nanoconversion. *Lansium domesticum* peels, a fruit with a common name as longkong carries plenteous metabolites like terpenoids (lansionic acid, triterpene glycosides) and organic compounds (lansionic acid and methyl ester), accomplish reduction of metal ions to stable NPs by capping it. These metabolites possess strong reducing potential (Shankar et al. 2014). Likewise, pomegranate peels are enriched with polyphenols like condensed tannins, hydrolysable tannins, ellagic and gallic acids (Goudarzi et al. 2016). According to Kuppusamy et al. (2015), polyphenols are the superior candidate for green NPs fabrication due to following betterment (a) non-toxic, (b) control the size, morphology of NPs, (c) simple synthesis mechanisms, (d) can be scaled up with minimum cost, (e) environmental friendly. They explained the three-fold working of polyphenol mediated NPs formation as polyphenol complex formation is the first step, followed by bioreduction of metal ion to nascent ion, and finally capping of fabricated NPs by phenolics (oxidized).

Other peels like papaya are loaded in metabolites like chymopapain and papain (Kokila et al. 2016), mango peel is endowed with polysaccharides, lignins, flavonoids, hemicelluloses, and pectins (Yang and Li 2013), Orange peels loaded with limonene, citral, neohesperidin, naringin, rutin, rhamnose, eriocitrin, and vitamin C (Rao et al. 2017a). Banana peels are bloated with polymers like lignins, hemicelluloses, and pectins (Bankar et al. 2010) and watermelon rind comprises of metabolites like cellulose, citrulline, pectins, proteins, and carotenoids (Patra et al. 2016).

Table 11.4 List of nanomaterial/nanoparticles mediated environmental pollutants remediation

Nanomaterials	Environmental pollutants
Lead, nickel, zinc, copper	Alumina
Radioactive metal toxin	Biophosphonate-modified magnetite
Polycyclic aromatic hydrocarbon, trihalomethanes, atrazine, lead, cadmium, chromium, zinc, polycyclic aromatic compound, chlorophenols, herbicides, dichlorodiphenyltrichloroethane	Carbon nanotube
Copper, lead, cadmium, zinc	Carbon nanotube/ hydroxyquinoline
Crocein orange G, acid green 25	Chitosan-based Fe_3O_4
Cyanine acid blue	Copper/iron oxide sawdust
Arsenic, chromium	Cerium oxide
Arsenic	Cerium oxide supported carbon nanotube
Arsenic and chromium	Flower-like iron oxide nanostructure
Cadmium, cobalt	Graphene nanosheets
Acid black 24, nitrate, barium, humic acid, arsenic, cadmium, chromium, nickel, lead, nitrate, perchlorate, organic compounds, 2,2′ dichlorobiphenyl	Iron oxide
Phosphate	Iron/granular-activated carbon
Chromium	Iron oxide coated polypyrrole
Arsenic	Magnetite/graphene
Nickel	Magnesium hydroxide/ alumina nanocomposite membrane
Organic dyes	Manganese oxide films
Nickel	Magnesium oxide/nickel oxide/alumina
Toluene, benzene	Multi-walled carbon nanotubes/iron
Acid green 27	Nanochitosan
Nickel	Nanocrystalline calcium hydroxyapatite
Benzene, toluene, xylene, ethylbenzene	Nanoporous-activated carbon fiber
Azo dyes and phenols	Nitrogen- and iron-doped titanium dioxide
Chromium	Palladium
Nitrate	Palladium/copper/alumina
Trichloroethylene	Palladium, palladium/gold
Heavy metals	Polyrhodamine magnetic
Copper, lead	Polylayered silicate nanocomposite

(continued)

Table 11.4 (continued)

Nanomaterials	Environmental pollutants
Organic dyes	Silver/amidoxime fiber
4 nitrophenol	Sulfate ion/titanium dioxide
Acid orange 7, reactive orange 16, polychlorinated biphenyls, benzene, chlorinated alkanes	Titanium dioxide
Malachite green	Tin-doped titanium dioxide double layer thin films
Cadmium	Titanate nanoflowers
Arsenic	Titanium dioxide/ magnetite
Chromium	Titanium dioxide/graphene sheet
Polychlorinated biphenyls, benzene, chlorinated alkanes	Titanium dioxide
Total organic content	Titanium dioxide and alumina
Toluene	Titanium nanorods
4- chlorocatechol	Zinc oxide
Thiophene	Zinc-based nanocrystalline aluminum oxide (Zn/ Al_2O_3)

Table 11.5 Fraction of consumed non-vegetarian food from the source

Non-vegetarian (Jha and Prasad 2016)	
Animals (each)	% Utilization
Cow	50–54
Sheep/goat	52
Pig	60–62
Chicken	68–72
Turkey	78

11.5.1.2 Poultry and Slaughters Waste

Not anyone in the world will deny possessing food drive we all drool towards yummy food with the choice to plum on plant based or animal food. However, the point is to bring into the limelight about the food processing waste generated in tons globally. This section ratifies poultry and slaughterhouse waste and their possible exorcism. Animal food subsumes both edible and non-edible parts (choice-based). Table 11.5 illustrate fraction of food that is only consumed. From production to processing, slaughter house and poultry procreate refusals like hatchery waste, sawdust, wood shavings, straws, peanuts, feathers, entrails, organs of murdered birds, processing wastewater, bio-solids, blood, hides, bones, and most of these are constructed with organic and inorganic nutrients. The above consequences diminish the essence of water, soil, and air as well (Williams 2008). A few options like applicability of poultry waste by-products to land application as crop nutrients, animal re-feeding, and biogas production from slaughterhouse waste were implemented to discern the natural terrain (Williams 2008; Ek et al. 2011; Ahmad and Ansari 2012; Malav et al. 2018). A group of experimenter propounded a safe method to decompose broiler slaughterhouse waste verging on to coir pith together with caged layer manure from poultry farms (Bharathy et al. 2012). Another major concern is the increasing population, as will impose hefty demand on the devastation of poultry and animal meat and inasmuch the waste generation. Therefore, dependency wholly on the above-mentioned management techniques would not rather be effective. From ended few years, researchers are putting effort to harness NPs from poultry-cum-slaughters waste pauperizing the load on other management modes.

11.5.1.3 Chicken Offals

11.5.1.3.1 Eggshells

A couple of researchers passed on a neoteric approach to use waste eggshells as such serving as a reactor system for controlled synthesis of amorphous $Co(OH)_2$ nanorod arrays accompanying on Ni foam, Ti foam, or glass as substrate and also swept off the fact of not being synthesized conventionally by direct mixing of the parent constituents. The metabolism of embryo creates carbonate ions, upon reaction with calcium ions coming out of uterus step up to sturdy, protective eggshell. They have utilized $CoSO_4$ and NaOH solution, the former one withheld in the emptied eggshell and the later one outside the shell. Being porous in nature, bestowed protein membrane is permeable and thus permitting the diffusion of ions slowly is definitely a stringent condition to procure $Co(OH)_2$ nanorods (Meng and Deng (2017)). Natural offal-like eggshell implied to be quite promising for manipulating materials at nanoscale. In another work, group of experimenters reported calcium carbonate nanoparticles fabrication using waste eggshells, majorly sourced to have 94% of calcite and other components. Figure 11.5 portrays various components present in eggshell. They achieved nanosize by virtue of mechanochemical cum sonochemical method, ball milled precisely in wet condition (polypropylene glycol) to avoid iron contamination through steel and superseded with high intensity ultrasonic horn in propinquity of N,N-dimethylformamide (DMF), decalin, and tetrahydrofuran. An irregular, 10 nm sized $CaCO_3$ NPs were established after characterization (Hassan et al. 2013). According to Wang et al. (2012a, b), microwave-assisted concept was operated to contrive fluorescent and water soluble C-Dots due to efficient and intensive energy treatment by microwave. They have devised a green, rapid, and eco-friendly method to fabricate C-Dot from easily procured protein-rich eggshell membrane trash that showed excellent fluorescent property detected by photoluminescence emission spectra.

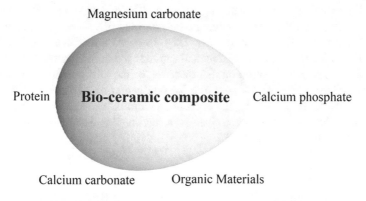

Magnesium carbonate

Protein   **Bio-ceramic composite**   Calcium phosphate

Calcium carbonate   Organic Materials

Fig. 11.5 Components of eggshell convoluted in nanotransformation

Fig. 11.6 Pictorial representation of constituents of feather

Keratin

Water   Lipid

11.5.1.3.2   Feather

Poultry processing plants not only serves us with flesh but also creates heap of feathers each year. Sundaram et al. (2015) demonstrated green synthesis of keratin NPs from chicken feather. In addition, they studied morphology of designed keratin NPs by scanning electron microscope and X-Ray diffraction. Feathers are protein rich. Figure 11.6 represents the constituent elements of feathers. In cosmetic products, keratins are primarily used. In another work, Gao et al. (2014) developed a strategy to perform pyrolysis of chicken feather offal to Ni_3S_2-carbon coaxial nanofiber, in the subsistence of the catalyst nickel acetate tetrahydrate at 650 °C. The product then acidified to obtain nitrogen-doped carbon nanotube. They reported 41.8% chicken feather carbon conversion to N-CNT and confirmed via characterization through FE-SEM, TEM, XPS, and XRD.

11.5.1.3.3   Fish Offal

Jha and Prasad (2014) accounted a new technique to cut down fish discard in stipulation to fabricate nanomaterial to overcome dirtiness. Fish discard from kitchen and fish market are comparatively quite lesser than coastal processing sector. These are propertied with highly complex molecules (proteins) and essential fatty acids. They made it happen, as combo of an aggregates and individual silver NPs were fabricated ranging from 8 to 40 nm from gut waste of the fish (*Labeo rohita*) and were further characterized for confirmation. They proposed a mechanism of silver NPs biosynthesis from fish discard extract and summarized in Schematics (Fig. 11.7a).

In an another investigation on fish discards, a couple of researcher utilized the potential of fish scales of *Lobeo rohita* for fabrication of copper NPs (25–37 nm) with no outsourced stabilizing and reducing materials. They also contemplated the mechanism for the fabricated copper NPs, outlined in Schematics (Fig. 11.7b) (Sinha and Ahmaruzzaman 2015).

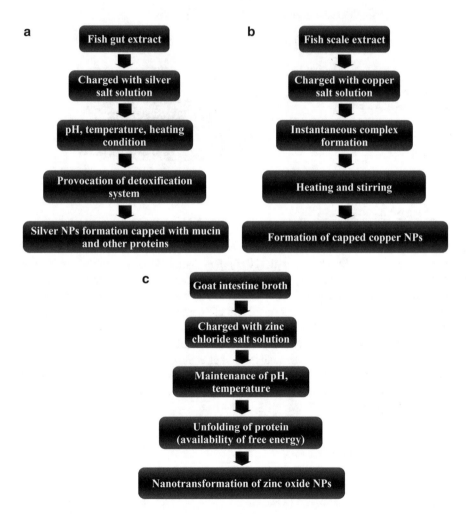

Fig. 11.7 (**a**) Contemplated mechanism of silver NPs from fish *Labeo rohita* gut. (**b**) Contemplated mechanism of Cu NPs fabrication from scales of fish *Labeo rohita*. (**c**) Contemplated mechanism of zinc oxide NPs fabricated from goat intestine

11.5.1.3.4 Goat Offal

In a study, goat intestine waste was utilized for the artifact of zinc oxide NPs through green, eco-friendly protocol in a concern to intercept environmental pollution. The particle size of zinc oxide NPs fabricated were in the range of 3–11 nm. Employing UV-visible spectroscopic technique, complete reaction decorum was monitored. A contemplated mechanism for biological synthesis of zinc oxide NPs has been proffered and is summarized in the Schematics (Fig. 11.7c) (Jha and Prasad 2016).

Chamundeeswari et al. (2011) reported the artifact of magnetic iron NPs from goat blood by incineration method. In her another experimental work, she applied a simple wet-precipitation mode to fabricate another nanobiocomposite employing the iron NPs with different coating material (collagen) (Chamundeeswari et al. 2013a). Further, she extended her work to fabricate bionanocomposite using iron NPs with a mixture of chitosan and gelatin with varying size that ranges from 80 to 300 nm. The so-formed material possesses magnetic properties having magnetic saturation of 18.97 emu/g (Chamundeeswari et al. 2013b).

11.5.2 Nanomaterials from Combustible Waste

Literally, combustion is a phenomena wherein burning occurs by virtue of the reaction between the reductant and an oxidant to exonerate light and heat. The material that bear up combustion are expounded as combustible material and so as combustible waste are defined. On the basis of origin, combustible waste could be either of two types (a) natural combustible waste (NCW) and (b) synthetic combustible waste (SCW). NCW comprises of fallen leaves, waste paper, wood, bagasse, etc., wherein SCW includes cigarette butt, diapers, plastic bottles, ceramics, porcelain, rubber, leather, clothes, etc. Numberless researchers have taken effort to utilize specified combustible waste to value-added products as nanomaterials (Table 11.6). Researchers have fabricated nanomaterials through green, chemical and physical protocol. They synthesized graphene from various combustible wastes (Akhavan et al. 2014; Suryawanshi et al. 2012; Adolfsson et al. 2015; Gong et al. 2014; Sharma et al. 2014; Essawy et al. 2017).

Akhavan et al. (2014) explained in their work about the fabrication of graphene from cheap progenitor source like bagasse, wood, leaf, fruit waste, bone, cow dung, and newspaper through chemical method. They worked to conquer the limitation of using notably pure graphite as the initial source for fabrication. The sources they have chosen are well provided in carbon content. The graphene sheets they have synthesized were exclusive of the initial material and purity was propped similar to the graphene sheets obtained through Hummers' method.

Another group of investigators employed carbon source as spent neem leaves to fabricate graphene quantum dots. The dried neem leaves were submitted to pyrolysis to procure black powder, and further ball milled to obtain fine powder. The so-obtained ball milled powder considered as precursor for graphene quantum dots

Table 11.6 Catalogue of trash sourced nanomaterials

Waste sources	Concerned nanomaterial	Synthesis approach	References
Kitchen waste			
Banana peel	Silver NPs	Bioinspired	Bankar et al. (2010)
Orange peel	Silver NPs	Biosynthesis	Kaviya et al. (2011)
Punica granatum peel	Silver NPs Gold NPs	Biosynthesis	Ahmad et al. (2012)
Punica granatum peel	Silver NPs	Biogenesis	Edison and Sethuraman (2013)
Mango peel	Silver NPs	Green synthesis	Yang and Li (2013)
Orange peel	Silver NPs Platinum NPs	Biosynthesis	Castro et al. (2015)
Food waste	Carbon nanodot	Ultrasound irradiation	Park et al. (2014)
Lansium domesticum fruit peel	Gold NPs	Green synthesis	Shankar et al. (2014)
Rambutan peel	Zinc oxide nanocrystals	Green synthesis	Yuvakkumar et al. (2014)
Banana peel	Silver NPs	Biosynthesis	Ibrahim (2015)
Cavendish banana peel	Silver NPs	Biological	Kokila et al. (2015)
Vegetable waste	Silver NPs	Green synthesis	Kumar et al. (2015)
Citrullus lantus rind	Gold NPs	Biosynthesis	Patra and Baek (2015)
Waste vegetable peel	Silver NPs	Green synthesis	Sharma et al. (2016)
Arachis hypogaea	Silver NPs	Green synthesis	Velu et al. (2015)
Citrus sinensis	Carbon NPs	Green synthesis	Adedokun et al. (2016)
Pomegranate peel	Silver NPs	Facile thermal decomposition	Goudarzi et al. (2016)
Carica papaya peel	Silver NPs	Green synthesis	Kokila et al. (2016)
Banana peel extract	Silver NPs	Green synthesis	Narayanamma et al. (2016)
Rind of watermelon	Silver NPs	Biosynthesis	Patra et al. (2016)
Mangosteen pericarp waste	Silver NPs Gold NPs	Green synthesis	Park et al. (2017)
Orange peel	Carbon dots	Hydrothermal carbonization	Prasannan and Imae (2013)
Orange peel	Silver NPs	Biogenesis	Rao et al. (2017a)
Orange peel	Silver NPs	Biosynthesis	Saratale et al. (2017)
Poultry and slaughters waste			
Goat blood	Iron NPs	Incineration	Chamundeeswari et al. (2011)
Eggshells membrane	Carbon nanodots	Microwave-assisted	Wang et al. (2012a, b)
Goat blood	Iron NPs coated with collagen	Wet-precipitation	Chamundeeswari et al. (2013a)

(continued)

Table 11.6 (continued)

Waste sources	Concerned nanomaterial	Synthesis approach	References
Goat blood	Iron NPs coated chitosan and folic acid	Incineration	Chamundeeswari et al. (2013b)
Eggshells	Calcium carbonate NPs	Mechanochemical cum sonochemical	Hassan et al. (2013)
Chicken feather (including quills and barbs)	Nitrogen- doped carbon nanotubes	Biological synthetic protocol	Gao et al. (2014)
Gut of fish *Lobeo rohita*	Silver NPs	Biological synthetic protocol	Jha and Prasad (2014)
Fish scales of *Lobeo rohita*	Copper NPs	Green protocol	Sinha and Ahmaruzzaman (2015)
Chicken feathers	Keratin NPs	Biological synthetic protocol	Sundaram et al. (2015)
Goat slaughters waste (intestine)	Zinc oxide NPs	Biological synthetic protocol	Jha and Prasad (2016)
Ferrocene-chicken oil mixture	Vertically aligned carbon nanotube	Thermal chemical vapor deposition	Suriani et al. (2016)
Eggshells	Cobalt hydroxide NPs	Bioinspired	Meng and Deng (2017)
Combustible waste			
Wood wool	Carbon nano-onions	Pyrolysis	Sonkar et al. (2012)
Wood, leaf, bagasse, newspaper, soot powder	Graphene	Chemical method	Akhavan et al. (2014)
Fallen willow leaves	Carbon nanosphere	Thermal decomposition	Qu et al. (2014)
Dead neem leaves	Graphene quantum dots	Pyrolysis	Suryawanshi et al. (2012)
Mineral water bottles(used)	Nanocarbon tube	Arc-discharge method	Berkmans et al. (2014)
Waste polypropylene	Graphene flakes	Pyrolysis	Gong et al. (2014)
Tyre rubber	Silicon carbide NPs	Pyrolysis	Rajarao et al. (2014)
Solid waste plastic	Graphene crystals	Pyrolysis	Sharma et al. (2014)
Waste paper	Graphene oxide quantum dots	Microwave-assisted thermal degradation/ sonication	Adolfsson et al. (2015)
Corrugated container	Cellulose nanocrystal	Enzymatic hydrolysis and sonication	Tang et al. (2015)

(continued)

Table 11.6 (continued)

Waste sources	Concerned nanomaterial	Synthesis approach	References
Polyethylene terephthalate bottle waste	Graphene	Thermal decomposition	Essawy et al. (2017)
Cigarette butt	Silver nanostructure	Green protocol	Murugan et al. (2017)
Lard oil	Graphene sheets	Inductively coupled plasma-assisted chemical vapor deposition	Wu et al. (2017)
Ashes			
Coal fly ash	Carbon nanofiber	Catalytic chemical vapor deposition	Hintsho et al. (2014)
Coal fly ash	Carbon nanomaterial	Catalytic chemical vapor deposition	Hintsho et al. (2016)
Pure rice husk ash	Nanosilica	Precipitation	Sinyoung et al. (2017)
Waste microorganisms			
Bacteria	Lithium metal phosphate NPs	Biological	Zhou et al. (2014)
Bacterial cellulose	3D carbon nanomaterials	Pyrolysis	Wu et al. (2016)
Metarhizium robertsii	Silver NPs	Photo-induced biosynthesis	Rożalska et al. (2016)
Pharmaceutical waste			
Norfloxacin tinidazole combination	Gold, zirconia oxide, cadmium sulfide	Green protocol	Jha and Prasad (2012)
Agricultural waste			
Avena sativa biomass	Gold NPs	Green protocol	Armendariz et al. (2004)
Annona squamosa peel	Silver NPs	Green protocol	Kumar et al. (2012)
Coconut shell	Porous graphene nanosheet	Simultaneous activation-graphitization	Sun et al. (2013)
Cotton wood	Magnetic biochar	Pyrolysis	Zhang et al. (2013)
Punica granatum peel	Platinum NPs	Green protocol	Dauthal and Mukhopadhyay (2014)
Grape stem, stalk, seeds	Gold NPs	Green protocol	Krishnaswamy et al. (2014)
Oil palm leaves	Porous carbon NPs	Pyrolysis	Kumar et al. (2014)
Rice bran	Gold NPs	Green protocol	Malhotra et al. (2014)
Waste cotton	Cellulose NPs	Green method	Meyabadi et al. (2014)
Sugarcane bagasse	Graphene oxide	Muffled atmosphere	Somanathan et al. (2015)

(continued)

Table 11.6 (continued)

Waste sources	Concerned nanomaterial	Synthesis approach	References
Bilberry waste, spent coffee grounds	Silver NPs	Green protocol	Baiocco et al. (2016)
Coffee silverskin	Lipid NPs	Homogenization/sonication	Rodrigues et al. (2016)
Bagasse	Magnetic biochar	Pyrolysis	Thines et al. (2017)

fabrication. Upon diluting the precursor with water, mixture was dispensed to filtration to disarticulate the unreacted carbon and the bulkier particles. The filtrate showed strong fluorescence thus resembling graphene quantum dots (Suryawanshi et al. 2012). Similarly, through pyrolysis Gong et al. (2014), Rajarao et al. (2014) and Sharma et al. (2014) harnessed the potential of waste polypropylene, tyre rubber, solid waste plastic, wood wool to procure graphene flakes, silicon carbide NPs, graphene crystal, carbide nano-onions, respectively.

Through microwave-assisted thermal degradation coupled with sonication to procure graphene oxide quantum dots were taken up by Adolfsson et al. (2015). They fabricated nanomaterial from cellulose through the embodiment of an intermediate as carbon nanosphere. Upon degradation of catalyst by means of H_2SO_4 carbon nanosphere is obtained. Further, the intermediate were sonicated and heated in the aura of HNO_3 to stockpile graphene oxide quantum dots.

Murugan et al. (2017) worked on green protocol to fabricate silver NPs using cigarette butt waste. Cigarette butt comprises of polycyclic aromatic compounds, heavy metals nicotine, and ethyl phenol. They prepared two different extracts with and without tobacco. The varying extracts were charged with silver nitrate salt solution under the reaction conditions. Further, the fabricated silver NPs were characterized.

In the majority of research work, mechanism of creation of NPs from agricultural waste is not revealed. Malhotra et al. (2014) in their study investigated the mechanism involved in biomineralization of gold NPs from rice agro-waste (rice bran). They deciphered that ferulic acid, as the function molecule is accountable for the bioconversion gold salt to gold metal. They also monitored the heat modification in the bioconversion process and concluded the process to be exothermic in nature. The reaction between the ferulic acid and the gold salt was of first-order kinetics.

11.5.3 Nanomaterials from Ash

Thermal power plant produces offshoot as fly ash upon ignition of coal. Fly ash is removed from combustible gases before being set free into the surroundings. The morphological feature of fly ash is spherical with diameter ranging from 1 to 150 μm. The chemical makeup of fly ash depends upon the nonexplosive matter in the coal (Ramezanianpour 2014). Hintsho et al. (2016) that out of 800 million tons of coal fly ash only 15% are being recycled have reported it. Recyclability of fly ash surely is the fruitful outcome as they stated about its high disposal cost and as an environmental threat. They also explained its present applications (mine fill, mine site remediation, adsorbent for heavy metals, catalyst), with none of them commercially being employed. In this background, nanotechnology is at budding stage to unburden the tons of fly ash load aids in creating pollution.

According to Sinyoung et al. (2017), nanosilica is created through established method from sodium silicate upon heating sodium carbonate with quartz at very high temperature (1300 °C). High-energy requirement made this process a bit costlier. Therefore, to overcome this cost issue nanosilica produced from rice husk ash is advantageous, apart from cost it also lower the risk imposed on the environment pollution. They also explained the groundwork of nanosilica (5–60 nm) from rice husk ash having porous structure. The nanosilica was calcined in the laboratory to decontaminate K_2O, CaO, and MgO.

In another abstraction, carbonaceous nanostructures were fabricated from South African fly ash. Fly ash is a waste product consummated from the thermogenesis of coal, and its constituents were found to be Al_2O_3, SiO_2, FeO, Fe_2O_3, MnO, MgO, Na_2O, CaO, and K_2O after examining it through X-ray diffraction. Above all, it acts as catalyst for the proliferation of carbon nanomaterials. The carbon nanofibers were fabricated through catalytic chemical vapor deposition method in shade of acetylene gas extremely elevated temperature (400–700 °C). Further carbonaceous were characterized by transmission electron microscopy, laser Raman spectroscopy, Brunauer–Emmett–Teller analysis to measure surface area, and thermogravimetric analysis to quantify carbon content (Hintsho et al. 2014).

Hintsho et al. (2016) in their research article studied about the gaseous (nitrogen and hydrogen) significance on the assembly of carbonaceous nanomaterials from acetylene in the presence of fly ash catalysts. They came up with the matter that hydrogen gas procured higher yield of carbonaceous nanomaterial than nitrogen, reasoned by virtue of the multifunctional roles of hydrogen. Further, the materials formed were featured by the methods like Raman spectroscopy and Mössbauer spectroscopy.

11.5.4   Nanomaterials from Pharmaceutical Waste

Pharmaceutical waste is consolidated outcome of expired medicines/products, contaminated drugs, and discarded/unused chemicals. Figure 11.8 represents various progenitors of pharmaceuticals that aids in enhancing environment pollution. Not much research work has been absorbed in this area, so nanomaterials fabrication could be a more desirable option for pharma companies to recycle leftover medicines.

According to Kadam et al. (2016), only the functioning part of the engulfed pharmaceuticals is metabolized by the biological system of ours and the non-metabolized

Fig. 11.8 Representation of major progenitor of pharmaceutical waste that pollutes environment

part enters the environment through excretions. The final terminus of excrete is lakes and other natural water sources. They documented that health center generates approximately 40,000 tons of waste per day in India.

Jha and Prasad (2012) in their unique research work opened up a pathway to recycle waste drugs to fabricate nanomaterials like metal (gold), oxide (zirconia oxide), and chalcogenide (cadmium sulfide) NPs. This proposal could be a boon to pharma companies in recycling waste as they have implemented green protocol to achieve nanosynthesis. They utilized quinolone group of antibiotic (norfloxacin) and tinidazole combination to fabricate above-mentioned nanomaterials. They prepared reaction broth utilizing 4 month expired Norflox-TZ dissolved initially in sterile distilled water. Further, HCl was added to achieve complete dissolution of the tablet, and the mixture was heated over steam bath at 40 °C. The prepared broth was utilized for NPs fabrication. They explained molecular structure of compounds norfloxacin and tinidazole facilitates redox reaction. The ketonic group and fluorine atom in norfloxacin indicates redox potential (Fig. 11.9). The zwitterionic form of norfloxacin makes it a suitable candidate for nanosynthesis. The cationic form upon addition of metal ions generates oxide/chalcogenides NPs wherein the anionic form upon reaction with metal ion produces metal NPs.

11.5.5   Nanomaterials from Waste Microorganisms

Microbes are not only meant to cause diseases rather are of prime importance, carried from industrial importance to health insurance to environmental concern and many more. Undoubtedly, microbes work very well for us as antagonist and protagonist as well. In this section, a brief description about the utilization of waste biomass to nanomaterial genesis is focused, as its disposal is definitely a huge concern. Only limited research work has been taken up by the researcher.

Rożalska et al. (2016) fabricated silver NPs from *Metarhizium robertsii* (filamentous fungi) biomass after being employed for decontamination of nonylphenol. They have used multiple biomass extract dilutions to synthesize photo-induced silver NPs, wherein obtained better results with 20 and 50% diluted sourced samples

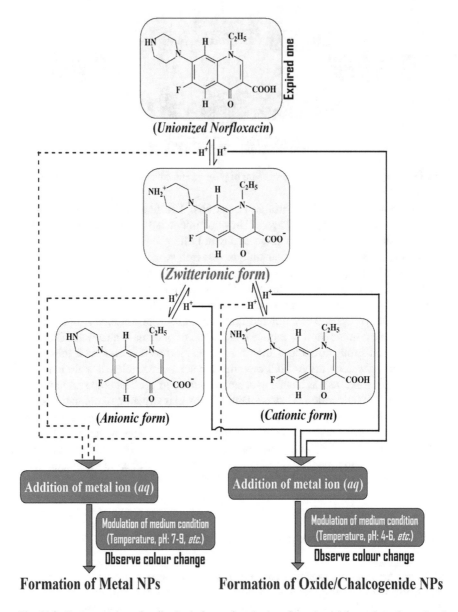

Fig. 11.9 Representation of zwitterionic form of expired norfloxacin and formation of metal and oxide/chalcogenide NPs upon addition of metal ions

within 3 h. The so-obtained nanomaterials were homogeneous in size and shape after being characterized by UV-Visible spectrophotometer, FT-IR, TGA, XPS spectroscopy, STEM, and DLS. FT-IR and XPS analysis confirmed the involvement of protein in the nanofabrication and further capping agents were quantified by thermogravimetric analyses.

Another group of researcher demonstrated the conversion of polyphosphate accumulated rich recombinant *E.coli* and wild type *E.coli* into lithium monophosphate nanostructure. Initially, the two categories of cells were harvested from phosphate-rich media with 20 percent more media depleted by recombinant strain. Further cells were treated with metal ions ($FeCl_3$) followed by lithium acetate solution and then annealed at 600 °C. Nanostructures formed were characterized by XRD, TEM, and EDX. TGA analysis reveals 20 wt% carbon content derived from the bacteria (Zhou et al. 2014).

Wu et al. (2016) mentioned in their article about bacterial cellulose, a biomass matter and its conversion to high value-added three-dimensional carbon nanomaterials. Bacterial cellulose is a three-dimensional nanoporous cellulose network that makes it an effective reaction system. They accounted on preparation of various nanomaterials like CNF aerogels.

11.5.6 Nanomaterials from Agricultural Waste

Agricultural wastes are generally the remainder of agricultural whirl. It comprises of crop residue, peels, pulp, stem, stalk, seeds, leaves, bagasse, fertilizers, and manures. Zhang et al. (2012) characterized agricultural waste based on the following five criteria: (a) Energy content of the waste (fruit waste), (b) Biomass combustion (grassland fires, crop residue burning forest), (c) Agricultural and natural fibers (cereals/insoluble fiber residue), (d) Bacterial and fungal communities (grown over the leftovers), (e) Decomposition (fruits and vegetables), (f) Agricultural water pollution (fertilizer/manure) (Sangeetha et al. 2017a, b, c). With increasing population, there is high demand on the food and so as the increment in agricultural waste (Sangeetha et al. 2017c). According to Castro et al. (2015), plant broths are economically useful for fabricating NPs following "green chemistry" approach with an additional advantage of undergoing scale up of the designed process. They elucidated that plant extracts endowed with biomolecules like enzymes, proteins, carbohydrates, organic acids, and amino acids, participate in NPs formation with dual function as reducing and capping negotiator is environmentally safe (Prasad 2014). Many researchers have taken effort to fabricate NPs from agricultural waste (Table 11.6) through green method, pyrolysis.

According to Dauthal and Mukhopadhyay (2014), agro-industrial waste *Punica granatum* peels effectively aided in fabricating platinum NPs. They elucidated that these peels are rich in ellagic tannins, ellagic acid, quercetin, gallic acid, punicalagin and endowed with antimutagenic, chemo-preventive, and antioxidant potential. They followed green protocol to fabricate platinum NPs through charging the extract by platinum salt solution under defined reaction conditions. Further, biofabricated platinum NPs was characterized. Armendariz et al. (2004), Kumar et al.

(2012), Krishnaswamy et al. (2014), Malhotra et al. (2014), Meyabadi et al. (2014), and Baiocco et al. (2016) also followed green protocol to fabricate NPs.

Zhang et al. (2013) fabricated magnetic/c-Fe_2O_3 composite through pyrolysis from cottonwood (pre-treated with $FeCl_3$). The biomass was kept in the $FeCl_3$ solution for couple of hours then air dried for 2 h at 80 °C. The prepared biomass was pyrolyzed in a furnace at particular reaction condition. Then, the composite procured were crushed and sieved accordingly.

Sun et al. (2013) for the first time fabricated graphene nanosheets from agricultural biomass waste coconut shell. They elucidated that coconut shell as a promising carbon source to fabricate activated carbon as these shells are enriched mainly with cellulose fibers. Above all coconut shells are cheap, environmental friendly, easily available in abundant quantity and most importantly a sustainable source. In the study, they have fabricated porous graphene-like nanosheets through effective simultaneous activation-graphitization route. The concept they employed is the synchronous ingression of the catalyst required for graphitization ($FeCl_3$) and activation ($ZnCl_2$) in coalition with the carbon source with metal precursor. They explained that after carbonization. They illustrated that the catalyst can form carburized phase during the heating process followed by fabrication of graphene-like nanosheets through organization cum disintegration phenomena of carburized phase. Finally, Fe components were discarded leaving behind the pure graphene-like nanosheets. Table 11.6 summarizes the synthesis of different nanomaterials synthesized from various waste sources.

11.6 Application of Nanomaterials Recycled from Waste

Nanomaterials procured from various waste sources are already revealed in the chapter. The comprehensive view of the applications of WMNPs has been elucidated below.

Zhang et al. (2013) prepared magnetic biochar and found its potential in arsenic removal. In addition, platinum NPs as a green catalyst for removal of pollutant like 3 nitrophenol in association with $NaBH_4$ were reported by Dauthal and Mukhopadhyay (2014). In another experimentation, silver NPs and nitrogen-doped carbon nanoparticle acts as catalyst to remove 4- nitrophenol (Edison and Sethuraman 2013; Gao et al. 2014; Rożalska et al. 2016). Further researchers have employed copper NPs, fluorescent carbon NPs to remove pollutant methylene blue dye (Sinha and Ahmaruzzaman (2015); Adedokun et al. 2016). Essawy et al. (2017) demonstrated removal of methylene blue and acid blue 25 by graphene NPs. Using carbon nanospheres, Rhodamine B and heavy metals were removed (Qu et al. 2014).

Biological applications of WMNPs include antibacterial (Kaviya et al. 2011; Yuvakkumar et al. (2014); Ibrahim 2015; Kokila et al. 2015; Patra and Baek 2015; Aziz et al. 2015, 2016; Patra et al. 2016; Rożalska et al. 2016; Sharma et al. 2016) antioxidant efficacy (Kokila et al. 2015; Patra and Baek 2015; Patra et al. 2016), proteasome inhibitory (Patra and Baek 2015), anticancer (Saratale et al. 2017), anticandidal (Patra et al. 2016), larvicidal and pupicidal (Velu et al. 2015; Murugan et al. 2017), growth promoter (Sonkar et al. 2012; Park et al. 2014), DNA detection (Godavarthi et al. 2017), imaging and targeted delivery (Chamundeeswari et al. 2013a, b; Park et al. 2014), pharmaceutical and biomedical practice (Shankar et al. 2014) were also demonstrated by vssarious researchers. Other applications of WMNPs include high power super capacitors (Sun et al. 2013) and cosmetic formulations (Rodrigues et al. 2016).

11.7 Conclusion

Waste management through conventional method is valuable but suffers from tremendous setback. A neoteric technology named "nanotechnology" has expanded its aura to most probably engulf the waste-related concern in near future. This chapter gives a rundown of the general waste management protocol with their related troubles, nanotechnology role in various stages of waste management, nanoconversions from various waste sources, and their budding applications. Despite of being fortunate technology in waste recycling, the uppermost entanglement is the scale up of waste conversion to nanomaterials, as maximum outcome is still executed at laboratory scale. Nanotechnology with the promise to shoot up with its heights in controlling waste in coming future will definitely be effectuated.

References

Adedokun O, Roy A, Awodugba AO, Devi PS (2016) Fluorescent carbon nanoparticles from Citrus sinensis as efficient sorbents for pollutant dyes. Luminescence 32:62–70. https://doi.org/10.1002/bio.3149

Adolfsson KH, Hassanzadeh S, Hakkarainen M (2015) Valorization of cellulose and waste paper to graphene oxide quantum dots. Royal Soc Chem Adv 5:26550–26558. https://doi.org/10.1039/C5RA01805F

Ahmad J, Ansari TA (2012) Biogas from slaughterhouse waste: towards an energy self-sufficient industry with economical analysis in India. J Microbial Biochem Technol S12:001–004. https://doi.org/10.4172/1948-5948.S12-001

Ahmad N, Sharma S, Rai R (2012) Rapid green synthesis of silver and gold nanoparticles using peels of Punicagranatum. Adv Mater Lett 3:376–380

Akhavan O, Bijanzad K, Mirsepah A (2014) Synthesis of graphene from natural and industrial carbonaceous wastes. Royal Soc Chem Adv 4:20441–20448. https://doi.org/10.1039/c4ra01550a

Alhassan FH, Rashid U, Taufiq-Yap YH (2015) Synthesis of waste cooking oil based biodiesel via ferric-manganese promoted molybdenum oxide/zirconia nanoparticle solid acid catalyst: influence of ferric and manganese dopants. J Oleo Sci 64:505–514

Alturaif HA, ALOthman ZA, Shapter JG, Wabaidur SM (2014) Use of carbon nanotubes (CNTs) with polymers in solar cells. Molecules 19:17329–17344

Amin MT, Alazba AA, Manzoor U (2014) A review on removal of pollutants from water/wastewater using different types of nanomaterials. Adv Mater Sci Eng 190:190–208. https://doi.org/10.1155/2014/825910

Armendariz V, Herrera I, Peralta-Videa JR, Jose-Yacaman M, Troiani H, Santiago P, Gardea-Torresdey JL (2004) Size controlled gold nanoparticle formation by Avena sativa biomass: use of plants in nanobiotechnology. J Nanopart Res 6:377–382

Arshadi M, Foroughifard S, Gholtash JE, Abbaspourrad A (2015) Preparation of iron nanoparticles-loaded Spondias purpurea seed waste as an excellent adsorbent for removal of phosphate from synthetic and natural waters. J Coll Inter Sci 452:69–77

AWM Handbook (2014) Waste to resources. Teri Press, New Delhi

Aziz N, Faraz M, Pandey R, Sakir M, Fatma T, Varma A, Barman I, Prasad R (2015) Facile algae-derived route to biogenic silver nanoparticles: synthesis, antibacterial and photocatalytic properties. Langmuir 31:11605–11612. https://doi.org/10.1021/acs.langmuir.5b03081

Aziz N, Pandey R, Barman I, Prasad R (2016) Leveraging the attributes of Mucor hiemalis-derived silver nanoparticles for a synergistic broad-spectrum antimicrobial platform. Front Microbiol 7:1984. https://doi.org/10.3389/fmicb.2016.01984

Baiocco D, Lavecchia R, Natali S, Zuorro A (2016) Production of metal nanoparticles by agro-industrial wastes: a green opportunity for nanotechnology. Chem Eng Trans 47:67–72. https://doi.org/10.3303/CET1647012

Bankar A, Joshi B, Kumar AR, Zinjarde S (2010) Banana peel extract mediated novel route for the synthesis of silver nanoparticles. Colloid Surf A Physicochem Eng Asp 368:58–63

Basel Convention (1989) Basel convention on the control of transboundary movements of hazardous wastes protocol on liability and compensation Basel convention protocol on liability and compensation. http://www.basel.int

Berkmans J, Jagannatham M, Priyanka S, Haridoss P (2014) Synthesis of branched, nano channeled, ultrafine and nano carbon tubes from PET wastes using the arc discharge method. Waste Manag 11:2139–2145. https://doi.org/10.1016/j.wasman.2014.07.004

Bharathy N, Sakthivadivu R, Sivakumar K, Saravanakumar VR (2012) Disposal and utilization of broiler slaughter waste by composting. Vet World 5:359–361

Bhat N (2007) Nanoelectronics era: novel device. J Indian Int Sci 87:61–74

Castro L, Blázquez ML, González F, Muñoz JA, Ballester A (2015) Biosynthesis of silver and platinum nanoparticles using orange peel extract: characterisation and applications. IET Nanobiotechnol 9:252–258

Çelebi O, Üzüm Ç, Shahwan T, Erten HN (2007) A radiotracer study of the adsorption behavior of aqueous Ba2+ ions on nanoparticles of zero-valent iron. J Hazard Mater 148:761–767

Chamundeeswari M, Senthil V, Kanagavel M, Chandramohan SM, Sastry TP (2011) Preparation and characterization of nanobiocomposites containing iron nanoparticles prepared from blood and coated with chitosan and gelatin. Mater Res Bull 46:901–904

Chamundeeswari M, Kumar BS, Muthukumar T, Muthuraman L, Sai KP, Sastry TP (2013a) Iron nanoparticles from blood coated with collagen as a matrix for synthesis of nanohydroxyapatite. Bull Mater Sci 36:1165–1170

Chamundeeswari M, Sastry TP, Lakhsmi BS, Senthil V, Agostinelli E (2013b) Iron nanoparticles from animal blood for cellular imaging and targeted delivery for cancer treatment. Biochim Biophys Acta Gen Subj 1830:3005–3010

Dauthal P, Mukhopadhyay M (2014) Biofabrication, characterization, and possible bio-reduction mechanism of platinum nanoparticles mediated by agro-industrial waste and their catalytic activity. J Ind Eng Chem 25:185–191. https://doi.org/10.1016/j.jiec.2014.07.009

Demirbas A (2011) Waste management, waste resource facilities and waste conversion processes. Energ Conv Manag 52:1280–1287

Dermatas D, Mpouras T, Panagiotakis I (2018) Application of nanotechnology for waste management: challenges and limitations. Waste Manag Res 36:197–199

Dutta N, Mukhopadhyay A, Dasgupta AK, Chakrabarti K (2014) Improved production of reducing sugars from rice husk and rice straw using bacterial cellulase and xylanase activated with hydroxyapatite nanoparticles. Bioresour Technol 153:269–277

Edison TJI, Sethuraman MG (2013) Biogenic robust synthesis of silver nanoparticles using Punica granatum peel and its application as a green catalyst for the reduction of an anthropogenic pollutant 4-nitrophenol. Spectrochim Acta A Mol Biomol Spectrosc 04:262–264

Ek AEW, Hallin S, Vallin L, Schnürer A, Karlsson M (2011) Slaughterhouse waste co-digestion—experiences from 15 years of full-scale operation. World Renewable Energy Congress, Linköping

Essawy NAE, Ali SM, Farag HAC, Konsowa AH, Elnouby M, Hamad HA (2017) Green synthesis of graphene from recycled PET bottle wastes for use in the adsorption of dyes in aqueous solution. Ecotoxicol Environ Safety 145:57–68

Fa W, Gong C, Tian L, Peng T, Zan L (2011) Enhancement of photocatalytic degradation of poly(vinylchloride) with Perchlorinated Iron (II) Phthalocyanine modified Nano-TiO2. J Appl Poly Sci 122:1823–1828

Gade A, Ingle A, Whiteley C, Rai M (2010) Mycogenic metal nanoparticles: progress and applications. Biotechnol Lett 32:593–600

Gangadhar G, Maheshwari U, Gupta S (2012) Application of nanomaterials for the removal of pollutants from effluent streams. Nanosci Nanotechnol Asia 2:140–150

Gao L, Li R, Sui X, Li R, Chen C, Chen Q (2014) Conversion of chicken feather waste to N-doped carbon nanotubes for the catalytic reduction of 4-Nitrophenol. Environ Sci Technol 48:10191–10197

Godavarthi S, Kumara KM, Véleza EV, Eligioc AH, Mahendhir M, N. Comoe NH, Alemane M, Gomeza LM (2017) Nitrogen doped carbon dots derived from Sargassum fluitans as fluorophore for DNA detection. J Phytochem Phytobiol, B: Biol 172:36–41.

Gong J, Liu J, Wen X, Jiang Z, Chen X, Mijowska E, Tang T (2014) Upcycling waste polypropylene into graphene flakes on organically modified montmorillonite. Ind Eng Chem Res 53:4173–4181

Goudarzi M, Mir MN, Mousavi-Kamazani M, Bagheri S, Salavati-Niasari M (2016) Biosynthesis and characterization of silver nanoparticles prepared from two novel natural precursors by facile thermal decomposition methods. Sci Rep 6:32539–32513. https://doi.org/10.1038/srep32539

Hassan TA, Rangari VK, Rana RK, Jeelani S (2013) Sonochemical effect on size reduction of CaCO3 nanoparticles derived from waste eggshells. Ultrason Sonochem 20:1308–1315

Hintsho N, Shaikjee A, Masenda H, Naidoo D, Billing D, Franklyn P, Durbach S (2014) Direct synthesis of carbon nanofibers from South African coal fly ash. Nanoscale Res Lett 9:1–11

Hintsho N, Shaikjee TPK, Masenda H, Naidoo D, Franklyn DS (2016) Effect of nitrogen and hydrogen gases on the synthesis of carbon nanomaterials from coal waste fly ash as a catalyst. J Nanosci Nanotechnol 16:4672–4683

Ibrahim HMM (2015) Green synthesis and characterization of silver nanoparticles using banana peel extract and their antimicrobial activity against representative microorganisms. J Rad Res Appl Sci 8:265–275. https://doi.org/10.1016/j.jrras.2015.01.007

Jha AK, Prasad K (2012) Synthesis of nanomaterials using expired medicines: an ecofriendly option. Nanotechnol Dev 2:36–39

Jha AK, Prasad K (2014) Synthesis of silver nanoparticles employing fish processing discard: an eco-amenable approach. J Chin Adv Mater Soc 2:179–185

Jha AK, Prasad K (2016) Synthesis of ZnO nanoparticles from goat slaughter waste for environmental protection. Int J Curr Eng Technol 6:147–151

Joo SH, Cheng IF (2006) Nanotechnology for environmental remediation. Springer, New York

Joseph T, Morrison M (2006) Nanoforum report: nanotechnology in agriculture and food. pp 1–14. www.nanoforum.org

Kadam A, Patil S, Patil S, Tumkur A (2016) Pharmaceutical waste management an overview. Ind J Pharma Pract 9:2–8

Karimi M, Keyhani A, Akram A, Rahman M, Jenkins B, Stroeve P (2013) Hybrid response surface methodology-genetic algorithm optimization of ultrasound-assisted transesterification of waste oil catalysed by immobilized lipase on mesoporous silica/iron oxide magnetic core-shell nanoparticles. Environ Technol (UK) 34:2201–2211

Kaviya S, Santhanalakshmi J, Viswanathan B, Muthumary J, Srinivasan K (2011) Biosynthesis of silver nanoparticles using citrus sinensis peel extract and its antibacterial activity. Spectrochim Acta A Mol Biomol Spectrosc 79:594–598

Kokila T, Ramesh PS, Geetha D (2015) Biosynthesis of silver nanoparticles from Cavendish banana peel extract and its antibacterial and free radical scavenging assay: a novel biological approach. Appl Nanosci 5:911–920

Kokila T, Ramesh PS, Geetha D (2016) Biosynthesis of AgNPs using *Carica Papaya* peel extract and evaluation of its antioxidant and antimicrobial activities. Ecotoxicol Environ Safety 134:467–473

Krishnaswamy K, Vali H, Orsat V (2014) Value-adding to grape waste: green synthesis of gold nanoparticles. J Food Eng 142:210–220. https://doi.org/10.1016/j.jfoodeng.2014.06.014

Kumar R, Roopan SM, Prabhakarn A, Khanna VG, Chakroborty S (2012) Agricultural waste Annona squamosa peel extract: biosynthesis of silver nanoparticles. Spectrochim Acta A Mol Biomol Spectrosc 90:173–176

Kumar A, Hegde G, Manaf SABA, Ngaini Z, Sharma KV (2014) Catalyst free silica templated porous carbon nanoparticles from bio-waste materials. Chem Commun 50:12702–12705

Kumar V, Vermab S, Choudhuryc S, Tyagid M, Chatterjee S, Variyara PS (2015) Biocompatible silver nanoparticles from vegetable waste: its characterization and bio-efficacy. Int J Nano Mater Sci 4:70–86

Kumar U, Sikarwar S, Sonker RK (2016) Carbon nanotube: synthesis and application in solar cell. J Inorg Organomet Polym Mater 26:1231–1242. https://doi.org/10.1007/s10904-016-0401-z

Kuppusamy S, Thavamani P, Megharaj M, Naidu R (2015) Bioremediation potential of natural polyphenol rich green wastes: a review of current research and recommendations for future directions. Environ Technol Innov 4:17–28. https://doi.org/10.1016/j.eti.2015.04.001

Li Z, Zheng L, Saini V, Bourdo S, Dervishi E, Biris AS (2013) Solar cells with graphene and carbon nanotubes on silicon. J Exp Nanosci 8:565–572

Lunge SS, Singh S, Sinha A (2014) Magnetic nanoparticle: synthesis and environmental applications. International Conference on Chemical Civil and Environmental Engineering. https://doi.org/10.15242/IICBE.C1114009

Luther W (2008) Application of nano-technologies in the energy sector, vol 9 of the series Aktionslinie Hessen-nanotech of the Hessian Ministry of Economy, Transport, Urban and Regional development Www.Hessen-Nanotech.De

Lymer J (2016) Waste classification guide. Royal Society of Chemistry Environmental Chemistry Group. p 4624

Malav OP, Birla R, Virk KS, Sandhu HS, Mehta N, Kumar P, Wagh RV (2018) Safe disposal of slaughter house waste. Appro Poult Dairy Vet Sci 2:3–5

Malhotra A, Sharma N, Navdezda KN, Dolma K, Sharma D, Nandanwar HS, Choudhury AR (2014) Multi-analytical approach to understand biomineralization of gold using rice bran: a novel and economical route. Roy Soc Chem 74:39484–39490. https://doi.org/10.1039/C4RA05404K

Meng X, Deng D (2017) Trash to treasure: waste eggshells as chemical reactors for the synthesis of amorphous co(OH)$_2$ nanorod arrays on various substrates for applications in rechargeable alkaline batteries and electrocatalysis. ACS Appl Mater Interf 9:5244–5253

Meyabadi TF, Dadashian F, Sadeghi GMM, Asl HEZ (2014) Spherical cellulose nanoparticles preparation from waste cotton using a green method. Powder Technol 261:232–240. https://doi.org/10.1016/j.powtec.2014.04.039

Murugan K, Suresh U, Panneerselvam C, Rajaganesh R, Roni M, Aziz AT, Hwang J, Sathishkumar K, Rajasekar A, Kumar S, Alarfaj AA, Higuchi A, Benelli G (2017) Managing wastes as green resources: cigarette butt-synthesized pesticides are highly toxic to malaria vectors with little impact on predatory copepods. Environ Sci Pollut Res 11:10456–10470. https://doi.org/10.1007/s11356-017-0074-3

Narayanamma MK, Rani A, Raju ME (2016) Natural synthesis of silver nanoparticles by banana peel extract and as an antibacterial agent. Int J Sci Res 5:1431–1441

Nikalje AP (2015) Nanotechnology and its applications in medicine. Med Chem 5:81–89

Nogueira V, Lopes I, Freitas AC, Rocha-Santos TAP, Gonçalves F, Duarte AC, Pereira R (2015) Biological treatment with fungi of olive mill wastewater pre-treated by photocatalytic oxidation with nanomaterials. Ecotoxicol Environ Safety 115:234–242

Park SY, Lee HU, Park ES, Lee SC, Lee JW, Jeong SW, Kim CH, Lee YC, Huh YS, Lee J (2014) Photoluminescent green carbon nanodots from food-waste- derived sources: large-scale synthesis, properties, and biomedical applications. Appl Mater Sci Inter 6:3365–3370

Park JS, Ahn EY, Park Y (2017) Asymmetric dumbbell-shaped silver nanoparticles and spherical gold nanoparticles green- synthesized by mangosteen (*Garcinia mangostana*) pericarp waste extracts. Int J Nanomed 12:6895–6908

Patra JK, Baek KH (2015) Novel green synthesis of gold nanoparticles using Citrullus lanatus rind and investigation of proteasome inhibitory activity, antibacterial, and antioxidant potential. Int J Nanomed 10:7253–7264

Patra JK, Das G, Baek KH (2016) Phyto-mediated biosynthesis of silver nanoparticles using the rind extract of watermelon (*Citrullus lanatus*) under photo-catalyzed condition and investigation of its antibacterial, anticandidal and antioxidant efficacy. Photochem Photobiol 161:200–210. https://doi.org/10.1016/j.jphotobiol.2016.05.021

Prasad R (2014) Synthesis of silver nanoparticles in photosynthetic plants. J Nanopart 963961. https://doi.org/10.1155/2014/963961

Prasannan A, Imae T (2013) One-pot synthesis of fluorescent carbon dots from orange waste peels. Ind Eng Chem Res 52:15673–15678

Qu J, Zhang Q, Xia Y, Cong Q, Luo C (2014) Synthesis of carbon nanospheres using fallen willow leaves and adsorption of rhodamine B and heavy metals by them. Environ Sci Pollut Res 2:1408–1419. https://doi.org/10.1007/s11356-014-3447-x:2014

Rahman FA (2000) Reduce, reuse, recycle: alternatives for waste management. pp 1–4

Rajarao R, Ferreira R, Sadi SHF, Khanna R, Sahajwalla V (2014) Synthesis of silicon carbide nanoparticles by using electronic waste as carbon source. Mater Lett 120:65–68. https://doi.org/10.1016/j.matlet.2014.01.018

Ramezanianpour AA (2014) Cement replacement materials. Spring Geochem/Mineralo Verlag, Berlin. https://doi.org/10.1007/978-3-642-36721-2_2

Rao M, Jha B, Jha AK, Prasad K (2017a) Fungal nanotechnology: a Pandora to agricultural science and engineering. In: Prasad R (ed) Fungal nanotechnology. Fungal biology. Springer, Berlin

Rao M, Jha B, Jha AK (2017b) Nanoparticles from kitchen waste (Orange peels): an avenue for conversion of green waste to value added product. Int J Plant Res 30(supl):21–24

Rodrigues F, Alves AC, Nunes C, Sarmento B, Amaral MH, Reis S, Oliveira MBPP (2016) Permeation of topically applied caffeine from a food by-product in cosmetic formulations: is nanoscale in vitro approach an option? Int J Pharm 513:496–503. https://doi.org/10.1016/j.ijpharm.2016.09.059

Rożalska S, Soliwoda K, Długoński J (2016) Synthesis of silver nanoparticle by *Metarhizium robertsii* waste biomass extract after nonylphenol degradation and its antimicrobial and catalytic potential. Roy Soc Chem 6:21475–21485. https://doi.org/10.1039/C5RA24335A

Sangeetha J, Thangadurai D, Hospet R, Harish ER, Purushotham P, Mujeeb MA, Shrinivas J, David M, Mundaragi AC, Thimmappa AC, Arakera SB, Prasad R (2017a) Nanoagrotechnology for soil quality, crop performance and environmental management. In: Prasad R, Kumar M, Kumar V (eds) Nanotechnology. Springer Nature Singapore Pte Ltd, Singapore, pp 73–97

Sangeetha J, Thangadurai D, Hospet R, Purushotham P, Karekalammanavar G, Mundaragi AC, David M, Shinge MR, Thimmappa SC, Prasad R, Harish ER (2017b) Agricultural nanotechnology: concepts, benefits, and risks. In: Prasad R, Kumar M, Kumar V (eds) Nanotechnology. Springer Nature Singapore Pte Ltd, Singapore, pp 1–17

Sangeetha J, Thangadurai D, Hospet R, Purushotham P, Manowade KR, Mujeeb MA, Mundaragi AC, Jogaiah S, David M, Thimmappa SC, Prasad R, Harish ER (2017c) Production of bionanomaterials from agricultural wastes. In: Prasad R, Kumar M, Kumar V (eds) Nanotechnology. Springer Nature Singapore Pte Ltd, Singapore, pp 33–58

Saratale RG, Shin H, Kumar G (2017) Exploiting fruit byproducts for eco-friendly nanosynthesis: Citrus × Clementina peel extract mediated fabrication of silver nanoparticles with high efficacy against microbial pathogens and rat glial tumor C6 cells. Environ Sci Pollut Res 11:10250–10263. https://doi.org/10.1016/j.matlet.2014.08.122

Shan G, Surampalli RY, Tyagi RD, Zhang TC (2009) Nanomaterials for environmental burden reduction, waste treatment, and nonpoint source pollution control: a review. Front Environ Sci Eng China 3:249–264

Shankar S, Jaiswal L, Aparna RSL, Prasad RGSV (2014) Synthesis, characterization, in vitro biocompatibility, and antimicrobial activity of gold, silver and gold silver alloy nanoparticles prepared from Lansium domesticum fruit peel extract. Mater Lett 137:75–78

Sharma S, Kalita G, Hirano R, Shinde SM, Papon R, Ohtani H, Tanemura M (2014) Synthesis of graphene crystals from solid waste plastic by chemical vapor deposition. Carbon 72:66–73

Sharma K, Kaushik S, Jyoti A (2016) Green synthesis of silver nanoparticles by using waste vegetable peel and its antibacterial activities. J Pharma Sci Res 8:313–316

Sinha T, Ahmaruzzaman M (2015) Green synthesis of copper nanoparticles for the efficient removal (degradation) of dye from aqueous phase. Environ Sci Pollut Res 22:20092–20100

Sinyoung S, Kunchariyakun K, Asavapisit S, Mackenzie KJD (2017) Synthesis of belite cement from nano-silica extracted from two rice husk ashes. J Environ Manag 90:53–60

Somanathan T, Prasad K, Ostrikov K, Saravanan A, Krishna VM (2015) Graphene oxide synthesis from agro waste. Nanomaterials 5:826–834

Sonkar SK, Roy M, Babar DG, Sarkar S (2012) Water soluble carbon nano-onions from wood wool as growth promoters for gram plants. Nanoscale 4:7670–7675

Sun L, Tian C, Li M, Meng X, Wang L, Wang R, Yin J, Fu H (2013) From coconut shell to porous graphene-like nanosheets for high-power supercapacitors†. J Mater Chem 1:6462–6470

Sundaram M, Banu AN et al (2015) A study on anti-bacterial activity of keratin nanoparticles from chicken feather waste against *Staphylococcus aureus* (bovine mastitis Bacteria) and its antioxidant activity. Eur J Biotechnol Biosci 3:1–5

Suriani AB, Dalila AR, Mohamed A, Rosmi MS, Mamat MH, Malek MF, Ahmad MK, Hashim N, Isa IM, Soga T, Tanemura M (2016) Parametric study of waste chicken fat catalytic chemical vapour deposition for controlled synthesis of vertically aligned carbon nanotubes. Cogent Phys 3:1–18

Suryawanshi A, Biswal M, Mhamane D, Gokhale R, Patil S, Guin D, Ogale S (2012) Large scale synthesis of graphene quantum dots (GQDs) from waste biomass and their use as an efficient and selective photoluminescence on-off-on probe for Ag+ ions. Nanoscale 20:11664–11670. https://doi.org/10.1039/x0xx00000x

Tang Y, Shen X, Zhang J, Guo D, Kong F, Zhang N (2015) Extraction of cellulose nano-crystals from old corrugated container fiber using phosphoric acid and enzymatic hydrolysis followed bysonication. Carbohydr Polym 125:360–366

Thines KR, Abdullah EC, Mubarak NM, Ruthiraan M (2017) Synthesis of magnetic biochar from agricultural waste biomass to enhancing route for waste water and polymer application: a review. Renew Sustain Energy Rev 67:257–276

Tran DL, Le VH, Pham HL, Hoang TMN, Nguyen TQ, Luong TT, Ha PT, Nguyen XP (2010) Biomedical and environmental applications of magnetic nanoparticles. Adv Nat Sci Nanosci Nanotechnol 1:045013–045015. https://doi.org/10.1088/2043-6262/1/4/045013

Velu K, Elumalai D, Hemalatha P, Janaki A, Babu M, Hemavathi M, Kaleena PK (2015) Evaluation of silver nanoparticles toxicity of *Arachis hypogaea* peel extracts and its larvicidal activity against malaria and dengue vectors. Environ Sci Pollut Res 22:7769–17779

Wang Q, Liu X, Zhang L, Lv Y (2012a) Microwave-assisted synthesis of carbon nanodots through an eggshell membrane and their fluorescent application. Analyst 137:5392–5397

Wang X, Guo Y, Yang L, Han M, Zhao J, Cheng X (2012b) Nanomaterials as sorbents to remove heavy metal ions in wastewater treatment. J Environ Anal Toxicol 2:1–7. https://doi.org/10.4172/2161-0525.1000154

Williams CM (2008) Poultry waste management in developing countries. Poult Dev Rev 4

Wu ZY, Liang HW, Chen LF, Hu BC, Yu SH (2016) Bacterial cellulose: a robust platform for design of three dimensional carbon-based functional nanomaterials. Acc Chem Res 49:96–105

A, Li X, Yang J, Du C, Shen W, Yan J (2017) Upcycling waste lard oil into vertical graphene sheets by inductively coupled plasma assisted chemical vapor deposition. Nano 1:96–105. https://doi.org/10.3390/nano7100318

Yang N, Li WH (2013) Mango peel extract mediated novel route for synthesis of silver nanoparticles and antibacterial application of silver noparticles loaded onto non-woven fabrics. Ind Crop Prod 48:81–88

Yang K, Zhu LZ, Xing BS (2006) Adsorption of polycyclic aromatic hydrocarbons by carbon nanomaterials. Environ Sci Technol 40:1855–1861

Yao Y, Zhang J, Wu G, Wang S, Hu Y, Su C, Xu T (2017) Iron encapsulated in 3D N-doped carbon nanotube/porous carbon hybrid from waste biomass for enhanced oxidative activity. Environ Sci Pollut Res 24:7679–7692

Yu CY, Huang LY, Kuan IC, Lee SL (2013) Optimized production of biodiesel from waste cooking oil by lipase immobilized on magnetic nanoparticles. Int J Mol Sci 14:24074–24086

Yuvakkumar R, Suresh J, Nathanael AJ, Sundrarajan M, Hong SI (2014) Novel green synthetic strategy to prepare ZnO nanocrystals using rambutan (*Nephelium lappaceum* L.) peel extract and its antibacterial applications. Mater Sci Eng C 41:17–27

Zhang Z, Gonzalez AM, Davies EGR, Liu Y (2012) Agricultural wastes. Water Environ Res 84:1386–1406

Zhang M, Gao B, Varnoosfaderani S, Hebard A, Yao Y, Inyang M (2013) Preparation and characterization of a novel magnetic biochar for arsenic removal. Bioresour Technol 130:457–462

Zhou Y, Yang D, Zeng Y, Zhou Y, Ng WJ, Yan Q, Fong E (2014) Recycling bacteria for the synthesis of LiMPO$_4$ (M = Fe, Mn) nanostructures for high-power lithium batteries. Small 10:3997–4002

Chapter 12
Synthesis of Nanomaterials Involving Microemulsion and Miceller Medium

Santosh Kumar, Mohammad Y. Wani, and Joonseok Koh

12.1 Introduction

Nanotechnology brings in new techniques to control single atoms and molecules. Nanotechnology comes out from the diverse scientific fields like physics, chemistry, biology, materials, and engineering sciences. Nanomaterials in nanotechnology are defined as small objects that behave as a whole unit in terms of its transport properties and potential application. To control the size and shape of nanoparticles, colloidal self-assembly as confined reaction media offers good advantages. Nanomaterials have been proving attractive due to their important catalytic, electrical, optical, and drug delivery properties accompanied with improved physical properties like mechanical, thermal stability, or chemical passivity. Microemulsion medium is one of the important synthetic methods which control the particle properties like size, geometry, morphology, and surface area (Hu et al. 2009; Pileni 2003; Richard et al. 2017). Many reviews about different aspects of microemulsions have already been written (Malik et al. 2012; Pileni 2008; Lopez-Quintela et al. 2004; Cushing et al. 2004; Shervani et al. 2006). Our intention is to summarize the most recent work carried out in this area, using microemulsion and other micellar methods to control the particle size and morphology of nanoparticles for the synthesis of organic and inorganic nanomaterials.

S. Kumar · J. Koh (✉)
Department of Organic and Nano System Engineering, Konkuk University,
Seoul, Republic of Korea
e-mail: ccdjko@konkuk.ac.kr

M. Y. Wani
Faculty of Science, Chemistry Department, University of Jeddah, Jeddah,
Kingdom of Saudi Arabia

© Springer Nature Switzerland AG 2018
R. Prasad et al. (eds.), *Exploring the Realms of Nature for Nanosynthesis*,
Nanotechnology in the Life Sciences, https://doi.org/10.1007/978-3-319-99570-0_12

12.2   Microemulsions

The word microemulsion was given by Schulman and co-workers in 1959 and mostly he used four component systems like hydrocarbons, ionic surfactants, co-surfactants, and an aqueous phase in his work. The basic observation made by Schulman and co-workers (Schulman et al. 1959) during the titration of a co-surfactant into a coarse microemulsion composed of a mixture of water/surfactant in a sufficient quantity was that a low viscosity, transparent, isotropic, and very stable micro droplet system was formed. These spherical micro droplets had a diameter between 600 and 8000 nm. Since that day microemulsions have found a wide range of applications in pharmaceutics, cosmetics, detergency, lubrication, and oil recovery.

Microemulsions are multicomponent isotropic, homogeneous, optically transparent and thermodynamically stable dispersion of two immiscible liquids [polar phase (water), a nonpolar phase (oil)] in the presence of an emulsifier or surfactant. On a microscopic level, the surfactant molecules are arranged in different forms to give microstructures in which droplets of oil are dispersed in a continuous water phase (O/W microemulsion) over a bicontinuous "sponge" phase and water droplets dispersed in a continuous oil phase (W/O microemulsion). Microemulsion droplet size is generally 100 times smaller than in traditional emulsions and varies from 1 to 100 nm and therefore the microemulsion system behaves as a nanoreactor for the synthesis of nanoparticles with a low polydispersity. The resulting nanomaterials can be characterized by different techniques like: microstructure optical methods, electron microscopy (SEM, TEM), small-angle X-ray and neutron scattering, rheological measurements, electron paramagnetic resonance (EPR), infrared (IR) spectroscopy, nuclear magnetic resonance (NMR), ellipsometry, and electrical conductivity (Destree et al. 2008; Zhong et al. 2007). Microemulsions have different types, such as water-in-oil (W/O), oil-in-water (O/W), and water-in-CO_2 (W/C).

12.3   Micelles

Micelles are liquid dispersions containing surfactant aggregates in micellar dispersions, the aggregates are made of surfactant only and are usually dispersed in water. Spheroidal aggregates can be formed when surfactant molecules are dissolved in organic solvents that are called as reverse micelles (Pileni 1989). It can be formed in the presence or absence of water where large surfactant aggregates are formed in the presence of water and very small aggregates in the absence of water. Water instantly solubilizes in the polar core to form "water pool" contents, which are characterized by water-surfactant molar ratio. The aggregates having a small amount of water ratio ($W° < 15$) are usually called reverse micelles, whereas the aggregates corresponding to droplets with large amount of water molecules ($W° > 15$) are called microemulsions (Luisi et al. 1986). The micro droplets can be formed in diverse ways in which the oil-swollen micelles are dispersed in water as oil-in-water (O/W) microemulsion or water swollen micelles dispersed in oil as for water-in-oil (W/O) microemulsion, also known as reverse microemulsion (Fig. 12.1). The formation of these microemulsions depends on the proportion of various components in the system and also on the hydrophilic-lipophilic balance of the surfactant used.

Reverse micelles also known as "Water-in-oil microemulsion are the microemulsion systems in which polar head groups of surfactant molecules are attracted by aqueous core and directed towards inside and hydrocarbon chain i.e. a polar part is attracted by non-aqueous phase and directed towards outside." By varying the water content, the size and shape of the aqueous core can be easily controlled and uniform size nanoreactors (5–10 nm) can be obtained at a particular ratio of the aqueous phase to the surfactant (Luisiet al. 1986), and it is possible to precipitate the inorganic and organic materials in such nanoreactors (Malik et al. 2012).

12.4 Water-in-Oil (W/O) Microemulsions

Dispersion of water in a hydrocarbon-based continuous phase results in the formation of a "water-in-oil" microemulsion that is mostly situated at the oil apex of a water/oil/surfactant triangular phase diagram. Thermodynamically driven surfactant self-assembly produces aggregates known as reverse micelles (Ekwall et al. 1970); spherical reverse micelles are the most common form which minimizes the surface energy. The polar or ionic components get sequestered into the central cores of these reversed micelles and thereby resulting in the fine dispersion of inorganic and organic materials in oil (Fig. 12.2). It is important to notice that these systems are dynamic in nature, i.e., they frequently collide via random Brownian motion and unite to form dimers, which may swap the contents and then break apart again. Clearly, any encapsulated inorganic reagents will get mixed inside the micelles.

This process is fundamental to nanoparticle synthesis inside reversed micellar "templates," allowing different reactants solubilized in separate micellar solutions to react upon mixing. Micelles in these systems can be depicted as "nanoreactors" providing a proper environment for the controlled nucleation and growth. In

Fig. 12.1 Typical structure of reverse micelle

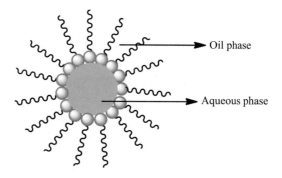

Oil phase

Aqueous phase

Fig. 12.2 Water-in-oil
(W/O) microemulsion

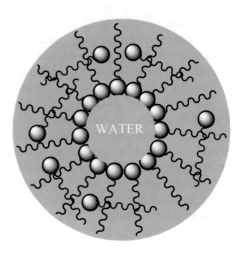

addition, at the latter stages of growth, steric stabilization provided by the surfactant layer prevents the nanoparticles from aggregating (Solanki and Murthy 2011; Lopez-Quintela et al. 2004). Generally, two methods are involved in the synthesis of nanoparticles by using microemulsion techniques. The first method is called the one microemulsion method. This method includes "energy triggering" and the "one microemulsion plus reactant" method. In the energy triggering method, the reaction is initiated by implementing a triggering agent into the single microemulsion which contains a reactant precursor. This fluid system is activated to initiate the reactions that finally lead to the particle formation. For example, to trigger the preparation of nano size gold techniques like pulse radiolysis and laser photolysis have been used (Kurihara et al. 1983). However, in one microemulsion plus reactant method the initiation of the reaction involves the direct addition of a pure reactant (liquid or gaseous phase) into the microemulsion of another reactant. The ions, e.g., metals, are initially dissolved in the aqueous phase of a W/O microemulsion and then the aqueous solution of precipitating agents, e.g., salt NaOH or a gas phase, e.g., NH_3 (g) and $CaCO_3$ (g), are added to the microemulsion solution. Microemulsion method has been widely used to synthesize various nanoscale crystallines in the last several decades since it is a powerful approach to control the particle size and particle distribution. The synthesis of $Li_4Ti_5O_{12}$/C composite was reported by Wang et al. (2014) by using oleic acid as carbon precursor and particle size controller in the microemulsion method.

12.5   Oil-in-Water (O/W) Microemulsions

Microemulsions can normally be prepared in a narrow temperature range (10–20 °C), but this could be overcome by using a mixture of ionic and non-ionic surfactants, which are effective in a wide range of temperatures and exhibit great solubilizing capacity (Sanchez-Dominguez et al. 2012). By using a suitable mixture of ionic and non-ionic surfactants one could increase the temperature range for producing O/W microemulsions. By increasing the pH of the microemulsion more carboxylic acid groups are neutralized, and the negative charge at the interface provides the double layer which boosts the formation of O/W microemulsion. The driving force for producing O/W microemulsion is the presence of the charged head group. The hydrocarbon core of the normal micelles solubilizes more oil, thus forming swollen micelles which are oil-in-water (O/W) microemulsions. The size of droplets can be tuned in the range of 1–100 nm by using varied concentration of the dispersed phase and surfactant. The use of an aqueous continuous phase lets the incorporation of organic droplets (micelles) and thereby declining the use and human exposure to the organic solvents. Moreover, the micelles (oil-in-water) (Fig. 12.3) can be employed as transporters of a large number of organic compounds or particles, e.g., optical limiting units. The reversed micelles (water-in-oil) can be utilized to carry inorganic compounds. The O/W microemulsions were prepared by following the phase diagram reported by Wang et al. (1995). The exact amounts of surfactant and co-surfactant were dissolved in water, and their compositions were calculated from the phase diagram. Addition of a small volume of heptane on stirring mechanically and then sonicating for 5 min in an ultrasonic bath resulted in the formation of oil-in-water microemulsion. Different microemulsions with various droplet sizes were prepared by varying the concentration of surfactant and dispersed phase. The droplets constituting dispersed phase of an oil-in-water microemulsion can act as transporters of lipophilic solutes across an aqueous environment. Sanchez-Dominguez et al. (2015) have reported formation of TiO_2 and Zn-doped TiO_2 nanoparticles by O/W microemulsion. They used a water/Synperonic91/5/ Isooctane microemulsion system. The microemulsions containing the organometallic precursors (Ti-2EH alone or Ti-2EH and Zn-2 EH) were prepared

Fig. 12.3 Oil-in-water (O/W) microemulsion

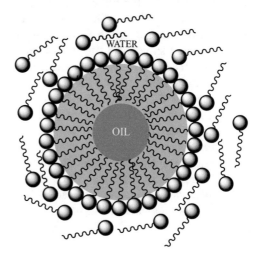

by mixing the surfactant, oil component (isooctane plus organometallic precursors), and water. The atomic ratio of the metal precursors was varied from 100:0 to 90:10 of Ti:Zn. The components were mixed by gentle stirring at the appropriate temperature (26 °C), until microemulsions became transparent, fluid, and isotropic. A certain amount of precipitating agent (ammonium hydroxide 25%) was added under vigorous stirring and continued overnight, followed by centrifugation and washing cycles (first with water, then isopropanol, followed by chloroform and finally isopropanol). Finally, it was dried at room temperature for 2 days, ground in an agate pestle and mortar, and calcined at 400 °C (Sanchez-Dominguez et al. 2015).

12.6 Bicontinuous Microemulsions

Bicontinuous structure or sponge phase plays a key role in colloidal and interfacial science. It is a complex structure, in which water and oil are used as continuous phases. The sponge exhibits a continuous structure, but it is likely to "fill" the sponge with a liquid. The liquid forms a continuous phase and the material of sponge also form a continuous phase. Bicontinuous structures are more frequently observed in microemulsions, in mesophases, in relatively dilute surfactant solutions, and as structures that cannot be depicted in terms of particles, and in nanophilic systems: zeolites, copolymer molecular metals, volcanic minerals, surfactant-templated synthetic mesoporous oxides (Anderson et al. 2006), and composite media. They are usually not found in discrete micellar aggregates. A bicontinuous arrangement of oil and water channels exists instead of isolated droplets. A comparable amount of water and oil is present in most of the microemulsions and those that are stabilized by lesser amounts of surfactants are of great interest (Kresge et al. 1992). A microemulsion with a mean curvature on the average of zero is a strong indication of bicontinuity. There is no long-range order in the structure of microemulsions and also the mean curvature over the dividing surface is not homogeneous at any given instant. However, we can visualize bicontinuous microemulsion structures using well-defined models (continuous paths between interconnected spheres, distorted lamellae, tubule structures with one of the solvents confined to branched tubes with negligible surface type structures similar to that established for bicontinuous cubic phase) but portray them as being thermally disrupted or melted. The thermodynamic model for bicontinuous microemulsions was first provided by Schulreich et al. (2013) reported bicontinuous microemulsions with extremely high temperature stability based on skin friendly oil, sugar surfactant, and co-surfactant benzyl alcohol. Steudle et al. (2015) demonstrated biocatalysis in bicontinuous microemulsions. The activity of the squalene-hopenecyclase from *Alicyclobacillus acidocaldarius* (AacSHC) towards its natural substrate squalene in bicontinuous microemulsions.

12.7   Water-in-CO$_2$ (W/C) Microemulsions

Water-in-CO$_2$ microemulsions have received great attention for the synthesis of nanoparticles (Lim et al. 2002). Carbon dioxide is a smart organic solvent as it is nontoxic, nonflammable, highly volatile, cheap, and environmentally benign in nature. A typical water-in-CO$_2$ microemulsion system is shown in Fig. 12.4. Water-in-CO$_2$ (W/C) microemulsions are formed with especially designed surfactants containing "CO$_2$-philic" fluorocarbon moieties (Mohamed et al. 2016). TiO$_2$-nanoparticles can be formed by the controlled hydrolysis of titanium tetraisopropoxide (TTIP) in water-in-CO$_2$ (W/C) microemulsions and stabilized with the surfactants ammonium carboxylate perfluoropolyether and poly(dimethyl amino ethyl methacrylate-block-1H,1H,2H,2H-perfluorooctyl methacrylate) (Lim et al. 2002). The effects of a homologous series of sodium *p-n*-alkylbenzoate hydrotropes in water-in-CO$_2$ (W/C) microemulsions have been investigated, by comparing the phase behavior and droplet structures (Yan et al. 2015). The W/C microemulsions appeared to be generally stable upon addition of hydrotropes, however, on increasing the alkyl chain length of the hydrocarbon and fluorocarbon moieties of the surfactants, different effects on stability were observed. Yan et al. (2015) research findings were relevant to the understanding of self-assembly of co-adsorbed species in supercritical CO$_2$, as the hydrotrope layers potentially have significant effects on surfactant packing, and can modify the physico-chemical properties of scCO$_2$ through formation of worm-like micellar assemblies.

Fig. 12.4   Water-in-CO$_2$ microemulsions

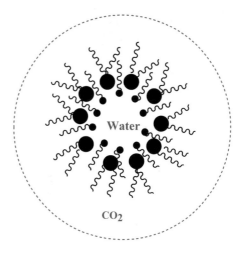

12.8 Mechanism of Nanoparticle Synthesis

In a microemulsion, the aqueous droplets continuously collide and break apart, resulting in a continuous exchange of solution contents. The process of the formation of nanoparticles has been described through the La Mer burst nucleation and Ostwald ripening to describe the change in the particle size. The variation in the size of particles with precursor concentration and with the size of the aqueous droplets in the microemulsion could be explained based on two models. The first is based on La Mer diagram (La Mer and Dinegam 1950), which has been anticipated to explain the precipitation in an aqueous medium and thus is not explicit to the microemulsions. This diagram (Fig. 12.5) is established on the principle that nucleation is a limiting step in the precipitation reaction and demonstrates the variation of the concentration with time during a precipitation reaction. In the first step, there is a continuous increase in concentration with time and once the concentration approaches critical supersaturation value, nucleation occurs, leading to a decrease of the concentration. The nucleation occurs within the concentrations C_{max}^* and C_{min}^* and by the growth of the particles by diffusion there occurs a decrease in concentration and the growth of particles continues till the concentration reaches the solubility value. This model has been applied to the microemulsion medium, i.e., firstly nucleation occurs and later only growth of particles occurs. Following this model, the size of the particles will increase continuously with concentration of the precursor or a minimum in the variation of the size with the concentration might be expected. This is because the number of nuclei is constant, and the increase of concentration favors increase in size of particles.

The second model is based on the thermodynamic stabilization of the particles by the surfactant (La Mer and Dinegam 1950) and the nucleation occurs continuously during nanoparticle formation.

The size of the particles remains constant as the precursor concentration and size of aqueous droplets varies. Both the models are limiting models: as the La Mer diagram does not take surfactant stabilization into account, and the thermo-

Fig. 12.5 A typical La Mer diagram

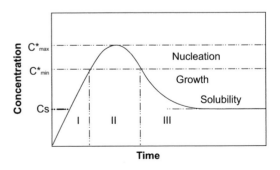

dynamic stabilization model does not take into account that the nucleation of particles are more difficult than the growth by diffusion. Because of the stochastic nature of nanoparticle formation, Monte Carlo simulations can better explain the mechanism of formation of microemulsions. But it is computationally exhaustive due to very large number of micelles considered in a simulation. A mean field approximation with Monte Carlo simulation is possibly required to account for all the probable combinations of various species in the micelles and outcomes of the fission fusion events. The population balance equations with mean field approximation, restricting the maximum number of molecules in a micelle, can be solved and the particle size distribution can be computed from the fraction of nucleated micelles. The several stages involved in the formation of nanoparticles within the water droplets are as: chemical reaction, the nucleation, and the particle growth. In these microemulsions, two different reactants (A and B) are introduced, e.g., in metal oxalate "A" should be a metal ion and "B" should be the oxalate ion. These two microemulsions are continuously stirred and the droplets collide resulting in the interchange of reactants. During this process, the reaction takes place inside the nanoreactor.

12.9 Synthesis of Various Nanomaterials

12.9.1 Metals

Metals display important catalytic properties and synthesis of metal nanoparticles is therefore very interesting. Different metal nanoparticles, including silver and gold nanoparticles have been prepared using inverse microemulsion (Barnickel and Wokaum 1990). Rivera-Rangel et al. (2017) have demonstrated a green synthesis of silver nanoparticles using microemulsion with castor oil as the oily phase, Brij 96V and 1,2-hexanediol as the surfactant and co-surfactant, respectively. Geranium (*P. hortorum*) leaf aqueous extract was employed as a reducing agent. The content and concentration of a metallic precursor and geranium leaf extract in the systems used makes it possible to obtain varied sizes of silver nanoparticles from 25 to 150 nm. The O/W synthesis method is more eco-friendly because the continuous phase is water. The most common method involved in the preparation of metal nanoparticles through microemulsions is the reduction method. Nanoparticles of platinum, palladium, rhodium, and Iridium have been prepared by reverse micelles (Boutonnet et al. 1982).

12.9.2 Synthesis of Nanoparticles of Metal Salts

Nanoparticles of silver bromide have been synthesized by mixing of two micro-emulsions containing the precursor salts $AgNO_3$ and KBr. The microemulsions are composed of AOT, n-heptane, and water. As shown in Fig. 12.6, the role of reaction cage is played by the inner water cores. The initial monomeric AgBr entity is stabilized by AOT adsorption and grows with the fast exchange between water cores (Monnoyer et al. 1995).

12.9.3 Synthesis of Metal Sulfide Nanoparticles

The preparation method is usually applied in the in situ preparation of metal sulfide particles (Chen et al. 2008). Mesoporous CdS nanoparticles have been synthesized using an O/W microemulsion by dissolving the precursor and the precipitating

Fig. 12.6 Model illustrating the role of reaction cage played by the inner water cores (adapted with permission from Elsevier Monnoyer et al. 1995)

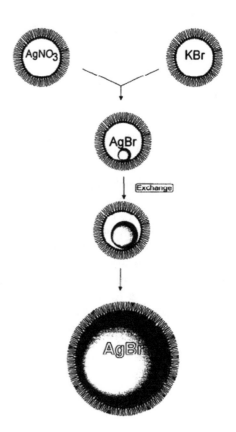

agent ($CdCl_2$ and thioacetamide) in continuous aqueous phase with precipitation of nanoparticles by gamma ray irradiation (Chen et al. 2008). The average diameter of the particles was found to depend on the relative amount of Cd^{2+} and S^{2-}.

12.9.4 Metal Oxides Nanoparticles

CeO_2, ZrO_2, $Ce_{0.5}Zr_{0.5}O_2$, and TiO_2 nanoparticles are synthesized by using oil-in-water microemulsions in contrast to the typically used water-in-oil microemulsion method (Sanchez-Dominguez et al. 2010; Boutonnet and Sanchez-Dominguez 2017). The specific surface area of the nanomaterials was 200–370 m^2 g^{-1} and the particle size was very small (~2–3 nm). CeO_2, ZrO_2, $Ce_{0.5}Zr_{0.5}O_2$, and TiO_2 nanoparticles can be prepared by water/Synperonic®10/6/hexane O/W microemulsions. The microemulsion containing an organometallic precursor has been prepared by mixing suitable amounts of the surfactant, the oil component, and deionized water with composition as: 64.5% water, 21.5% surfactant, and 14% oil phase for TiO_2, ZrO_2, and $Ce_{0.5}Zr_{0.5}O_2$, and 60% water, 20% surfactant, and 20% oil phase for CeO_2. To obtain a transparent, homogeneous, and fluid isotropic phase, the mixture was magnetically stirred at 25 °C (for TiO_2) or 35 °C (for CeO_2, ZrO_2, and $Ce_{0.5}Zr_{0.5}O_2$) and then the microemulsion was kept at 35 °C. A turbid yellow color for CeO_2 and $Ce_{0.5}Zr_{0.5}O_2$ nanoparticles and turbid white for ZrO_2 and TiO_2 nanoparticles appeared on adding 25 wt% NH_3, till pH 10–11 was achieved. The reaction mixture was stirred overnight, followed by centrifugation and washed several times with ethanol and chloroform and finally dried at 70 °C. Calcination was done at a heating rate of 5 °C min^{-1} and keeping at 400 °C during 2 h in an air atmosphere (Sanchez-Dominguez et al. 2010).

12.9.5 Synthesis of Magnetic Nanoparticles

Iron oxide nanoparticles are of great interest because of their unique magnetic properties. The magnetic nanoparticles were first formed in micelles from the oxidation of Fe^{2+} salts to form Fe_3O_4 and Fe_2O_3 (Inouye et al. 1982). This reaction was carried out in a sodium bis (2-ethylhexyl) sulfosuccinate (AOT)/isooctane system to form spherical nanoparticles with surprisingly tight size distributions of less than 10%. Okoli et al. (2012) demonstrated the synthesis of magnetic iron oxide nanoparticles for protein binding and separation, obtained from W/O and O/W microemulsions with sizes ranging from 2 to 10 nm. For synthesis via O/W microemulsion reaction method, the microemulsion consisted of a non-ionic surfactant (Synperonic 10/6), oil phase containing the iron precursor (hexane plus iron(III) 2-ethylhexanoate) and aqueous phase. It was prepared by the addition of organometallic precursor solution to the three components and stirring at 30 °C until a homogeneous, transparent brownish isotropic phase was obtained. The precipitating agent (NH$_3$) was added to the stirring mixture till pH 11 was achieved to precipitate the magnetic iron oxide nanoparticles. The reaction mixture was kept stirring at 30 °C for 48 h. The magnetic nanoparticles were obtained by centrifugation and later washed and dried. For synthesis via the W/O microemulsion method, it consisted of a cationic surfactant (CTAB), co-surfactant (1-butanol), oil phase (n-octane), and aqueous phase. The aqueous phase contained the Fe salt precursors in a mole ratio of 2:1 [FeCl$_3$/FeCl$_2$]. The addition of the precursor solution to the mixture of CTAB/1-butanol/n-octane gave a way for the formation of a microemulsion. A magnetic nanoparticle was achieved by adding the precipitating agent (NH$_3$) to the microemulsion containing the precursor upon vigorous stirring until pH 11 was achieved. The obtained magnetic nanoparticles were separated by centrifugation, washed, and dried at 70 °C (Okoli et al. 2012). Recently, Hu et al. (2017) developed a novel method to synthesize iron oxide nanoparticles in situ in O/W microemulsions for improving oil recovery. Iron oxide nanoparticles were in situ synthesized in microemulsions containing brine, n-hexane, mixture of SDS and Span 80 as the surfactants, and propyl alcohol as the co-solvent.

12.9.6 Synthesis of Mn-Zn Ferrite Nanoparticles

Pemartin et al. (2014) have synthesized Mn-Zn ferrite magnetic nanoparticles via oil-in-water (O/W) microemulsion reaction method. By variation of the precipitating agent and the oil phase concentration, distinctive characteristics were found. The microemulsions containing organometallic precursors (Fe-2EH, Zn-2EH, Mn-2EH, 2 wt% solution of metal in hexane) were synthesized by mixing the surfactant, oil component, and water. The surfactant: water weight ratio was 25:75 and the concentration of oil component was 12 or 20 wt%; this corresponds to a water:surfactant:oil weight ratio of 66:22:12 and 60:20:20, respectively. The atomic ratio of the metal precursors was 2:0.5:0.5 of Fe:Mn:Zn. The components were mixed by stirring gently at appropriate temperature (25 °C for sample with 12 wt% oil and 40 °C for sample with 20 wt% oil), until microemulsions became brown transparent, fluid and isotropic. Then, a certain amount of precipitating agent (aqueous solution of 0.5 M TMAH or 2.5 M NaOH) was added under vigorous stirring up to pH = 12.5. Due to the high toxicity of TMAH, and its large tendency to adsorb onto the surface of iron oxides, a lower concentration to reach the same final pH was used to facilitate its removal from the system as the obtained nanoparticles find potential application in biomedicine. The reaction mixture was stirred overnight, followed by centrifugation and repeated washings with a mixture of water/ethanol and dried in the oven at 70 °C for 2 days. Aubery et al. (2013) have reported microemulsions as reaction media for synthesizing mixed oxide nanoparticles and the relationships between microemulsion structure, reactivity, and nanoparticle characteristics. To determine the effect on the formation of Mn-Zn ferrite nanoparticle, phase behavior, dynamics, and structure of W/O microemulsions of the aqueous solution/Synperonic13-6.5/1-hexanol/isooctane system were studied. The co-surfactant 1-hexanol caused a decrease in the microemulsion regions in comparison to the systems without co-surfactant; however, overlap of microemulsion regions with precursor salts (PS) and precipitating agent (PA) was attained at lower S/O ratios in the systems, compared to the system without co-surfactant. PA microemulsions at 50 °C are non-percolated while the PS microemulsions are percolated.

12.9.7   Nanowires

The nanomaterials fabricated in the reverse micelles are spherical particles in most cases. However, the variations in the properties of nanoparticles are affected by the shape of nanoparticles. For example, cubic Pt nanoparticles have been synthesized and showed extremely good catalysis, selectivity, and activity (Ahmad et al. 1996). Platinum nanowire with high surface area (32.4 ± 3.6 m^2 g^{-1}) and average diameter (2.2 nm) has been synthesized by the reduction of a platinum complex by NaBH$_4$ in the presence of cetyltrimethylammonium bromide in a two-phase water-chloroform system (Song et al. 2007). Qi et al. (1997) synthesized cubic BaSO$_4$ nanoparticles by using reverse micelles of TX-100/hexanol-cyclohexane/water. They confirmed that the water content significantly affected the shape of the nanoparticles in the reverse micelles. In contrast, in the non-ionic reverse micelle 0.1 M BaCl$_2$ and Na$_2$CO$_3$ aqueous solution was added to 0.2 MCl$_2$E$_4$/cyclohexane solution, and mixing the two reverse micelles, BaCO$_3$ nanowires were obtained.

12.9.8   Nanorods

The reverse micellar route was adopted for the preparation of zinc oxalate nanorods for which the microemulsion was prepared with CTAB acts as a surfactant, 1-butanol as co-surfactant, isooctane as nonpolar phase, and 0.1 M aqueous solution of Zn^{2+} and C$_2$O$_4^{2-}$. Microemulsion I contained 0.1 M zinc nitrate solution while microemulsion II contained 0.1 M ammonium oxalate solution, both were mixed slowly and stirred overnight on a magnetic stirrer. The weight fraction of the components in the microemulsion was 16.76% of CTAB, 13.9% of n-butanol, 59.29% of isooctane, and 10.05% of aqueous phase. The precursor was separated from the surfactant and nonpolar phase by centrifugation at an interval of every 2 h of reaction. The precursor was then washed with a mixture of 1:1 chloroform/methanol to remove the surfactant and then dried at room temperature. These zinc oxalate nanorods synthesized at different time intervals were then calcined at 300 °C for 6 h to obtain the resulting zinc oxide (ZnO) nanostructures (Sharma and Ganguli 2014).

12.9.9   Nanocomposites

Reverse micelle is an important method for synthesizing the composite nanoparticles. Composite nanoparticles like CdS-ZnS and CdS-Ag$_2$S have been successfully synthesized by reverse micelles (Han et al. 1998). For core shell nanoparticles synthesis, two steps are involved; in first step, the core nanoparticles are formed in the reverse micelles and the second step involves the growth of shell particle on the core. CdS/ZnS is a typical example of shell-core type composite nanoparticles (where CdS is the core and the ZnS is the shell) synthesized by mixing the reverse micelles containing Cd^{2+} and S^{2-} in a 1:2 ratio resulting in the formation of a core CdS reverse micelle solution. In this reverse micelle S^{2-} is in excess. After several minutes, Zn^{2+} containing reverse micelle is added, which results in the precipitation of ZnS inside the core of CdS nanoparticles and a shell-core type CdS/ZnS composite nanoparticle is obtained. Using this method, CdS/ZnS and ZnS/CdS composite nanoparticles have been synthesized. Another type of composite nanoparticle has been reported to contain two metals but not in the 1:1 ratio. Chen and Chen (2002) have reported the synthesis of Au-Ag bimetallic nanoparticles in water/AOT/isooctane microemulsions (Chen and Chen 2002). Wang et al. (2014) have synthesized ultrafine Li$_4$Ti$_5$O$_{12}$/C composite by microemulsion with oleic acid as carbon precursor and particle size controller. Recently, Pal et al. (2017) have demonstrated fiber formation in emulsion electrospinning by revealing the viscoelastic interaction between dispersed and continuous phase. Composite electrospun matrices of poly(ε-caprolactone) with or without hydroxyapatite were developed from an oil-in-water emulsion. The fiber formation and uniformity were clearly governed by the viscoelastic interaction between the continuous and dispersed phase. Caging of droplets by optimal quantity of poly(vinyl alcohol) (PVA) in continuous phase resulted in uniform stretching and coalescence of droplets (Pal et al. 2017).

12.10   Synthesis of Medium-Chain-Length poly-3-Hydroxyalkanoates (mcl-PHA) Nanoparticle

Ishak and Annuar (2017) demonstrated the effect of temperature on the phase behavior of medium-chain-length poly-3-hydroxyalkanoates nanoparticle. The O/W emulsion was used as template to produce mcl-PHA-incorporated nanoparticle. Water-to-organic phase, oil-to-surfactants, and cremophor EL-to-Span 80 ratio were optimized using response surface methodology with the size of oil droplets produced (20–30 nm).

12.11   Synthesis of Organic Nanomaterials

Synthesis of organic nanoparticles is usually done oil-in-water microemulsions, also known as microemulsion polymerization (Vaucher et al. 2002). However, this technique didn't make much progress due to the phase separation occurring during the process. Margulis-Goshen and Magdassi (2012) have reviewed the methods used to produce organic nanomaterials from microemulsions. Atik and Thomas (1981) reported the use of microemulsion polymerization for the preparation of organic nanoparticles, using a CTAB/styrene/hexanal/water O/W microemulsion system. They obtained monodisperse latex nanoparticles of diameters 35 and 20 nm. The reaction was carried out either thermally using azobisisobutyronitrile (AIBN) or radiolytically. The particles formed were transparent up to 60% of water in the microemulsion systems. Several other organic nanoparticles of cholesterol, retinol rhodiarome, and rhovanil have been synthesized using different microemulsion systems such as AOT/heptane/water, triton/decanol/water, and CTABr/hexanol/water microemulsions (Destree and Nagy 2006). Generally, the active organic compounds are directly precipitated in the aqueous cores of the microemulsion and are detected and observed with iodine vapor and transmission electron microscope. The mechanism of the formation of nanoparticles consists of several steps. The solution of the active compound in an appropriate solvent penetrates inside the aqueous cores by crossing the interfacial film. The solvent plays a key role in carrying the active compound into the aqueous cores, where it gets precipitated due to its insolubility in water and thereby resulting in the formation of nuclei. The growth of nuclei can occur due to the exchange of an active compound between the aqueous cores, and finally these nanoparticles are stabilized by the surfactants (Nagy and Mittal 1999). Lavaud et al. (2017) have developed a simple procedure to prepare a new Prussian blue (PB)-lecithin reverse micellar system comprising PB nanoparticles within the Aonys formulation containing peceol, β-sitosterol, lecithin, ethanol, and water. The lecithin-containing reverse micellar systems including ultra-small PB nanoparticles which are stable over time, while lecithin-free reverse micellar systems show aggregation of the nanoparticles, even for freshly mixed solutions, leading to system destabilization and PB precipitation with time. Using PB nanoparticles within a reverse microemulsion for in vivo Cs^+ (caesium) elimination has great advantages (Lavaud et al. 2017).

12.12   Conclusion

Nanotechnology today is considered to be a mature interdisciplinary scientific field, where the applications are enormous. However, the growing demand of new advanced nano-technologies necessitates the development of new methods to synthesize nanoparticles of desired size, morphology, and properties. To this end, various methods have been successfully developed and used to fine tune the particle size and morphology of the nanoparticles. Microemulsion method is developed long time back, holds promise even today because it allows controlling the particle size and morphology, offers thermodynamic stability, monodispersity, homogeneity, and easy synthetic procedure. The particle size and morphology can be controlled by judicious choice of the various components of the microemulsion and other micellar media. However, the use of microemulsion method in pharmaceutical industry has some concerns like the toxicity of the surfactant and stability of the microemulsion which depends on temperature and pH, and these parameters change when the microemulsion is delivered inside the body. If these concerns are addressed, this method would be greatly helpful in the pharmaceutical industry and drug delivery field.

References

Ahmad TD, Wang ZL, Hengiein A, Sayed MA (1996) Cubic colloidal platinum nanoparticles. Chem Mater 8:1161–1163

Anderson MT, Martin JE, Odinek J, Newcomer P (2006) Synthesis of surfactant-templated mesoporous materials from homogeneous solutions. In book: Access in Nanoporous Matererials p 29–37

Atik SS, Thomas JK (1981) Polymerized microemulsions. J Am Chem Soc 103:4279–4280

Aubery C, Solans C, Prevost S, Gradzielski M, Sanchez-Dominguez M (2013) Microemulsions as reaction media for the synthesis of mixed oxide nanoparticles: relationships between microemulsion structure, reactivity, and nanoparticle characteristics. Langmuir 29:1779–1789

Barnickel P, Wokaum A (1990) Synthesis of metal colloids in inverse microemulsions. Mol Phys 69:1–9

Boutonnet M, Sanchez-Dominguez M (2017) Microemulsion droplets to catalytically active nanoparticles. How the application of colloidal tools in catalysis aims to well-designed and efficient catalysts. Catal Today 285:89–103

Boutonnet M, Kizling J, Stenius P, Maire G (1982) The preparation of monodisperse colloidal metal particles from microemulsions. Colloid Surf 5:209–225

Chen DH, Chen CJ (2002) Formation and characterization of Au-Ag bimetallic nanoparticles in water-in-oil microemulsions. J Mater Chem 12:1557–1562

Chen J, Wang X, Zhang Z (2008) In situ fabrication of mesoporous CdS nanoparticles in microemulsion by gamma ray irradiation. Mater Lett 62:787–790

Cushing BL, Kolesnichenko VL, Connor CJO (2004) Recent advances in the liquid-phase syntheses of inorganic nanoparticles. Chem Rev 104:3893–3946

Destree J, Nagy B (2006) Mechanism of formation of inorganic and organic nanoparticles from microemulsions. Adv Colloid Interf Sci 123:353–367

Destree C, Debuigne F, George S, Champagne B, Guillaume M, Ghijsen J, Nagy JB (2008) J complexes of retinol formed within the nanoparticles prepared from microemulsions. Colloid Polym Sci 286:1463–1470

Ekwall P, Mandell L, Solyom P (1970) The solution phase with reversed micelles in the cetyltrimethylammonium bromide-hexanol-water-system. J Colloid Interf Sci 35:266–272

Han MY, Huang W, Chew CH, Gan LM, Zhang XJ, Ji W (1998) Large nonlinear absorption in coated Ag$_2$S/CdS nanoparticles by inverse microemulsion. J Phys Chem B 102:1884–1887

Hu A, Yao Z, Yu X (2009) Phase behavior of a sodium dodecanol allyl sulfosuccinicdiester/n-pentanol/methyl acrylate/butyl acrylate/water microemulsion system and preparation of acrylate latexes by microemulsion polymerization. J Appl Polym Sci 113:2202–2208

Hu Z, Nourafkan E, Gao H, Wen D (2017) Microemulsions stabilized by in-situ synthesized nanoparticles for enhanced oil recovery. Fuel 210:272–281

Inouye K, Endo R, Otsuka Y, Miyashiro K, Kaneko K, Ishikawa T (1982) Oxygenation of ferrous ions in reversed micelle and reversed microemulsion. J Phys Chem 86:1465–1469

Ishak KA, Annuar MSM (2017) Temperature-induced three-phase equilibrium of medium-chain-length poly-3-hydroxyalkanoates-incorporated emulsion system for production of polymeric nanoparticle. J Disper Sci Technol 39(3):375–383. https://doi.org/10.1080/01932691.2017.1320563

Kresge CT, Leonowicz ME, Roth WJ, Virtula JC, Beck JS (1992) Ordered mesoporous molecular sieves synthesized by a liquid-crystal template mechanism. Nature 359:710–712

Kurihara K, Kizling J, Stenius P, Fendler JH (1983) Laser and pulse radiolytically induced colloidal gold formation in water-in-oil microemulsions. J Am Chem Soc 105:2574–2579

La Mer VK, Dinegam RH (1950) Theory, production and mechanism of formation of monodispersed hydrosols. J Am Chem Soc 72:4847–4854

Lavaud C, Kajdan M, Compte E, Maurel J, Him JLK, Bron P, Oliviero E, Long J, Larionova J, Guari Y (2017) In situ synthesis of Prussian blue nanoparticles within a biocompatible reverse micellar system for in vivo Cs$^+$ uptake. New J Chem 41:2887–2890

Lim KT, Hwang HS, Lee MS, Lee GD, Hong SS, Johnston KP (2002) Formation of TiO$_2$ nanoparticles in water-in-CO$_2$ microemulsions. J Chem Soc Chem Commun 14:1528–1529

Lopez-Quintela MA, Tojo C, Blanco MC, Garcia Rio L, Leis JR (2004) Microemulsion dynamics and reactions in microemulsions. Curr Opin Colloid Interf Sci 9:264–278

Luisi PL, Majid LJ, Fendler JH (1986) Solubilization of enzymes and nucleic acids in hydrocarbon micellar solution. Crit Rev Biochem 20:409–474

Malik MA, Wani MY, Hashim MA (2012) Microemulsion method: a novel route to synthesize organic and inorganic nanomaterials. Arab J Chem 5:397–417

Margulis-Goshen K, Magdassi S (2012) Organic nanoparticles from microemulsions: formation and applications. Curr Opin Colloid Interface Sci 17:290–296

Mohameda A, Ardyani T, Bakar SA, Sagisaka M, Ono S, Narumi T, Kubota M, Brown P, Eastoe J (2016) Effect of surfactant headgroup on low-fluorine-content CO$_2$-philichybrid surfactants. J Supercrit Fluids 116:148–154

Monnoyer P, Fonseca A, Nagy JB (1995) Preparation of colloidal AgBr particles from microemulsions. Colloid Surf A Physicochem Eng Asp 100:233–243

Nagy JB, Mittal KL (1999) Handbook of microemulsion science and technology. Marcel Dekker, New York, p 499

Okoli C, Sanchez-Dominguez M, Boutonnet M, Järås S, Civera C, Solans C, Kuttuva GR (2012) Comparison and functionalization study of microemulsion-prepared magnetic iron oxide nanoparticles. Langmuir 28:8479–8485

Pal J, Wu D, Hakkarainen M, Srivastava RK (2017) The viscoelastic interaction between dispersed and continuous phase of PCL/HA-PVA oil-in-water emulsion uncovers the theoretical and experimental basis for fiber formation during emulsion electrospinning. Eur Polym J 96:44–54

Pemartin K, Solans C, Alvarez-Quintana J, Sanchez-Dominguez M (2014) Synthesis of Mn-Zn ferrite nanoparticles by the oil-in-water microemulsion reaction method. Colloids Surf A Physicochem Eng Aspects 451:161–171

Pileni MP (ed) (1989) Structure and reactivity in reverse micelles. Elsevier, Amsterdam

Pileni MP (2003) Nanocrystals: fabrication, organization and collective properties. C R Chimie 6:965–978

Pileni MP (2008) Supracrystals of inorganic nanocrystals: an open challenge for new physical properties. Acc Chem Res 41:1799–1809

Qi LM, Ma J, Chen H, Zhao Z (1997) Reverse micelle based formation of BaCO$_3$ nanowires. J Phys Chem B 101:3460–3463

Richard B, Lemyre JL, Ritcey AM (2017) Nanoparticle size control in microemulsion synthesis. Langmuir 33:4748–4757

Rivera-Rangel RD, González-Muñoz MP, Avila-Rodriguez M, Razo-Lazcano TA, Solans C (2017) Green synthesis of silver nanoparticles in oil-in-water microemulsion and nanoemulsion using geranium leaf aqueous extract as a reducing agent. Colloids Surf A 536:60–67. https://doi.org/10.1016/j.colsurfa.2017.07.051

Sanchez-Dominguez M, Liotta LF, Carlo GD, Pantaleo G, Venezia AM, Solans C, Boutonnet M (2010) Synthesis of CeO_2, ZrO_2, Ce0.5Zr0.5O2, and TiO_2 nanoparticles by a novel oil-in-water microemulsion reaction method and their use as catalyst support for CO oxidation. Catal Today 158:35–43

Sanchez-Dominguez M, Pemartin K, Boutonnet M (2012) Preparation of inorganic nanoparticles in oil-in-water microemulsions: a soft and versatile approach. Curr Opin Colloid Interface Sci 17:297–305

Sanchez-Dominguez M, Morales-Mendoza G, Rodriguez-Vargas MJ, Ibarra-Malo CC, Rodriguez-Rodriguez AA, Vela-Gonzalez AV, Perez-Garcia SA, Gomez R (2015) Synthesis of Zn-doped TiO_2 nanoparticles by the novel oil-in-water (O/W) microemulsion method and their use for the photocatalytic degradation of phenol. J Environ Chem Eng 3:3037–3047

Schulman JH, Stoekenius W, Prince LM (1959) Mechanism of formation and structure of microemulsions by electron microscopy. J Phys Chem 63:1677–1680

Schulreich C, Angermann C, Höhn S, Neubauer R, Seibt S, Stehle R, Lapp A, Richardt A, Diekmann A, Hellwega T (2013) Bicontinuous microemulsions with extremely high temperature stability based on skin friendly oil and sugar surfactant. Colloids Surf A Physicochem Eng Aspects 418:39–46

Sharma S, Ganguli AK (2014) Spherical-to-cylindrical transformation of reverse micelles and their templating effect on the growth of nanostructures. J Phys Chem B 118:4122–4131

Shervani Z, Ikushima Y, Hakuta Y, Kunieda H, Aramaki K (2006) Effect of cosurfactants on water solubilization in supercritical carbon dioxide microemulsions. Colloid Surf A Physicochem Eng Aspects 289:229–232

Solanki JN, Murthy ZVP (2011) Controlled size silver nanoparticles synthesis with water-in-oil microemulsion method: a topical review. Ind Eng Chem Res 50(22):12311–12323

Song Y, Garcia RM, Dorin RM, Wang H, Qiu Y, Coker EN, Steen WA, Miller JE, Shelnutt JA (2007) Synthesis of platinum nanowire networks using a soft template. Nano Lett 7(12):3650–3655

Steudle AK, Nestl BM, Hauer B, Stubenrauch C (2015) Activity of squalene-hopenecyclases in bicontinuous microemulsions. Colloids Surf B Biointerf 135:735–741

Vaucher S, Fielden J, Li M, Dujardin E, Mann S (2002) Molecule-based magnetic nanoparticles: synthesis of cobalt hexacyanoferrate, cobalt pentacyanonitrosyl ferrate and chromium hexacyanochromate coordination polymers in water-in-oil microemulsions. Nano Lett 2:225–229

Wang LN, Zhang Y, Muhammed M (1995) Synthesis of nanophase oxalate precursors of YBaCuO superconductor by coprecipitation in microemulsions. J Mater Chem 5:309–314

Wang Y, Rong H, Li B, Xing L, Li X, Li W (2014) Microemulsion-assisted synthesis of ultrafine $Li_4Ti_5O_{12}$/C nanocomposite with oleic acid as carbon precursor and particle size controller. J Power Sources 246:213–218

Yan C, Sagisaka M, James C, Rogers SE, Peach J, Hatzopoulos MH, Eastoe J (2015) Action of hydrotropes in water-in-CO_2 microemulsions. Colloids Surf A Physicochem Eng Aspects 476:76–82

Zhong O, Hiroshi Y, Keisaku K (2007) Preparation and optical properties of organic nanoparticles of porphyrin without self-aggregation. J Photochem Photobiol A Chem 189:7–14

Chapter 13
Hydrothermal Nanotechnology: Putting the Last First

Sumit K. Roy and Kamal Prasad

13.1 Nanotechnology: An Introduction

Nanoparticles: Last in the line above the atom, the most difficult to find, and the hardest to learn from. Like atoms nanoparticles are present around us everywhere and are produced by variety of sources like vacuum cleaners, laser printers, candle flames and many more. But unlike atoms we can see the structure of nanoparticles and the breakthrough came with the advent of electron microscope in 1931, when German scientists Ernst Ruska and Max Knoll built the first Transmission Electron Microscope, or TEM. Scientists could now explore the structure of organic molecules that make up the human body, such as proteins, and inorganic materials, such as metals, by examining a cross section of a sample (Boysen and Boysen 2011).

As the matter at the nanoscale behave differently than bulk matter (Rao et al. 2006), nanotechnology has evolved as a recognized branch of science and has broad reach across several scientific disciplines and many industries. Some consider that the general concept of nanotechnology started with a talk that Richard P. Feynman gave in 1959 at the California Institute of Technology. In his talk titled "There is plenty of room at the bottom" the American theoretical physicist asked the question "what would happen if we could arrange the atoms one by one the way we want them?" opened up the possibilities of the nanoscale world, controlling the atoms and molecules that make up matter.

In spite of the fact that the nanotechnology has been around us since 50 years, there is no unanimity about its definition. The word nano comes from the Greek

S. K. Roy
Department of Physics, St. Xavier's College, Ranchi, India

K. Prasad (✉)
Department of Physics, Tilka Manjhi Bhagalpur University, Bhagalpur, Bihar, India

© Springer Nature Switzerland AG 2018
R. Prasad et al. (eds.), *Exploring the Realms of Nature for Nanosynthesis*,
Nanotechnology in the Life Sciences, https://doi.org/10.1007/978-3-319-99570-0_13

word "*nanos*", meaning dwarf and is used as a scientific prefix that stands for 10^{-9} or 1 billionth; the most commonly accepted definition is probably "*Nanotechnology is the study and use of structures between 1 nanometer (nm) to 100 nanometer (nm) in size*". The European Commission offers the following definition "*Nanotechnology is the study of the phenomena and fine tuning of materials at atomic, molecular and macromolecular scale, where properties differ significantly from those at the larger scale*" (Boysen and Boysen 2011). The National Nanotechnology Initiative offers the following definition "Nanotechnology is the understanding and control of matter at dimensions between approximately 1 and 100 nanometers, where unique phenomena enable novel applications. Encompassing nanoscale science, engineering, and technology, nanotechnology involves imaging, measuring, modeling, and manipulating matter at this length scale". These two are among the many definitions available from different scientific and non-scientific communities like Foresight institute, National Nanotechnology Initiative, IBM Watson Research Center and is based on the fact that materials at the nanoscale have novel properties.

13.2   Synthesis of Nanomaterials

Nanoparticle synthesis refers to methods for creating nanoparticles. The synthesis of nanoparticles with a purpose of having better control over particles size distribution, morphology, purity, quantity and quality by employing environment-friendly economical processes has always been a challenge for the researchers.

Two historical evidences—medieval stained glasses and the Lycurgus cup—indicate that humans have been making and using nanoparticles for nearly 2000 years (Kuno 2012). The Roman Lycurgus cup, a relic from the fourth century AD is a well-preserved example of production of metal nanoparticles. What make the cup special are its unique colours that change depending on viewing angle. The light directly transmitted through the cup appears red whereas that transmitted at an angle appears green. This multicolour phenomenon stems from the presence of metal nanoparticles embedded within it as studied by Barber and Freestone (1990) through TEM. Their study on a small fragment of the cup revealed the presence of 50–100 nm diameter gold/silver nanoparticles. The other interesting observation is by Fredrickx et al. (2002) who confirmed the presence of gold nanoparticles in a red glass fragment found in the excavated workshop of Johann Kunckle, a well-known seventeenth century glass maker. The modern era of scientific probing to nanoparticles begins with Faraday's work (1857) on gold nanoparticles.

Nanoparticles synthesis has taken a giant leap since Faraday's time when colloidal metal nanoparticles and semiconductor quantum dots were first time made to the formation of buckyball by vapourizing graphite using a laser in an atmosphere of helium gas by Kroto et al. (1985). The buckyball is also known as fullerene or C60 for the 60 carbon atoms it contains. It is spherical in shape having interlocking hexagons and pentagons of carbon atoms, similar to those in a soccer ball. After the discovery of buckyballs, came the discovery of tiny needles of carbon that run anywhere from 1 to 100 nm in diameter, by Sumio Iijima, who named them carbon nanotubes (Ijima and Ichihashi 1993). These tubes are made of carbon atoms connected in hexagons and pentagons like buckyballs but in a cylindrical shape. Figure 13.1 shows a representative diagram of a graphene sheet, a single wall carbon nanotube and a buckyball.

Nanomaterials come in variety of flavours: Fractals (Mandelbrot 1983; Bunde and Havlin 1996), porous materials (Srivastava et al. 2002), zeolites (Dyer 1988), fullerenes (Kroto et al. 1985), micelles (Yin and Wang 1997; Sun and Murray 1999), self-assembled monolayers (Wang 1998), etc. A broader classification for the different types of nanomaterials can be organic nanoparticles (Dendrimers, micelles, liposomes and ferritin), inorganic nanoparticles (Metal and metal oxide based) and carbon-based nanoparticles (fullerenes, graphene, carbon nano tubes (CNT), carbon nanofibres and carbon black).

There are various nanomaterial synthesis techniques available nowadays (Fig. 13.2), namely mechanochemical, chemical, hydrothermal, flame combustion, emulsion precipitation, etc.; each method varies in effectiveness and has its own advantages and disadvantages. The performances of nanomaterials are closely related to the ways they are synthesized, but they are all based on one of the two *fundamental mechanisms*: *Energy maximization (Top-down approach) and Supra-Molecular Chemistry (Bottom-up approach)*.

13.2.1 Synthesis of Nanomaterials from Above and Below

Nanoparticles can be derived from larger molecules, or synthesized by "bottom-up" methods. In general, top-down and bottom-up are the two main approaches for nanomaterial synthesis.

- Top-down: size reduction from bulk materials.
- Bottom-up: material synthesis from atomic level.

Top-down approach aims at pushing hard and reaching down. It begins with a pattern generated on a larger scale, then reduced to nanoscale by breaking up larger

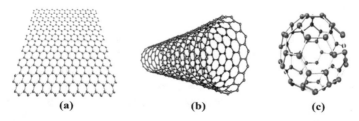

(a)   (b)   (c)

Fig. 13.1 Representative diagrams of (**a**) a graphene sheet, (**b**) a carbon nanotube and (**c**) a buckyball

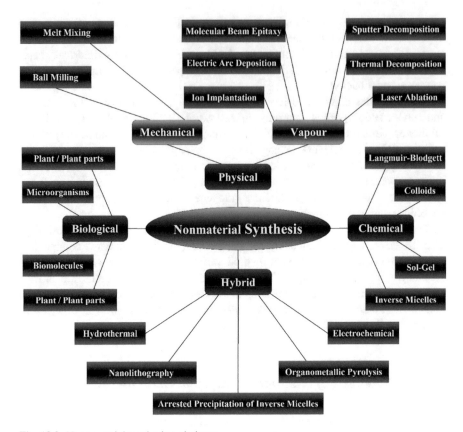

Fig. 13.2 Nanomaterial synthesis techniques

particles using physical processes like crushing, milling, grinding, etc. This method is currently used to manufacture computer chips as well as other products we use every day; however, the biggest problem with the top-down approach is the imperfection of the surface structure. Such imperfection would have a significant impact on physical properties and surface chemistry of nanostructures and nanomaterials. Usually this route is not suitable for preparing uniformly shaped materials, and it is very difficult to realize very small particles even with high energy consumption. Also this method is neither cheap nor quick to manufacture and thus not suitable for large-scale production.

Bottom-up approach refers to the build-up of a material from the bottom: atom-by-atom, molecule-by-molecule or cluster-by-cluster. This route is more suitable for preparing nanoscale materials of uniform size, shape and distribution. The bottom-up approach is relatively new approach, in the theoretical stage, with researchers continuously upgrading the experimental techniques to synthesize nanomaterials by this technique. In 1977, Eric Drexler, while still a student at MIT, first outlined the fundamental bottom-up approach to nanotechnology in which he described how we might manipulate individual atoms and thereby synthesize materials. Drexler who was the first to be granted Ph.D. in Molecular Nanotechnology postulated that we could use tiny machines called molecular assemblers, to build just about anything-using common chemicals (Drexler 1986, 1992; Bimberg et al. 1999). Few high-tech bottom-up techniques for synthesizing quantum wells and quantum dots are: Molecular Beam Epitaxy (MBE), colloidal growth, arrested precipitation, inverse micelles, etc. (Drexler 1986; Bimberg et al. 1999; Turkevich et al. 1951; Brust et al. 1994; Rossetti et al. 1985).

There are three generic types of nanostructures important to us. The classification is based on the dimensions in which a carrier (electrons and holes) is free to move. These structures are alternatively referred to as low dimensional systems and possess a given dimensionality, the schematic illustrations of which are given in Fig. 13.3:

- Quantum dots (zero-dimensional system)
- Quantum wires (one-dimensional system)
- Quantum wells (two-dimensional system)

In a particular low dimensional nanostructure, carrier confinement occurs in a particular direction if the physical thickness of the material in that direction becomes smaller than the characteristics or de-Broglie length of mobile carriers. In a quantum dot, the structure is narrow along all three directions of a co-ordinate system. It is usually depicted by a small spherical particle of radius smaller than the characteristic de-Broglie wavelength of carriers in that material (Bera et al. 2010). In a nanowire, the material is stretched along a single direction with other two orthogonal dimensions much narrower (Schmid 2005). The cross-sectional geometry of a nanowire may have different shapes such as circular, rectangular or square. Finally, in quantum well the carriers have one degree of confinement and two degree of freedom (Zory 1993; Rao et al. 1999).

The controlled synthesis of desirable composition, size, shape and crystal structure is of considerable interest because of its potential application in obtaining materials with wide technological applications. Among the various synthesis techniques mentioned in Fig. 13.2, hydrothermal technique has proven to be an excellent method for the synthesis of different chemical compounds.

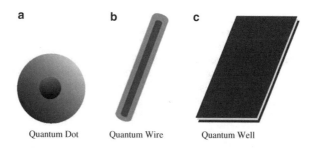

Fig. 13.3 Schematic illustration of (**a**) quantum dot, (**b**) quantum wire and (**c**) quantum well

a   b   c

Quantum Dot   Quantum Wire   Quantum Well

13.3 Hydrothermal Method

The term hydrothermal usually refers to any heterogeneous reaction in the presence of aqueous solvents under elevated pressure and temperature conditions to dissolve and recrystallize materials that are relatively insoluble under ordinary conditions. The word was first coined by British Geologist, Sir Roderick Murchison (1792–1871) in explaining the formation of various rocks and minerals in earth's crust. It all started with developing an understanding of the mineral formation process in nature that is how in the presence of water under elevated pressure and temperature formation of minerals and rocks takes place in earth's crust. The Greek prefix hydro (The Greek word for hydro is húdōr—ὕδωρ) means water and thermal (The Greek word for thermal is thérmē-θέρμη) means heat. In the last three decades hydrothermal technique has become a very popular means to simulate the natural conditions existing under the earth's crust, and synthesizing them in laboratory (Byrappa and Yoshimura 2001), still there is no unanimity about its definition, neither there is any unanimity about the critical values of temperature and pressure above which the process can be categorized as hydrothermal. There have been several definitions of hydrothermal technique since its advent and are listed in Table 13.1.

As mentioned research on hydrothermal technique started in the middle of the nineteenth century by geologists and the initial work related to hydrothermal synthesis of materials is attributed to R.W. Bunsen, who grew barium and strontium carbonate at temperatures above 200 °C and pressures above 100 bars in 1839 (Bunsen 1848). After that in 1845, C.E. Schafhautl observed the formation of small

Table 13.1 Definition of hydrothermal synthesis prescribed by different researchers

Author (year)	Definition
Morey and Niggli (1913)	In the hydrothermal method the components are subjected to the action of water, at temperatures generally near though often considerably above the critical temperature of water (~370 °C) in closed bombs, and therefore, under the corresponding high pressures developed by such solutions
Laudise and Parker (1970)	Hydrothermal growth means growth from aqueous solution at ambient or near-ambient conditions
Lobachev (1973)	Defined it as a group of methods in which crystallization is carried out from superheated aqueous solutions at high pressures
Rabenau (1985)	Defined hydrothermal synthesis as the heterogeneous reactions in aqueous media above 100 °C and 1 bar
Byrappa (1992)	Defines hydrothermal synthesis as any heterogenous reaction in an aqueous media carried out above room temperature and at pressure greater than 1 atm
Roy (1994)	Declares that hydrothermal synthesis involves water as a catalyst and occasionally as a component of solid phases in the synthesis at elevated temperature (>100 °C) and pressure (greater than a few atmospheres)
Yoshimura and Suda (1994)	Proposed the following definition: reactions occurring under the conditions of high-temperature–high-pressure (>100 °C, >1 atm) in aqueous solutions in a closed system

quartz crystals upon transformation of precipitated silicic acid in a steam digester (Schafhautl 1845). Thereafter, Wöhler in 1848 recrystallized apophyllite by heating apophyllite in water solutions at 180–190 °C under 10–12 atm pressure (Wöhler 1848). The next significant contribution was by H. De Senarmont, in the year 1856 where he synthesized six-sided quartz prism with pyramidal termination. Then it was French mineralogist, Daurree (1857), who first used a steel tube to synthesize quartz and wollastonite at about 400 °C with water as a solvent. Barrer in 1940 succeeded in synthesizing for the first time analcime, a member of zeolite group, using the hydrothermal technique (Barrer 1948). This initiated the flurry of artificial production of bulk single crystals of quartz and zeolites during late 1930s and 1940s. In 1950–1970, Roy and co-workers used the hydrothermal process to synthesize systematically for the first time the whole range of 7, 10 and 14 Å clay minerals and selected zeolite families with an enormous range of compositions (Roy and Roy 1955). The hydrothermal technique received a wide acceptance due to its technological efficiency in developing bigger, purer and dislocation-free single crystals and practically all inorganic species, starting from native elements to the most complex oxides, silicates, germanates, phosphates, chalcogenides, and carbonates have been obtained by this method. The year 1990 marks the beginning of the work on the processing of fine to ultra-fine particles with a controlled size and morphology. Nowadays hydrothermal researchers are able to understand the mechanism involved in hydrothermal synthesis owing to the advent of optical instruments which can probe up to the atomic and subatomic structures. Among the different chemical methods, hydrothermal synthesis is often preferred due to its simplicity, allowing the control of grain size, morphology and degree of crystallinity by easy changes in the experimental procedure. Thus hydrothermal technique can control the formation process and yield even sub-micrometre to nanosize crystalline products with desired properties.

Today, hydrothermal technique has found its place in several branches of science and technology, and is of use for people of several streams like material scientists, earth scientists, material engineers, metallurgists, physicists, chemists, biologists and others. A number of other techniques find its origin in the hydrothermal technique; at the core is the hydrothermal technique, and the other techniques have their roots attached to it as depicted in Fig. 13.4.

The other major reason for the success of hydrothermal technique is that it meets the global concern and awareness for the development of environment-friendly materials. In few other techniques such as solid-state reaction method, hazardous volatile gases are emanated during the high temperature sintering. Considering the fatal affects which these hazardous substances cause to living organism and in general to environment, the hydrothermal technique occupies a unique place in modern science and technology. Hydrothermal technique is in line with the development of ecologically benign, "green" compounds with the recent RoHS and WEEE legislation stipulated by the EU (Jha and Prasad 2010).

Fig. 13.4 Hydrothermal
processing: an
interdisciplinary technique

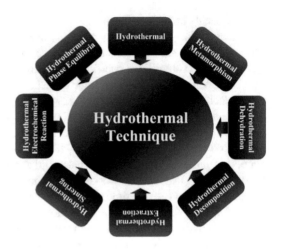

13.4   Exploring the Role of Solvent

In any hydrothermal system or reaction, the role played by the solvent under the
action of temperature and pressure is very important. Water is one of the most
important solvents used for the purpose and its properties such as vapour pressure,
density, surface tension and viscosity are greatly changed under hydrothermal
conditions (Meng et al. 2016). It was most popularly used as a hydrothermal
mineralizer in all the earlier experiments due to the advantages that water is cheaper
than other solvents, and it can act as a catalyst for the formation of desired materials
by tuning the temperature and the pressure. It is nontoxic, nonflammable,
noncarcinogenic, nonmutagenic, and thermodynamically stable. Another advantage
is that water is very volatile, so it can be easily removed from the product. However,
several compounds do not show high solubility for water even at supercritical
temperature, and hence the size of the crystals or minerals obtained in all the early
hydrothermal experiments did not exceed thousandths or hundredths of a millimetre.
Therefore, the search for other suitable mineralizers began. The other reason behind
the investigation of different solvents in the chemical reactions is to bring down the
pressure and temperature conditions close to ambient conditions and to increase the
solubility of the desired compound. A variety of aqueous and non-aqueous solutions
were tried to obtain highly crystalline products. Depending upon the type of solvent
used in such chemical reactions, a common term in use is solvothermal synthesis.
Specifying the solvent, solvothermal synthesis can be classified as: hydrothermal,
glycothermal, ammonothermal, many more.

Glycothermal process is a special class of solvothermal process in which organic
solvent, especially glycols, is used to prevent the formation of hard agglomerations
during crystallization. Though the reaction mechanism in non-aqueous solutions is
complex, then also combining glycols and hydrothermal treatment is a successful
method for preparing highly crystalline nanoparticles with narrow size distribution
and low aggregation. Very few investigations have been reported dealing with the

use of glycol to synthesize crystalline ceramic powders under hydrothermal conditions (Kil et al. 2015).

Ammonothermal synthesis uses ammonia instead of water to epitomize the hydrothermal synthesis as ammonia resembles water in its physical properties (Richter and Niewa 2014). Since ammonia is more reactive than water in terms of metal solubility, so less harsh temperature and pressure conditions are required to carry out the process. The first ammonothermal syntheses were carried out in the year 1960 by Juza and Jacobs and subsequently followed by Jacobs and co-workers (Juza and Jacobs 1966; Juza et al. 1966; Jacobs and Schmidt 1982). The chemists working in the supercritical region dealing with the materials synthesis, extraction, degradation, treatment, alteration, phase equilibria study, etc. prefer to use the term supercritical fluid technology (Byrappa and Adschiri 2007).

13.5   Hydrothermal Synthesis

Hydrothermal synthesis or crystal growth under hydrothermal conditions is typically carried out in a pressurized vessel called an autoclave. In an autoclave, nutrient is supplied along with solvent, which may be water or any other chemical. To enhance the process, some amount of surface modifiers is also mixed. A temperature gradient is maintained between the opposite ends of the growth chamber. At the hotter end the nutrient solute dissolves, while at the cooler end it is deposited on a seed crystal, growing the desired crystal. The block diagram of hydrothermal synthesis procedure is as shown in Fig. 13.5.

In hydrothermal method, the involved reaction time is long and highly corrosive salt is used to synthesize inorganic materials, so the autoclave must be capable of sustaining highly corrosive reaction system at high temperature and pressure for a longer time duration. Thus the major factors deciding the choice of a suitable autoclave are temperature, pressure and the ability of the autoclave material to resist corrosion in that pressure-temperature range for a given solvent. Other desirable conditions for autoclave are:

- Length should be sufficient enough to set a desired temperature gradient.
- The choice of autoclave material should be such that it is inert to acids/bases and oxidizing agents.
- High creep-rupture strength (is measured by measuring the span of time a material can withstand without breakdown or rupture when subjected to stress at a given temperature).

An autoclave satisfying the above conditions is practically hard to design. Most autoclaves are designed to satisfy the specific need of a given project. Stainless steel, cobalt-based and titanium-based alloys are mostly employed for making autoclave due to their superior corrosion resistance and high creep-rupture strength (Laudise and Nielsen 1961; Clauss 1969). Additional care must be taken to counter the fact that under various corrosion processes the actual strengths of alloys are less

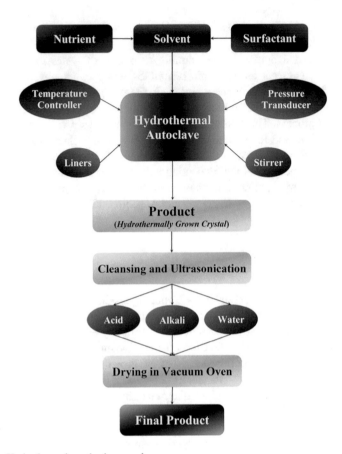

Fig. 13.5 Hydrothermal synthesis procedure

than their strengths under ideal conditions. In addition, it must be noted that the creep-rupture strengths decrease significantly with increasing temperature (Byrappa and Yoshimura 2001), thereby lowering the maximum permissible pressure of an experiment at higher temperature.

The corrosion which the autoclave material encounters at high temperature and pressure during hydrothermal process comes under either of the following three, namely, crevice corrosion, stress corrosion cracking and inter-crystalline corrosion. The crevice or crack can be taken care off by properly polishing the inner surface of the vessel. Inter-crystalline corrosion can be taken care off by using low carbon grade of stainless steels, by high-temperature solution heat treatment and alloying with small amounts of metals (Nb, Ta and Ti) forming very stable carbides. Stress corrosion can be retarded through proper annealing of the alloy used for making autoclaves, and through the use of molybdenum containing austenitic steels.

In addition, systems containing hydrogen cannot be synthesized by hydrothermal technique as hydrogen reacts badly at high temperature with the alloys used for preparing autoclaves and can reduce the strength of autoclaves either by hydrogen embrittlement, irreversible hydrogen damage, or metal-hydride formation. These problems could be overcome through careful selection of alloys containing small additives such as Ti, Mo and V; heating in H_2-free atmosphere and using alloys with low thermodynamic activity.

Obtaining a contamination-free, high purity crystal is one of the major requirements of any synthesis method. A major setback to this requirement in hydrothermal method is the use of highly corrosive mineralizer that reacts with the material of the autoclave causing a permanent decay. This can be prevented by suitable lining of the inner walls of the autoclave or by placing separate liners over the inner surface. Among very early users of liners are Daurree (1857) and Allen et al. (1912), who either used pure glass or some variants of glass accompanied with steel tube as host for the glass tube, then came the era of steel autoclave and noble metal lining. Different metal linings are used depending upon the solvent medium (Litvin and Tules 1973) (Table 13.2).

13.6   Types of Hydrothermal Method

In hydrothermal method, the physical parameters, which can be controlled externally, are temperature, pressure, time and heating rate. There is also control over the parameters of the solute such as its pH value, concentration and the type of solute. As there are a number of processing variables which can be controlled during hydrothermal synthesis, it can be hybridized with several other processes to obtain highly crystalline products with narrow size distribution, high purity and low

Table 13.2 Materials used as reactor linings

Material	Temperature (°C)	Solutions	Remarks
Silver	600	Hydroxides	Recrystallization (gradual) and dissolution (partial)
Gold	700	Hydroxides, sulphates	Dissolution (partial) in hydroxides
Platinum	700	Hydroxides, chlorides, sulphates	Get Blacken in chlorides in presence of sulphur ions. Dissolution (partial) in hydroxides
Copper	450	Hydroxides	Reduction of corrosion in presence of fluoride ions as well as organic compounds
Titanium	550	Chlorides, hydroxides, sulphates, sulphides	Corrosion in NaOH solution greater than 25% and in NH_4Cl solution greater than 10%
Tantalum	500	Chlorides	Corrosion starts in NH_4Cl solution 78%
Armco iron	450	Hydroxides	Oxidation (gradual) producing magnetite
Graphite	450	Sulphates	Pyrolytic graphite most suitable for linings
Teflon	300	Chlorides, hydroxides	Poor thermal conduction

aggregation. Lots of work has been reported where hydrothermal synthesis has been hybridized with microwaves, electrochemical process, ultrasound, mechanical milling, optical radiation, hot-pressing and many other processes to enhance the quality of product.

13.6.1 Microwave-Assisted Hydrothermal Processing (MAH)

Microwaves, a component of the Maxwell's rainbow are short wavelength radio-waves in the gigahertz (GHz) range. Under the influence of the oscillating electric field of the electromagnetic wave, the charges within the sample execute forced oscillation and are polarized. Microwaves enhance crystallization kinetics by one to two orders of magnitude with respect to standard hydrothermal processing. Different materials have energy levels differently spaced hence they interact with microwaves differently. Their loss factor or conversion efficiency determines the amount of electromagnetic energy that will be converted into heat. So for efficient and rapid heating, reaction medium with high microwave absorbing ability should be selected. The microwave-assisted hydrothermal method combines the merits of microwave and hydrothermal methods, thereby reducing the reaction time to few minutes or few hours as compared to that in a purely hydrothermal method where reaction time ranges from several hours to several days. Among many experimental observations, one is by Xu et al. (2004) who synthesized a high quality pure hydroxy-sodalite zeolite membrane on the support of α-Al$_2$O$_3$ by microwave-assisted hydrothermal method and reported the synthesis of hydroxy-sodalite zeolite membrane within 45 min compared to 6 h of synthesis time by the conventional hydrothermal method.

Thus compared with conventional hydrothermal method, microwave-assisted hydrothermal synthesis has emerged as a novel heating model in material science due to its rapid volumetric heating, low reaction temperature, homogeneous thermal transmission, low thermal gradients and the phase purity with better yield. This method is promising for fabricating nanomaterials with various morphology viz. nanowires, nanorods, nano-needles and it is generally a low-cost process for the bulk production of nano- and micro-structures.

This method is used mostly for synthesis of ceramic powders. Many groups have reported on the successful synthesis of ceramics by MAH technique like, Newalkar and Komarneni (2002) synthesized micropore-free hydrothermally stable mesoporous silica SBA-15 under microwave hydrothermal conditions at 373 K for 120 min. Bondioli et al. (2005) prepared nanocrystalline Pr-doped ceria powders with uniform sizes of 25–30 nm within 1 h by microwave-assisted hydrothermal route. Komarneni et al. (1993) prepared BaTiO$_3$, SrTiO$_3$, Ba$_{1/2}$Sr$_{1/2}$TiO$_3$, BaZrO$_3$, SrZrO$_3$, PbZrO$_3$, and Pb(Zr$_{0.62}$Ti$_{0.48}$)O$_3$; ceramics and ZnO rod-assembled microspheres were successfully synthesized by Zhu et al. (2011). They concluded that ZnO prepared by MAH technique had an excellent optical property and higher photocatalytic activity and displays good recyclability and stability than that of ZnO prepared by conventional hydrothermal technique. In most cases, ionic solvents are used as the medium for the hydrothermal reaction in the presence of microwaves because it helps to reduce the processing temperatures for various kinds of the ceramic materials. Thus, the synthesis of ferrites at low temperature is also possible by this process. $ZnFe_2O_4$ nanoparticles, iron hydroxyl phosphates $(NH_4Fe_2(PO_4)_2OH \cdot 2H_2O)$ nanostructures, α-FeOOH hollow spheres, β-FeOOH architectures and α-Fe_2O_3 nanoparticles were prepared successfully by Tadjarodi et al. (2014) and Cao et al. (2010), respectively. Other synthesized materials are SnO_2 microspheres (Dong et al. 2008) and Bi_2O_3 (Schmidt et al. 2018), etc.

13.6.2 Sol-Gel-Hydrothermal (SG-HT) Method

The sol-gel method is well known as one of the versatile methods to synthesize inorganic compounds, including metal oxides, complex oxides, chalcogenides and many more. The Royal Society of Chemistry explains the sol-gel method as the process of the settling of (nm sized) particles from a colloidal suspension onto a pre-existing surface, resulting in ceramic materials. The desired solid particles (*e.g.* metal alkoxides) are suspended in a liquid, forming the "sol", which is deposited on a substrate by spinning, dipping or coating, or transferred to a mould. The particles in the sol are polymerized by partial evaporation of the solvent, or addition of an initiator, forming the "gel", which is then heated at high temperature to give the final solid product.

The sol-gel method is a promising low-energy method; however, in the case of powder syntheses, a subsequent heating process is often needed to obtain the desired compounds, and agglomerated nanoparticles are usually obtained. On the other hand, the sol-gel method combined with hydrothermal method (SG-HT) is capable of synthesizing microstructure-controlled nanoparticles at a lower temperature. The experimental procedure, divided in two major heads, sol-gel preparation and hydrothermal treatments are as depicted in Fig. 13.6.

Few reports on SG-HT are listed here. Valencia et al. (2010) synthesized titanium dioxide nanoparticles by the sol-gel method followed by a hydrothermal treatment using tetraisopropyl orthotitanate (TIOT) and 2-propanol. Gong et al. (2008) reported the synthesis of carbon-coated Li_2FeSiO_4 material through hydrothermal-assisted sol-gel process by adding sucrose to the synthetic precursor. The Li_2FeSiO_4 so obtained showed large reversible capacity under high rate charge and discharge and excellent capacity retention. Ueno et al. (2015) synthesized silver-strontium titanate hybrid nanoparticles by sol-gel-hydrothermal method. Fuentes et al. (2010) also prepared strontium titanate, $SrTiO_3$, using the sol-gel-hydrothermal reaction of $TiCl_4$ and a $SrCl_2$ solution in an oxygen atmosphere.

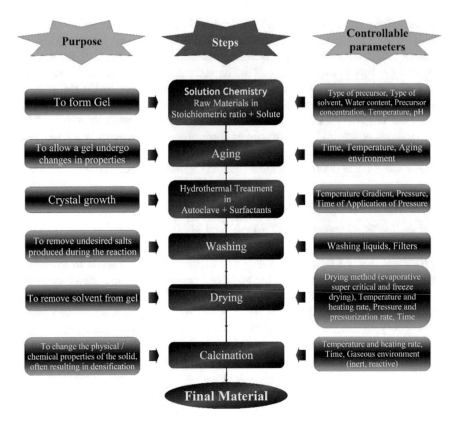

Fig. 13.6 Sol-gel combined hydrothermal method

13.6.3 Hydrothermal-Electrochemical Synthesis (HE)

Hydrothermal-electrochemical technique combines hydrothermal process with electrochemical method and results in deposition of crystalline thin films on metallic substrate directly from nanostructured products of hydrothermal reaction. It is particularly important in the case of crystalline oxide products, which cannot precipitate out of solution in the absence of an applied electric potential.

Hawkins and Roy (1962) introduced electrodes into a hydrothermal bomb and studied the influence of an electric field on the synthesis of kaolinite. They showed a very substantial effect (>100 °C) of lowering the crystallization temperature of kaolinite. Yoshimura popularized this concept by extending it to the synthesis of electroceramics, particularly perovskite-type oxides. Figure 13.7 shows the schematic diagram of the autoclave designed in Yoshimura's laboratory at the Tokyo Institute of Technology, Japan, for the hydrothermal-electrochemical reactions, using three-electrode technique of Bard et al. (1980).

Fig. 13.7 Schematics of
the electrochemical cell
with dc power supply for
the hydrothermal-
electrochemical method

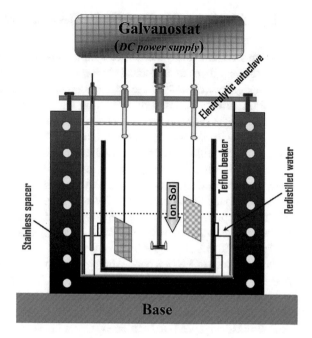

This has a provision to measure potentials of electrodes in aqueous solutions at high temperatures and pressures on a thermodynamically meaningful scale, regardless of the electrolysis conditions. Yoshimura's group has worked extensively on this and successfully prepared perovskite-type titanates such as $SrTiO_3$, $CaTiO_3$ and $BaTiO_3$ thin films up to several micrometres in thickness, almost linearly, at fairly lower temperatures (<90 °C) (Suchanek and Yoshimura 1998; Yoshimura et al. 1999). They have also studied the growing interface, the transport mechanism of the atoms across the growing film without concentration gradients. It also enables fabrication of ceramic super lattices. Villegas and Stickney (1992) and Colletti et al. (1998) grew monomolecular or atomic layer films of GaAs, CdTe, CdSe and CdS semiconductors, equivalent to molecular beam epitaxial, using aqueous solutions instead of a vapour phase for transport of growth species.

Basca et al. (1992) also crystallized polycrystalline films of cubic $BaTiO_3$ on Ti metal substrates by electrochemical, hydrothermal and electrochemical-hydrothermal reaction of a Ti metal with a saturated solution of $Ba(OH)_2$ in the temperature range 80–200 °C. The electrochemical-hydrothermal combination reduced the reaction time to only 30–45 min. The reduced reaction time is desirable as they observed that crystallinity decreased with extended time of reaction for a given temperature.

13.6.4   Mechanochemical-Hydrothermal Synthesis (MCH)

The mechanochemical-hydrothermal synthesis often termed as wet mechanochemical synthesis utilizes an aqueous solution as a milling medium. The aqueous solution actively participates in the mechanochemical reaction, thereby accelerating dissolution, diffusion, adsorption, reaction rate and crystallization (Yoshimura and Suchanek 1997). For the mechanochemical-hydrothermal processing, there is no need for a pressure vessel and external heating as pressure can be applied at room temperature by milling equipment, and energy required for reaction is provided in the form of mechanical energy. Thus, mechanochemical-hydrothermal technique jointly uses the mechanism of hydrothermal and mechanical alloying. Collision between the balls, between balls and wall (present inside milling instrument) and their mutual friction creates localized zones of high temperature and high pressure and the heat generated adiabatically raises the temperature of the slurries as high as 450–700 °C (Kosova et al. 1997).

Ceramics such as $PbTiO_3$ and hydroxyapatite have been made with this approach. Shuk et al. (2001), Riman et al. (2002) and Abdel-Aala et al. (2008) have synthesized nano-crystallite hydroxyapatite by this method. Although the mechanochemical method appears promising for the synthesis of ceramic precursors, it has some possible disadvantages like high energy consumption and possible contaminations from the milling media. In addition, control of particle size, morphology and aggregation is a challenge for this method since MC-HT method fails to regulate rates of nucleation, growth and ageing unlike conventional hydrothermal technologies.

13.6.5   Ultrasound-Assisted Hydrothermal Crystallization (UAH)

Under the ultrasound-assisted hydrothermal crystallization, often known as sonochemical technique, molecules undergo chemical reactions due to the application of powerful ultrasound radiation (20 kHz to 10 MHz). Ultrasound is known to accelerate the reaction kinetics by as much as two orders of magnitude owing to acoustic cavitation (Yamakov et al. 2004). Besides accelerating the reaction rate, UA-HT also yields polycrystalline thin films with smooth surface, thereby reducing the possibility of electrical breakdown when subjected to electric field. The sonochemical environment is also considered to alter molecular chemistry and enhance mass transport (Peters 1996). The schematic diagram of the ultrasonically assisted hydrothermal system has been illustrated in Fig. 13.8.

This method, initially proposed for the synthesis of iron nanoparticles, is nowadays used to synthesize different metal oxides. The sonochemical technique involves passing sound waves of fixed frequency through slurry or solution but this arrangement is difficult to achieve as hydrothermal method is carried out in a sealed

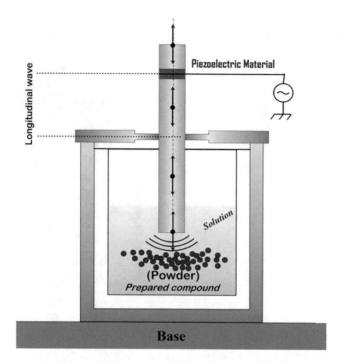

Fig. 13.8 Schematics of the ultrasonically assisted hydrothermal system

stainless-steal container under high pressure and high temperature. Ishikawa et al. (2008) constructed a reaction container with an ultrasonic transducer integrated into the lid of an autoclave for synthesis of $KNbO_3$ piezoelectric ceramics and ultrasonically assisted hydrothermal PZT thin films. They observed noticeable effectiveness—increase in the reaction rate of the $KNbO_3$ powder and smoothening in the surface of the PZT thin film. They further enhanced the device by improving the vibration speed even under high-temperature conditions and examined its effect on PZT thin film deposition and $(K,Na)NbO_3$ powder fabrication.

13.6.6   Hydrothermal Hot-Pressing (HHP)

Hydrothermal Hot-Pressing (HHP) technique is a processing route for producing a ceramic body at relatively low temperatures typically below 300 °C. This technique is applied to solidify many kinds of the materials such as silicate glass, titania, calcium carbonate and hydroxyapatite (Hosoi et al. 1998). The hydrothermal hot-pressing method (Fig. 13.9) has two characteristics: continuous compression of samples under hydrothermal conditions and space for water retreat. Compression accelerates compaction of starting powder and prevents development of cracks by shrinkage. Compaction depends on the rate at which water can be expelled from the

Fig. 13.9 Schematic of
autoclave for hydrothermal
hot-pressing

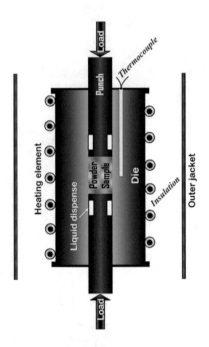

starting powder. The hydrothermal hot-pressing method requires the space for water retreat, into which water included in the starting powder is released. Without the space, water exists in pore space of the starting powder and hinders compaction of the powder. During the hydrothermal treatment, mass transport leading to densification occurs mostly by a dissolution precipitation mechanism. Low processing temperatures enable the incorporation of organic components that can further improve the mechanical strength of the ceramics.

13.6.7 Hydrothermal–Photochemical Synthesis (HP)

Photochemical synthesis is used to describe a chemical reaction caused by absorption of light. Photochemical reactions proceed differently than temperature-driven reactions. Photochemical paths access high-energy intermediates that cannot be generated thermally, thereby overcoming large activation barriers in a short period of time, and allowing reactions otherwise inaccessible by thermal processes. Adding optical irradiation to hydrothermal technique increases growth rates by an order of magnitude for ceramics and three orders of magnitude for metals (Tamir and Zahavi 1985; Itoh et al. 1991). Enhancement of the reaction rate can be attributed to temperature, diffusional enhancement due to light induced thermal gradients that micro-stir the solution and/or photo-chemistry. Examples of ceramics synthesized by the hydrothermal–photochemical include thin films of Ni-, Zn-, Co- and Mn-ferrites, Tl_2O_3 and Fe_3O_4.

13.7 Merits of Hydrothermal Methods

Like any other conventional synthesis techniques, the hydrothermal technique also has its merits and demerits. They are listed as under:

- The hydrothermal synthesis is particularly suitable for the growth of large good-quality crystals of desirable composition, size, shape and crystal structure. Crystals of controlled shape, size and composition are essential as they strongly influence magnetic, electric, optic and other properties of the sample. This method is especially of great importance to those industries (like pharmaceuticals, paint industry, etc.) whose starting material is powder in raw form because hydrothermal synthesis yields products with controlled size morphology.
- Under hydrothermal conditions, since the properties of water such as vapour pressure, density, surface tension, viscosity, and ionic product can be greatly changed, it can decrease the reaction temperature of systems; prepare crystalline products with narrow size distribution, high purity and low aggregation.
- Hydrothermal synthesis is very useful technique when it comes to defect (crystal defects: stress defects, Schottky defects) minimization. For example, tungstates of Ca, Ba and Sr synthesized at room temperature by a hydrothermal method do not contain Schottky defects usually present in similar materials prepared by solid-state reaction technique at high temperatures, which results in improved luminescent properties. Added advantage of hydrothermal technique is proper stoichiometry due to the non-attainment of volatilization temperature and microcrack-free samples, as phase transition is not involved at low temperatures.
- With this method, it is possible to grow crystals of compounds with high melting points at lower temperatures.
- Materials, which have a high vapour pressure near their melting points, can be grown by the hydrothermal method.
- Compounds with elements in oxidation states that are difficult to obtain (like compounds of transition metal) can be obtained inside a pressurized vessel by the hydrothermal method.
- Low temperature phases viz. quartz, berlinite and others can be easily achieved by hydrothermal technique.
- Under hydrothermal conditions as reaction takes place in a closed system, studies on phase equilibrium and the influence of temperature and pressure on phase boundary can be carried out.
- From the environmental perspective, hydrothermal methods are more environmentally benign than many other methods, thereby abiding by the rules imposed by European Union.
- A major advantage of hydrothermal synthesis is that this method can be hybridized with other processes like microwave, electrochemistry, ultrasound, mechano-chemistry, optical radiation and hot-pressing to gain advantages such as enhancement of reaction kinetics and increase ability to make new materials.

13.8 Demerits of Hydrothermal Methods

The hydrothermal technique is challenged at two major fronts:

- The need of expensive autoclaves.
- As reaction takes place in a closed system, it is impossible to observe and monitor the growth process of materials.

13.9 Comparison of Ba(Fe$_{1/2}$Nb$_{1/2}$)O$_3$ Synthesized by Solid-State Reaction Method, Hydrothermal Technique and Mechanical Alloying

With increasing working population, rising household incomes and increasing disposable income, the global consumer electronics market has thrived imposing high demand of microelectronic, opto-electric and magneto-electric devices. Most of these devices employ ceramics, which are usually lead based, and have a short service life, often ending up in a landfill within a few months or years. Attempts to recycle the lead in these products have not been successful. The disposed lead in the environment has subsequent effects on the ecosystem and needs to be addressed. A large body of work has been reported in the last decade on the development of lead-free piezoceramics in the quest to replace lead-based ceramics (Cross 2004; Rödel et al. 2009; Takenaka and Nagata 2005; Shrout and Zang 2007; Lau et al. 2008). It is observed from the literature that sodium potassium niobate (Saito et al. 2004; Fuyuno 2005; Kimura and Ando 1999; Wolny 2004), bismuth sodium titanate (Prasad et al. 2007; Sakata and Masuda 1974; Dorcet et al. 2008; Trolliard and Dorcet 2008; Nagata et al. 2004) and related materials are potential lead-free materials.

In recent years, complex perovskites with nominal chemical formula A(Fe$_{1/2}$B$_{1/2}$)O$_3$ (A: Ba, Sr and Ca; B: Nb, Ta and Sb) have attracted much attention because of their giant dielectric (10^3–10^5) response over a wide temperature and frequency interval (Wang et al. 2007a, b; Raevski et al. 2003; Chung et al. 2004; Homes et al. 2001; Bhagat and Prasad 2010; Roy et al. 2017). Among these perovskites barium iron niobate, Ba(Fe$_{1/2}$Nb$_{1/2}$)O$_3$ (abbreviated hereafter as BFN) and barium iron tantalate, Ba(Fe$_{1/2}$Ta$_{1/2}$)O$_3$ (abbreviated hereafter as BFT) bulk and nano have been synthesized by different researchers by variety of techniques, like solid-state reaction method, hydrothermal method and mechanical alloying. Some of the methods are biased to performance and neutral to the environment. The hydrothermal technique is one, which is in line with the development of ecologically benign, "green" compounds.

Reaney et al. (1994) first prepared BFN by solid-state reaction method. Saha and Sinha (2002), Raevski et al. (2003), Chung et al. (2004), Eitssayeam et al. (2006), Intatha et al. (2007, 2009) and Wang et al. (2007a, b) further investigated BFN but all synthesized it by high-temperature reaction method. Charoenthai et al. (2008) for the first time synthesized BFN powders by microwave synthesis. Microwave-assisted sol-gel synthesis of BFN ceramics for MLC (Multilayer Capacitor) applications was carried out by Sonia and Kumar (2017). Köferstein and Ebbinghaus (2017) prepared BFN by a low temperature aqueous synthesis. In a very recent work BFN ceramic powder with average size 12–20 nm were synthesized mechanochemically by Bochenek et al. (2018). A relatively new approach for obtaining nanosized particles of $Ba(Fe_{1/2}Ta_{1/2})O_3$ and $Ba_{0.06}(Na_{1/2}Bi_{1/2})_{0.94}TiO_3$ was reported by Mahto et al. (2016, 2018) which can be epitomized for BFN nanoparticles too. The advantage behind this method is the surety of phase formation. In this approach, the ceramics are first formed by solid-state reaction method ensuring proper phase and then milled to nanosize. As phase, formation takes through solid-state reaction method so reduction to nanoceramics involves less milling hour, thereby reducing energy consumption and the possibility of contamination.

Here we provide a comparative study of BFN prepared by solid-state reaction method, hydrothermal method and mechanochemical method. In a particular work polycrystalline BFN was prepared by Bhagat and Prasad by solid-state reaction method (Bhagat and Prasad 2010), and BFN nano powders were prepared by a low temperature aqueous synthesis by Köferstein and Ebbinghaus (2017), whereas BFN ceramic powder were synthesized mechanochemically by Bochenek et al. (2018). An attempt has been made to show the experimental parameters and involved steps by a flowchart (Fig. 13.10) in the above-mentioned synthesis techniques. Figure 13.11 shows the diffraction peaks and the SEM images for bulk and nano BFN ceramics.

BFN samples were prepared by Köferstein and Ebbinghaus (2017) by reaction of Ba^{2+}, Fe^{3+} and Nb^{5+} reactants in boiling NaOH solutions with different concentrations (2, 6, 8, 10 and 14 mol/L of NaOH). Detailed characterization was carried on the

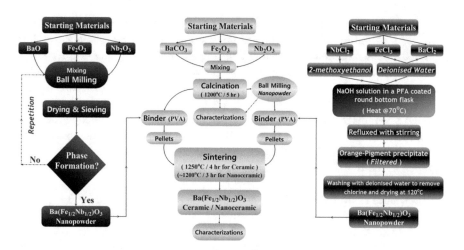

Fig. 13.10 Comprehensive synthesis mechanism of $Ba(Fe_{1/2}Nb_{1/2})O_3$ by different techniques

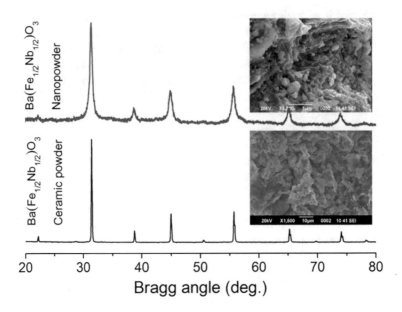

Fig. 13.11 Diffraction peaks and the SEM images for bulk and nano BFN ceramics

as-prepared powder BFN-8M sintered at 1000 °C, as below this concentration impurity % was more and above this concentration formation of pure phase took place but with increase in particle size. The average crystallite size was calculated to be 15 nm and the root-mean-square strain was found to be 1.7×10^{-3}. TEM images showed spherical particles in the range between 9 and 25 nm. The dielectric constant and loss tangent at room temperature at 1 kHz frequency for the BFN-8M sintered at 1200 °C is reported to be 5.8×10^3 and 0.5 (approximately) with activation energy ≈ 5 eV.

Bochenek et al. (2018) prepared BFN nanoceramics by high-energy ball milling. After 20 h of milling of the mixture of the starting powders pure BFN perovskite phase has been recorded. As the flowchart depicts BaO being one of the starting powder, no clacination step was required for phase formation. The average crystallite size was found to be nearly 16 nm (approximately) for the the 20 h milled BFN sample. The average grain size for the 20 h milled sample was found to be 12.8 μm through SEM images. The final parameters of the obtained ceramics suggested that the optimal high-energy milling time is 60 h. The dielectric constant and loss tangent at room temperature at 1 kHz frequency for the 60 h milled BFN nanoceramics is reported to be 6×10^4 and 0.5. The phase transition temperature at 1 kHz was reported to be 254 °C.

The values of average crystallite size and lattice strain, reported by Bhagat and Prasad (2010), respectively, are 84 nm and 0.0008. The low value of lattice strain in solid-state reaction method might be due to the fact that the synthesis procedure does not impose much constraint in the formation of compound as is generally found in case of extensive ball milling technique. Grains of unequal sizes (2–5 μm) were found to be distributed throughout the sample. Dielectric constant near to room temperature (20 °C) was found to be 3699. The value of activation energy was estimated to be 0.327 eV at 1 kHz.

Compared with BFN ceramics obtained by the mixed-oxide method, the soft-chemistry synthesis and high-energy ball milling lead to ceramics: with up to one order of magnitude, higher permittivity values exhibit lower temperature dependency and higher activation energy. The activation energy is higher in chemical route, which reflects the fact that the crystallite growth process starts only at high temperatures in these processes.

13.10   Conclusion

We started with the sentence "Nanoparticles: Last in the line above the atom, the most difficult to find, and the hardest to learn from". Indeed, a true nanoscale control of matter is more difficult than its conception. Then also our understanding of the laws of nature at nanoscale using quantum physics is still impressive. Keeping in mind the generations to come, hydrothermal synthesis is a suitable method for preparing nanoparticles which find its application in all streams of life, from healthcare to manufacturing, space travel to improving our environment. The hydrothermal technique can be easily hybridized with other synthesis techniques to reduce the reaction time and efforts, yield products having desired properties, enhance the quality of product, etc. In the present study a technologically important material, BFN nanopowder grown hydrothermally, was considered to establish these facts.

References

Abdel-Aala EA, El-Midanya AA, El-Shall E (2008) Mechanochemical–hydrothermal preparation of nano-crystallite hydroxyapatite using statistical design. Mater Chem Phys 112:202–207

Allen ET, Crenshaw JL, Johnston J (1912) The mineral sulphides of iron; with crystallographic study by ES Larsen. Am J Sci 195:169–236

Barber DJ, Freestone IC (1990) An investigation of the origin of the colour of Lycurgus cup by analytical transmission electron microscopy. Archaeometry 32:33–45

Bard AJ, Faulkner LR, Leddy LR, Zoski CG (1980) Electrochemical methods-fundamentals and applications, vol 2. Wiley, New York

Barrer RM (1948) Syntheses and reactions of mordenite. J Chem Soc 435:2158–2163

Basca R, Ravindranathan P, Dougherty JP (1992) Electrochemical, hydrothermal, and electrochemical-hydrothermal synthesis of barium titanate thin films on titanium substrates. J Mater Res 7:423–428

Bera D, Qian L, Tseng TK, Holloway PH (2010) Quantum dots and their multimodal applications: a review. Materials 3:2260–2345

Bhagat S, Prasad K (2010) Structural and impedance spectroscopy analysis of $Ba(Fe_{1/2}Nb_{1/2})O_3$ ceramic. Phys Status Solidi A 207:1232–1239

Bimberg D, Grundmann M, Ledentsov NN (1999) Quantum dot heterostructures. Wiley, Chichester

Bochenek D, Niemiec P, Adamczyk M, Szafraniak-Wiza I (2018) Physical properties of lead-free $BaFe_{1/2}Nb_{1/2}O_3$ ceramics obtained from mechanochemically synthesized powders. J Mater Sci:1–12. https://doi.org/10.1007/s10853-018-2254-z

Bondioli F, Ferrari AM, Lusvarghi L, Manfredini T, Nannarone S, Pasqualia L, Selvaggi G (2005) Synthesis and characterization of praseodymium-doped ceria powders by a microwave-assisted hydrothermal (MH) route. J Mater Chem 15:1061–1066

Boysen E, Boysen N (2011) Nanotechnology for dummies, 2nd edn. Wiley, Chichester

Brust M, Walker M, Bethell D, Schiffrin DJ, Whyman R (1994) Synthesis of thiol-derivatised gold nanoparticles in a two-phase liquid–liquid system. J Chem Soc Chem Commun 7:801–802

Bunde A, Havlin S (1996) Fractals and disordered systems. Berlin, Springer

Bunsen R (1848) Bemerkungen Zu Cinigen Einwürten Gegen Mehrere Ansichtenüber die Chemisch-geologischon Erscheinungen in Island. Eur J Org Chem 65:70–85

Byrappa K (1992) Hydrothermal growth of crystals. Prog Cryst Growth Charact Mater 21:R7–R10

Byrappa K, Adschiri T (2007) Hydrothermal technology for nanotechnology. Prog Cryst Growth Charact Mater 53:117–166

Byrappa K, Yoshimura M (2001) Handbook of hydrothermal technology. Norwich, William Andrew

Cao SW, Zhu YJ, Cui JB (2010) Iron hydroxyl phosphate microspheres: microwave-solvothermal ionic liquid synthesis, morphology control, and photoluminescent properties. J Solid State Chem 183:1704–1709

Charoenthai N, Traiphol R, Rujijanagul G (2008) Microwave synthesis of barium iron niobate and dielectric properties. Mater Lett 62:4446–4448

Chung CY, Chang YH, Chen GJ (2004) Effects of lanthanum doping on the dielectric properties of $Ba(Fe_{0.5}Nb_{0.5})O_3$ ceramic. J Appl Phys 96:6624–6628

Clauss FJ (1969) Engineer's guide to high-temperature materials. Addison-Wesley, Reading, 401P

Colletti LP, Flowers BH, Stickney JL (1998) Formation of thin films of CdTe, CdSe, and CdS by electrochemical atomic layer epitaxy. J Electrochem Soc 145:1442–1449

Cross LE (2004) Materials science: lead free at last. Nature 432:24–25

Daurree M (1857) Sur le Metamorphisme et Recherches Experimentales sur Quelques-uns. Ann Mines 12:289–326

Dong WS, Li MY, Liu C, Lin F, Liu Z (2008) Novel ionic liquid assisted synthesis of SnO_2 microspheres. J Colloid Interface Sci 319:115–122

Dorcet V, Trolliard G, Boullay P (2008) Reinvestigation of phase transitions in $Na_{0.5}Bi_{0.5}TiO_3$ by TEM. Part I: first order rhombohedral to orthorhombic phase transition. Chem Mater 20:5061–5073

Drexler KE (1986) Engines of creation: the coming era of nanotechnology. Anchor Books, Doubleday

Drexler KE (1992) Nanosystems: molecular machinery, manufacturing, and computation. Wiley, Chichester

Dyer A (1988) An introduction to zeolite molecular sieves. Wiley, New York

Eitssayeam S, Intatha U, Pengpat K, Tunkasiri T (2006) Preparation and characterization of barium iron niobate $(BaFe_{0.5}Nb_{0.5}O_3)$ ceramics. Curr Appl Phys 6:316–318

Faraday M (1857) The Bakerian Lecture: experimental relations of gold (and other metals) to light. Philos Trans R Soc Lond 147:145–181

Fredrickx P, Schryvers D, Janssens K (2002) Nanoscale morphology of a piece of ruby red knuckel glass. Phys Chem Glasses 43:176–183

Fuentes S, Zarate RA, Chavez E, Munoz P, Dıaz-Droguett D, Leyton P (2010) Preparation of $SrTiO_3$ nanomaterial by a sol–gel-hydrothermal method. J Mater Sci 45:1448–1452

Fuyuno I (2005) Toyota's production line leads from lab to road. Nature 435:1026–1027

Gong ZL, Li YX, He GN, Li J, Yang Y (2008) Nanostructured Li_2FeSiO_4 electrode material synthesized through hydrothermal-assisted sol–gel. Process Electrochem Solid State Lett 11:A60–A63

Hawkins DB, Roy R (1962) Electrolytic synthesis of kaolinite under hydrothermal conditions. J Am Ceram Soc 45:507–508

Homes CC, Vogt T, Shapiro SM, Wakimoto S, Ramirez AP (2001) Optical response of high-dielectric-constant perovskite-related oxide. Science 293:673–676

Hosoi K, Korenaga T, Hashida T, Takahashi H, Yamasaki N (1998) New synthesis technique for making hydroxyapatite ceramics using hydrothermal hot-pressing. Rev High Press Sci Technol 7:1405–1407

Ijima S, Ichihashi T (1993) Single shell carbon nanotubes of 1-nm diameter. Nature 363:603–605

Intatha U, Eitssayeam S, Pengpat K, MacKenzie KJ, Tunkasiri T (2007) Dielectric properties of low temperature sintered LiF doped $BaFe_{0.5}Nb_{0.5}O_3$. Mater Lett 61:196–200

Intatha U, Eitssayeam S, Tunkasiri T (2009) Giant dielectric behavior of $BaFe_{0.5}Nb_{0.5}O_3$ perovskite ceramic. Condens Matter Theor 23:429–435

Ishikawa M, Kadota Y, Takiguchi N, Hosaka H, Moritay T (2008) Synthesis of nondoped potassium niobate ceramics by ultrasonic assisted hydrothermal method. Jpn J Appl Phys 47:7673–7677

Itoh T, Hori S, Abe M, Tamaura Y (1991) Light-enhanced ferrite plating of $Fe_{3-x}M_xO_4$ (M = Ni, Zn, Co, and Mn) films in an aqueous solution. J Appl Phys 69:5911–5914

Jacobs H, Schmidt D (1982) High-pressure ammonolysis in solid-state chemistry. Curr Top Mater Sci 8:387–427

Jha AK, Prasad K (2010) Understanding biosynthesis of metallic/oxide nanoparticles: a biochemical perspective. In: Ashok Kumar S, Thiagarajan S, Wang S-F (eds) Biocompatible nanomaterials synthesis, characterization and applications. NOVA Science, New Yorks

Juza R, Jacobs H (1966) Ammonothermal synthesis of magnesium and beryllium amides. Angew Chem Int Ed 5:247–247

Juza R, Jacobs H, Gerke H (1966) Ammonothermal synthese von Metallamiden und Metallnitriden. Ber Bunsen Phys Chem 70:1103–1105

Kil HS, Jung YJ, Moon JI, Song JH, Lim DY, Cho SB (2015) Glycothermal synthesis and photocatalytic properties of highly crystallized anatase TiO_2 nanoparticles. J Nanosci Nanotechnol 15:6193–6200

Kimura M, Ando A (1999) Piezoelectric Ceramic Composition U.S. Patent, Patent No. 6083415, Murata Manufacturing, Japan

Köferstein R, Ebbinghaus SG (2017) Investigations of $BaFe_{0.5}Nb_{0.5}O_3$ nano powders prepared by a low temperature aqueous synthesis and resulting ceramics. J Eur Ceram Soc 37:1509–1516

Komarneni S, Li QH, Stefanson KM, Roy R (1993) Microwave-hydrothermal processing for synthesis of electroceramic powders. J Mater Res 8:3176–3183

Kosova NV, Khabibullin AK, Boldyrev VV (1997) Hydrothermal reactions under mechanochemical treating. Solid State Ion 101:53–58

Kroto HW, Heath JR, O'Brien SC, Curl RF, Smalley REC (1985) C_{60} Buckminsterfullerene. Nature 318:162–163

Kuno M (2012) Introductory nanoscience, physical and chemical concepts. Garland Science, Taylor and Francis, Boca Raton

Lau ST, Cheng CH, Choy SH, Lin DM, Kwok KW, Chan HLW (2008) Lead-free ceramics for pyroelectric applications. J Appl Phys 103:104105–104104

Laudise RA, Nielsen JW (1961) Hydrothermal crystals growth. Solid State Phys 12:149–222

Laudise RA, Parker R (1970) The growth of single crystals. Prentice-Hall, Englewood Cliffs

Litvin BN, Tules DA (1973) Apparatus for hydrothermal synthesis and growth of monocrystals. In: Lobachev AN (ed) Crystallization processes under hydrothermal conditions. Studies in Soviet Science. Consultant Bureau, New York, p 139

Lobachev AN (ed) (1973) Crystallization processes under hydrothermal conditions. Consultants Bureau, New York, pp 1–255

Mahto UK, Roy SK, Chaudhuri S, Prasad K (2016) Effect of milling on the electrical properties of $Ba(Fe_{1/2}Ta_{1/2})O_3$ ceramic. Adv Mat Res 5:181–192

Mahto UK, Roy SK, Prasad K (2018) High energy milled $Ba_{0.06}Na_{0.47}Bi_{0.47}TiO_3$ ceramic: structural and electrical properties. IEEE Trans Dielectr Insul 25:174–180

Mandelbrot BB (1983) The fractal geometry of nature. W.H. Freeman, New York

Meng LM, Wang B, Ma MG, Lin KL (2016) The progress of microwave-assisted hydrothermal method in the synthesis of functional nanomaterials. Mater Today Chem 1:63–83

Morey GW, Niggli P (1913) The hydrothermal formation of silicates, a review. J Am Chem Soc 35:1086–1130

Nagata H, Shinya T, Hiruma Y, Takenaka T, Sakaguchi I, Haneda H (2004) Developments in dielectric materials and electronic devices. Ceram Trans 167:213–221

Newalkar BL, Komarneni S (2002) Simplified synthesis of micropore-free mesoporous silica, SBA-15, under microwave-hydrothermal conditions. Chem Commun 16:1774–1775

Peters D (1996) Ultrasound in materials chemistry. J Mater Chem 6:1605–1618

Prasad K, Kumari K, Lily, Chandra KP, Yadav KL, Sen S (2007) Electrical conduction in $(Na1/2Bi1/2)TiO3$ ceramic: impedance spectroscopy analysis. Adv Appl Ceram 106:241–246

Rabenau A (1985) The role of hydrothermal synthesis in preparative chemistry. Angew Chem Int 24:1026–1040

Raevski IP, Prosandeev SA, Bogatin AS, Malitskaya MA, Jastrabik L (2003) High dielectric permittivity in $AFe_{1/2}B_{1/2}O_3$ nonferroelectric perovskite ceramics (A= Ba, Sr, Ca; B= Nb, Ta, Sb). J Appl Phys 93:4130–4136

Rao DS, Muraleedharan K, Dey GK, Halder SK, Bhagavannarayan G, Banerji P, Bose DN (1999) Transmission electron microscopy and X-ray diffraction studies of quantum wells. Bull Mater Sci 22:947–951

Rao CNR, Müller A, Cheetham AK (eds) (2006) The chemistry of nanomaterials: synthesis, properties and applications. Wiley-VCH, New York

Reaney IM, Petzelt J, Voitsekhovskii VV, Chu F, Setter N (1994) B-site order and infrared reflectivity in A (B′B′) O$_3$ Complex perovskite ceramics. J Appl Phys 76:2086–2092

Richter TMM, Niewa R (2014) Chemistry of ammonothermal synthesis. Inorganics 2:29–78

Riman RE, Suchanek WL, Byrappa K, Chen CW, Shuk P, Oakes CS (2002) Solution synthesis of hydroxyapatite designer particulates. Solid State Ion 151:393–402

Rödel J, Jo W, Seifert KTP, Anton EM, Granzow T (2009) Perspective on the development of lead-free piezoceramics. J Am Ceram Soc 92:1153–1177

Rossetti R, Hull R, Gibson JM, Brus LE (1985) Excited electronic states and optical spectra of ZnS and CdS crystallites in the $\cong$15 to 50 A size range: evolution from molecular to bulk semiconductor properties. J Chem Phys 82:552–559

Roy R (1994) Acceleration the kinetics of low-temperature inorganic syntheses. J Solid State Chem 111:11–17

Roy DM, Roy R (1955) Synthesis and stability of minerals in the system $MgO-Al_2O_3-SiO_2-H_2O$. Am Min 40:147–178

Roy SK, Singh SN, Mukherjee SK, Prasad K (2017) $Ba_{0.06}(Na_{1/2}Bi_{1/2})_{0.94}TiO_3-Ba(Fe_{1/2}Ta_{1/2})O_3$: giant permittivity lead-free ceramics. J Mater Sci Mater Electron 28:4763–4771

Saha S, Sinha TP (2002) Structural and dielectric studies of $BaFe_{0.5}Nb_{0.5}O_3$. J Phys Condens Matter 14:249–258

Saito Y, Takao H, Tani T, Nonoyama T, Takatori K, Homma T, Nagaya T, Nakamura M (2004) Lead-free piezoceramics. Nature 432:84–87

Sakata K, Masuda Y (1974) Ferroelectric and antiferroelectric properties of $(Na_{1/2} Bi_{1/2})TiO_3-SrTiO_3$ solid-solution ceramics. Ferroelectrics 7:347–349

Schafhautl CE (1845) Gelehrte Anzeigen. Akad Wiss Münchem 20:578

Schmid G (2005) Nanoparticles. Wiley VCH, New York

Schmidt S, Kubaski ET, Volanti DP, Sequinel T, Bezzon VD, Tabcherani SM (2018) Synthesis of acicular α-Bi_2O_3 microcrystals by microwave-assisted hydrothermal method. Particulate Sci Technol:1–5. https://doi.org/10.1080/02726351.2018.1457108

Shrout TR, Zang SJ (2007) Lead-free piezoelectric ceramics: alternatives for PZT? J Electroceram 19:111–124

Shuk P, Suchanek WL, Hao T, Gulliver E, Riman RE (2001) Mechanochemical-hydrothermal preparation of crystalline hydroxyapatite powders at room temperature. J Mater Res 16:1231–1234

Sonia CM, Kumar P (2017) Microwave assisted sol-gel synthesis of high dielectric constant CCTO and BFN ceramics for MLC applications. Proc Appl Ceram 11:154–159

Srivastava DN, Perkas N, Gedanken A, Felner I (2002) Sonochemical synthesis of mesoporous iron oxide and accounts of its magnetic and catalytic properties. J Phys Chem B 106:1878–1883

Suchanek WL, Yoshimura M (1998) Preparation of strontium titanate thin films by the hydrothermal-electrochemical method in a solution flow system. J Am Ceram Soc 81:2864–2868

Sun S, Murray CB (1999) Synthesis of monodisperse cobalt nanocrystals and their assembly into magnetic superlattices. J Appl Phys 85:4325–4330

Tadjarodi A, Salehi M, Imani M, Ebrahimi S, Pardehkhorram R (2014) Glycine assisted synthesis of $ZnFe_2O_4$ nanoparticles by one pot microwave heating route and organic pollutant adsorption for water treatment, The 18th international electronic conference on synthetic organic chemistry, Multidisciplinary Digital Publishing Institute

Takenaka T, Nagata H (2005) Current status and prospects of lead-free piezoelectric ceramics. J Eur Ceram Soc 25:2693–2700

Tamir S, Zahavi J (1985) Laser-induced gold deposition on a silicon substrate. J Vac Sci Technol A 3:2312–2315

Trolliard G, Dorcet V (2008) Reinvestigation of phase transitions in $Na_{0.5}Bi_{0.5}TiO_3$ by TEM. part II: second order orthorhombic to tetragonal phase transition. Chem Mater 20:5074–5082

Turkevich J, Stevenson PL, Hillier J (1951) A study in the nucleation and growth processes in the synthesis of colloidal gold. Discuss Faraday Soc 11:55–75

Ueno S, Nakashima K, Sakamoto Y, Wada S (2015) Synthesis of silver-strontium titanate hybrid nanoparticles by sol-gel-hydrothermal method. Nanomaterials 5:386–397

Valencia Hurtado SH, Marín Sepúlveda JM, Restrepo Vásquez GM (2010) Study of the bandgap of synthesized titanium dioxide nanoparticules using the sol-gel method and a hydrothermal treatment. Open Mater Sci J 4:9–14

Villegas I, Stickney JL (1992) Preliminary studies of GaAs deposition on Au (100), (110), and (111) surfaces by electrochemical atomic layer epitaxy. J Electrochem Soc 139:686–694

Wang AL (1998) Structural analysis of self-assembling nanocrystal superlattices. Adv Mater 10:13–30

Wang Z, Chen XM, Ni L, Liu XQ (2007a) Dielectric abnormities of complex perovskite $Ba(Fe_{1/2}Nb_{1/2})O_3$ ceramics over broad temperature and frequency range. Appl Phys Lett 90:022904–022903

Wang Z, Chen XM, Ni L, Liu YY, Liu XQ (2007b) Dielectric relaxations in $Ba(Fe_{1/2}Ta_{1/2})O_3$ giant dielectric constant ceramics. Appl Phys Lett 90:102905–102903

Wöhler F (1848) Recrystallization of apophyllite. Ann Chem Pharm 65:80–84

Wolny WW (2004) European approach to development of new environmentally sustainable electroceramics. Ceram Int 30:1079–1083

Xu XC, Bao YN, Song CS, Yang WS, Liu J, Lin LW (2004) Microwave-assisted hydrothermal synthesis of hydroxy-sodalite zeolite membrane. Microporous Mater 75:173–181

Yamakov V, Wolf D, Phillpot S, Mukherjee A, Gleiter H (2004) Deformation-mechanism map for nanocrystalline metals by molecular- dynamics simulation. Nat Mater 3:43–47

Yin JS, Wang ZL (1997) Ordered self-assembling of tetrahedral oxide nanocrystals. Phys Rev Lett 79:2570–2573

Yoshimura M, Suchanek W (1997) In situ fabrication of morphology-controlled advanced ceramic materials by soft solution processing. Solid State Ion 98:197–208

Yoshimura M, Suda H (1994) Hydrothermal processing of hydroxyapatite: past, present, and future. In: Brown PW, Constanz B (eds) Hydroxyapatite and related materials. CRC Press, Boca Raton, pp 45–72

Yoshimura M, Suchanek WL, Watanabe T, Sakurai B (1999) In situ fabrication of $SrTiO_3$-$BaTiO_3$ layered thin films by hydrothermal-electrochemical technique. J Eur Ceram Soc 19:1353–1359

Zhu Z, Yang D, Liu H (2011) Microwave-assisted hydrothermal synthesis of ZnO rod-assembled microspheres and their photocatalytic performances. Adv Powder Technol 22:493–497

Zory PS Jr (ed) (1993) Quantum well lasers. Academic, San Diego

Chapter 14
Hydrothermal Synthesis of Hybrid Nanoparticles for Future Directions of Renewal Energy Applications

G. P. Singh, Neha Singh, Ratan Kumar Dey, and Kamal Prasad

14.1 Introduction

The term "hydrothermal" was first coined by a geologist Sir Roderick Murchison (1792–1781) to illustrate the changes observed in the earth's crust by the formation of various rocks and minerals due to action of water at high temperature and pressure. The formation of beryl crystal in single crystal structure is the pertinent example of nature using hydrothermal origin whereas as the quartz is the example of manmade crystal using this technique. Upon the observation of great structural control and manipulation, this process was widely introduced in nineteenth century for the research purpose. R.W. Bunsen in 1839, first time used this process to prepare Barium (Ba) and Strontium carbonate at temperatures above 200 °C and pressures above 100 bars. Thereafter, E. Schafhaut in 1845 grew the quartz crystals of silicic acid in small dimension using autoclave hydrothermal decomposition (Lalena et al. 2008). This process was further comprehensively verified (Roy 1994; Roy and Tuttle 1956) on clay mineral and synthesis of zeolite families of various sizes ranging between 7 and 14 Å for oxide phase transformation study and creation of various vacancy disordering. Afterwards, this technique has been considered as one of the most versatile techniques which can produce a variety of fine particles of different materials in various shapes and structures.

In early 1990s, the hydrothermal process was extensively adopted by the researcher for processing of fine to ultra-fine particles with a controlled size and

G. P. Singh (✉)
Centre for Nanotechnology, Central University of Jharkhand, Ranchi, Jharkhand, India

N. Singh · R. K. Dey
Centre for Applied Chemistry, Central University of Jharkhand, Ranchi, Jharkhand, India

K. Prasad
Department of Physics, Tilka Manjhi Bhagalpur University, Bhagalpur, Bihar, India

© Springer Nature Switzerland AG 2018
R. Prasad et al. (eds.), *Exploring the Realms of Nature for Nanosynthesis*,
Nanotechnology in the Life Sciences, https://doi.org/10.1007/978-3-319-99570-0_14

morphology. The easy control over the sizes, shapes, and morphologies dragged it into the era of new technology, namely "Nanotechnology" as the major materials processing tools. The control over the particles manipulation from sub-micrometer to nanosize allows better understanding over the relevant technological applications. The phenomenological new terminology "Nanotechnology" was introduced by the Eric Drexler (1986) in his book Engines of Creation, where the emphasis was given on the manipulation of individual atoms and molecules to build structures to complex atomic specifications. Perhaps, the concept of nanotechnology extracted from the landmark lecture on *"There's Plenty of Room at the Bottom"* delivered by the Richard Feynman (1959), physicist.

The research interest in materials having small size close to nanosize (1–100 nm) in one of the spatial dimensions is grown due to the fact that small-sized system exhibits completely new properties from their bulk counterparts, and these properties vary with size and/or shape of the particles. The properties include optical, magnetic, and electronic which may be important applications in material technologies like microelectronics, catalytic systems, hydrogen storage, ferrofluids, chemical nanosensors, etc. Nanomaterials have been extensively investigated during the last decade due to their wide variety of applications. It is observed that field of nanomaterial synthesis is very dynamic. Many process such as gas condensation, chemical vapor synthesis, mechanical attrition, chemical precipitation, sol-gel technique, electrodeposition, and some other methods widely used are molecular beam epitaxy, ionized cluster beam, liquid metal ion source, consolidation, sputtering, and gas aggregation of monomers chemical precipitation in presence of capping agents, reaction in micro-emulsions, and auto-combustion are commonly used techniques for synthesis of nanophosphors. Table 14.1 shows the impact of the various processes on the morphology, quality of product, reaction control, etc., whereas Table 14.2 presents the advantages of the hydrothermal process over other processes.

Table 14.1 Comparison of various synthesis processes

S. no.	Synthesis techniques	Structural control	Morphology control	Reactivity	Particle size (nm)
1.	Solid state technique	Poor	Poor	Poor	≥ 20
2.	Co-precipitation technique	Good, high	Moderate	Good, high	≥ 10
3.	Spray technique	Excellent	Moderate	Good, high	≥ 10
4.	Emulsion synthesis	Excellent	Excellent	Good, high	≥ 10
5.	Spray pyrolysis technique	Excellent	Excellent	Good, high	≥ 10
6.	Sol-gel synthesis	Excellent	Moderate	Good, high	≥ 10
7.	Hydrothermal technique	Excellent	Good, high	Good, high	≤ 10

Table 14.2 Comparative analysis of the different synthesis process versus hydrothermal methods

S. no.	Hydrothermal process	Other process
1.	Hydrothermal process is the self-purifying process and chemicals required very low purity of acetates, isopropoxides, oxides, nitrates, and chlorides of the elements	Solid state process requires high purity chemical of the oxides, acetates, isopropoxides, and oxy-nitrates
2.	Inexpensive precursor materials are useful for this process. The product quality is better than any other processes	Other processing methods required high purity precursor chemicals to produce quality products
3.	This process works at very low temperature. In most cases, it works below 200 °C which minimizes the problems related to the stoichiometry control for the materials have volatile nature. Here, no calcination or annealing is required for the final product in various cases	Other methods produce amorphous materials as a product that needs the high temperature calcination for a couple of hours
4.	Hydrothermal process is easy and safe for samples preparation because it is an environment-friendly method compared to others. This method provides versatile options for the synthesis of the multifunctional semiconductor and ceramic materials like electronic ceramics, bioceramics, catalyst supports, membranes, and ceramics with electronic, magnetic, and optical properties	Other methods are not so versatile for the multifunctional semiconductor and ceramic materials like electronic ceramics, bioceramics, catalyst supports, membranes, and ceramics with electronic, magnetic, and optical properties
5.	Because of cost-effective and reproducible nature of production of high-quality ceramic powders on a large scale, it can be scaled up to meet the industrial requirement or demands	Low quality product restricts the scaling up of the sol-gel and co-precipitation techniques
6.	Hydrothermal process allows designing the particles in various shapes and sizes in shorter reaction times over the good control on the reaction process	Other methods such as sol-gel and co-precipitation have less control on the size and shape/morphology of the particles

Thanks are due to the advancement in the technologies which led the researchers to understand the hydrothermally grown nanosized particles, their formation process and their control in order to get the desired properties to such nanomaterials. It is therefore hydrothermal technology considered as one of the most apprised techniques in the field of nanotechnology. Presently, the research reports forecast that in coming future, the hydrothermal technology may take a key place in several interdisciplinary branches of science and technology as well as shall work as bridge among the important technologies including nanotechnology, biotechnology, sensor technology and geo-technology, and advanced materials technology. Thus, it enriches that the hydrothermal processing of advanced materials is a highly interdisciplinary technique that is commonly used by materials scientists, hydrometallurgists, physicists, engineers, chemists, ceramists, biologists, and geologists. Figure 14.1 shows various branches of science and technology which utilize the hydrothermal process as a source of materials development.

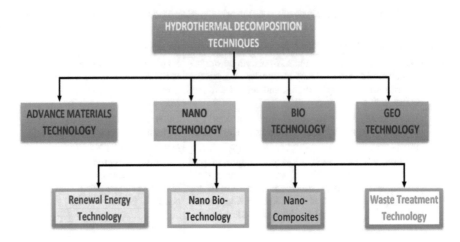

Fig. 14.1 Application of hydrothermal process in various science and technology fields

In addition to the processing of nanomaterials, it offers distinct advantages due to good control over the diffusivity in a strong solvent medium in a closed system. It is highly required for the case of nanomaterials especially of the functional materials because of the physico-chemical characteristics of nano-systems. As the size is reduced to the nanometer range, the materials exhibit peculiar and interesting mechanical and physical properties viz. improved mechanical strength, enhanced diffusivity, higher specific heat, and electrical resistivity compared to their conventional coarse grained counterparts due to a quantum size effect (Yamakov et al. 2004). With continuous growing demand for composite nanostructures, the hydrothermal technique also offers a unique option for coating of various polymeric compounds on metals, polymers, and ceramics as well as for the fabrication of powders or bulk ceramics. Therefore, it is now considered as a frontline technology for the processing of advanced materials for nanotechnology application. Nowadays, the conventional hydrothermal methods as well as its variants have emerged as a versatile synthesis option for the preparation of multifunctional ceramics materials for electronic, bio, magnetic, catalysts, optical, and other properties (Ortiz-Landeros et al. 2012).

14.2   Types of Hydrothermal Method

Unique advantage of hydrothermal synthetic process provides the numerous ways for technological hybridization in order to make new materials for high-technology applications. To this end, various upgradations in the hydrothermal process are done in combination with other advance processes. Presently, this process is combined or hybridized with the different sources of energies such as microwaves, sol-gel, electrochemistry, ultrasound, mechano-chemistry, optical radiation, and hot-pressing.

The working process of these hybrid techniques are briefly discussed in Chap. 13 for the synthesis of the advanced nanomaterials and their composites.

14.3   Hydrothermal Processing Chemistry of Advanced Materials for Nanotechnology Applications

Understanding of the chemistry of materials in hydrothermal processing is considered as the key features for the purpose of designing new materials for advance technology. The process chemistry includes the understanding of the solvent chemistry behavior such as structure at critical, dielectric constant, pH variation, viscosity, coefficient of expansion, and density of materials in hydrothermal process conditions with respect to temperature and pressure dependency. Using this process, mostly the solubility, stability, yield, dissolution, and precipitation reactions of materials are studied because the analysis of the change in intermediate kinetics of the intermediate phase formation is absent. Therefore, in 1950–60s, the synthesis of zeolites in presence of temperature and pressure storm the importance of the crystallization kinetics of the materials and the ability to empirically define the essential roles of various parameters such as temperature, pressure, precursor, and time involve in crystallization kinetics of various compounds (Byrappa and Adschiri 2007). Accurate understanding on the role of these parameters for the formation of solution species, solid phases, and the rate of their formation requires to explore the appropriate modeling approach. In this connection, the rational engineering approach is developed to understand the intermediate conditions of the process and also further improve the process development. The steps involved in this approach are (1) computing of thermodynamic equilibria as a function of chemical processing variables, (2) generation of equilibrium diagrams to access the process variable for the points of interest, (3) plan hydrothermal experiments to authenticate the computed diagrams, and (4) utilization of the processing variables for controlling the reactions and crystallization kinetics. Using such a modeling approach, theoretical stability field diagrams or yield diagrams are designed to obtain yield close to 100%. Considering the materials product is in pure phase, the yield (Y) of materials can be expressed in the form of

$$Y_i = \frac{100\left(m_i^x - m_i^y\right)}{m_i^x}\%$$

here m^x and m^y are the input and equilibrium molar concentrations of the species, respectively, for the ith atoms. As a result, this approach successfully provides the control over the phase purity, particle size, size distribution, and particle morphology by predicting the optimal synthesis conditions (Lencka and Riman 2003; Gersten 2003; Riman et al. 2002).

14.4 Instrumentation Requirement for the Hydrothermal Processing of Nanomaterials

As the reaction are being carried out at high temperature and pressure in a closed vessel in hydrothermal processing, the requirement or selection of materials to design the equipment known as autoclave, reactor, pressure vessel, or high pressure bomb is very important. Designing an ideal hydrothermal vessel is supposed to be a tough job and perhaps difficult because each experimental conditions has different objectives and tolerance limits. In most practical cases, the materials used for an ideal hydrothermal autoclave should have the following characteristics such as unresponsiveness to acids, bases and oxidizing agents, simplicity of assembly and disassembly, sufficient length for desired temperature gradient, leak-proof capabilities at operating temperature and pressure, and quality to sustain high pressure and temperature during experiments for long periods without damage. Considering these requirements, materials used for the fabrication are glass cylinder, quartz cylinder, and alloys, such as austenitic stainless steel, iron, nickel, cobalt-based super alloys, and titanium and its alloys. The commonly used reactors in the hydrothermal processing of advanced nanomaterials are general purpose autoclaves, morey type flat plate seal, stirred reactors, cold-cone seal tuttle-roy type, TZM autoclaves, batch reactors, flow reactors, microwave-hydrothermal reactors, mechano-chemical-hydrothermal, piston cylinder, opposed anvil, and opposed diamond anvil. The parameters to be considered for the selection of a suitable reactor are

1. Experimental temperature and pressure.
2. Corrosion resistance behavior for a given solvent or fluid at operating pressure-temperature.

14.5 Hydrothermal Synthesis of Advanced Nanomaterials

The nanomaterials are well recognized nowadays because of their unique properties such as mechanical, chemical, physical, thermal, electrical, optical, magnetic, and also specific surface area properties. Based on such properties, it is known by several names such as nanostructures, nano-electronics, nano-photonics, nano-biotechnology, and nano-analytics. In the last one decade, a large variety of nanomaterials and devices with new capabilities have been generated employing nanoparticles based on metals, metal oxides, and semiconductors including the II–VI and III–V compounds, silicates, sulfides, hydroxides, tungstates, titanates, carbon, zeolites, ceramics, and a variety of composites. Here, some of the forms of the nanomaterials are discussed.

14.5.1   Metal Nanoparticles

Owing to the fundamental technological properties, noble metals like Au, Ag, Pt, Co, Ni, and Fe, metal alloys such as FePt, CoPt, and multilayers of Cu/Co, Co/Pt have drawn considerable attention of the researchers to think in new ways of the utilization of these materials in different sizes and morphology (Forster and Antonietti 1998; Zhu et al. 2003; Puntes et al. 2001; Xie et al. 2005) because of the inherent properties of metal nanoparticles strongly governed by their size, shape, and structure. More often, the synthesis of these metals was done by using both the hydrothermal and hydrothermal-supercritical water techniques in order to study the crystal growth process and shape control. Zhu et al. 2003 have reported the mild hydrothermal reactions process for the synthesis of silver nanoparticles in dendrite nanostructures in presence of the anisotropic nickel nanotubes as a reducing agent. The morphologies changed from dendrite to compact crystals were observed during the progress of the reaction from non-equilibrium to quasi-equilibrium conditions. The use of PVP surfactant instead of the nickel, the nanostructures silver were swapped by bulk particles. Xie et al. (2005) and Liu et al. (2005) have reported the hydrothermal synthesis of cobalt nanorods and nanobelts with and without surfactants. Cobalt nanorods with hcp structures was obtained at 90 °C with an average particle size of 10 nm diameter and 260 nm length (Liu et al. 2005) by using micro-emulsion with hydrothermal process. Similarly, Co nanobelts via a surfactant-assisted hydrothermal reduction process at 160 °C or 20 h have been reported by Xie et al. (2005). Liu et al. (2005) have reported a complex-surfactant-assisted hydrothermal route to ferromagnetic nickel nanobelts at about 110 °C in 24 h.

Recently, the synthesis of Ag, Au, Pd, In, Pt, Si, Ge, Cu, etc. nanoparticles via supercritical conditions reactions are becoming very popular due to its fast kinetics and rapid particle production quality with the shortest dwelling time. There are several reports on the preparation of nanoparticles under supercritical conditions reactions (Reverchon and Adami 2006; Kameo et al. 2015; Shah et al. 2001; McLeod et al. 2004). Similarly, the coating of nanocrystalline films of Cu, Ni, Ag, Au, Pt, Pd, Rh, etc. on silicon wafers for microelectronics, data storage, etc. were also done (Blackburn et al. 2001). It has been further extended to numerous other materials. Accordingly, the hydrothermal-solvothermal and hydrothermal-supercritical conditions offer unique advantages over the preparation of the metal nanoparticles over other conventional methods.

14.5.2   Metal Oxide Nanoparticles

Recent demands of mono-dispersed nano-dimension metal oxide particles such as TiO_2, ZnO, Fe_2O_3, Fe_3O_4, CeO_2, ZrO_2, CuO, Al_2O_2, Dy_2O_3, In_2O_3, Co_3O_4, NiO, and its composites with graphene, CNT, or metals with control morphology drawn considerable attention towards the processing techniques especially for the case of hydrothermal process and its conditions for technological applications including catalytic, optical, drug delivery, magnetic imaging, hyperthermia, cancer therapy, photo-catalytic, luminescent, electronic, high-density information storage, etc. These applications mostly require particles of narrow size distribution with a high dispersibility in the appropriate dispersive medium depending on the technological requirement. Thus, to get the high dispersion of metal oxides mono-dispersed particles required a variety of modifications in the hydrothermal technique.

Adschiri et al. (Adschiri et al. 1992a, b, 2000, 2001; Hakuta et al. 1998a, b) described a continuous synthesis of narrow size metal oxide particles using metal nitrates and in presence of the supercritical water as the reacting medium. Here, the production of small size fine particles is because water rapidly dehydrates the metal hydroxides before the occurrence of the substantial growth of particles. The overall reactions involve a two-step process, i.e., hydrolysis of metal salts to metal oxides and its dehydration

$$M(NO_3)_2 + x.H_2O \rightarrow M(OH)_x + x.HNO_3$$

$$M(OH)_x \rightarrow MO_{x/2} + (1/2)x.H_2O.$$

In this process, rate of dehydration is controlled in such a way that the rate of reaction is hardly affected by diffusion growth process, resulting in small size particle formation. Additionally, the process also brought the negligible mass transfer because of the gas like viscosity and diffusivity of water in the critical region. Thus, the overall synthesis rate is significantly high. Moreover, if the temperature is introduced in the process, the reaction rate is also affected.

Recently, TiO_2 and its composites are extensively developed by a large number of researchers using the hydrothermal process for the energy harvesting applications because TiO_2 is non-toxic and chemically inert and produces maximum light scattering. In a most recent work, Singh et al. (2014) successfully prepared the ternary phase plasmonic photo-catalyst using hydrothermal process. The photocatalyst includes graphene-Au-TiO_2 nano-system and is used for the production of hydrogen from photo-catalytic water splitting and control of environmental pollution through dye degradation in presence of simulated solar spectrum. The X-ray diffraction patterns of these nanocomposites are shown in Fig. 14.2. It is observed that the peaks shifted towards higher angles, which imply the reduction in interplaner spacing after incorporation of Au. The Au peaks are observed in the sample containing Au contents 8.0 wt% in rGO-TiO_2. Transmission electron microscope (TEM) images of these composites are shown in Fig. 14.3.

The environmental pollution control through dye degradation process involves firstly the absorption of photons by TiO_2 that transfer electrons from the valence band of TiO_2 to the conduction band through electron-hole pair formation. The conduction band surface electron is consumed by reaction with O_2 and simultaneously the holes in the valence band react with OH or H_2O species in turn gives the hydroxyl radical. This hydroxyl radical destroys the available organic chemical contaminants

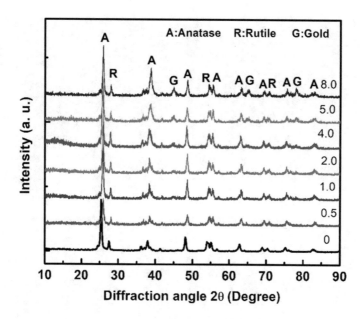

Fig. 14.2 X-ray diffraction patterns of rGO-TiO$_2$ with different Au wt%

in air, water, and soil through photo-catalytic oxidation. Further, this technological process is also used for the treatment of polluted surface and groundwater, waste water, and drinking water.

MoS$_2$/CuO heterostructural nanoflowers (Li et al. 2015a, b) and MoS$_2$/VO$_2$ hybrids (Chen et al. 2018) were developed by using two-step hydrothermal process, which significantly enhanced the photo-catalytic activity as compared to only MoS$_2$ nanoflower. Recently, bimetallic Co-Mn oxides/carbon hybrid based on CoMn$_2$O$_4$ anchored onto reduced graphene oxide and N-doped reduced graphene oxide via a hydrothermal method for the application in Zn-air batteries. Both catalysts displayed ORR half-wave potential of 110 and 60 mV lower than that of the state of art Pt/C catalyst, respectively (Gebremariam et al. 2018).

14.5.3 Carbon Nanomaterials

Presently, the development of carbon polymorphs such as graphite, diamond, amorphous carbon or diamond-like carbon, fullerenes, and carbon nanotubes has considered as a vital ingredients of the technological materials. But the phase stabilities of these polymorphs and their physico-chemical phenomena for the formation is still unknown. Hence, a large number of efforts were put to synthesize them with varied condition. Therefore, it is observed that the stabilities of graphite and diamond were mainly controlled by pressure-temperature tuning in the system

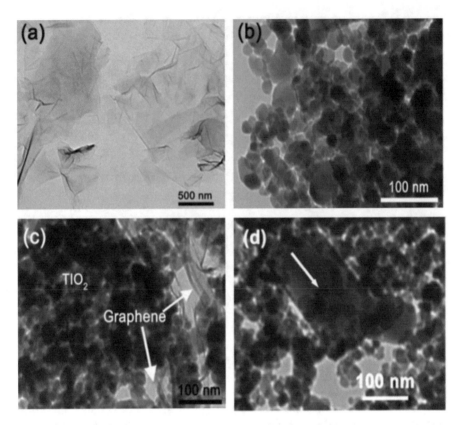

Fig. 14.3 TEM micrographs of (**a**) graphene oxide, (**b**) TiO₂, (**c**) rGO-TiO₂, and (**d**) Au-rGO-TIO₂

(Melto et al. 1974; Navon 1991; Heggerty 1986). The hydrothermal (Vasilev et al. 1968; Orlov 1973) process provoked the material researchers to explore the possibility of synthesizing them at fairly low pressure and temperature conditions. Because it is highly promising for reactions involving volatiles, as they attain the supercritical fluid state and supercritical fluids are known for their greater ability to dissolve non-volatile solids (Orlov 1973). The formation of various carbon polymorphs was reported by Gogotsi et al. (1995, 1996) and Roy et al. (1996) in hydrothermal conditions by decomposition of silicon carbide in supercritical water.

Further, Basavalingu et al. (2001a, b) have explored the possibilities of producing carbon polymorphs under hydrothermal conditions through decomposition of silicon carbide in the presence of organic compounds instead of pure water at above 700 °C and 100 MPa. Here, the decomposition of the silicon carbide occurs either to quartz or to cristobalite with high yield carbon particles. The shape of the carbon particles formed in this process are discrete or linked spherical having elongated and irregular with pore diameter of 20–30 nm. Further, according to Schmidt and Benndorf (1998), dissociation of organic molecules in presence of higher concentration of atomic hydrogen and C_1H_x radicals is assumed an ideal environment for sp^3-carbon phase stabilization in stimulated natural environments. It is also found that this stabilizer not only increased the yield of the carbon particles but also allows the control over the tailoring of the shape of the particles. Later on, the growth of the of diamond/sp^3 hybridized carbon were also successfully done without using diamond precursor seed or the metal catalysts in hydrothermal conditions (Basavalingu et al. 2006, 2007).

14.5.4 Carbon Nanotubes

The development of the materials in tube structure in nano-dimension was the beginning of the new era of the materials science for the technology point of view. Materials designed in the nanotube structure are carbon nanotubes (CNTs), barium titanate (BTNT), strontium titanate nanotubes (STNTs), titania nanotube (TNT), silicon nanotube (SNT), antimony nanotubes (SbNTs), and gallium nitride nanotube (GNT) were developed (Ijima 1991; Seo et al. 2001; Wang et al. 2003; Goldberger et al. 2003; Tsai and Teng 2004; Zhao et al. 2005; Qui et al. 2005). Among these materials, the CNTs got huge popularity within the scientific and technological research due to outstanding electrical, optical, catalytic, and mechanical properties. It is considered as the potential materials for designing MEMS and NEMS devices. The benefits of the use of CNTs for such applications are due to ease of alignment in different orientation as per the need of the devices such as field effect transistors, electron-field emitters, and many more (Singh et al. 2014; Katayama et al. 1998). The CNTs exist in two different structure, namely single wall carbon nanotubes (SWCNTs) and multiwall carbon nanotubes (MWCNTs). The structure of the carbon networks such as carbon-carbon single and double bonds available in CNTs and their conjugated structure not only provides the essence of the polymeric but also the retaining some analogous optoelectronic characteristics of other conjugated polymers. Because of owing the conjugated polymer nature, it is also considered as quantum wires (Yakobson and Smalley 1997; Ajayan and Ebbesen 1997). Further, the unusual coupling of the molecular symmetries, the CNTs became very attractive for many other potential applications such as single molecular transistors, scanning probe microscope tips, electrochemical energy storage, photo-catalyst, protein/DNA supports, filtration membranes, and artificial muscles (Frank et al. 1998; Dai et al. 1996; Che et al. 1998; Guo et al. 1998).

Therefore, the selection of the methods for the synthesis of the CNTs is an important task. The commonly known process for synthesis of the CNTs is the CVD method and pyrolysis of hydrocarbons. Both processes require metal as a catalyst. The yield is very low in this process. The observation of appearance of the natural MWCNTs with rocks of hydrothermal origin (DeVries et al. 1994) drawn possibility that the hydrothermal process may be a reproducible fabrication method of it. Gogotsi and his group (1995) developed hydrothermal CNTs using a polyethylene (PE) or ethylene glycol (EG) and water with Ni powder in a sealed gold capsule at high pressure and temperature. They found that CNTs can be synthesized by suing of any liquid, solid, or gaseous carbon source. Basavalingu et al. 2006 have further studied the transformation of SWCNTs to MWCNTs under hydrothermal conditions using SWCNTs in the Tuttle-type autoclaves filled with double distilled water as a solvent with varying temperature from 200 to 800 °C at a pressure of 100 MPa for 30 m to 48 h. Initially, the SWCNTs slowly transformed into polyhedral graphitic nanoparticles above 550 °C and, then finally into MWCNTs after 800 °C and 48 h of treatment. Thereafter, several groups (Motiei et al. 2001; Lee et al. 2004; Chang et al. 2002) have synthesized the CNTs using different types of carbon-based precursor materials such as CO_2 and supercritical toluene and appropriate catalyst like Mg, Fe, or FePt nanocrystals, under hydrothermal techniques. The hydrothermal technique may offer special advantages for the processing of materials in terms of high quality with control over the tube diameter and structures.

14.6   Applications for Renewal Energy Harvesting

Energy crisis is the worldwide main concern since fossil fuels are facing rapid depletion and its consumption contributes to the rise in the average global temperature. Among the challenges to be embedded lately with agricultural activities is to explore clean and renewable energy resources.

14.6.1   Photo-Catalytic Hydrogen Evolution from Water Splitting

Considering an ideal fuel for the future, hydrogen can be produced from clean and renewable energy sources like water. First time, the photo-electrochemical hydrogen production was reported by Fujishima and Honda (1972) and after that many report were published on the semiconductor photo-catalysis (Asahi et al. 2001; Khaselev and Turner 1998; Zou et al. 2001; Fox and Dulay 1993). Mostly emphasize the semiconductor photo-catalytic water and environmental purification (Hoffmann et al. 1995; Herrmann 1999), but very few of them are related to photo-catalytic hydrogen production (Hoffmann et al. 1995; Herrmann 1999; Wu et al. 2016; Zhu et al. 2018). Table 14.3 presents the photo-catalytic semiconductor materials used for the hydrogen production from water.

Recently, Singh et al. (2014) prepared reduced graphene oxide (rGO)-TiO_2 nanocomposite hydrothermally and its photo-catalytic properties for H_2 evolution values were demonstrated in Fig. 14.4. The H_2 values were recorded after irradiation of solar spectrum in UV-Visible region in presence of methanol as a scavenger. The TiO_2 samples produce very low H_2, i.e., 0.56 mmol (i.e., 112 μmol h^{-1}), whereas

Table 14.3 Recent photo-catalysts for water splitting H_2 production

S. no.	Photo-catalysts	H_2-evolution (μmol h^{-1} g^{-1})	References
1.	Pt, Cr, Ta loaded TiO_2	11.7	Tanigawa and Irie (2016)
2.	Cu/Ga/In/S/TiO_2	50.6	Kandiel and Takanabe (2016)
3.	1 wt% Pt/C-HS-TiO_2	5713.6	Zhu et al. (2016)
4.	Platinized sub-10 nm rutile TiO_2 (1 wt% Pt)	932	Li et al. (2015)
5.	Rh- and La-Co doped $SrTiO_3$	84	Wang et al. (2014)
6.	$Cu_{1.94S}$-Zn_xCd_1 ($0 \leq x \leq 1$)	7735	Chen et al. (2016)
7.	MoS_2/Co_2O_3/poly(heptazine imide)	0.67	Dontsova et al. (2015)
8.	Bi_4NbO_8Cl	6.25	Fujito et al. (2016)
9.	CdS nanorod/ZnS nanoparticle	239	Jiang et al. (2016)
10.	Ni/CdS/g-C_3N_4	1258.7	Yue et al. (2016)
11.	CdS/WS/graphene	1842	Xiang et al. (2016)
12.	V-doped TiO_2/RGO	160	Agegnehu et al. (2016)
13.	Pt/g-C_3N_4 conjugated polymers	1.2	Zhang et al. (2016)
14.	Au-TiO C_3N_4 Nanohybrids	647	Zhang et al. (2016)
15.	$SrTiO_3$:La,Rh/Au/$BiVO_4$:Mo	90	Wang et al. (2016)
16.	CoOx-B/TiO_2-TaON	40	Gujral et al. (2016)
17.	MoS_2/$CuInS_2$	202	Yuan et al. (2016)
18.	Copper-organic framework; H_2PtCl_6	30	Wu et al. (2016)
19.	BP/$BiVO4$	160	Zhu et al. (2018)
20.	Bi_4TaO_8X (X = Cl, Br)	0.15	Tao et al. (2017)
21.	$ZnIn_2S_4$/$In(OH)_3$	1030.1	Zhao et al. (2018)

these values are dramatically enhanced after the addition of rGO with TiO_2. Small amount of rGO, i.e., 0.5 wt% produces 1.1 mole of H_2 which is about 1.8-fold more than the TiO_2. A maximum of 1.34 mmol H_2 (i.e., 268 μmol h^{-1}), i.e., 2.4-fold was obtained in case of rGO amount of 1.0 wt% mixed with TiO_2. The enhancement in the H_2 evolution in presence of rGO is due to excellent electron accepting and transferring capability, and the capacity to restrain the recombination of electrons-holes effectively due to two-dimensional π-conjugation structure of graphene (Zhang et al. 2010; Yu et al. 2009). More than 1.0 wt% rGO reduces the photo-catalytic activity results reduction in H_2 production value.

Further, the Au was loaded on the rGO-TiO_2 nanocomposites to see the plasmonic effect on the H_2 evolution. Figure 14.5 shows the photo-catalytic activity of the Au co-catalyst loaded on 1 wt% rGO-TiO_2. The 0.5 wt% Au loading shows the H_2 evolution value of 7 mmol (i.e., 1.4 mmol h^{-1}) which is 5.2-fold higher than 1 wt% rGO-TiO_2. The maximum value of H_2 production was achieved 12 mmol (i.e., 2.4 mmol h^{-1}) in case of 2.0 wt% Au loading. More than 2.0 wt% Au contents loading gradually decreases the H_2 evolution activity.

The charge transfer mechanism is projected in Fig. 14.6 and the photo-catalysts reaction steps involved in water splitting H_2 evolution are summarized as follows:

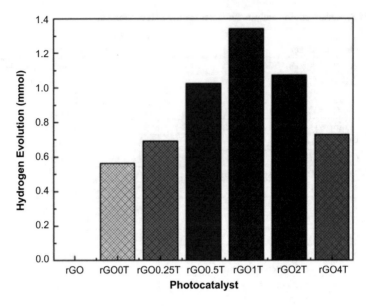

Fig. 14.4 Photo-catalytic activity of photo-catalyst for H$_2$ evolution from methanol aqueous solution under UV-Visible light irradiation (Singh et al. 2014)

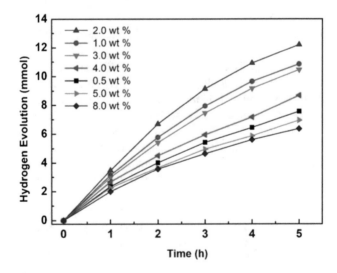

Fig. 14.5 Hydrogen evolution through plasmonic active Au loaded rGO1T photo-catalysts from methanol aqueous solution under UV-visible light (Singh et al. 2014)

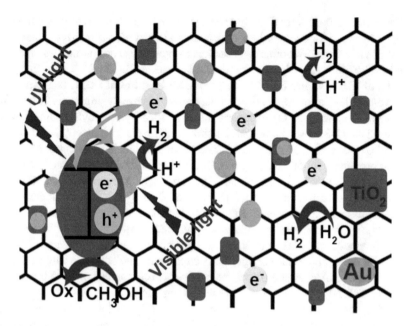

Fig. 14.6 Charge transfer process in Au/graphene/TiO₂ photo-catalyst (Singh et al. 2014)

$$\text{Reduce Graphene}(rGO)\,/\,Au\,/\,TiO_2 \xrightarrow{h\nu} rGO\,/\,Au\left(e^-\right)/\,TiO_2\left(h^+\right)$$

$$rGO\,/\,Au\left(e^-\right) + 2H^+ \rightarrow rGO\,/\,Au + H_2 \uparrow$$

$$n\,TiO_2\left(h^+\right) + CH_3OH + 6OH^- \rightarrow n\,TiO_2 + CO_2 + 5H_2O$$

14.6.2 Dye-Sensitized Solar Cells (DSSC)

Photovoltaic (PV) technology is also considered as one of the feasible green resources for electrical energy generation via solar technology. In this regard, the dye-sensitized solar cell (DSSC) is of countless importance due to several attractive features such as designing and fabrication of DSSC is inexpensive, flexible, transparent, and sensitive to low light levels. Besides its easiness to be used in larger applications, the production of the materials used for making DSSC an ideal candidate needs to explore the techniques which are able to increase the materials efficiency of solar to energy conversion. The hydrothermal techniques is also considered as one of the viable means for the production of materials in comparison to other available methods.

Wang et al. (2009) were directly grown large-scale macroporous TiO$_2$ nanowires on spiral-shaped titanium wires as photoanodes of dye-sensitized solar cells (DSSCs) via a facile hydrothermal reaction without any seeds, templates, and TiO$_2$ powder. The efficiency of 0.86% for the DSSC was obtained with a J_{sc} of 2.30 mA/cm^2, V_{oc} of 616 mV, and FF of 0.61. This MNT-based mini DSSC is a promising photovoltaic device for applications in the fields of high-integrated microelectronic equipment. Zn$_2$SnO$_4$ (ZTO) nanoparticles developed by employing same process were used as a DSSC with commercial photo sensitizers such as N719 and N3 dye molecules (Das et al. 2016; 2018a, b). The N3 dye molecules give better dye sensitization property than N719 dye molecules. The solar devices designed with ZTO-N719 showed improved photovoltaic performance compared to the ZTO-N3 devices. The overall conversion efficiency of 2.56% is achieved under 1SUN 1.5 AM illumination and sensitization for 12 h. Rajamanickam et al. (2017) implemented the hydrothermally TiO$_2$ nanorods prepared at low temperature for solar cell as a photoanode materials and found the power conversion efficiency of 5.42% along with the open-circuit voltage, short-circuit current density and fill factor of t 0.647 V, 13.64 mA/cm^2, and 61.5%, respectively. After the use of 2500 h runs of it in a solar cell, J-V characteristics were remaining constant.

Further, the perovskite-based solar cell were designed by Dunlap-Shohl et al. (2018) using delafossite oxides (CuCrO$_2$, i.e., CCO) as promising hole transport layer because of wide band gap and favorable energy band alignment relative to the perovskite absorber. Pure Phase CCO films were deposited by spin casting suspensions of hydrothermally synthesized CCO nanoparticles in the form of glass/ITO/CCO/CH$_3$NH$_3$PbI$_3$/C60/BCP/Ag as solar cell device structure. It exhibits highly stabile power conversion efficiencies exceeding 14%.

14.7 Future Scope of Hydrothermal Synthesis Process

The futuristic scope of hydrothermal synthesis process for advanced technological application depends upon its operation simplicity with nanometer resolution, control over the reaction kinetics, suitability with the theoretical approaches to the growth of materials from solutions, development of in situ mapping of techniques, in situ surface modification and attaining high dispersion of materials. It offers many advantages over conventional and non-conventional synthetic methods. All forms such as powders, fibers, films of metal, ceramics, and polymers can be synthesized by hydrothermal synthesis. Present continuous demands of this technology lead it not only within the materials growth but also it explores towards the various branches of science and technology. For instance, the hydrothermal technique is regarded as the most appropriate technique to formulate the materials for advanced technology like solar energy conversion, hydrogen generation from water, drug delivery, hyperthermia, neutron therapy, bio-imaging, fluorescent labeling, and more. The beauty of this method is to ease the control of the size and shape of the nanocrystals, and significantly reduced aggregation levels which is not possible in other processing techniques. As continuous increasing demand for composite nano-structures, it also offers a unique method for designing polymer-ceramic composites or textured ceramic-ceramic composites with anisotropic properties.

14.8 Conclusion

Hydrothermal technology for the processing of advanced materials has gained lots of advantages due to easy adaptability and being environmental friendly. The great advantages of hydrothermal technology for nanomaterials processing are the production of particles that are mono-dispersed with total control over their shape and size in addition to their chemical homogeneity with the highest dispersibility. A great variety of advanced nanomaterials whether nanoparticles, or nanocomposites covering metals, metal oxides, semiconductors including II–VI and III–V compounds, silicates, sulfides, hydroxides, tungstates, titanates, carbon, zeolites ceramics, composites, etc. have been processed using hydrothermal technology. The use of multi-energy systems like microwave-hydrothermal, electro-chemical-hydrothermal, or mechano-chemical-hydrothermal drives this technology to a new and totally unexplored avenue in present scenario. The use of capping agents, surfactants, and other organic molecules contribute greatly to the surface modification of these nanocrystals to obtain the desired physico-chemical characteristics. Hydrothermal technology has the ability to significantly accelerate the kinetics of synthesis, to model the theoretical approaches from solutions, and to develop in situ observation techniques. The combination of hydrothermal technology and nano-technology can answer most of the problems associated with advanced materials processing.

References

Adschiri T, Kanaszawa K, Arai K (1992a) Rapid and continuous hydrothermal synthesis of boehmite particles in subcritical and supercritical water. J Am Ceram Soc 75:2615–2618

Adschiri T, Kanaszawa K, Arai K (1992b) Rapid and vontinuous hydrothermal crystallization of metal oxide particles in supercritical water. J Am Ceram Soc 75:1019–1033

Adschiri T, Hakuta Y, Arai K (2000) Hydrothermal synthesis of metal oxide fine particles at supercritical conditions. Ind Eng Chem Res 39:4901–4907

Adschiri T, Hakuta Y, Sue K, Arai K (2001) Hydrothermal synthesis of metal oxide nanoparticles at supercritical conditions. J Nanopart Res 3:227–235

Agegnehu AK, Pan CJ, Tsai MC, Rick J, Su WN, Lee JF, Hwang BJ (2016) Visible light responsive noble metal-free nanocomposite of V-doped TiO_2 nanorod with highly reduced graphene oxide for enhanced solar H_2 production. Int J Hydrog Energy 41:6752–6762

Ajayan PM, Ebbesen TW (1997) Nanometre-size tubes of carbon. Rep Prog Phys 60:1025–1062

Asahi R, Morikawa T, Ohwaki T, Aoki K, Taga Y (2001) Visible-light photocatalysis in nitrogen-doped titanium oxides. Science 293:269–267

Basavalingu B, Byrappa K, Yoshimura M (2001a) Advances in high pressure science and technology. Tata McGraw Publishers, 417 pp

Basavalingu B, Jose M, Moreno C, Byrappa K, Gogotsi YG (2001b) Decomposition of silicon carbide in the presence of organic compounds under hydrothermal conditions. Carbon 39:1763–1766

Basavalingu B, Byrappa K, Madhusudan P, Dayananda AS, Yoshimura M (2006) Hydrothermal synthesis and characterization of micro to nano sized carbon particles. J Mater Sci 41:1465–1469

Basavalingu B, Byrappa K, Madhusudan P, Yoshimura M (2007) Hydrothermal synthesis of nano-sized crystals of diamond under sub-natural conditions. J Geo Soc India 69:665

Blackburn JM, Long DP, Cabanas A, Watkins JJ (2001) Deposition of conformal copper and nickel films from supercritical carbon dioxide. Science 294:141–145

Byrappa K, Adschiri T (2007) Hydrothermal technology for nanotechnology. Prog Cryst Growth Charact Mater 53:117–166

Chang JY, Ghule A, Chang JJ, Tzing SH, Ling YC (2002) Opening and thinning of multiwall carbon nanotubes in supercritical water. Chem Phys Lett 363:583–590

Che G, Lakshmi BB, Fisher IR, Martin CR (1998) Carbon nanotubule membranes for electrochemical energy storage and production. Nature 393:346–349

Chen Y, Zhao S, Wang X, Peng Q, Lin R, Wang Y, Shen R, Cao X, Zhang L, Zhou G (2016) Synergetic integration of $Cu_{1.94}S$-$Zn_xCd_{1-x}S$ heteronanorods for enhanced visible-light-driven photocatalytic hydrogen production. J Am Chem Soc 138:4286–4289

Chen G, Zhang X, Guan L, Zhang H, Xie X, Chen S, Tao J (2018) Phase transition promoted hydrogen evolution performance of MoS_2/VO_2 hybrids. J Phys Chem C 122:2618. https://doi.org/10.1021/acs.jpcc.7b12040

Dai GH, Hafner JH, Rinzler AG, Colbert DT, Smalley RE (1996) Nanotubes as nanoprobes in scanning probe microscopy. Nature 384:147–150

Das P, Roy A, Devi PS (2016) Zn_2SnO_4 as an alternative photoanode for dye sensitized solar cells: current status and future scope. Trans Ind Ceram Soc 75:1–8

Das PP, Roy A, Agarkar S, Devi PS (2018a) Hydrothermally synthesized fluorescent Zn_2SnO_4 nanoparticles for dye sensitized solar cells. Dyes Pigments 154:303–313

Das P, Roy A, Devi PS (2018b) Hydrothermally synthesized fluorescent Zn_2SnO_4 nanoparticles for dye sensitized solar cells. Dyes Pigments 154:11–22

DeVries RC, Roy R, Somiya S, Yamada S (1994) A review of liquid phase systems pertinent to diamond synthesis. Trans Mater Res Soc Japan 14B:641

Dontsova D, Fettkenhauer C, Papaefthimiou V, Schmidt J, Antonietti M (2015) 1,2,4-Triazole-based approach to noble-metal-free visible-light driven water splitting over carbon nitrides. Chem Mater 28:772–778

Dunlap-Shohl WA, Daunis TB, Wang X, Wang J, Zhang B, Barrera D, Yan Y, Hsu JWP, Mitzi DB (2018) Room-temperature fabrication of a delafossite $CuCrO_2$ hole transport layer for perovskite solar cells. J Mater Chem A 6:469–477

Eric Drexler K (1986) Engines of Creation: The Coming Era of Nanotechnology, Doubleday Publisher, United States

Feynman RP (1959) There's plenty of room at the bottom. Annual meeting of the American Physical Society

Forster S, Antonietti M (1998) Amphiphilic block copolymers in structure-controlled nanomaterial hybrids. Adv Mater 10:195–217

Fox MA, Dulay M (1993) Heterogeneous photocatalysis. Chem Rev 93:341–357

Frank S, Poncharal P, Wang ZI, De Heer WA (1998) Carbon nanotube quantum resistors. Science 280:1744–1746

Fujishima A, Honda K (1972) Electrochemical photolysis of water at a semiconductor electrode. Nature 238:37–38

Fujito H, Kunioku H, Kato D, Suzuki H, Higashi M, Kageyama H, Abe R (2016) Layered perovskite oxychloride Bi_4NbO_8Cl: a stable visible light responsive photocatalyst for water splitting. J Am Chem Soc 138:2082–2085

Gebremariam TT, Chen F, Wang Q, Wang J, Liu Y, Wang X, Qaseem A (2018) Bimetallic Mn-Co oxide nanoparticles anchored on carbon nanofibers wrapped in nitrogen doped carbon for application in Zn-air batteries and supercapacitors. ACS Appl Energy Mater 1:1612. https://doi.org/10.1021/acsaem.8b00067

Gersten B (2003) In: Byrappa K, Ohachi T (eds) Handbook of crystal growth technology. William Andrew Publications, New York

Gogotsi YG, Nickel KG, Kofstad PJ (1995) Hydrothermal synthesis of diamond from diamond-seeded β-SiC powder. Mater Chem 5:2313–2314

Gogotsi YG, Kofstad P, Yoshmura M, Nickel KG (1996) Formation of sp3-bonded carbon upon hydrothermal treatment of SiC. Diam Relat Mater 5:151

Goldberger J, He R, Zhang Y, Lee S, Yan H, Choi HJ, Yang P (2003) Single-crystal gallium nitride nanotubes. Nature 422:599–602

Gujral SS, Simonov AN, Higashi M, Fang XY, Abe R, Spiccia L (2016) Highly dispersed cobalt oxide on TaON as efficient photoanodes for long-term solar water splitting. ACS Catal 6:3404–3417

Guo Z, Sadler PJ, Tsang SC (1998) Immobilization and visualization of DNA and proteins on carbon nanotubes. Adv Mater 10:701–703

Hakuta Y, Adschiri T, Suzuki T, Chida T, Seino K, Arai K (1998a) Flow method for rapidly producing barium hexaferrite particles in supercritical water. J Am Ceram Soc 81:2461–2464

Hakuta Y, Onai S, Terayama H, Adschiri T, Aria K (1998b) Production of ultra-fine ceria particles by hydrothermal synthesis under supercritical conditions. J Mater Sci Lett 17:1211–1213

Herrmann JM (1999) Heterogeneous photocatalysis: fundamentals and applications to the removal of various types of aqueous pollutants. Catal Today 53:115–129

Heggerty S (1986) Diamond genesis in a multiply-constrained model. Nature 320:34

Hoffmann MR, Martin ST, Choi WY, Bahnmann DW (1995) Environmental applications of semiconductor photocatalysis. Chem Rev 95:69–96

Ijima S (1991) Helical microtubules of graphitic carbon. Nature 354:56–58

Jiang D, Sun Z, Jia H, Lu D, Du P (2016) A cocatalyst-free CdS nanorod/ZnS nanoparticle composite for high-performance visible-light-driven hydrogen production from water. J Mater Chem A Mater Energy Sustain 4:675–683

Kameo A, Yoshimura T, Esumi K (2015) Preparation of noble metal nanoparticles in supercritical carbon dioxide. Colloid Surf A Physicochem Eng Aspect 215:181–189

Katayama K, Yao H, Nakanishi F, Doi H, Saegusa A, Okuda N, Yamala T (1998) Lasing characteristics of low threshold ZnSe-based blue/green laser diodes grown on conductive ZnSe substrates. Appl Phys Lett 73:102

Kandiel TA, Takanabe K (2016) Solvent-induced deposition of Cu-Ga-In-S nanocrystals onto a titanium dioxide surface for visible-light-driven photocatalytic hydrogen production. Appl Catal B Environ 184:264–269

Khaselev O, Turner JA (1998) A monolithic photovoltaic-photo-electrochemical device for hydrogen production via water splitting. Science 280:425–427

Lalena JN, Cleary DA, Carpenter E, Dean NF (2008) Nanomaterials synthesis. In: Lalena N, Cleary DA, Carpenter E, Dean NF (eds) Inorganic materials synthesis and fabrication. Wiley, Hoboken

Lee DC, Mikulec FV, Korgel BA (2004) Carbon nanotube synthesis in supercritical toluene. J Am Chem Soc 126:4951–4957

Lencka MM, Riman RE (2003) In: Byrappa K, Ohachi T (eds) Handbook of crystal growth technology. William Andrew Publications, New York

Li L, Yan J, Wang T, Zhao ZJ, Zhang J, Gong J, Guan N (2015a) Sub-10 nm rutile titanium dioxide nanoparticles for efficient visible-light-driven photocatalytic hydrogen production. Nat Commun 6:5881–5810

Li H, Yu K, Lei X, Guo B, Chao L, Hao F, Zhu Z (2015b) Synthesis of MoS$_2$@CuO heterogeneous structure with improved photocatalysis performance and H$_2$O adsorption analysis. Dalton Trans 44:10438–10447

Liu W, Zhong W, Wu X, Tang N, Du Y (2005) Hydrothermal microemulsion synthesis of cobalt nanorods and self-assembly into square-shaped nanostructures. J Cryst Growth 284:446–452

McLeod MC, Gale WF, Roberts CB (2004) Metallic nanoparticle production utilizing a supercritical carbon dioxide flow process. Langmuir 20:7078–7082

Motiei M, Hacohen YR, Calderon-Moreno JM, Gedanken A (2001) Preparing carbon nanotubes and nested fullerenes from supercritical CO_2 by a chemical reaction. J Am Chem Soc 123:8624–8625

Melto CE, Giardini AA (1974) The composition and significance of gas released from natural diamonds from Africa and Brazil. Am Mineral 59:775

Navon O (1991) High internal pressures in diamond fluid inclusions determined by infrared absorption. Nature 353:746–748

Ortiz-Landeros J, Gómez-Yáñez C, López-Juárez R, Dávalos-Velasco I, Pfeiffer H (2012) Synthesis of advanced ceramics by hydrothermal crystallization and modified related methods. J Adv Ceram 1:204–220

Orlov Yu L (1973) Mineralogy of diamond. Nauka, Moscow in Russian

Puntes VF, Drishnan KM, Alivisatos AP (2001) Colloidal nanocrystal shape and size control: the case of cobalt. Science 291:2115–2117

Qui T, Wu XL, Mei YF, Wan GJ, Chu PK, Siu GG (2005) From Si nanotubes to nanowires: synthesis, characterization, and self-assembly. J Cryst Growth 277:143–148

Rajamanickam G, Narendhiran S, Muthu SP, Mukhopadhyay S, Perumalsamy R (2017) Hydrothermally derived nanoporous titanium dioxide nanorods/nanoparticles and their influence in dye-sensitized solar cell as a photoanode. Chem Phys Lett 689:19–25

Reverchon E, Adami R (2006) Nanomaterials and supercritical fluids. J Supercrit Fluid 37:1–22

Riman RE, Suchanek WL, Byrappa K, Chen CW, Shuk P, Oakes CS (2002) Solution synthesis of hydroxyapatite designer particulates. Solid State Ionics 151:393–402

Roy R, Tuttle OF (1956) Investigations under hydrothermal conditions. Phys Chem Earth 1:138–180

Roy R (1994) Accelerating the kinetics of low-temperature inorganic syntheses. J Solid State Chem 111:11–17

Roy R, Ravichandran D, Ravindranathan P, Badzian A (1996) Evidence for hydrothermal growth of diamond in the C-H-O and C-H-O halogen system. J Mater Res 11:1164–1168

Seo DS, Lee JK, Kim H (2001) Preparation of nanotube-shaped TiO_2 powder. J Cryst Growth 229:428–432

Shah PS, Husain S, Johnston KP, Korgel BA (2001) Nanocrystal arrested precipitation in supercritical carbon dioxide. J Phys Chem B 105:9433–9440

Singh GP, Shrestha KM, Nepal A, Klabunde KJ, Sorensen CM (2014) Graphene supported plasmonic photocatalyst for hydrogen evolution in photocatalytic water splitting. Nanotechnology 25:265701 (11pp)

Schmidt I, Benndorf C (1998) Mechanisms of low temperature growth of diamond using halogenated precursorgases. Diam Relat Mater 7:266

Tanigawa S, Irie H (2016) Visible-light-sensitive two-step overall water-splitting based on band structure control of titanium dioxide. Appl Catal B Environ 180:1–5

Tao X, Zhao Y, Mu L, Wang S, Li R, Li C (2017) Bismuth tantalum oxyhalogen: a promising candidate photocatalyst for solar water splitting. Adv Energy Mater 8:1701392–1701397

Tian Z, Liu J, Voigt JA, Xu H, Mcddermott MJ (2003) Dendritic growth of cubically ordered nanoporous materials through self-assembly. Nano Lett 3:89

Tsai CC, Teng H (2004) Regulation of the physical characteristics of titania nanotube aggregates synthesized from hydrothermal treatment. Chem Mater 16:4352–4358

Vasilev VG, Kovalski VP, Cherski NV (1968) Origin of diamond. Nedra, Moscow (in Russian)

Wang D, Yu D, Peng Y, Meng Z, Zhang S, Qian Y (2003) Formation of antimony nanotubes via a hydrothermal reduction process. Nanotechnology 14:748–751

Wang H, Liu Y, Li M, Huang H, Zhong M, Shen H (2009) Hydrothermal growth of large-scale macroporous TiO2 nanowires and its application in 3D dye-sensitized solar cells. Appl Phys A Mater Sci Process 97:25–29

Wang Q, Hisatomi T, Mab SSK, Li Y, Domen K (2014) Core/shell structured La- and Rh-co-doped SrTiO$_3$ as a hydrogen evolution photocatalyst in Z-scheme overall water splitting under visible light irradiation. Chem Mater 26:4144–4150

Wang Q, Hisatomi T, Jia Q, Tokudome H, Zhong M, Wang C, Pan Z, Takata T, Nakabayashi M, Shibata N (2016) Scalable water splitting on particulate photocatalyst sheets with a solar-to-hydrogen energy conversion efficiency exceeding. Nat Mater 15:611–615

Wu ZL, Wang CH, Zhao B, Dong J, Lu F, Wang WH, Wang WC, Wu GJ, Cui JZ, Cheng P (2016) A semi-conductive copper-organic framework with two types of photocatalytic activity. Angew Chem Int Ed 55:4938–4942

Xiang Q, Cheng F, Lang D (2016) Hierarchical layered WS$_2$/Graphene-modified CdS nanorods for efficient photocatalytic hydrogen evolution. ChemSusChem 9:996–1002

Xie Q, Dai Z, Huang W, Liang J, Jiang C, Qian YT (2005) Synthesis of ferromagnetic single-crystalline cobalt nanobelts via a surfactant-assisted hydrothermal reduction process. Nanotechnology 16:2958–2962

Yamakov V, Wolf D, Phillpot S, Mukherjee A, Gleiter H (2004) Deformation-mechanism map for nanocrystalline metals by molecular-dynamics simulation. Nat Mater 3:43–47

Yakobson BI, Smalley RE (1997) Fullerene nanotubes: C1,000,000 and beyond: some unusual new moleculeslong, hollow fibers with tantalizing electronic and mechanical properties-have joined diamonds and graphite in the carbon family. Am Sci 85:325

Yu JG, Wang WG, Cheng B, Su BL (2009) Enhancement of photocatalytic activity of mesoporous TiO$_2$ powders by hydrothermal surface fluorination treatment. J Phys Chem C 113:6743–6750

Yuan YJ, Chen DQ, Huang YW, Yu ZT, Zhong JS, Chen TT, Tu WG, Guan ZJ, Cao DP (2016) MoS$_2$ nanosheet-modified CuInS$_2$ photocatalyst for visible-light-driven hydrogen production from water. ChemSusChem 9:1003–1009

Yue X, Yi S, Wang R, Zhang Z, Qiu S (2016) Cadmium sulfide and nickel synergetic co-catalysts supported on graphitic carbon nitride for visible-light-driven photocatalytic hydrogen evolution. Sci Rep 6:22268–22269

Zhang H, Lv XJ, Li YM, Wang Y, Li JH (2010) P25-Graphene composite as a high performance photocatalyst. ACS Nano 4:380–386

Zhang J, Jin X, Morales-Guzman PI, Yu X, Liu H, Zhang H, Razzari L, Claverie JP (2016a) Engineering the absorption and field enhancement properties of Au-TiO$_2$ nanohybrids via whispering gallery mode resonances for photocatalytic water splitting. ACS Nano 10:4496–4503

Zhang G, Lan ZA, Lin L, Lin S, Wang X (2016b) Overall water splitting by Pt/g-C$_3$N$_4$ photocatalysts without using sacrificial agents. Chem Sci 7:3062–3066

Zhao J, Wang X, Chen R, Li L (2005) Synthesis of thin films of barium titanate and barium strontium titanate nanotubes on titanium substrates. Mater Lett 59:2329

Zhao J, Yan X, Zhao N, Li X, Lu B, Zhang X, Yu H (2018) Cocatalyst designing: a binary noble-metal-free cocatalyst system consisting of ZnIn$_2$S$_4$ and In(OH)$_3$ for efficient visible-light photocatalytic water splitting. RSC Adv 8:4979–4986

Zhu Y, Zheng H, Li Y, Gao L, Yang Z, Qian YT (2003) Synthesis of ag dendritic nanostructures by using anisotropic nickel nanotubes. Mater Res Bull 38:1829–1834

Zhu Z, Chen JY, Su KY, Wu RJ (2016) Efficient hydrogen production by water-splitting over Pt-deposited C-HS-TiO$_2$ hollow spheres under visible light. J Taiwan Inst Chem Eng 60:222–228

Zhu M, Sun Z, Fujitsuka M, Majima T (2018) Z-scheme photocatalytic water splitting on a 2D heterostructure of black phosphorus/bismuth vanadate using visible light. Angew Chem Int Ed Engl 57:2160–2164

Zou ZG, Ye JH, Sayama K, Arakawa H (2001) Direct splitting of water under visible light irradiation with an oxide semiconductor photocatalyst. Nature 414:625–627

Chapter 15
Biomedical Applications and Characteristics of Graphene Nanoparticles and Graphene-Based Nanocomposites

S. Rajeshkumar and P. Veena

15.1 Introduction

Graphene is the nanomaterial consisting of a single layer of carbon arranged in an allotropic form. Graphene is chemically, mechanically, and thermally stable compound (Wang et al. 2017a). It has various other properties such as large specific surface area and high flexibility (Hsieh et al. 2015). Graphene oxide, reduced graphene oxide, graphene nanosheets, graphene nanocrystals, graphene nanocubes, graphene nanotubes, graphene quantum dots, and others are also widely studied due to its exceptional properties (Fig. 15.1). Different methods are carried out to prepare graphene such as chemical vapor deposition, chemical or electrochemical reduction, and mechanical exfoliation (Kim et al. 2017). The graphene layer is formed by chemical vapor deposition. Graphene oxide (GO) is prepared by oxidation of purified graphite by Hummer's method and reduced graphene oxide (rGO) is prepared by chemical reduction of reduced graphene oxide (Devi and Kumar 2018).

Graphene is stronger than many other metals and it is highly flexible. It is the strongest material than any other material known. Graphene has many such notable features that allow the inorganic nanoparticle to accumulate in it to form nanocomposites (Wei et al. 2017). Graphene nanoparticles are decorated with other nanoparticles on its surface for more beneficial property (Luan et al. 2018). Graphene is mixed with several other nanoparticles such as metals (Ag, Au, Cu, Zn), non-metals (chitosan, polymers, epoxy), or metal oxides (TiO_2, ZnO) in order to produce nanocomposites (Acar Bozkurt 2017) (Fig. 15.2).

S. Rajeshkumar (✉)
Department of Pharmacology, Saveetha Dental College and Hospitals, Saveetha Institute of Medical and Technical Sciences, Chennai, TN, India

P. Veena
Nanotherapy Laboratory, School of Bio-Sciences and Technology, Vellore Institute of Technology, Vellore, TN, India

© Springer Nature Switzerland AG 2018   341
R. Prasad et al. (eds.), *Exploring the Realms of Nature for Nanosynthesis*,
Nanotechnology in the Life Sciences, https://doi.org/10.1007/978-3-319-99570-0_15

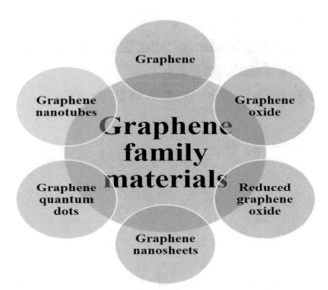

Fig. 15.1 Graphene materials

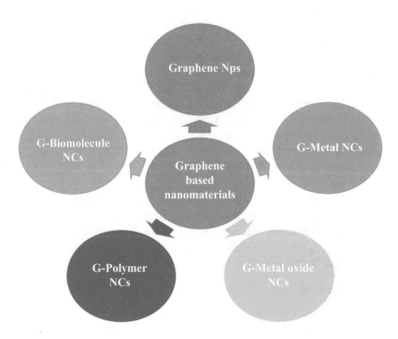

Fig. 15.2 Graphene-based nanocomposites

Graphene nanocomposites have tremendous application in various fields such as antimicrobial, anticancer, biosensing, bioimaging, and drug delivery (Khalil et al. 2017) (Fig. 15.3).

In this review article, we focused and discussed the major application of graphene nanoparticles and graphene-based nanocomposites such as graphene-gold, graphene-silver, graphene-zinc, and graphene-nickel in various biomedical aspects such as a biosensor, bioimaging, antimicrobial, anticancer drug delivery, etc. (Fig. 15.4).

15.2 Application of Graphene Nanoparticles and Graphene-Based Nanocomposites

15.2.1 Graphene Nanoparticles

Graphene nanoparticles (GNPs) have gained noticeable attention in the biomedical field such as tissue engineering, bone regeneration, bio-conductivity, drug delivery, etc. Low oxygen graphene nanoparticles have an exceptional property of osteoinduction and osteoconduction which helps in the bone regeneration of mesenchymal stem cells. They are used to treat trauma in the bone tissue and retaining the stem cell activity (Elkhenany et al. 2017). Graphene has a 2D and unique structure

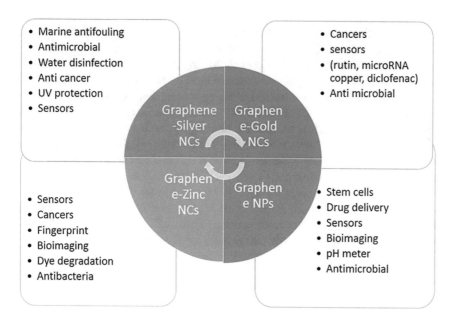

Fig. 15.3 Graphene NPs and graphene-based NCs applications

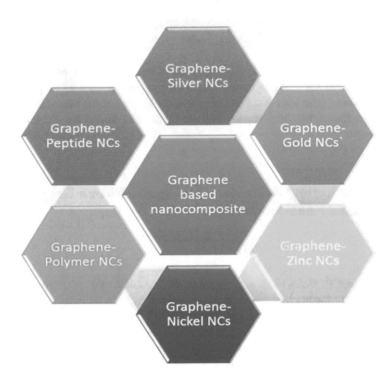

Fig. 15.4 Widely used graphene-based nanocomposite

containing various special properties such as high specific area, modified easily and
it is a strong material. This property allows the graphene to act as a carrier for a
multifunctional drug delivery (Zhao et al. 2017b). A single layer graphene intensi-
fies the property of the sensors. Graphene is used widely in sensors and biosensors
such as enzymatic and non-enzymatic glucose sensors, hydrogen peroxide sensors,
and immuno-sensors (Bollella et al. 2017). Nanographene oxide has a photoluminescence property which is utilized for bioimaging technique such as fluorescent
imaging and 2 photon fluorescent imaging (Lin et al. 2016). Remarkable properties
of the graphene such as larger surface area provide the pH meter a higher sensitivity
and flexibility compared to a glass electrode. Optical graphene oxide has an insulat-
ing property which is usually involved in optical sensors (Salvo et al. 2018). An
antimicrobial property of the graphene oxide is attained by the sharp edges of the
GO (graphene oxide) which causes damages to the cell membrane of the bacteria
that decreases the drug resistivity (Yousefi et al. 2017) (Table 15.1).

Table 15.1 Biomedical application of graphene nanoparticles

S. no.	Application	Function	Characteristics	Reference
1	Stem cell	Osteoinductive and osteoconductive platform for stem cell and bone regeneration	Graphene thickness is 1.0–1.2 nm	Elkhenany et al. (2017)
2	Drug delivery	The unique structure of graphene acts as a carrier for multifunctional drug delivery	Graphene nanoparticles size is 5 nm	Zhao et al. (2017a, b)
3	Sensors	Sensors and biosensors are made effective using graphene, GO, and rGO	500 nm of nanoflakes size and 1 nm of the thickness of graphene	Bollella et al. (2017)
4	Bioimaging	Graphene used in fluorescent imaging and 2 photon fluorescent imaging	Graphene nanoparticles size of 10–60 nm	Lin et al. (2016)
5	pH meter	Fabrication of pH meter with graphene increases its sensitivity and quality	3–20 nm usually between 100 nm of graphene quantum dots	Salvo et al. (2018)
6	Antimicrobial	Exceptional properties of graphene oxide contain high antimicrobial property and control drug-resistant microbes	100 nm, 200 nm, and 50 nm of graphene is used for different effect on bacteria	Yousefi et al. (2017)

15.2.2   Graphene-Based Nanocomposites

15.2.2.1   Graphene-Gold Nanocomposites

Gold is an excellent thermal and electrical conductor. This property helps these nanocomposites in electrochemical sensing. Graphene and gold nanoparticles are coated on an acupuncture needle for sensing rutin. Acupuncture needle has a small surface area which acts as an electrode for sensing the lower quantity of sample (Niu et al. 2018). Another sensor for detecting rutin is made of carbon screen-printed electrode which is coated with the graphene, gold NPs, and chitosan (Apetrei and Apetrei 2018). Graphene and gold nanoparticles are used for fabricating the PEC aptasensor (photo-electrochemical aptasensor) which is used to sense the diclofenac, an anti-inflammatory drug (Okoth et al. 2018). Reduced graphene oxide (rGO) and gold nanoparticles are used for fabricating the electrode for detection of copper in high salt electrolyte substrate. Reduced graphene oxide has a large surface area which has many active sites and electrical transfer is promoted by gold. Nafion is a polymer that is used as cation exchanger to bind the reduced graphene oxide and gold together tightly. Five to 10 nm diameter of silver nanoparticles is used to decorate the graphene on the electrode surface (Liu et al. 2017). SPR biosensor (sensitive surface plasmon resonance) has two different layers, the bottom layer made of graphene oxide to attain maximum surface area for microRNA to get adhere onto the surface and the top layer made of gold for providing signal amplification. The detection limit of microRNA is measured as 0.1 fM (Li et al. 2017). Graphene-based sensor plays a vital role in cancer detection. Graphene oxide and gold nanoparticles are fabricated on a glass electrode for detecting the cancer biomarkers. Graphene oxide helps to detect the cancer biomarkers and gold nanoparticles provide the signaling (Ali et al. 2017). Graphene and graphene-based nanocomposites possess a promising application in cancer disease. Graphene-gold based nanocomposite with its exceptional properties it acts as therapeutic tools and detection of cancer disease. Graphene-silver nanocomposite acts as various sensors such as gene biosensor, immune-sensor, and enzyme biosensor. Various therapeutic techniques also involve G-Au NPs such as phototherapy, combination therapy, and chemotherapeutic drug delivery (Al-ani et al. 2017) (Table 15.2).

15.2.2.2 Graphene-Silver Based Nanocomposites

An excellent antimicrobial property of silver is utilized and widely used in several applications. The reduced graphene oxide which is ornamented with the silver nanoparticles shows higher antibacterial activity by showing the enhanced zone of inhibition. The large surface area of the graphene promotes the activity of the "attacking and killing" mechanism (Ganguly et al. 2017). Graphene decorated with reduced silver is used as a sensor to detect the hydrogen sulfide gas. Doped graphene allows fabricating the sensor to achieve high resistance and more sensitivity to hydrogen sulfide gas (Ovsianytskyi et al. 2018). Composite G/Ag NPs also

Table 15.2 Biomedical application of graphene-gold nanocomposite

S. no.	Application	Function	Characterization	Reference
1	Cancer	Early detection of breast cancer biomarkers using graphene oxide-gold coated electrodes with detection limit 0.16 and 0.23 nM	The thickness of graphene sheet is 0.34 nm and gold nanoparticles size is 10–50 nm	Ali et al. (2017)
		Graphene-Au nanoparticle sensors for detecting and tools for therapy	Gold nanoparticles are determined to be having a particle size of 100 nm	Al-Ani et al. (2017)
2	Sensors	Detection of rutin with detection limit 1.1×10^{-8} M	Nanoflakes structure of graphene decorated with gold on its surface	Niu et al. (2018) and Apetrei and Apetrei (2018)
		Detection of diclofenac		Okoth et al. (2018)
		Detection of copper from seawater	Gold nanoparticles diameter is about 5–10 nm and thickness of graphene is 0.8–1.2 nm	Liu et al. (2017)
		Detection of microRNA	The average size of gold nanoparticles is 16 nm to 22 nm.	Li et al. (2017)

applied as water disinfectant due to its elasticity and large porosity. Graphene nanoparticles and silver nanoparticles are ornamented with the melamine sponge which is then submerged in a suspension of bacteria. The bacterial membrane is destructed by the composite providing antibacterial property to the sponge (Deng et al. 2017). Graphene/silver nanocomposites are coated with a polyester fabric to enhance the quality such as UV and mechanical protection. Fabric coated with graphene/silver nanocomposites is also highly durable, stretchable, foldable, and flexible (Ouadil et al. 2017). Reduced graphene oxide-silver nanoparticles are synthesized by the green method and reduced using vitamin C to the composite. The cytotoxicity study of the synthesized nanocomposites towards human lung cancer is significantly high compared to silver nanoparticles and graphene oxide. The cytotoxicity of the rGO-Ag NPs is IC50 of 30 lg/mL (Kavinkumar et al. 2017). Methimazole can be detected sensitively by the electrode that is fabricated by the graphene and the silver NPs. Graphene nanosheets are synthesized by Hummer's method from graphite. The surface of the graphene is decorated with the Ag NPs. The composite is fabricated on the carbon paste electrode. Due to the large surface area which leads to higher active sites the adsorption of the methimazole is high and sensitive than the carbon paste electrode (Alaqad et al. 2017). Antimicrobial property of silver assists in the process of antifouling of the submerged surfaced undersea. Graphene material decorated with silver nanoparticles is fabricated on the submerged surfaces. The antibacterial property of the silver and structure of graphene together act as an antifouling agent. The diameter of the silver nanoparticle is 2–5 nm (Yee et al. 2016) (Table 15.3).

15.2.2.3 Graphene-Zinc Based Nanocomposite

Chemically converted graphene decorated with Au and ZnO is synthesized in-situ by one-pot low temperature. Size of the Zn nanoparticles is 30–60 nm and Au particles are −11.3 nm. AZG (Au-ZnO-Graphene) nanocomposite is immobilized with BSA (bovine serum albumin) which has high biocompatibility and shows maximum cell viability for human ovarian cancer cells (Naskar et al. 2018). Graphene nanoparticles with zinc are used to sense hydrogen gas at 200 ppm. Graphene-zinc oxide nanocomposite increases the sensitivity and decreases the minimum operable temperature. The prepared composite powder is covered on the alumina substrate for gas sensing. The XRD reveals that the graphene-zinc formed a crystalline structure and the FTIR value is 1564 cm^{-1} (Anand et al. 2014). Highly sensitive DNA biosensor made using graphene and zinc oxide is effective in detecting the H5 gene of Avian Influenza. The sensitivity of geno sensor is increased to ($P < 0.05$). The sensitivity of this biosensor is 3.2580 m AmM À1 and the detection limit is 7.4357 mM (Low et al. 2016). RNA sequence can also be detected effectively using biosensors fabricated with the zinc-graphene nanocomposite because of its high specific area and good biocompatibility. A single-strandard DNA sequence is synthesized using coconut cadang cadang viroid (CCCVd) complementary RNA sequence. The synthesized sequence is immobilized on the surface of the biosensor

Table 15.3 Biomedical application of graphene-silver nanocomposite

S. no.	Application	Function	Characteristic	Reference
1	Antibacterial	"Capturing and killing" mechanism carried out by silver nanoparticles to kill *E.coli* with inhibition zone 2.1 cm	Crystal structure with 0.44 nm interspace in reduced graphene oxide and silver nanocomposites	Ganguly et al. (2017)
2	Sensor	Hydrogen sulfide gas sensor electrode	Noncrystalline smooth structure with a 10–30 nm particles size of silver nanoparticles	Ovsianytskyi et al. (2018)
		Detection of doxorubicin with the detection limit of 2 nM	XRD shows dendritic structure and the nanoparticles distributed on the surface of the reduced GO	Guo et al. (2017)
		Methimazole detection	Silver particle size is 8–12 nm	Alaqad et al. (2017)
		Methylene blue detection	Silver-reduced graphene oxide is quasi-spherical in shape and the diameter is 28–32 nm	Chettri et al. (2017)
3	Water disinfection	Preparation of melamine sponge with graphene-silver NPs	Silver nanoparticles distributed on the MS is 38.41 ± 13.83 nm	Deng et al. (2017)
4	Anticancer activity	Higher cytotoxicity level of rGO-gold nanoparticles towards lung cancer cell	Size of the silver nanoparticles is greater than 5 nm	Kavinkumar et al. (2017)
6	UV protection fabrics	Ag-G NPs are coated on a fabric for UV protection	Graphene oxide has to interspace between 0.9 nm	Ouadil et al. (2017)
6	Marine antifouling agent	Antimicrobial property of silver acts against the bacterial growth on submerged surfaces	Silver nanoparticles have a size of 76–82 nm	Yee et al. (2016)

fabricated with GZO nanocomposite which provides good detection limit and sensitivity. The detection limit is 4.3×10^{-12} M. The zinc oxide and graphene nanocomposites are formed into a clear crystalline structure. Detection of ammonia is achieved successfully at room temperature using zinc oxide nanowire and reduced graphene oxide. Zinc transfers the electron to the graphene which increases the sensitivity to the sensor more than bare graphene sensor (Wang et al. 2017b). Bioimaging can be successfully obtained by using AgInZnS and graphene oxide nanocomposite. Mice organs and cancer cells are effectively imaged using this nanocomposite. The absence of toxic cadmium reduces the cytotoxicity of the cells. The synthesized nanocomposites emit four colors red, yellow, green, and orange. The nanocomposites prepared shows highly crystalline structure with the diameter of 4–5 nm. This is effectively applied in bioimaging of breast cancer cells

SK-BR-3 (Zang et al. 2017). A study discovered a new drug carrier made of zinc-clinoptilolite/graphene which has high drug load capacity and possesses a property to release the drug slowly into the cell medium. The research is successfully carried out with a cancer drug—doxorubicin (Khatamian et al. 2016). Zinc oxide is decorated with the graphene oxide nanoparticles. The size of the zinc oxide is 20–25 nm with a spherical shape. The synthesized nanocomposite is used for photodegrading the dyes such as methyl red, congo red, crystal violet, and neutral red. The ZnO-GO nanocomposite effectively photodegrades azo dye and bare ZnO nanoparticles (Atchudan et al. 2017) (Table 15.4).

Table 15.4 Graphene-zinc based nanocomposite

S. no.	Application	Function	Characteristics	Reference
1	Latent Fingerprint detection	TiO_2, iron oxide, and ZnO are used as developing fingerprint powders		Amrutha et al. (2017)
2	Cancer	Gold-zinc-graphene immobilized with BSA (bovine serum albumin) has maximum CV in human ovarian cancer cells	The average particle size of zinc oxide is −11.30 nm and gold is 30–60 nm	Naskar et al. (2018)
3	Sensors	Hydrogen gas detection by ZnO and graphene	Flower-like structure distributed on the graphene sheet	Anand et al. (2014)
		Avian Influenza H5 gene detection by DNA biosensor made of graphene-ZnO	Spherical in shape and the average particle size is 200 nm	Low et al. (2016)
		Single-stranded RNA detection genosensor—CCCVd sequence	Zinc oxide is observed to be Qazi spherical shape	Low et al. (2017)
		Glucose sponsor using nickel and zinc is effective in detecting glucose from human serum sample	Size of nickel NPs is 18 nm and thickness of GO is 0.34 nm	Mazaheri et al. (2017)
4	Drug delivery	Zinc incorporated with graphene as a drug carrier of cancer drug—doxorubicin	XRD shows 20–30 nm thick graphene	Khatamian et al. (2016)
5	Bioimaging	Imaging of breast cancer cells SK-BR-3 using AgInZnS and graphene oxide	Average size is 10 nm	Zang et al. (2017)
6	Dye degradation	ZnO and GO nanocomposite are used effectively in the photodegradation of azodye MO, CV, CR, NR	Zinc oxide nanoparticles have 20–25 nm of diameter and spherical shape	Atchudan et al. (2017)
7	Antibacterial activity	Effective against *Bacillus subtilis, Escherichia coli* with 18.9 mm and 23.8 nm	XRD shows 25.9–56.2 nm of particle size	Alswat et al. (2017)

15.2.3 Other Graphene-Based Nanocomposite

Apart from graphene, gold, silver and zinc nanocomposite, there are several other graphene-based nanocomposites that possess effective application in various fields. Graphene and nickel nanocomposite is used in degrading radionuclides such as uranium and thorium. Uranium (VI) and thorium (IV) are adsorbed more when the temperature of the system increases. Graphene has the high surface area, pore size and density provide high absorption than another nanocomposite (Lingamdinne et al. 2017). Graphene oxide and lead possess an excellent catalytic property which is used to degrade organic dyes such as congo red, methyl orange, and methylene blue with use of reducing agent $NaBH_4$ (Omidvar et al. 2017). Bacterial cellulose and graphene oxide nanocomposite effectively act as a drug carrier for ibuprofen. It has high cell viability and it is effective than graphene oxide (Luo et al. 2017). When a graphene is modified with macromolecule such as peptides, DNA, and proteins in biosensors, it enhances the sensitivity and selectivity of the system. Peptide-modified graphene nanocomposite is used in many biosensors such as fluorescent biosensors, spectroscopic biosensors, electrochemical biosensors, and electronic biosensors. Among these biosensors, fluorescent biosensors and electrochemical biosensors are effective than spectroscopic biosensors and electronic biosensors. Peptide-modified graphene nanocomposite increases the biorecognition and biocompatibility of the sensors (Wang et al. 2017a). Graphene oxide modified with hydroxyapatite nanocomposite has less cytotoxicity to the human skin cancer cells (A431). The property of the nanocomposite is analyzed using BSA (bovine serum albumin). GO-HAp nanocomposite has good biocompatibility which is used in dentistry, drug delivery, and other biomedical application (Ramadas et al. 2017) (Table 15.5).

15.3 Conclusion

In summary, the graphene nanoparticles and graphene-based nanocomposites are involved in many applications in the biomedical field. Graphene is highly effective because of its exceptional properties such as high specific surface area, good physiochemical property, mechanical and electrical stability. Graphene consists of 2D structure which is used to decorate other nanoparticles such as metals, metal oxides, polymers, and biomacromolecules. Our review concludes that the graphene nanoparticles and graphene-based nanocomposite such as G-Silver NCs, G-Zinc NCs, G-Gold NCs, G-Nickel NCs, G-Polymer NCs, and G-Peptide NCs are effective in many applications such as cancer studies, antimicrobial activity, drug delivery, biosensors, marine antifouling agent, bioimaging, UV protection, and dye degradation. In future, graphene nanoparticles decorated with a suitable metal or non-metal nanoparticles can enhance the effect on graphene nanoparticles and will be useful in many biomedical fields.

Conflict of Interest The authors declared that there is no conflict of interest.

Table 15.5 Other graphene-based nanocomposite

S. no.	Nanocomposite	Application	Function	Characteristics	Reference
1	Nickel-graphene	Removing radionuclides	Nickel ferrite and reduced GO is used to remove radionuclides such as uranium and thorium from wastewater	41.41 nm size of graphene oxide and 32.16 nm of reduced graphene oxide. It is noncrystalline cube structure	Lingamdinne et al. (2017)
2	Graphene-bioglass	Bone tissue engineering	–	–	Pazarçeviren et al. (2017)
3	Graphene-lead	Dye degradation	Organic dyes such as Congo red, methyl orange, and methylene blue are degraded from water using graphene oxide and lead	11 nm particle size and spherical shape	Omidvar et al. (2017)
4	Graphene-cellulose	Drug delivery	Ibuprofen is loaded on the graphene oxide and bacterial cellulose nanocomposite for drug delivery	500 nm particle size and thick fiber porous structure	Luo et al. (2017)
5	Graphene-peptide	Biosensors	Biosensors with biomacromolecules enhance the efficiency of the sensors. It increases the selectivity and sensitivity of the sensors	Crystalline honeycomb structure of graphene with surface modified peptide	Wang et al. (2017a)
6	Graphene-hydroxyapatite	Cancer	Cytotoxicity of graphene oxide and hydroxyapatite is examined using human skin cancer cell	Diameter of Hap is −32 nm and length −60 to 85 nm	Ramadas et al. (2017)
7	Graphene oxide-PEG-folic acid	Cancer	Anticancer drug camptothecin is loaded in a graphene for its large surface area	Width of the PEG is 11.5 nm and GO is 10.5 nm	Deb and Vimala (2018)
8	Graphene-Iron-SnO$_2$	Sensor	Used as humid sensor by humid air measured at room temperature	Iron nanoparticles have diameter of 17 nm	Toloman et al. (2017)
9	Graphene-nickel	Sensor	Enzymeless detection of glucose with detection limit 388.4 1A mM	Nickel particle size is 99.2 nm	Ji et al. (2017)

References

Acar Bozkurt P (2017) Sonochemical green synthesis of Ag/graphene nanocomposite. Ultrason Sonochem 35:397–404. https://doi.org/10.1016/j.ultsonch.2016.10.018

Al-ani LA, Alsaadi MA, Kadir FA, Hashim NM, Julkapli NM, Yehye WA (2017) Graphene gold based nanocomposites applications in cancer diseases; efficient detection and therapeutic tools. Eur J Med Chem 139:349–366. https://doi.org/10.1016/j.ejmech.2017.07.036

Alaqad KM, Abulkibash AM, Charles O, Hamouz SA, Saleh TA (2017) Silver nanoparticles decorated graphene modified Carbon paste electrode for molecular methimazole determination. Chem Data Collection 12:168–182. https://doi.org/10.1016/j.cdc.2017.09.003

Ali A, Lluís J, Sánchez A, Sullivan CKO, Nooredeen M (2017) DNA biosensors based on gold nanoparticles-modified graphene oxide for the detection of breast cancer biomarkers for early diagnosis. Bioelectrochemistry 118:91–99. https://doi.org/10.1016/j.bioelechem.2017.07.002

Alswat AA, Ahmad MB, Saleh TA (2017) Preparation and characterization of zeolite\zinc oxide-copper oxide nanocomposite: antibacterial activities. Colloids Interface Sci Commun 16:19–24. https://doi.org/10.1016/j.colcom.2016.12.003

Amrutha VS, Anantharaju KS, Prasanna DS, Rangappa D, Shetty K, Nagabhushana H, Darshan GP (2017) Enhanced sunlight driven photocatalytic performance and visualization of latent fingerprint by green mediated $ZnFe_2O_4$–RGO nanocomposite. Arab J Chem., in print. https://doi.org/10.1016/j.arabjc.2017.11.016

Anand K, Singh O, Singh MP, Kaur J, Singh RC (2014) Hydrogen sensor based on graphene/ZnO nanocomposite. Sensors Actuators B Chem 195:409–415. https://doi.org/10.1016/j.snb.2014.01.029

Apetrei IM, Apetrei C (2018) A modified nanostructured graphene-gold nanoparticle carbon screen- printed electrode for the sensitive voltammetric detection of rutin. Measurement 114:37–43. https://doi.org/10.1016/j.measurement.2017.09.020

Atchudan R, Edison TNJI, Perumal S, Shanmugam M, Lee YR (2017) Direct solvothermal synthesis of zinc oxide nanoparticle decorated graphene oxide nanocomposite for efficient photodegradation of azo-dyes. J Photochem Photobiol A Chem 337:100–111. https://doi.org/10.1016/j.jphotochem.2017.01.021

Bollella P, Fusco G, Tortolini C, Sanzò G, Favero G, Gorton L, Antiochia R (2017) Beyond graphene: electrochemical sensors and biosensors for biomarkers detection. Biosens Bioelectron 89:152–166. https://doi.org/10.1016/j.bios.2016.03.068

Chettri P, Vendamani VS, Tripathi A, Singh MK, Pathak AP, Tiwari A (2017) Green synthesis of silver nanoparticle-reduced graphene oxide using Psidium guajava and its application in SERS for the detection of methylene blue. Appl Surf Sci 406:312–318. https://doi.org/10.1016/j.apsusc.2017.02.073

Deb A, Vimala R (2018) Camptothecin loaded graphene oxide nanoparticle functionalized with polyethylene glycol and folic acid for anticancer drug delivery. J Drug Deliv Sci Technol 43:333–342. https://doi.org/10.1016/j.jddst.2017.10.025

Deng C, Gong J, Zhang P, Zeng G, Song B, Liu H (2017) Preparation of melamine sponge decorated with silver nanoparticles-modified graphene for water disinfection. J Colloid Interface Sci 488:26–38. https://doi.org/10.1016/j.jcis.2016.10.078

Devi M, Kumar A (2018) Structural, thermal and dielectric properties of in-situ reduced graphene oxide—polypyrrole nanotubes nanocomposites. Mater Res Bull 97:207–214. https://doi.org/10.1016/j.materresbull.2017.09.010

Elkhenany H, Bourdo S, Hecht S, Donnell R, Gerard D, Abdelwahed R, Dhar M (2017) Graphene nanoparticles as osteoinductive and osteoconductive platform for stem cell and bone regeneration. Nanomed Nanotechnol Biol Med 13:2117–2126. https://doi.org/10.1016/j.nano.2017.05.009

Ganguly S, Das P, Bose M, Kanti T, Mondal S (2017) Sonochemical green reduction to prepare ag nanoparticles decorated graphene sheets for catalytic performance and antibacterial application. Ultrason Sonochem 39:577–588. https://doi.org/10.1016/j.ultsonch.2017.05.005

Guo H, Jin H, Gui R, Wang Z, Xia J, Zhang F (2017) Electrodeposition one-step preparation of silver nanoparticles/carbon dots/reduced graphene oxide ternary dendritic nanocomposites for sensitive detection of doxorubicin. Sensors Actuators B Chem 253:50–57. https://doi.org/10.1016/j.snb.2017.06.095

Hsieh SH, Chen WJ, Yeh TH (2015) Effect of various amounts of graphene oxide on the degradation characteristics of the ZnSe/graphene nanocomposites. Appl Surf Sci 358:63–69. https://doi.org/10.1016/j.apsusc.2015.08.220

Ji Z, Wang Y, Yu Q et al (2017) One-step thermal synthesis of nickel nanoparticles modified graphene sheets for enzymeless glucose detection. J Colloid Interface Sci 506:678–684. https://doi.org/10.1016/j.jcis.2017.07.064

Kavinkumar T, Varunkumar K, Ravikumar V (2017) Anticancer activity of graphene oxide-reduced graphene oxide-silver nanoparticle composites. J Colloid Interface Sci 505:1125–1133. https://doi.org/10.1016/j.jcis.2017.07.002

Khalil I, Rahmati S, Muhd Julkapli N, Yehye WA (2017) Graphene metal nanocomposites—recent progress in electrochemical biosensing applications. J Ind Eng Chem 59:1–15. https://doi.org/10.1016/j.jiec.2017.11.001

Khatamian M, Divband B, Farahmand-Zahed F (2016) Synthesis and characterization of zinc (II)-loaded zeolite/graphene oxide nanocomposite as a new drug carrier. Mater Sci Eng C 66:251–258. https://doi.org/10.1016/j.msec.2016.04.090

Kim TH, Lee D, Choi JW (2017) Live cell biosensing platforms using graphene-based hybrid nanomaterials. Biosens Bioelectron 94:485–499. https://doi.org/10.1016/j.bios.2017.03.032

Li Q, Wang Q, Yang X, Wang K, Zhang H, Nie W (2017) High sensitivity surface plasmon resonance biosensor for detection of microRNA and small molecule based on graphene oxide-gold nanoparticles composites. Talanta 174:521–526. https://doi.org/10.1016/j.talanta.2017.06.048

Lin J, Chen X, Huang P (2016) Graphene-based nanomaterials for bioimaging. Adv Drug Deliv Rev 105:242–254. https://doi.org/10.1016/j.addr.2016.05.013

Lingamdinne LP, Choi YL, Kim IS, Yang JK, Koduru JR, Chang YY (2017) Preparation and characterization of porous reduced graphene oxide based inverse spinel nickel ferrite nanocomposite for adsorption removal of radionuclides. J Hazard Mater 326:145–156. https://doi.org/10.1016/j.jhazmat.2016.12.035

Liu M, Pan D, Pan W, Zhu Y, Hu X, Han H, Shen D (2017) Talanta in-situ synthesis of reduced graphene oxide/gold nanoparticles modified electrode for speciation analysis of copper in seawater. Talanta 174:500–506. https://doi.org/10.1016/j.talanta.2017.06.054

Low SS, Tan MTT, Loh HS, Khiew PS, Chiu WS (2016) Facile hydrothermal growth graphene/ZnO nanocomposite for development of enhanced biosensor. Anal Chim Acta 903:131–141. https://doi.org/10.1016/j.aca.2015.11.006

Low SS, Loh HS, Boey JS, Khiew PS, Chiu WS, Tan MTT (2017) Sensitivity enhancement of graphene/zinc oxide nanocomposite-based electrochemical impedance genosensor for single stranded RNA detection. Biosens Bioelectron 94:365–373. https://doi.org/10.1016/j.bios.2017.02.038

Luan VH, Bae D, Hun J, Lee W (2018) Mussel-inspired dopamine-mediated graphene hybrid with silver nanoparticles for high performance electrochemical energy storage electrodes. Compos Part B 134:141–150. https://doi.org/10.1016/j.compositesb.2017.09.070

Luo H, Ao H, Li G et al (2017) Bacterial cellulose/graphene oxide nanocomposite as a novel drug delivery system. Curr Appl Phys 17:249–254. https://doi.org/10.1016/j.cap.2016.12.001

Mazaheri M, Aashuri H, Simchi A (2017) Three-dimensional hybrid graphene/nickel electrodes on zinc oxide nanorod arrays as non-enzymatic glucose biosensors. Sensors Actuators B Chem 251:462–471. https://doi.org/10.1016/j.snb.2017.05.062

Naskar A, Bera S, Bhattacharya R, Roy SS, Jana S (2018) Effect of bovine serum albumin immobilized Au–ZnO–graphene nanocomposite on human ovarian cancer cell. J Alloys Compd 734:66–74. https://doi.org/10.1016/j.jallcom.2017.11.029

Niu X, Wen Z, Li X, Zhao W, Li X, Huang Y (2018) Fabrication of graphene and gold nanoparticle modified acupuncture needle electrode and its application in rutin analysis. Sensors Actuators B Chem 255:471–477. https://doi.org/10.1016/j.snb.2017.07.085

Okoth OK, Yan K, Feng J, Zhang J (2018) Chemical label-free photoelectrochemical aptasensing of diclofenac based on gold nanoparticles and graphene-doped CdS. Sensors Actuators B Chem 256:334–341. https://doi.org/10.1016/j.snb.2017.10.089

Omidvar A, Jaleh B, Nasrollahzadeh M (2017) Preparation of the GO/Pd nanocomposite and its application for the degradation of organic dyes in water. J Colloid Interface Sci 496:44–50. https://doi.org/10.1016/j.jcis.2017.01.113

Ouadil B, Cherkaoui O, Safi M, Zahouily M (2017) Surface modification of knit polyester fabric for mechanical, electrical and UV protection properties by coating with graphene oxide, graphene and graphene/silver nanocomposites. Appl Surf Sci 414:292–302. https://doi.org/10.1016/j.apsusc.2017.04.068

Ovsianytskyi O, Nam Y, Tsymbalenko O, Lan P (2018) Sensors and Actuators B : Chemical Highly sensitive chemiresistive H$_2$S gas sensor based on graphene decorated with Ag nanoparticles and charged impurities. Sensors Actuators B Chem 257:278–285. https://doi.org/10.1016/j.snb.2017.10.128

Pazarçeviren AE, Tahmasebifar A, Tezcaner A, Keskin D, Evis Z (2017) Investigation of bismuth doped bioglass/graphene oxide nanocomposites for bone tissue engineering. Ceram Int 44:3791. https://doi.org/10.1016/j.ceramint.2017.11.164

Ramadas M, Bharath G, Ponpandian N, Ballamurugan AM (2017) Investigation on biophysical properties of hydroxyapatite/graphene oxide (HAp/GO) based binary nanocomposite for biomedical applications. Mater Chem Phys 199:179–184. https://doi.org/10.1016/j.matchemphys.2017.07.001

Salvo P, Melai B, Calisi N, Paoletti C, Bellagambi F, Kirchhain A, Di Francesco F (2018) Graphene-based devices for measuring pH. Sensors Actuators B Chem 256:976–991. https://doi.org/10.1016/j.snb.2017.10.037

Toloman D, Popa A, Stan M, Socaci C, Biris AR, Katona G, Iacomi F (2017) Reduced graphene oxide decorated with Fe doped SnO2 nanoparticles for humidity sensor. Appl Surf Sci 402:410–417. https://doi.org/10.1016/j.apsusc.2017.01.064

Wang L, Zhang Y, Wu A, Wei G (2017a) Designed graphene-peptide nanocomposites for biosensor applications: a review. Anal Chim Acta 985:24–40. https://doi.org/10.1016/j.aca.2017.06.054

Wang T, Sun Z, Huang D, Yang Z, Ji Q, Hu N, Zhang Y (2017b) Studies on NH$_3$ gas sensing by zinc oxide nanowire-reduced graphene oxide nanocomposites. Sensors Actuators B Chem 252:284–294. https://doi.org/10.1016/j.snb.2017.05.162

Wei T, Sun J, Zhang F, Zhang J, Chen J, Li H, Zhang XM (2017) Acetylene mediated synthesis of Au/graphene nanocomposite for selective hydrogenation. Catal Commun 93:43–46. https://doi.org/10.1016/j.catcom.2017.01.029

Yee MSL, Khiew PS, Chiu WS, Tan YF, Kok YY, Leong CO (2016) Green synthesis of graphene-silver nanocomposites and its application as a potent marine antifouling agent. Colloids Surf B Biointerf 148:392–401. https://doi.org/10.1016/j.colsurfb.2016.09.011

Yousefi M, Dadashpour M, Hejazi M, Hasanzadeh M, Behnam B, de la Guardia M, Mokhtarzadeh A (2017) Anti-bacterial activity of graphene oxide as a new weapon nanomaterial to combat multidrug-resistance bacteria. Mater Sci Eng C 74:568–581. https://doi.org/10.1016/j.msec.2016.12.125

Zang Z, Zeng X, Wang M, Hu W, Liu C, Tang X (2017) Tunable photoluminescence of water-soluble AgInZnS–graphene oxide (GO) nanocomposites and their application in-vivo bioimaging. Sensors Actuators B Chem 252:1179–1186. https://doi.org/10.1016/j.snb.2017.07.144

Zhao H, Ding R, Zhao X, Li Y, Qu L, Pei H, Zhang W (2017a) Graphene-based nanomaterials for drug and/or gene delivery, bioimaging, and tissue engineering. Drug Discov Today 22:1302–1317. https://doi.org/10.1016/j.drudis.2017.04.002

Zhao J, Wu L, Zhan C, Shao Q, Guo Z (2017b) Overview of polymer nanocomposites : computer simulation understanding of physical properties. Polymer 133:272–287. https://doi.org/10.1016/j.polymer.2017.10.035

Chapter 16
Nanodiagnostics Tools for Microbial Pathogenic Detection in Crop Plants

Sandra Pérez Álvarez, Marco Antonio Magallanes Tapia,
Jesús Alicia Chávez Medina, Eduardo Fidel Héctor Ardisana,
and María Esther González Vega

16.1 Physiopathology of Pathogenic Infections in Plants

The annual loss of an important part of crops is, must of the time, because pest incidence. This is the reason why the nature of the mechanisms of pathogenic infection and resistance in the host is still under study at present. One of the most discussed issues is the coevolution capacity of the pathogen and the host, understood as the ability of the pathogen to infect an unknown host, and the ability of the host to defend against this new pathogen (Anderson et al. 2010).

For years, it has been investigated, without special success, if the introduction of plant species in a locality significantly reduces the infection by a certain pathogen present there. The hypothesis has been that invasive species leave behind their natural enemies, and that they do not receive special pressure from established pathogens in the areas they reach, since they are adapted to live on native species. However, the evidence on this was contradictory (Wolfe 2002; Agrawal et al. 2005; Parker and Gilbert 2007; van Kleunen and Fischer 2009) until Gilbert and Parker (2010) studied the infection of *Stemphylium solani* isolated in California on species of native and European *Trifolium*, finding a strong increase of the infection as the pathogen was confronted to the new hosts.

This evolution must occur with greater speed and amplitude in pathogens, given the speed with which they reproduce and give rise to new generations, compared with

S. P. Álvarez (✉) · M. A. M. Tapia · J. A. C. Medina
Instituto Politécnico Nacional, CIIDIR Unidad Sinaloa, Depto. de Biotecnología Agrícola,
Guasave, Sinaloa, Mexico

E. F. H. Ardisana
Facultad de Ingeniería Agronómica, Universidad Técnica de Manabí, Portoviejo, Ecuador

M. E. G. Vega
Instituto Nacional de Ciencias Agrícolas (INCA), Carretera a Tapaste,
San José de las Lajas, Mayabeque, Cuba

© Springer Nature Switzerland AG 2018
R. Prasad et al. (eds.), *Exploring the Realms of Nature for Nanosynthesis*,
Nanotechnology in the Life Sciences, https://doi.org/10.1007/978-3-319-99570-0_16

that of plants. Flor's theory about the existence of a gene for a gene (complementarity between the host and the pathogen) is being modified from the understanding that the interaction between them is determined by a complex set of genes in the host and the pathogen (Gilbert and Parker 2010).

Until today, a subject of deep interest in host–pathogen interactions is how plants are able to perceive the attack of pathogens. To date, two proposals are known; the first is based on Flor (1956) theory of gene-to-gene correspondence, and led to the discovery of pathogenesis-related proteins (PRPs), which are proteins produced by plants as a result of pathogen attack (Loon 1985). The PRPs are divided into two large groups, according to Tang et al. (2017):

- Cytoplasmic nucleotide binding domain leucine-rich repeat domain-containing receptors, known as NLRs (Jones et al. 2016).
- Receptors localized in cell surface, which are proteins belonging to large families of receptor-like kinases (RLKs) and receptor-like proteins (RLPs) (Jones and Dangl 2006).

The second way is based on the fact that elicitors and the molecular patterns of damage generated by microorganisms can trigger defense signals in plants, leading to the identification of the abovementioned RLKs and RLPs as recognition patterns (PRRs) of the attacks of pathogens (Boller and Felix 2009). In this way, the NLRs would be responsible for the recognition of the effects at the cytoplasm level, and the PRRs would detect the effects on the apoplast (Tang et al. 2017).

As mentioned above, the constitutive or innate immunity of the plant depends on three events: the first is that the plant can detect the attack of the pathogen; the second, that a complex system of signals is activated, and the third, the expression of defense genes (Tang et al. 2017).

Truman et al. (2013) point out that two types of processes are involved in the detection of attack, which have been called protein-triggered immunity (PTI) (Truman et al. 2013) and effector-triggered immunity (ETI) (Jones and Dangl 2006). In the first of them, pathogenic proteins are recognized by the plant through their coupling to receptors directly involved in the defense system. In the second, pathogens produce effectors that are recognized by plant proteins, unleashing defensive mechanisms.

In the PTI pathway, the recognition of the infection triggers several rapid response processes, and after a few hours the synthesis of phytohormones (ethylene, conjugates of jasmonic acid, salicylic acid) is observed, signaling for the expression of a large group of genes (Fig. 16.1).

Although the recognition of the pathogen through ETI results in a defense mediated by proteins inhibiting invasion, this occurs only in resistant or immune hosts. If the pathogen does not produce effectors capable of triggering the ETI, or if it is capable of reprogramming the transcriptome of the host thanks to other effectors, an effector-triggered susceptibility is produced (ETS) (Mengiste 2012).

Today there is abundant evidence showing that ethylene, which was originally described as "the ripening hormone," participates in other processes of plant growth and development, and also in the response to biotic and abiotic stress (Liu et al. 2016). In Arabidopsis, a linear pathway of ethylene signaling has been discovered, leading

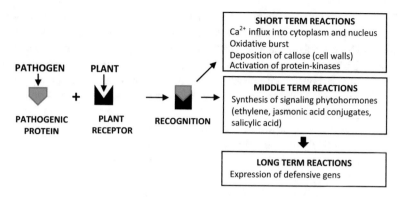

Fig. 16.1 Protein-triggered immunity (data from several authors, cited by Truman et al. 2013)

from its perception in the endoplasmic reticulum to its transcriptional regulation in the nucleus (Yang et al. 2015). However, the ethylene pathway shows diffuse patterns in terms of its participation in resistance: the ethylene insensitive mutants *ein1* and *ein2* are resistant to the biotrophic bacterium *Pseudomonas syringae* and are instead susceptible to the necrotrophic fungi *Botrytis cinerea* and *Plectosphaerella cucumerina* (van Loon et al. 2006; Chen et al. 2009).

Calmodulins are proteins containing four Ca^{2+} fixative sites. Through these bonds, they are responsible for the transport of calcium, which constitutes one of the first signs of recognition of pathogens, the subsequent production of salicylic acid, and the expression of immunity genes mediated by this phytohormone (Wang et al. 2009). Truman et al. (2013) found that three members of the CBP60 family are involved in signaling the resistance of Arabidopsis to *P. syringae* (one of them has a repressor effect and two have inductive effects). Apparently, these two types of actions are related to the repression of the response in the absence of the pathogen and its rapid activation when the attack begins. However, the CBP60 family is also involved in the response to abiotic stresses such as drought (Wan et al. 2012) and low temperatures (Kim et al. 2013).

It should be borne in mind that there are two large groups of pathogenic microorganisms for plants: biotrophic and necrotrophic (Glazebrook 2005). The former attack living tissues, and require them to continue alive to obtain the substances the pathogens require for their growth and reproduction. Instead, the latter need to kill the tissues to release the nutrients that their metabolism demands.Consequently, the defense strategy of the plants is different in each case and completely opposite to that of the pathogens: biotrophic pathogens induce the death of infected cells and confines the infection to a restricted area to maintain cellular life and counteract the attack of necrotrophic organisms (Glazebrook 2005).

Phytohormones participate in the defense against both types of pathogenic microorganisms, and everything indicates that salicylic acid (SA) is the main regulator of the response against biotrophic organisms, while jasmonic acid (JA) and ethylene play a greater role against necrotrophic organisms (Farmer et al. 2003; Vlot et al. 2009).

In response to these functions of phytohormones, pathogens have developed mechanisms to subvert their normal functioning, taking advantage of the existing antagonism between them, such as that between SA and JA. As SA contributes to resistance to biotrophic pathogens, necrotrophic pathogens can activate this path, attenuating that of JA, which gives them advantages of attack. Conversely, biotrophic pathogens trigger the activation of the JA pathway, inhibiting that of SA (Kazan and Lyons 2014).

On the other hand, abscisic acid (ABA), which is known to play an important role in tolerance to abiotic stress, assumes an ambivalent role, interfering at different levels—positively and negatively—with pathogenic signals (Fig. 16.2).

It is believed that this interference in both directions is linked to events as different as the suppression of the resistance conferred by salicylic acid (Mohr and Cahill 2007), the ethylene/JA balance (Anderson et al. 2004), the suppression of Reactive Oxygen Species (ROS), stomatal closure, and the accumulation of callose (Asselbergh et al. 2008). Ton et al. (2009) proposed a model according to which ABA stimulates the defense of the plants in the initial stages of the infection and acts as a suppressor in the advanced stages; however, the large number of contradictory reports on the role of this plant hormone does not yet provide solid explanations about its mode of action.

Consequently, the recognition and response to necrotrophic pathogens are still not sufficiently known, although all these evidences suggest that it is complex in nature and multiple genes and mechanisms are involved, as occurs in the interaction between *Arabidopsis thaliana* x *B. cinerea* (Rowe and Kliebenstein 2008).

The detection of the pathogen's DNA in the host or its products has been useful to diagnose the infection and monitor its progress. Real time-PCR (RT-PCR) (Lievens et al. 2005a; Pasquali et al. 2006) has been useful as a tool for this purpose. However, the presence of the DNA of a fungus in the host is not always an indicator of the infection. For example, the DNA of three strains of *Fusarium oxysporum* (two of them non-pathogenic for tomato) could be detected in plants of this species, even though the symptoms of the infection only appeared where the pathogenic race was present (Validov et al. 2011). However, the amount of fungal DNA detected was several times higher in plants inoculated with the pathogenic race than in plants treated with the two non-pathogenic races, revealing a quantitative relationship that favors the pathogenic microorganism over non-pathogens.

One of the transcription factors of the WRKY family (WRKY33) is involved in the resistance of *Arabidopsis thaliana* to *Alternaria brassicicola* and *B. cinerea* (Zheng et al. 2006; Birkenbihl et al. 2012). The signaling pathway involved is that of ROS; the WRKY family plays an important role in the resistance to numerous types of stress in plants—including the attack of pathogens—(Bakshi and Oelmüller 2014) but also to other physiological phenomena of abiotic origin such as water stress in Arabidopsis (Scarpeci et al. 2013), Chinese cabbage (Tang et al. 2014), and cassava (Wei et al. 2016).

Pathogens penetrate the host through stomata (Grimmer et al. 2012), through wounds or by using specialized mechanisms; the accumulation of molecules such as chitin (fungi), flagellin (bacteria), and others such as peptidoglycans and lipopolysaccharides is associated with pathogen attack and proliferation (Grant et al. 2013).

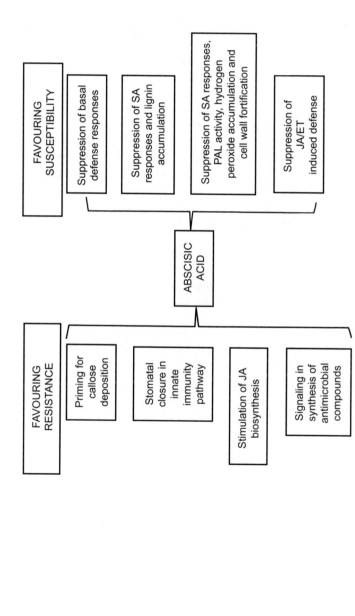

Fig. 16.2 Effects of ABA on plant–pathogen interactions (resumed from results published by Asselbergh et al. 2008)

The penetration of pathogens through stomata is counteracted by the so-called stomatal immunity (Melotto et al. 2006), which is a rapid stomatal closure to restrict the entry of microorganisms. ABA takes part in this closing of the stoma through the control of the hydric balance in the plant (Chater et al. 2014). It has been shown that the mechanism linking ROS and ABA with stomatal closure (Jammes et al. 2009) is related to resistance to *P. syringae* pv tomato DC3000 (Pst) (Jammes et al. 2011). In response to this, some pathogens have mechanisms to overcome the barrier of stomatal immunity. Coronatine, a phytotoxin secreted by some lines of *P. syringae* (Brooks et al. 2004), inhibits stomatal closure (Melotto et al. 2006) promoting the catalysis of SA and preventing its accumulation (Du et al. 2014). Other pathogenic metabolites behave like cytokinins, stimulating the production of ethylene, inhibiting the synthesis of ABA and stomatal closure (Tanaka et al. 2006). Pathogens can also inhibit the ABA-regulated stomatal closure by inducing metabolic processes involving oxalate (Guimaraes and Stotz 2004), triacylglycerol (McLachlan et al. 2016), or starch (Azoulay-Shemer et al. 2016).

As is known, Cl^-, NO_3^-, and malate are the main negative ions balancing K^+ to cause stomatal closure (Kim et al. 2010). In this way, the pathogenic strategy of causing the conversion of malate to oxalate leads to an ionic imbalance favoring potassium and keeping stomata open, facilitating the penetration of the pathogen. On the other hand, a main source of malate is starch, whose accumulation in the cells is related to the stomatal closure. Both events are directly related to each other, and the disturbance of one or the other by the presence of a pathogen could favor the penetration of this through the stoma that would remain open (Azoulay-Shemer et al. 2016).

The studies of McLachlan et al. (2016) in Arabidopsis and *Selaginella* (div. *Lycophyta*) showed that for both higher and lower plants, there is a decrease in the content of the lipid droplets present in the guard cells during the stomatal opening. These lipid droplets contain triacylglycerol (TAG), which is presumed to serve as a substrate for obtaining ATP necessary for the process of opening the stomata; in numerous species the presence of TAG in the guard cells has been detected. Lipid compounds play an important role in the defense against the invasive advancement of pathogens from diseased tissues to still healthy tissues (Shimada and Hara-Nishimura 2015). A possible pathway of pathogenic disturbance would be to alter the metabolism of TAG in favor of ATP production, keeping stomata open and favoring penetration.

The losses caused by pathogens are greater, among other factors, because their presence is always detected when the infection is already established in the plantation, through the characteristic symptoms of the disease. If early diagnosis methods could be established, it would be feasible to take much more effective preventive measures. According to Mazarei et al. (2008), and to the response ways that have been discussed above, there are four routes or elements that can be useful in this regard:

- Salicylic acid-induced response.
- Ethylene-induced response.
- Jasmonic acid-induced response.

- The PR1 gene, which in Arabidopsis is expressed in the presence of pathogenic infection or the application of substances such as salicylic acid itself (Lebel et al. 1998).

The picture is complex because there are pathogens that do not produce specific toxins, such as necrotrophic fungi and some bacteria. These organisms are characterized by the secretion of a large number of substances that degrade cell walls or counteract the action of host enzymes. Against these actions of the pathogen, effective resistance is low, as well as recognition through the PTI and ETI processes. Some toxins, in addition, can be recognized by the host, but this recognition does not go in the sense of resistance but in that of susceptibility (Friesen et al. 2008).

Finally, attention should be paid to the existence of beneficial microorganisms that act symbiotically with the plant. This phenomenon is particularly interesting because both pathogenic and symbiotic microbes have conserved sites of similar microbial molecular patterns (MAMPs) (Zamioudis and Pieterse 2012). The rhizobacteria have developed mutations of the MAMPs that prevent their recognition by the defensive systems of the plant (Trdá et al. 2014). In any case, the particular characteristics of the interaction between the host and the pathogen, and the nature of the molecular signals that are exchanged for mutual recognition and the establishment of defensive reactions in the host, must be taken into account to establish diagnostic techniques using nanotechnology.

16.2 Nanodiagnostics for Fungal Plant Pathogen

In global agriculture, phytopathogenic fungi are responsible for pre- and post-harvest diseases in several crops and they are responsible for quantitative losses; the damage they cause not only refers to economic losses of production but also to losses in biological production, that means, the alteration produced in the growth and development of the host plants attacked by these microorganisms (Agrios 2005).

Both bacteria and fungi cause high crop losses; however, fungal species are often the cause of pathological deterioration of fruits, leaves, stems, roots, tubers, among others. Some authors estimate that these losses are between 5 and 25% in developed countries and between 20 and 50% in developing countries (FHIA 2007).

The effects of fungi on plants can be local, when they affect a small portion of the tissue, or general if the damage is in the whole plant. The damage caused by the fungi is primarily necrosis (death of the tissue they infect), atrophy of the whole plant or parts, or overgrowth (García 2004).

The protection of economical crops is important to meet the demand of a growing world population for the production of food (Strange and Scott 2005). Plant pathogens (fungi, viruses, bacteria, nematodes) are responsible for direct yield losses ranging between 20 and 40% of global agricultural productivity (Oerke et al. 1994; Oerke 2006); for this reason the early detection of diseases is important to avoid disease spread with minimal loss to crop production (Sankaran et al. 2010; Martinelli et al. 2014).

Traditionally, the methods for the identification of plant pathogens are based on visual symptoms in the field and in vitro culture of the pathogen for morphological identification, these lead to mistake because they are not sensitive and they consume a lot of time and money. These methods depend on the experience of the person that makes the diagnosis (Fig. 16.3) (Sankaran et al. 2010; Kashyap et al. 2011). Some other identification methods more sensitive and useful for diagnosis include:

16.2.1 Polymerase Chain Reaction (PCR)

In 1983, Kary Mullis, PhD, a scientist at the Cetus Corporation, conceived of PCR as a method to copy DNA and synthesize large amounts of a specific target DNA. This technology allows the amplification of an original DNA strand and as a result, many copies are obtained. By the use of specific primers, pathogens can be identified quickly and precisely. In addition, a reverse transcription PCR has been used for pathogen detection in a very sensitive way (López et al. 2003). PCR can be affected by the quality of the DNA extracted, primers design that must be specific for the diseases or pathogen and necessary to initiate the reaction, enzyme activity (polymerase), buffer and concentration of deoxynucleoside triphosphate (dNTPs) (Schaad and Frederick 2002; Van der Wolf et al. 2001).

- Usually for primer design conserved genes are selected to amplify conserved regions, for fungi analysis the ITS (internal transcribed spacer) region of the RNAr is widely used. This region has conserved areas suitable for primer designing, also it has variable areas that allow discrimination over a wide range of taxonomic levels (ITS region) (White et al. 1990). The ITS is an ubiquitous region and it is present in all eukaryotic organisms. In some studies about fungal taxonomy, this ITS region has been used showing difference between species such as *Pythium ultimum* and *P. helicoides* (Kageyama et al. 2007); *Peronospora arborescens* and *P. cristata* (Landa et al. 2007); *Colletotrichum gloeosporioides* and *C. acutatum* (Kim et al. 2008), and within species e.g., allowing differentiation of

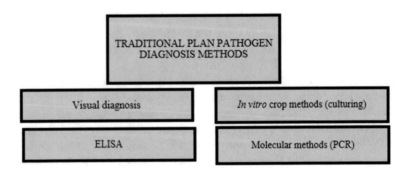

Fig. 16.3 Some traditional methods used for plant pathogen diagnosis

Puccinia striiformis f. sp. tritici (Zhao et al. 2007). Some other examples where the ITS region was used to design primers for fungi identification are *Sclerotium rolfsii* (Jeeva et al. 2010) and *Colletotrichum capsici* (Torres-Calzada et al. 2011).

- Some mitochondrial genes such as cytochrome oxidase genes (*cox*) can be used for fungal identification (Martin and Tooley 2003; Powers 2006) because similar to ribosomal genes they are present in high copy number and present in both rapidly evolving and conserved regions. Some pathogen species cannot be identified with certain primers. For example, certain *Phytophthora* species are impossible to identify using ITS sequences and others are not distinguished using mitochondrial cox II genes (Cooke et al. 2000; Martin and Tooley 2003).

16.2.2 Enzyme-Linked Immunosorbent Assay (ELISA)

The ELISA is an immunological test used to measure antibodies, antigens, proteins, and glycoproteins in biological samples (Clark and Adams 1977) with target antigens from the viruses, bacteria, and fungi that will bind specifically with antibodies conjugated to an enzyme. The detection is based on changes of color due to the interaction between the substrate and the immobilized enzyme (Gorris et al. 1994; López et al. 2001).

- The identification of *Cladosporium fulvum*, a potentially serious fungal pathogen of tomato, can be made based in the expression of ß-glucuronidase gene (GUS) that can be measured for detention and quantification of biomass in tomato leaves. This method was compared with plate-trapped antigen (PTA)-ELISA showing that both methods could detect the pathogen in a low level (<1 mg g^{-1}) (Karpovich-Tate et al. 1998).
- The fungal pathogen *P. ultimatum* causes the damping-off and root rot diseases of hundreds of diverse plant hosts including corn, soybean, potato, wheat, fir, and many ornamental species (Farr and Rossman 2014), and 188 isolates were identified as *P. ultimum* at the asexual stage based on morphological characteristics and their positive reaction with E5 antibody in ELISA (Yuen et al. 1993).

16.2.3 Immunofluorescence

It is a technique used for light microscopy with a fluorescence microscope and is used primarily on microbiological samples. This technique uses the specificity of antibodies to their antigen to target fluorescent dyes to specific biomolecule targets within a cell, and therefore allows visualization of the distribution of the target molecule through the sample. This technique is used to analyze microbiological samples, also to identify or detect pathogen infections in plant tissues (Ward et al. 2004). One example is the detection of *B. cinerea* that induce infection in onion seed heads and cause brown stain (Dewey and Marshall 1996).

16.2.4   Flow Cytometry (FCM)

It is an optical technique laser-based with several applications including the counting of cell and sorting, the detection of some biomarkers, and protein engineering. This technique is interesting because several parameters can be detected simultaneously. FCM was first used in bacteria to study the kinetics cycle of several cells, antibiotic susceptibility, for counting, for DNA characterization, also to characterize fungal spores, but still is a new methodology for plant pathogen identification (Chitarra and van den Bulk 2003).

For plant fungal diseases, FCM is used to characterize genome sizes of fungal and oomycete populations, several pathogen detection, and the monitoring of the viability, cultivability, and gene expression (D'Hondt et al. 2011).

16.2.5   Fluorescence Imaging

It is based in the fluorescence of chlorophyll on leaves and when fluorescence parameters change the pathogen infection can be analyzed based in the photosynthetic apparatus and photosynthetic electron transport reactions (Kuckenberg et al. 2009; Bürling et al. 2011). In plant fungal pathogen leaf rust and powdery mildew infections in wheat leaves at 470 nm were precisely detected with the analysis of temporal and spatial variations of chlorophyll fluorescence (Kuckenberg et al. 2009). At present the practical application of this technique directly in the field is limited (Chaerle et al. 2009; Cséfalvay et al. 2009).

16.2.6   Hyperspectral Techniques

These techniques are highly robust and use spectrum in a range of 350 to 2500 nm to obtain important information about plant condition. In addition, with these techniques the study of plant phenotype and the identification of crop diseases can be made (Mahlein et al. 2012). In the case of plant disease detection, measuring the changes in reflectance resulting from the biophysical and biochemical characteristic changes upon infection has been applied to identified *Magnaporthe grisea* infection of rice, *Phytophthora infestans* infection of tomato, and *Venturia inaequalis* infection of apple trees (Zhang et al. 2003; Delalieux et al. 2007).

The combination of molecular diagnosis and nanotechnology is a great technology for the rapid and precise identification of different plant pathogens. At present for diagnosis, several nanodevices and nanosystems have been used, as well as sequencing of DNA individual molecules. The recently developed nanomaterials with special nanoscale characteristics represent a tremendous advance in the technology of detection and diagnosis of plant pathogens (Khiyami et al. 2014).

The combination of biology and nanotechnology is called nanobiosensors (Yang et al. 2008), and they can increase greatly sensitivity and consequently when a disease problem is present in a crop they can diminish response time (Small et al. 2001). For these reasons, nanosensors can improve production and food safety in agriculture. In this context, Dubas and Pimpan (2008) investigated the particular optical properties of silver nanoparticles, the interaction between silver nanoparticles and sulphurazon-ethyl herbicide. The authors found that these nanoparticles increased concentrations of herbicide in a solution and induced a variation in color of the nanoparticles from yellow to orange red and finally to purple; this variation is useful for the detection of several contaminants such as organic pollutants and microbial pathogens in water bodies and in the environment (Pal et al. 2008).

16.2.7   Nucleic Acid Based Affinity

Biosensor is a new method for pathogen detection that allows the detection at molecular level of the diseases before any symptoms appear and this has been widely used for detection of bacteria, fungi, and genetically modified organisms. A single-stranded DNA (ssDNA) is the most used DNA probe on electrodes to measure hybridization between probe DNA and the complementary DNA analyses (Eun and Wong 2000). There are four major types of DNA-based biosensors depending on their mode of transduction: optical, piezoelectric, strip type, and electrochemical DNA biosensors. The optical DNA-based biosensors can be further classified to three subtypes—molecular beacons (MB), surface plasmon resonance (SPR), and quantum dots (Eun and Wong 2000).

(a) The molecular beacons (MB) is a new fluorescence probe based on DNA.
(b) The surface plasmon resonance (SPR).
(c) Quantum dots (QD) are semiconductors that use nanocrystals and light of specific wavelengths based in a luminescent property of the infrared light emitted (Fig. 16.4) (Edmundson et al. 2014). This technology is better than other fluorescent dyes because luminescence is more efficient, the spectra emission is small, and great photostability QD can be excited to all colors with only one light source (Warad et al. 2004). QD have been used for diagnosis of plant pathogens such as fungi. In yeast was first reported the mycosynthesis of semiconductor nanomaterials because yeast can produce cadmium sulfide (CdS) crystallites in response to cadmium salt stress (Dameron et al. 1989), but yeast is not the only microorganism that can produce CdS; nevertheless, the studies on their fluorescent properties are not enough (Yadav et al. 2015). *Fusarium oxysporum* Schltdl. can produce a myco-mediated synthesis of highly fluorescent CdTe quantum dots when reacted with a mixture of $CdCl_2$ and $TeCl_2$ (Jain 2003; Kashyap et al. 2013; Alghuthaymi et al. 2015). Another example is the use of QD-FRET (Fluorescence resonance energy transfer) mechanism (Grahl and Märkl 1996) for detection and identification of *Aspergillus amstelodami* in concentrations as low as 103 spores/mL in 5 min or less (Kattke et al. 2011).

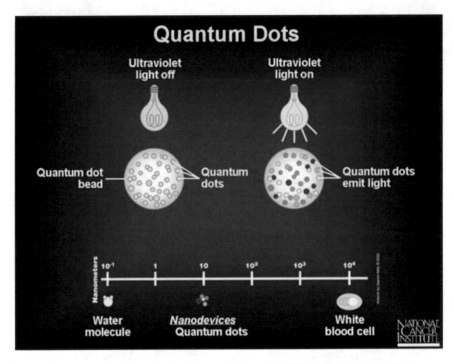

Fig. 16.4 Quantum dot schematization. (www.ubooks.pub/BOOKS/BO/E20R2020/MAIN/images/image014jpg)

Nanomaterials have been used for sensor development because they offer a friendly platform that can be utilized for the assembly of bio-recognition elements, the high surface area, high electronic conductivity and plasmonic properties of nanomaterials that enhance the limit of detection. Several metal and metal oxide nanoparticles, quantum dots, carbon nanomaterials such as carbon nanotubes and graphene as well as polymeric nanomaterials are used for biosensor construction (Singh et al. 2010).

Nanosensor is another technique used for the detection of plant pathogens in a very precise way (Fig. 16.5) (Esker et al. 2008; Rai and Ingle 2012). Several polymers such as polyaniline, polythiophene, and polypyrrole are used for sensors fabrication for the detection of molecular signal (Sekhon 2010), so these bio-nanosensors can improve sensitivity and reduce significantly the response time to discover potential disease problems and they were used in agriculture and food system to detect and quantify small amounts of pathogens like viruses, bacteria, and fungi (Rai and Ingle 2012). These nanosensors can be used in real time by the connection to a GPS, monitoring not only diseases but also soil conditions and crop health (Clarke et al. 2009; Rai and Ingle 2012). Nanosensors can allow scientists and farmers to identify plant diseases before visible symptoms appear and this will facilitate their control with enough time to avoid economic losses (Rai and Ingle 2012).

Fig. 16.5 Nanosensors.
Swierczewska et al.,
(2012)

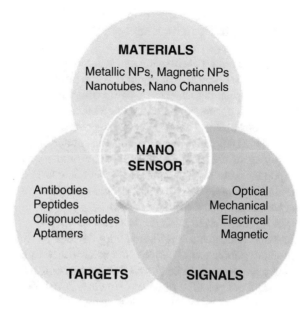

Nanosensors also can be used to detect some chemicals released during food spoilage and some of them are based on microfluidic devices (Baeummer 2004), so they can be used to identify pathogens proficiently in short time with high sensitivity.

A sensor with gold electrode and copper nanoparticles has been used to monitor the levels of salicylic acid in oil seeds to detect *Sclerotinia sclerotiorum* (Wang and Li 2010).

Biosensors combined with copper oxide (CuO) nanoparticles were used for detecting *Aspergillus niger* (Etefagh et al. 2013). *A. niger* causes black mold of onions and ornamental plants. Infection of onion seedlings can become systemic, manifesting only when conditions are conducive. *A. niger* causes a common post-harvest disease of onions, in which the black conidia can be observed between the scales of the bulb. The fungus also causes disease in peanuts and grapes (Schuster et al. 2002).

Several biosensors have been developed and described for mycotoxin analysis, some of them have the potential of multi-array mycotoxins identification with the principal role of decreasing the time for fungal pathogen detection (Baeummer 2004).

Nanoparticles have been used for the detection/diagnosis of fungal plant pathogens by several authors (Boonham et al. 2008; Yao et al. 2009). Some examples are gold nanoparticles, copper oxide, titanium dioxide, tin dioxide, and some others.

Partial charcoal is the wheat and triticale disease caused by the fungus *Tilletia indica* Mitra, also known as *Neovossia indica* (Mitra) Mundkur. Typically, the fungus infects only a few grains per spike and the disease only develops in a part of the infected grains. The disease is also known as Carnal Charcoal, and it was discovered in 1931 in India. Subsequently, it has been found in large areas of northern India, Pakistan, Nepal, Iran, and in some parts of Mexico and South of the United States (Vernon and Vizcarra-Mendoza 2000). Gold nanoparticles have been developed for detecting *T. indica* using surface plasmon resonance (Singh et al. 2010). These nanoparticles have a high electroactivity and electronic conductivity (Cao et al. 2011).

In plants infected with pathogenic fungus *Phytophthora cactorum*, nanoparticles of TiO_2 or SnO_2 on screen-printed carbon electrodes have been used for the identification of p-ethylguaiacol, a volatile compound present in this plants (Fang et al. 2014). *P. cactorum* was first identified on cacti in 1870 by Lebert and Cohn (Erwin and Ribeiro 1996). This oomycete is capable of infecting an extremely large number of hosts (Erwin and Ribeiro 1996; Benson and Jones 2001) and is problematic in low-lying or wet field conditions. It can limit production for many economically important crops such as apple, pear, rhododendron, azalea, and strawberry (Fujita 1990; Erwin and Ribeiro 1996). *P. cactorum* can cause root, collar, and crown rots, as well as foliar and fruit infections.

A briefcase-sized kit is used as another method of nanodiagnosis directly in the field and it contains some measuring devices, reagents, power supply, and other products (Goluch et al. 2006). By using nanodiagnostic kit equipment potential plant pathogens could be detected easily and quickly and this will allow to prevent farmers from potential diseases (Pimentel 2009). For example, 4mycosensor is an antibody-based assay for the real-time detection of ZEA, T-2/HT-2, DON, and FB1/FB2 mycotoxins on the same single strip in corn, wheat, oat, and barley samples at or below their respective European maximum residue limits (MRLs) (Lattanzio et al. 2012), so this immunoassay is fast, cheap, easy-to-use, and suitable for the purpose of quick screening of mycotoxins in cereals.

Another tools used for nanodiagnostic in plant pathology are magnetic nanoparticles (MNPs) consisting of magnetite (Fe_3O_4) and they were produced with hydrothermal protocols. The size and distribution of these particles were characterized by transmission electron microscopy (TEM) and dynamic light scattering (DLS), which revealed that the synthesized MNPs were highly monodispersed. The MNPs were used by Alghuthaymi (2017) for the molecular identification of two novels *Cladosporium cladosporioides* from tomato phylloplane; for this reason fungal DNA was extracted using MNPs in comparison with the conventional sodium dodecyl sulfate (SDS) method in the context of quality, quantity, and timing process. For PCR the universal primers ITS for fungi were used together with actin genes as housekeeping ones and the amplification result was 100% demonstrating the high quality of the isolated DNA. In this study, 47 isolates of *C. cladosporioides* complex were phylogenetically evaluated based on DNA sequences of the internal transcribed spacer regions ITS1 and ITS2 together with partial actin and translation elongation factor 1-α gene sequences. The *Cladosporium* strains were isolated in Saudi Arabia and Egypt from tomato crop and were identified as *C. asperulatum* and *C. myrtacearum* based on their molecular phylogenetic characteristics. The combination of magnetic nanoparticles and DNA isolation could be used commonly both in plant pathology laboratories and in the nanobiotechnology industry (Alghuthaymi 2017). Methods relating with magnetic separation have been extensively studied, but still there is a need for an efficient technique to isolate DNA for diagnostic applications with high sensitivity and MNP techniques use less steps, microcentrifuge tubes and time; for all these reasons, it is more cost-effective when compared with traditional techniques (Maeda et al. 2016).

16.3 Plant Viruses: Nanodiagnostic

16.3.1 Viruses

Viruses are pathogens that have one or two molecules of nucleic acids, surrounded by a protective layer, consisting of proteins. Ninety percent of the viruses have an RNA genome and the remaining 10% have a DNA genome, single-stranded or double-stranded. These biological entities are considered obligate parasites and they have the capacity to cause diseases in all living organisms, including man. Plant viruses cause the damage inside plant cells by intervening the allocation of resources that the plant has produced through photosynthesis, causing a delay in growth, and shortening the life of the host; however, other observed symptoms include chlorosis, yellowing, crinkling, mosaics, mottled, necrosis, deformation of leaves and fruits, and flower abortion, resulting in a severe impact on production. However, some viruses, under appropriate conditions, can infect a plant without producing symptoms of disease (Bové et al. 1988; Hull 2002; Agrios 2005; van der Want and Dijkstra 2006; Schuman and D'Arcy 2010).

Viruses are ranked as the second most important plant pathogens following fungi. Economic loss has been estimated more than several billion dollars per year worldwide in many crops such as barley, corn, potato, rice, and wheat. The crop damages owing to viral diseases are difficult to predict, because they depend on virus strain, host plant cultivar/variety, region, and time of infection (Hull 2002; Agrios 2005; Strange and Scott 2005; Ellis et al. 2008).

In addition, plants can develop symptoms similar to those produced by this group of phytopathogens when they are affected by biotic or abiotic factors such as nutritional imbalances, unfavorable weather, and pathogens caused by pests and other agents. Therefore, the diagnosis of viral diseases based on the observed symptoms is more difficult than those caused by other pathogens (Lievens et al. 2005b; van der Want and Dijkstra 2006).

16.3.2 Nanodiagnosis Applied to Virology

The diagnosis is the basis to manage plant diseases and to predict the crop loss by infection of plant pathogens. The main problem in the productivity of a crop lies mainly in the early detection of the disease or infectious agent. Most of the time, pesticides are applied preventively, causing residual toxicity in the crop in addition to environmental risks. On the other hand, the application of pesticides once the disease is presented cause an extra economic investment in crop management. Within the diseases, the viral ones are the most complicated to control due to the emphasis in the control of their vectors and consequently their propagation. Therefore, the key to success in the control of diseases based on a crop management system, particularly of viral origin, is to figure out the causes by the detection in early stages such as the replication of viral DNA or the initial production of its viral protein (Pearson et al. 2006; van der Want and Dijkstra 2006; Aboul-Ata et al. 2011).

The viral diagnosis based on nanomaterials and utilization of biomarkers including the development of multiple diagnostic kits has been of great importance to detect the variants and identify differential proteins production in healthy and sick states during the infectious cycle to stop the disease. These nano-based diagnostics not only increase the speed of detection but also increase the power of the detection (Prasanna 2007; Mousavi and Rezaei 2011; Rai and Ingle 2012).

Methods for detection and identification of viruses are critical in the management of these phytopathogens (Aboul-Ata et al. 2011). Therefore, detection methods should be more convenient, effective, specific and permit the use for detecting plant pathogens (McCartney et al. 2003). Nanotechnology is the creation and utilization of materials, devices, and systems through the control of matter on the nanometer-length scale, at the level of atoms, molecules, and supramolecular structures. It is the popular term for the construction and utilization of functional structures with at least one characteristic dimension measured in nanometers (Jain 2003; Tan et al. 2004). A rapid diagnosis of the disease often depends on the sensitivity of the genes used for the detection, identification, and control of smaller and less abundant targets (mRNA, DNA, proteins, and peptides) (Wang et al. 2006).

16.3.2.1 Field Sensing Systems Based in Nanosensors

This system of precision agriculture is used to monitor the appropriate time of planting and harvesting in the field, climatic conditions, efficient use of water and soil nutrition, as well as the detection of pests, including viruses. Field system makes use of nanomaterials such as polyaniline, polythiophene, and polypyrrole that function as a network of wireless sensors with global positioning system for real-time monitoring and satellite imagery in the field (Joseph and Morrison 2006; Derosa et al. 2010; Chen and Yada 2011; Ingale and Chaudhari 2013; Prasad et al. 2014, 2017).

Nanosensors identify plant diseases before visible symptoms appear and thus facilitate their control. Precision farming will allow improved agriculture production by providing precise data, helping growers to make better decisions (Rai and Ingle 2012).

16.3.2.2 Nanobiosensors as a Diagnostic System

Nanobiosensors are nanosensors with immobilized bioreceptor probes that are selective for target molecules which offer the advantage of being portable, small, quantitative, and reliable in the identification of potential disease problems. These systems are used in agriculture for the detection and quantification in minutes of pathogens such as viruses (Rai and Ingle 2012; Khiyami et al. 2014; Srinivasan and Tung 2015). These sensors can be linked to a GPS and distributed throughout the field for real-time monitoring of disease, soil conditions, and crop health (Nezhad 2014; Prasad et al. 2014, 2017).

Nanowire biosensors are a class of nanobiosensors, of which the major sensing components are made of nanowires coated by biological molecules such as DNA molecules, polypeptides, fibrin proteins, and filamentous bacteriophages. Since their surface properties are easily modified, nanowires can be decorated with virtually any potential chemical or biological molecular recognition unit, making the wires themselves analyte independent. Nanomaterials, due to their unique characteristics, including stable emissions and size-dependent properties, have attracted great attention as they transduce the chemical bonding event on their surface to a change in the conductance of the nanowire in an extremely sensitive, real-time and quantitative manner (Rai and Ingle 2012; Kashyap et al. 2017).

Nanowires were used as a transducer biosensor to detect cucumber mosaic virus (CMV) and papaya ring spot virus (PRSV), indicating that nanowires are good candidate materials for the manufacture of nanoscale biosensors for the diagnosis of viral diseases (Ariffin et al. 2014). Also, 11-mercaptoundecanoic acid was self-assembled on gold surface and then crosslinked with anti-maize chlorotic mottle virus (anti-MCMV) for MCMV disease detection. SPR response to MCMV solutions was evaluated over time with different concentrations from 1 to 1000 ppb. The limit of detection was evaluated to be approximately 1 ppb (Zeng et al. 2013).

Biosensors based on living cells are characterized by low limit of detection, high specificity, and rapid response time. A novel portable cell biosensor system for detection of potato virus Y (PVY), CMV, and tobacco rattle virus (TRV) was fabricated by immobilizing the vero cells carrying virus specific antibodies on their membranes. This study demonstrated an important step towards the development of a portable plant virus detection system suitable for on-field application (Perdikaris et al. 2011).

16.3.2.3 Nanodiagnosis Using Quantum Dots (QDs)

QDs, as described above in detail, are semiconductors made out of the elements from groups II and VI or groups III and V in periodic table that fluoresce when stimulated by an excitation light source. Furthermore, QDs are inorganic fluorophores presenting major advantages over traditional organic fluorophores used as markers on nucleic acids or proteins for visual detection and these nanomaterials have been most widely used for disease diagnosis (Wang et al. 2006; Adams and Barbante 2013; Holzinger et al. 2014). In the most of the cases, viral nucleic acids have been used for hybridization assays as well as antigen-antibody affinity for building immunoassays as mechanism, which will enable FRET follow their detection. FRET is an acronym for Förster resonance energy transfer and involves non-radiative transfer of energy between donor and acceptor fluorophores (Uhl et al. 2011).

Quantum Dots FRET-based biosensors are applied in the detection of viruses like necrotic yellow vein virus (BNYVV), a plant virus responsible for rhizomania disease in sugar beet and transmitted by *Polymyxa beta* (Keskin). BNYVV sensor FRET occurred between thioglycolic acid-modified cadmium telluride QD conjugated with anti-glutathione-S-transferase antibody (anti-GST) via electrostatic interaction as energy donor and fluorescent dye rhodamine attached to GST as energy acceptor (Safarpour et al. 2012).

Likewise, the nanodiagnostic of *citrus tristeza virus* (CTV) was reported where the cadmium telluride-quantum dots (CdTe-QDs) were conjugated with a specific antibody against coat protein (CP) of CTV, and the CP was immobilized on the surface of gold nanoparticles (AuNPs) to develop a specific and sensitive fluorescence resonance energy transfer (FRET) based nanobiosensor (Shojaei et al. 2016a). This was also done in another study; however, the conjugated was immobilized on the surface of carbon nanoparticles (CNPs). The limit of detection was measured at about 220 ng mL^{-1} of CTV (Shojaei et al. 2016b).

In the same way, the conjugated CdTe-QDs-CP of CTV was achieved in a parallel reaction, rhodamine dye molecules were attached to the purified recombinant CTV-CP. Two independent approaches were explored for detection of the infected plants. First, in fluorescence resonance energy transfer (FRET) based assay, the quenching ability of rhodamine molecules was applied for altering the QDs light emission. More specifically, donor-acceptor complexes (Ab-QD + CP-Rd) were created based on the affinity of antibody-antigen molecules. The resulting assembly brought Ab-QD (the donor) and the Rd-CP (the acceptor) into a close proximity and resulted in a substantial decrease in the intensity of QD light emission. Addition of free antigen into the solution resulted in the replacement of CP-Rd with free CP and a subsequent increase in the emission of QDs. In the second approach, a non-FRET based assay was performed through the addition of free antigen to the Ab-QD solution, which led to the aggregation of the Ab-QD conjugates and consequently to a significant increase in the light intensity emission of the QD (Safarnejad et al. 2017).

16.3.2.4 Gold Nanoparticles

Gold nanoparticles (GNP) are used in biosensors due to their optical, electronic, compatibility and simple modification properties. These nanomolecules have been in active use in the identification of chemical and biological agents (Li et al. 2010; Biju 2014). Predominantly, transmission electron microscopy has historically remained the predominant means to detect biospecific interactions using colloidal gold particles (Hayat 1989). In addition to the conventional colloidal gold with quasi-spherical particles (nanospheres), also nonspherical particles, such as nanorods, nanoshells, nanocages, nanostars, among other particles can be used (Khlebtsov and Dykman 2010).

In this regard, it was reported the use of DNA molded to self-assembly of gold nanoparticles grouped in different configurations for the colorimetric detection of the DNA of the Tomato leaf curl New Delhi virus (ToLCNDV), using a bifunctional oligonucleotide probe conjugated with gold nanoparticles. This study provided new insight into the preparation of gold nanoparticle clusters in defined shapes and sizes using a single type of GNP (Dharanivasan et al. 2016). Likewise, another report indicated the detection of tobacco mosaic virus (TMV) through immunochromatography technique for express diagnosis of plant viral infections on polycomposite test strips using colloidal gold nanoparticles as a visual label. In this work, polyclonal antibodies were used whose molecules can differ significantly in chemical composition and immunochemical and physicochemical properties (Drygin et al. 2009).

In other work, a highly practical and rapid lateral-flow assay (LFA) was developed for the detection of soybean mosaic virus (SMV). The LFA is an immunoassay method based on immuno-chromatography. In this research, they used gold nanoparticles for colloidal solution. The assay can be used to detect SMV in infected leaf samples and soybean seeds (Zhu et al. 2016).

16.3.2.5   Nanorod Technology

Gold nanorods (AuNRs) have larger absorption and scattering cross sections in the l-mode PPR wavelength region than AuNSs in their PPR wavelength region (Parab et al. 2010; Lin et al. 2012). The l-mode PPR wavelength is tunable from the visible to near-infrared regions (650–900 nm) by tailoring the aspect ratio of AuNRs. The spectral region is particularly useful for biosensing because within the range the background absorption and scattering of endogenous chromophores from biological mixtures (e.g., hemoglobins in blood) are minimal (Nusz et al. 2008). In particular, AuNRs have inherently higher sensitivity to the local dielectric environment than similar sized nanospheres (McFarland and Van Duyne 2003). Moreover, the l-mode PPR of AuNRs has higher bulk refractive index (RI) sensitivity and also narrower line width than the t-mode one (Ni et al. 2008).

Due to the unique optical properties of AuNRs, they have been recently developed to be highly sensitive optical transducers for nanodiagnosis, such as being probes for detecting target DNA of pathogenic bacteria (Parab et al. 2010; Wang et al. 2012) and Raman labels in sandwich immunoassay for detection of virus (Baniukevic et al. 2013). Based in the AuNR functionalized by antibodies, the detection of *cymbidium mosaic virus* (CymMV) or *odontoglossum ringspot virus* (ORSV) using fiber optic particle plasmon resonance immunosensor was recorded (Lin et al. 2014). These viruses were also detected in crude saps using the quartz crystal microbalance (QCM) technique with gold electrodes on either side immunosensor which measures the change in frequency of a quartz crystal resonator (Eun et al. 2002).

16.3.2.6 Nanoribbon Technology

The chemiresistive immunosensor based on antibody-functionalized polypyrrole (PPy) nanoribbons was used to detect viral plant pathogen (James 2013). In the study, CMV detection was reported, where lithographically patterned nanowire electrodeposition (LPNE) technique was employed to batchfabricate the PPy sensor by integrating patterned microelectrodes. For it, N-(3-Dimethylaminopropyl)-N'-ethylcarbodiimide (EDC)/N-hydroxysuccinimide (NHS) was used to surface functionalize antibodies for CMV onto the nanostructured PPy. It is important to mention that in this study the detection was optimized by adjusting the electrical conductivity of PPy nanoribbons, the dimensions of the nanostructure, and the ionic strength of the pH buffer (Chartuprayoon et al. 2013).

16.3.2.7 Immunosensors

This technique consists of antibodies (Ab) or antigens (Ag) coupled to a transducer that generate an analytical response, and it can provide an alternative to used detection system. Depending on the method of signal transduction, immunosensors may be divided into four basic groups: electrochemical, optical, piezoelectric, and thermometric. Among these different types of immunosensors, the electrochemical ones show more potential thanks to their higher sensitivity, higher speed, and permanent control (Ricci et al. 2007; Hassen et al. 2011; Sun et al. 2011).

In relation to it, a portable electromechanical immunosensor system was performed for the detection of CMV, based on immobilized CMV specific antibodies conjugated with gold nanoparticles. This immunosensor provides a promising successful tool for real-time and sensitive CMV detection in the nanophytopathology (Rafidah et al. 2016). Also, recently a label-free and sensitive electrochemical immunosensor for efficient and rapid detection of CTV was performed. The specific antibody against coat protein (CP) of CTV was successfully immobilized on 11-mercaptoundecanoic acid (MUA) and 3-mercapto propionic acid (MPA) modified gold electrode via carbodiimide coupling reaction using N-(3-dimethylaminopropyl)-N-ethylcarbodiimide hydrochloride (EDC) and N-HydroxySuccinimide (NHS) (Haji-Hashemi et al. 2017).

Capsicum chlorosis virus (CaCV) was detected efficiently in bell pepper by sensitive label-free amperometric-based biosensor. Antigen was immobilized over the surface of gold nanoparticle/multi-walled carbon nanotube (Nano-Au/CMWCNT) screen-printed electrodes using 1-Ethyl-3- (3- dimethylaminopropyl) carbodiimide (EDC)/N-hydroxysuccinimide (NHS) crosslinking chemistry followed by interaction with groundnut bud necrosis virus (GBNV)/CaCV specific polyclonal antibody (Sharma et al. 2017).

16.3.2.8 Nanochips

Based on an electronically addressable electrode, the nanochips are kinds of micro-arrays, which consist of fluorescent oligo capture probes that are used to identify hybridization. This nanotechnology is highly specific and sensitive to detect the single nucleotide changes occurring in bacteria and viruses. A particular feature of this system is that biotinylated immobilized molecules can be either oligo capture probes or amplified PCR samples (Sosnowski et al. 1997; López et al. 2009). Due to their ability to discriminate single nucleotide changes, nanochips have shown high specificity and accuracy to diagnose PVY, potato virus X (PVX), and potato leafroll virus (PLRV) in potato. In addition, the Nanogen system allows differentiating between the PVY, PVY^0, and $PVY^{N/NTN}$ races (Ruiz-García et al. 2004).

In the management of viral diseases, preventive control is efficient when the crop is free of viruses; therefore, a quick, accurate and effective diagnosis is of vital importance. In this sense, the diagnosis based on symptoms is not entirely effective, since these can be associated with nutritional deficiencies and interactions between hosts and viral co-infections. For this reason, the different nanodiagnostic technologies can be applied depending on the purpose of the investigation. Under this premise, it is necessary to continue conducting research to optimize these tests in order to obtain reliable results when the pathogen is in low titers in its host.

References

Aboul-Ata AE, Mazyad H, El-Attar AK, Soliman AM, Anfoka G, Zeidaen M, Gorovits R, Sobol I, Czosnek H (2011) Diagnosis and control of cereal viruses in the Middle East. Adv Virus Res 81:33–61

Adams FC, Barbante C (2013) Nanoscience, nanotechnology and spectrometry. Spectrochim Acta B 86:3–13

Agrawal AA, Kotanen PM, Mitchell CE, Power AG, Godsoe W, Klironomos J (2005) Enemy release? An experiment with congeneric plant pairs and diverse above-and belowground enemies. Ecology 86:2979–2989

Agrios GN (2005) Plant pathology, 5th edn. Academic Press, New York, p 803

Alghuthaymi MA (2017) Nanotools for molecular identification two novels *Cladosporium clado-sporioides* species (Cladosporiaceae) collected from tomato phyloplane. J Yeast Fungal Res 8(2):11–18

Alghuthaymi MA, Almoammar H, Rai M, Said-Galiev E, Abd-Elsalam KA (2015) Myconanoparticles: synthesis and their role in phytopathogens management. Biotechnol Biotechnol Equip 29(2):221–236

Anderson JP, Badruzsaufari E, Schenk PM, Manners JM, Desmond OJ, Ehlert C, Maclean DJ, Ebert PR, Kazan K (2004) Antagonistic interaction between abscisic acid and jasmonate-ethylene signaling pathways modulates defense gene expression and disease resistance in Arabidopsis. Plant Cell 16(12):3460–3479

Anderson JP, Gleason CA, Foley RC, Thrall PH, Burdon JB, Singh KB (2010) Plants versus pathogens: an evolutionary arms race. Funtc Plant Biol 39(6):499–512

Ariffin SAB, Adam T, Hashim U, Faridah S, Zamri I, Uda MNA (2014) Plant diseases detection using nanowire as biosensor transducer. Adv Mater Res 832:113–117

Asselbergh B, de Vleesschauwer D, Höfte M (2008) Global switches and fine-tuning-ABA modulates plant pathogen defense. Mol Plant Mic Interact 21(6):709–719

Azoulay-Shemer T, Bagheri A, Wang C, Palomares A, Stephan AB, Kunz HH, Schroeder JI (2016) Starch biosynthesis in guard cells but not in mesophyll cells is involved in CO2-induced stomatal closing. Plant Physiol 171:788–798

Baeummer A (2004) Nanosensors identify pathogens in food. Food Technol 58:51–55

Bakshi M, Oelmüller R (2014) WRKY transcription factors. Plant Signal Behav 9:2

Baniukevic J, Hakki Boyaci I, Goktug Bozkurt A, Tamer U, Ramanavicius A, Ramanaviciene A (2013) Magnetic gold nanoparticles in SERS-based sandwich immunoassay for antigen detection by well oriented antibodies. Biosens Bioelectron 43:281–288

Benson DM, Jones RK (2001) Diseases of woody ornamentals and trees in nurseries. APS Press, St. Paul, MN

Biju V (2014) Chemical modifications and bioconjugate reactions of nanomaterials forsensing, imaging, drug delivery and therapy. Chem Soc Rev 43:744–764

Birkenbihl RP, Diezel C, Somssich IE (2012) Arabidopsis WRKY33 is a key transcriptional regulator of hormonal and metabolic responses toward *Botrytis cinerea* infection. Plant Physiol 159:266–285

Boller T, Felix GA (2009) Renaissance of elicitors: perception of microbe-associated molecular patterns and danger signals by pattern-recognition receptors. Ann Rev Plant Biol 60:379–406

Boonham N, Glover R, Tomlinson J, Mumford R (2008) Exploiting generic platform technologies for the detection and identification of plant pathogens. In: European Journal of Plant Pathology. Springer, Dordrecht, pp 355–363

Bové JM, Vogel R, Albertini D, Bové JM (1988) Discovery of a strain of Tristeza virus (K) inducing no symptoms in Mexican lime. Proceedings of the 10th Conference of IOCV. Spain 1988. International Organization of Citrus Virologists, Riverside, CA, pp 14–16

Brooks DM, Hernandez-Guzman G, Kloek AP, Alarcon-Chaidez F, Sreedharan A, Rangaswamy V, Penaloza-Vazquez A, Bende CL, Kunkel BN (2004) Identification and characterization of a well-defined series of coronatine biosynthetic mutants of *Pseudomonas syringae* pv. tomato DC3000. Mol Plant Microbe Interact 17:162–174

Bürling K, Hunsche M, Noga G (2011) Use of blue-green and chlorophyll fluorescence measurements for differentiation between nitrogen deficiency and pathogen infection in winter wheat. J Plant Physiol 168:1641–1648

Cao X, Ye Y, Liu S (2011) Gold nanoparticle-based signal amplification for biosensing. Anal Biochem 417(1):1–16

Chaerle L, Lenk S, Leinonen I, Jones HG, Van Der Straeten D, Buschmann DC (2009) Multisensor plant imaging: towards the development of a stress catalogue. Biotechnol J 4:1152–1167

Chartuprayoon N, Rheem Y, Ng J, Nam J, Chen W, Myung N (2013) Polypyrrole nanoribbon based chemiresistive immunosensors for viral plant pathogen detection. Anal Methods 5(14):3497–3502

Chater CC, Oliver J, Casson S, Gray JE (2014) Putting the brakes on: abscisic acid as a central environmental regulator of stomatal development. New Phytol 202:376–391

Chen H, Yada R (2011) Nanotechnologies in agriculture: new tools for sustainable development. Trends Food Sci Technol 22:585–594

Chen H, Xue L, Chintamanani S, Germain H, Lin H, Cui H, Cai R, Zuo J, Tang X, Li X, Guo H, Zhou JM (2009) Ethylene insensitive3 and ethylene insensitive3-like1 repress salicylic acid induction deficient2 expression to negatively regulate plant innate immunity in *Arabidopsis*. Plant Cell 21:2527–2540

Chitarra LG, van den Bulk RW (2003) The application of flow cytometry and fluorescent probe technology for detection and assessment of viability of plant pathogenic bacteria. Eur J Plant Pathol 109:407–417

Clark MF, Adams A (1977) Characteristics of the microplate method of enzyme-linked immunosorbent assay for the detection of plant viruses. J Gen Virol 34:475–483

Clarke J, Wu H-C, Jayasinghe L, Patel A, Reid S, Bayley H (2009) Continuous base identification for single-molecule nanopore DNA sequencing. Nat Nanotechnol 4(4):265–270

Cooke DEL, Drenth A, Duncan JM, Wagels G, Brasier CM (2000) A molecular phylogeny of Phytophthora and related oomycetes. Fungal Genet Biol 30:17–32

Cséfalvay L, Gaspero GD, Matous K, Bellin D, Ruperti B, Olejnickova J (2009) Pre-symptomatic detection of Plasmopara viticola infection in grapevine leaves using chlorophyll fluorescence imaging. Eur J Plant Pathol 125:291–302

D'Hondt L, Höfte M, Van Bockstaele E, Leus L (2011) Applications of flow cytometry in plant pathology for genome size determination, detection and physiological status. Mol Plant Pathol 12(8):815–828

Dameron CT, Reeser RN, Mehra RK, Kortan AR, Carroll PJ, Steigerwaldm ML, Brus LE, Winge DR (1989) Biosynthesis of cadmium sulphide quantum semiconductor crystallites. Nature 338(6216):596–597

Delalieux S, van Aardt J, Keulemans W, Schrevens E, Coppin P (2007) Detection of biotic stress (Venturia inaequalis) in apple trees using hyperspectral data: non-parametric statistical approaches and physiological implications. Eur J Agron 27:130–143

DeRosa MC, Monreal C, Schnitzer M, Walsh R, Sultan Y (2010) Nanotechnology in fertilizers. Nat Nanotechnol 5:91

Dewey F, Marshall G (1996) Production and use of monoclonal antibodies for the detection of fungi. In: Proceeding of British Crop Protection Council Symposium, Farnham, UK, pp 18–21

Dharanivasan G, Mohammed Riyaz SU, Jesse DMIT, Muthuramalingam R, Rajendran G, Kathiravan K (2016) DNA templated self-assembly of gold nanoparticle clusters in the colorimetric detection of plant viral DNA using a gold nanoparticle conjugated bifunctional oligonucleotide probe. RSC Adv 6:11773

Drygin YF, Blintsov AN, Osipov AP, Grigorenko VG, Andreeva IP, Uskov AI, Varitsev YA, Anisimov BV, Novikov VK, Atabekov JG (2009) High-sensitivity express immunochromatographic method for detection of plant infection by Tobacco mosaic virus. Biochem Mosc 74:986–993

Du M, Zhai Q, Deng L, Li S, Li H, Yan L, Huang Z, Wang B, Jiang H, Huang T, Chang L, Jia W, Kang L, Jing L, Chuan L (2014) Closely related NAC transcription factors of tomato differentially regulate stomatal closure and reopening during pathogen attack. Plant Cell 26:3167–3184

Dubas ST, Pimpan V (2008) Green synthesis of silver nanoparticles for ammonia sensing. Talanta 76(1):29–33

Edmundson MC, Capeness M, Horsfall L (2014) Exploring the potential of metallic nanoparticles within synthetic biology. New Biotechnol 31(6):572–578

Ellis SD, Boehm MJ, Qu F (2008) Agriculture and natural resources: viral diseases of plants (PP401.05) [Fact Sheet]. Ohio State Univ. http://www.learnnc.org/lp/media/uploads/2010/11/viral-disease-fact-sheet.pdf

Erwin DC, Ribeiro OK (1996) Phytophthora: diseases worldwide. APS Press, St. Paul, MN

Esker PD, Sparks AH, Campbell L, Guo Z, Rouse M, Silwal SD, Tolos S, Van Allen B, Garrett KA (2008) Ecology and epidemiology in R: disease forecasting and validation. [Online]. Plant Health Instructor. https://doi.org/10.1094/PHIA_029-01

Etefagh R, Azhir E, Shahtahmasebi N (2013) Synthesis of CuO nanoparticles and fabrication of nanostructural layer biosensors for detecting Aspergillus niger fungi. Sci Iranica 20(3):1055–1058

Eun AJ-C, Wong S-M (2000) Molecular beacons: a new approach to plant virus detection. Phytopathology 90:269–275

Eun AJ-C, Huang L, Chew F-T, Li SF-Y, Wong S-M (2002) Detection of two orchid viruses using quartz crystal microbalance (QCM) immunosensors. J Virol Methods 99:71–79

Fang Y, Umasankar Y, Ramasamy RP (2014) Electrochemical detection of p-ethylguaiacol, a fungi infected fruit volatile using metal oxide nanoparticles. Analyst 139:3804–3810

Farmer EE, Alméras E, Krishnamurthy V (2003) Jasmonates and related oxylipins in plant responses to pathogenesis and herbivory. Curr Opin Plant Biol 6:372–378

Farr DF, Rossman AY (2014) Fungal databases, systematic mycology and microbiology laboratory. ARS, USDA, Washington, DC http://nt.ars-grin.gov/fungaldatabases/

FHIA (2007) Deterioro poscosecha de las frutas y hortalizas frescas por hongos y bacterias. 4:2-5. http://fhia.org.hn/dowloads/fhiainfdic2007.pdf

Flor HH (1956) The complementary genic systems in flax and flax rust. Adv Genet 8:29–54

Friesen TL, Faris JD, Solomon PS, Oliver RP (2008) Host-specific toxins: effectors of necrotrophic pathogenicity. Cell Microbiol 10:1421–1428

Fujita DB (1990) In: Jones AL, Aldwinkle HS (eds) Crown, collar, and root rot. Compendium of apple and pear diseases. APS Press, St. Paul, MN

García CV (2004). Introducción a la microbiología. Segunda Edición. Editorial EUNED, Costa Rica, pp 103–107

Gilbert GS, Parker IM (2010) Rapid evolution in a plant-pathogen interaction and the consequences for introduced host species. Evol Appl 3:144–156

Glazebrook J (2005) Contrasting mechanisms of defense against biotrophic and necrotrophic pathogens. Annu Rev Phytopathol 43:205–227

Goluch ED, Nam JM, Georganopoulou DG, Chiesl TN, Shaikh KA, Ryu KS, Barron AE, Mirkin CA, Liu C (2006) A biobarcode assay for on-chip attomolar-sensitivity protein detection. Lab Chip 6(10):1293–1299

Gorris MT, Alarcon B, Lopez M, Cambra M (1994) Characterization of monoclonal antibodies specific for Erwinia carotovora subsp. atroseptica and comparison of serological methods for its sensitive detection on potato tubers. App Environ Microbiol 60:2076–2085

Grahl T, Märkl H (1996) Killing of microorganisms by pulsed electric fields. Appl Microbiol Biotechnol 45(1–2):148–157

Grant MR, Kazan K, Manners JM (2013) Exploiting pathogens' tricks of the trade for engineering of plant disease resistance: challenges and opportunities. Microb Biotechnol 6(3):212–222

Grimmer MK, John Foulkes M, Paveley ND (2012) Foliar pathogenesis and plant water relations: a review. J Exp Bot 63:4321–4331

Guimaraes RL, Stotz HU (2004) Oxalate production by Sclerotinia sclerotiorum deregulates guard cells during infection. Plant Physiol 136:3703–3711

Haji-Hashemi H, Norouzia P, Safarnejadc MR, Ganjalia MR (2017) Label-free electrochemical immunosensor for direct detection of Citrus tristeza virus using modified gold electrode. Sensors Actuators B 244:211–216

Hassen WM, Duplan V, Frost E, Dubowski JJ (2011) Quantitation of influenza a virus in the presence of extraneous protein using electrochemical impedance spectroscopy. Electrochim Acta 56:8325–8328

Hayat MA (1989) Colloidal gold: principles, methods, and applications, vol 1. Academic Press, San Diego, CA, 538 p

Holzinger M, Le Goff A, Cosnier S (2014) Nanomaterials for biosensing applications: a review. Front Chem 2:63 p

Hull R (2002) Matthews' plant virology, 4th edn. Academia Press, San Diego, CA, 1001 p

Ingale AG, Chaudhari AN (2013) Biogenic synthesis of nanoparticles and potential applications: an eco-friendly approach. J Nanomed Nanotechol 4:165

Jain K (2003) Nanodiagnostics: application of nanotechnology (NT) in molecular diagnostics. Expert Rev Mol Diagn 3(2):153–161

James C (2013) Polypyrrole nanoribbon based chemiresistive immunosensors for viral plant pathogen detection. Anal Methods 5:3497–3502

Jammes F, Song C, Shin D, Munemasa S, Takeda K, Gu D, Cho D, Lee S, Giordo R, Sritubtim S, Leonhard N, Ellis BE, Murata Y, Kwak JM (2009) MAP kinases MPK9 and MPK12 are preferentially expressed in guard cells and positively regulate ROS-mediated ABA signaling. Proc Natl Acad Sci U S A 106:20520–20525

Jammes F, Yang X, Xiao S, Kwak JM (2011) Two Arabidopsis guard cell-preferential MAPK genes, MPK9 and MPK12, function in biotic stress response. Plant Signal Behav 6:1875–1877

Jeeva ML, Mishra AK, Vidyadharan P, Misra RS, Hegde V (2010) A species-specific polymerase chain reaction assay for rapid and sensitive detection of *Sclerotium rolfsii*. Aust Plant Pathol 39(6):517–523

Jones JD, Dangl JL (2006) The plant immune system. Nature 444:323–329

Jones JD, Vance RE, Dangl JL (2016) Intracellular innate immune surveillance devices in plants and animals. Science 354:aaf6395. https://doi.org/10.1126/science.aaf6395

Joseph T, Morrison M (2006) Nanotechnology in agriculture and food. A Nanoforum report, Institute of Nanotechnology. www.nanoforum.org

Kageyama K, Senda M, Asano T, Suga H, Ishiguro K (2007) Intra-isolateheterogeneity of the ITS region of rDNA in *Pythium helicoides*. Mycological Res 111:416–423

Karpovich-Tate N, Spanu P, Dewey FM (1998) Use of monoclonal antibodies to determine biomass of *Cladosporium fulvum* in infected tomato leaves. Mol Plant Pathog Interact 11:710–716

Kashyap PL, Kaur S, Sanghera GS, Kang SS, Pannu PPS (2011) Novel methods for quarantine detection of Karnal bunt (Tilletia indica) of wheat. Elixir Agric 31:1873–1876

Kashyap PL, Kumar S, Srivastava AK, Sharma AK (2013) Myconanotechnology in agriculture: a perspective. World J Microbiol Biotechnol 29(2):191–207

Kashyap PL, Kumar S, Srivastava AK (2017) Nanodiagnostics for plant pathogens. Environ Chem Lett 15:7–13

Kattke MD, Gao EJ, Sapsford KE, Stephenson LD, Kumar A (2011) FRET-based quantum dot immunoassay for rapid and sensitive detection of *Aspergillus amstelodami*. Sensors 11(6):6396–6410

Kazan K, Lyons R (2014) Intervention of phytohormone pathways by pathogen effectors. Plant Cell 26:2285–2309

Khiyami MA, Almoammar H, Awad YM, Alghuthaym MA, Abd-Elsalam KA (2014) Plant pathogen nanodiagnostic techniques: forthcoming changes? Biotechnol Biotechnol Equip 28(5):775–785

Khlebtsov NG, Dykman LA (2010) Optical properties and biomedical applications of plasmonic nanoparticles. J Quant Spectrosc Radiat Transf 111:1–35

Kim JT, Park SY, Choi W, Lee YH, Kim HT (2008) Characterization of Colletotrichum isolates causing anthracnose of pepper in Korea. Plant Pathol J 24(1):17–23

Kim TH, Bohmer M, Hu H, Nishimura N, Schroeder JI (2010) Guard cell signal transduction network: advances in understanding abscisic acid, CO_2, and Ca^{2+} signaling. Ann Rev Plant Biol 61:561–591

Kim Y, Park S, Gilmour SJ, Thomashow MF (2013) Roles of CAMTA transcription factors and salicylic acid in configuring the low-temperature transcriptome and freezing tolerance of Arabidopsis. Plant J 75:364–376

Kuckenberg J, Tartachnyk I, Noga G (2009) Temporal and spatial changes of chlorophyll fluorescence as a basis for early and precise detection of leaf rust and powdery mildew infections in wheat leaves. Precis Agric 10:34–44

Landa BB, Montes-Borrego M, Muñoz-Ledesma FJ, Jiménez-Díaz RM (2007) Phylogenetic analysis of downy mildew pathogens of opium poppy and PCRBased in planta and seed detection of *Peronospora arborescens*. Phytopathology 97(11):1380–1390

Lattanzio VMT, Nivarlet N, Lippolis V, Gatta SD, Huet AC, Delahaut P, Granier B, Visconti A (2012) Multiplex dipstick immunoassay for semi-quantitative determination of fusarium mycotoxins in cereals. Anal Chim Acta 718:99–108

Lebel E, Heifetz P, Thorne L, Uknes S, Ryals J, Ward E (1998) Functional analysis of regulatory sequences controlling PR-1 gene expression in Arabidopsis. Plant J 16(2):123–133

Li Y, Schluesener H, Xu S (2010) Gold nanoparticle-based biosensors. Gold Bull 43:2941. https://doi.org/10.1007/BF03214964

Lievens B, Brouwer M, Vanachter ACRC, Levesque CA, Cammue BPA, Thomma BPHJ (2005a) Quantitative assessment of phytopathogenic fungi in various substrates using a DNA macroarray. Environ Microbiol 7:1698–1710

Lievens B, Grauwet TJMA, Cammue BPA, Thomma BPHJ (2005b) Recent developments in diagnostics of plant pathogens: a review. Recent Res Dev Microbiol 9:57–79

Lin H-Y, Huang C-H, Huang C-C, Liu Y-C, Chau L-K (2012) Multiple resonance fiber-optic sensor with time division multiplexing for multianalyte detection. Opt Lett 37(19):3969–3971

Lin H-Y, Huang C-H, Lu S-H, Kuo I-T, Chau L-K (2014) Direct detection of orchid viruses using nanorod-based fiber optic particle plasmon resonance immunosensor. Biosens Bioelectron 51:371–378

Liu J, Zhang T, Jia J, Sun J (2016) The wheat mediator subunit TaMED25 interacts with the transcription factor TaEIL1 to negatively regulate disease resistance against powdery mildew. Plant Physiol 170:1799–1816

Loon LC (1985) Pathogenesis-related proteins. Plant Mol Biol 4(2–3):111–116

López MM, Llop P, Cubero J, Penyalver R, Caruso P, Bertolini E, Penalver J, Gorris MT, Cambra M (2001) Strategies for improving serological and molecular detection of plant pathogenic bacteria. In: Plant pathogenic bacteria. Springer, Berlin, pp 83–86

López MM, Bertolini E, Olmos A, Caruso P, Corris MT, Llop P, Renyalver R, Cambra M (2003) Innovative tools for detection of plant pathogenic viruses and bacteria. Int Microbiol 6:233–243

López MM, Llop P, Olmos A, Marco-Noales E, Cambra M, Bertolini E (2009) Are molecular tools solving the challenges posed by detection of plant pathogenic bacteria and viruses? Curr Issues Mol Biol 11:13–46

Maeda Y, Toyoda T, Mogi T, Taguchi T, Tanaami T, Yoshino T, Matsunaga T, Tanak T (2016) DNA recovery from a single bacterial cell using charge-reversible magnetic nanoparticles. Colloids Surf B Biointerfaces 139:117–122

Mahlein AK, Oerke E, Steiner U, Dehne H (2012) Recent advances in sensing plant diseases for precision crop protection. Eur J Plant Pathol 133:197–209

Martin FN, Tooley PW (2003) Phylogenetic relationships among Phytophthora species inferred from sequence analysis of mitochondrially encoded cytochrome oxidase I and II genes. Mycologia 95:269–284

Martinelli F, Scalenghe R, Davino S, Panno S, Scuderi G, Ruisi P, Villa P, Stroppiana D, Boschetti M, Goulart LR, Davis CE, Dandekar AM (2014) Advanced methods of plant disease detection: a review. Agron Sustain Dev 35(1):1–25

Mazarei M, Teplova I, Hajimorad MR, Stewart CN Jr (2008) Pathogen phytosensing: plants to report plant pathogens. Sensors 8:2628–2641

McCartney AH, Foster SJ, Fraaige BA, Ward E (2003) Molecular diagnostics for fungal plant pathogens. Pest Manag Sci 59:129–142

McFarland AD, Van Duyne RP (2003) Single silver nanoparticles as real-time optical sensors with zeptomole sensitivity. Nano Lett 3:1057–1062

McLachlan DH, Lan J, Geilfus CM, Dodd AN, Larson T, Baker A, Horak H, Kollist H, He Z, Graham I, Mickelbart MV, Hetherington AM (2016) The breakdown of stored triacylglycerols is required during light-induced stomatal opening. Curr Biol 26:707–712

Melotto M, Underwood W, Koczan J, Nomura K, He SY (2006) Plant stomata function in innate immunity against bacterial invasion. Cell 126:969–980

Mengiste T (2012) Plant immunity to necrotrophs. Annu Rev Phytopathol 50:267–294

Mohr PG, Cahill DM (2007) Suppression by ABA of salicylic acid and lignin accumulation and the expression of multiple genes, in Arabidopsis infected with *Pseudomonas syringae* pv. tomato. Funct Integr Genomics 7:181–191

Mousavi SE, Rezaei M (2011) Nanotechnology in agriculture and food production. J Appl Environ Biol Sci 1:414–419

Nezhad AS (2014) Future of portable devices for plant pathogen diagnosis. Lab Chip 14:2887–2904

Ni W, Chen H, Kou X, Yeung W, Wang J (2008) Optical fiber-excited surface plasmon resonance spectroscopy of single and ensemble gold nanorods. J Phys Chem C 112(22):8105–8109

Nusz GJ, Marinakos SM, Curry AC, Dahlin A, Höök F, Wax A, Chilkoti A (2008) Anal Chem 80:984–989

Oerke EC (2006) Crop losses to pests. J Agric Sci 144:31–43

Oerke EC, Dehne HW, Schönbeck F, Weber A (1994) Crop production and crop protection. Estimated losses in major food and cash crops. Elsevier, Amsterdam

Pal S, Ying W, Alocilja EC, Downes FP (2008) Sensitivity and specificity performance of a direct-charge transfer biosensor for detecting *Bacillus cereus* in selected food matrices. Biosyst Eng 99(4):461–468

Parab HJ, Jung C, Lee JH, Park HG (2010) A gold nanorod-based optical DNA biosensor for the diagnosis of pathogens. Biosens Bioelectron 26:667–673

Parker IM, Gilbert GS (2007) When there is no escape: the effects of natural enemies on native, invasive, and noninvasive plants. Ecology 88:1210–1224

Pasquali M, Piatti P, Gullino ML, Garibaldi A (2006) Development of a real-time polymerase chain reaction for the detection of *Fusarium oxysporum* f. Sp *basilica* from basil seed and roots. J Phytopathol 154:632–636

Pearson MN, Clover GRG, Guy PL, Fletcher JD, Beever RE (2006) A review of the plant virus, viroid and mollicute records for New Zealand. Australas. Plant Pathol 35:217–252

Perdikaris A, Vassilakos N, Yiakoumettis I, Kektsidou O, Kintzios S (2011) Development of a portable, high throughput biosensor system for rapid plant virus detection. J Virol Methods 177(1):94–99

Pimentel D (2009) Invasive plants: their role in species extinctions and economic losses to agriculture in the USA. In: Inderjit (ed) Management of invasive weeds, invading nature, Springer Series in invasion ecology, vol 5. Springer, Dordrecht, pp 1–7

Powers T (2006) Nematode molecular diagnostics: from bands to barcodes. Annu Rev Phytopathol 42:367–383

Prasad R, Kumar V, Prasad KS (2014) Nanotechnology in sustainable agriculture: present concerns and future aspects. Afr J Biotechnol 13(6):705–713

Prasad R, Bhattacharyya A, Nguyen QD (2017) Nanotechnology in sustainable agriculture: recent developments, challenges, and perspectives. Front Microbiol 8:1014. https://doi.org/10.3389/fmicb.2017.01014

Prasanna BM (2007) Nanotechnology in agriculture. ICAR National Fellow, Division of Genetics, I.A.R.I., New Delhi http://www.iasri.res.in/ebook/EBADAT/6-Other Useful Techniques/10-nanotech_in_Agriculture__BM_Prasanna__1.2.2007.pdf

Rafidah AR, Faridah S, Shahrul AA, Mazidah M, Zamri I (2016) Chronoamperometry measurement for rapid cucumber mosaic virus detection in plants. Proc Chem 20:25–28

Rai M, Ingle A (2012) Role of nanotechnology in agriculture with special reference to management of insect pests. Appl Microbiol Biotechnol 94(2):287–293

Ricci F, Volpe G, Micheli L, Palleschi G (2007) A review on novel developments and applications of immunosensors in food analysis. Anal Chim Acta 605:111–127

Rowe HC, Kliebenstein DJ (2008) Complex genetics control natural variation in *Arabidopsis thaliana* resistance to *Botrytis cinerea*. Genetics 180:2237–2250

Ruiz-García AB, Olmos A, Arahal DR, Antúnez O, Llop P, Pérez-Ortín JE, López MM, Cambra M (2004) Biochip electrónico para la detección y caracterización simultánea de los principales virus y bacterias patógenos de la patata. XII Congreso de la Sociedad Española de Fitopatología. Lloret de Mar. 12 p

Safarnejad MR, Samiee F, Tabatabie M, Mohsenifar A (2017) Development of quantum dot-based Nanobiosensors against Citrus Tristeza virus (CTV). Sensors & Transducers Published by IFSA Publishing, S. L. http://www.sensorsportal.com

Safarpour H, Safarnejad MR, Tabatabaei M, Mohsenifar A, Mohsenifar A, Rad R, Basirat M, Shahryari F, Hasanzadeh F (2012) Development of a quantum dots FRET-based biosensor for efficient detection of *Polymyxa betae*. Can J Plant Pathol 34:507–515

Sankaran S, MishraA ER, Davis C (2010) A review of advanced techniques for detecting plant diseases. Comput Electron Agric 72:1–13

Scarpeci TE, Zanor MI, Mueller-Roeber B, Valle EM (2013) Overexpression of *AtWRKY30* enhances abiotic stress tolerance during early growth stages in *Arabidopsis thaliana*. Plant Mol Biol 83(3):265–277

Schaad NW, Frederick RD (2002) Real-time PCR and its application for rapid plant disease diagnostics. Can J Plant Pathol 24:250–258

Schuman GL, D'Arcy CJ (2010) Essential plant pathology, 2nd edn. The American Phytopathological Society, St. Paul, MN, 369 p

Schuster E, Dunn-Coleman N, Frisvad JC, Van Dijck PW (2002) On the safety of Aspergillus niger: a review. Appl Microbiol Biotechnol 59(4–5):426–435

Sekhon BS (2010) Food nanotechnology: an overview. J Nanotechnol Sci Appl 3:1–15

Sharma A, Kaushal A, Kulshrestha S (2017) A nano-Au/C-MWCNT based label free amperometric immunosensor for the detection of capsicum chlorosis virus in bell pepper. Arch Virol 162:2047–2052

Shimada TL, Hara-Nishimura I (2015) Leaf oil bodies are subcellular factories producing antifungal oxylipins. Curr Opin Plant Biol 25:145–150

Shojaei TR, Salleh MAM, Sijam K, Rahim RA, Mohsenifar A, Safarnejad R, Tabatabaei M (2016a) Detection of Citrus tristeza virus by using fluorescence resonance energy transfer-based biosensor. Spectrochim Acta A Mol Biomol Spectrosc 169:216–222

Shojaei TR, Salleh MAM, Sijam K, Rahim RA, Mohsenifar A, Safarnejad R, Tabatabaei M (2016b) Fluorometric immunoassay for detecting the plant virus Citrus tristeza using carbon nanoparticles acting as quenchers and antibodies labeled with CdTe quantum dots. Microchim Acta 183:2277

Singh S, Singh M, Agrawal VV, Kumar A (2010) An attempt to develop surface plasmon resonance based immuno sensor for Karnal bunt (Tilletia indica) diagnosis based on the experience nano-gold based lateral flow immune dipstick test. Thin Solid Films 519(3):1156–1159

Small J, Call DR, Brockman FJ, Straub TM, Chandler DP (2001) Direct detection of 16S rRNA in soil extracts by using oligonucleotide microarrays. Appl Environ Microbiol 67(10):4708–4716

Sosnowski RG, Tu E, Butler WF, O'Connell JP, Heller MJ (1997) Rapid determination of single base mismatch mutations in DNA hybrids by direct electric field control. Proc Natl Acad Sci U S A 94:1119–1123

Srinivasan B, Tung S (2015) Development and applications of portable biosensors. J Lab Autom 20:365–389

Strange RN, Scott PR (2005) Plant disease: a threat to global food security. Annu Rev Phytopathol 43:83–116

Sun X, Du S, Wang X, Zhao W, Li Q (2011) A label-free electrochemical immunosensor for carbofuran detection based on a sol-gel entrapped antibody. Sensors 11:9520–9531

Swierczewska M, Liu G, Chen X (2012). High-sensitivity nanosensors for biomarker detection. Chemical Society Review 41:2641–2655

Tan W, Wang K, He X, Zhao XJ, Drake T, Wang L, Bagwe RP (2004) Bionanotechnology based on silica nanoparticles. Med Res Rev 24(5):621–638

Tanaka Y, Sano T, Tamaoki M, Nakajima N, Kondo N, Hasezawa S (2006) Cytokinin and auxin inhibit abscisic acid induced stomatal closure by enhancing ethylene production in Arabidopsis. J Exp Bot 57:2259–2266

Tang J, Wang F, Hou X, Wang Z, Huang Z (2014) Genome-wide fractionation and identification of WRKY transcription factors in chinese cabbage (Brassica rapa ssp. pekinensis) reveals collinearity and their expression patterns under abiotic and biotic stresses. Plant Mol Biol Rep 32(4):781–795

Tang D, Wang G, Zhou J (2017) Receptor kinases in plant-pathogen interactions: more than pattern recognition. Plant Cell 29:618–637

Ton J, Flors V, Mauch-Mani B (2009) The multifaceted role of ABA in disease resistance. Trends Plant Sci 14(6):310–317

Torres-Calzada C, Tapia-Tussell R, Quijano-Ramayo A, Martin-Mex R, Rojas-Herrera R, Higuera Ciapara I, Perez-Brito D (2011) A species-specific polymerase chain reaction assay for rapid and sensitive detection of Colletotrichum capsici. Mol Biotechnol 49(1):48–55

Trdá L, Fernandez O, Boutrot F, Héloir MC, Kelloniemi J, Daire X, Adrian M, Clément C, Zipfel C, Dorey S, Poinssot B (2014) The grapevine flagellin receptor VvFLS2 differentially recognizes flagellin-derived epitopes from the endophytic growth-promoting bacterium *Burkholderia phytofirmans* and plant pathogenic bacteria. New Phytol 201:1371–1384

Truman W, Sreekanta S, Lu Y, Bethke G, Tsuda K, Katagiri F, Glazebrook J (2013) The calmodulin-binding protein60 family includes both negative and positive regulators of plant immunity. Plant Physiol 163:1741–1751

Uhl J, Tang Y, Cockerill ER (2011) Fluorescence resonance energy transfer. In: Persing D, Tenover F, Tang Y, Nolte F, Hayden R, van Belkum A (eds) Molecular microbiology. ASM Press, Washington, DC, pp 231–244. https://doi.org/10.1128/9781555816834.ch14

Validov SZ, Kamilova FD, Lugtenberg BJJ (2011) Monitoring of pathogenic and non-pathogenic *Fusarium oxysporum* strains during tomato plant infection. Microb Biotechnol 4(1):82–88

van der Want JPH, Dijkstra J (2006) A history of plant virology. Arch Virol 51:1467–1498

van der Wolf J, van Bechhoven JRCM, Bonants PJM, Schoen CD (2001) New technologies for sensitive and specific routine detection of plant pathogenic bacteria. In: Plant pathogenic bacteria. Springer, Berlin, pp 75–77

van Kleunen M, Fischer M (2009) Release from foliar and floral fungal pathogen species does not explain the geographic spread of naturalized north American plants in Europe. J Ecol 97:385–392

van Loon LC, Geraats BP, Linthorst HJ (2006) Ethylene as a modulator of disease resistance in plants. Trends Plant Sci 11:184–191

Vernon C, Vizcarra-Mendoza M (2000) Separation kinetics of Karnal bunt (*Tilletia indica*) infected wheat (*Triticum aestivum*) grains in a batch operated fluidized bed. Food Sci Technol Int 6(2):137–143

Vlot AC, Dempsey DA, Klessig DF (2009) Salicylic acid, a multifaceted hormone to combat disease. Annu Rev Phytopathol 47:177–206

Wan D, Li R, Zou B, Zhang X, Cong J, Wang R, Xia Y, Li G (2012) Calmodulin binding protein CBP60g is a positive regulator of both disease resistance and drought tolerance in Arabidopsis. Plant Cell Rep 31:1269–1281

Wang L, Li PC (2010) Gold nanoparticle-assisted single base-pair mismatch discrimination on a microfluidic microarray device. Anal Biochem 400(2):282–288

Wang L, O'Donoghue MM, Tan W (2006) Nanoparticles for multiplex diagnostics and imaging. Nanomedicine 1(4):413–426

Wang L, Tsuda K, Sato M, Cohen JD, Katagiri F, Glazebrook J (2009) Arabidopsis CaM binding protein CBP60g contributes to MAMP induced SA accumulation and is involved in disease resistance against *Pseudomonas syringae*. PLoS Pathog 5(2):e1000301

Wang J, Wang X, Li Y, Yan S, Zhou Q, Gao B, Peng J, Du J, Fu Q, Jia S, Zhang J, Zhan L (2012) A novel, universal and sensitive lateral-flow based method for the detection of multiple bacterial contamination in platelet concentrations. Anal Sci 28:237–241

Warad HC, Ghosh SC, Thanachayanont C, Dutta J, Hilborn JG (2004) Highly luminescence manganese doped ZnS quantum dots for biological labeling. In: Proceedings of the International Conference on Smart Materials/Intelligent Materials, Chiang Mai, Thailand, 1–3 December 2004, pp 203–206

Ward E, Foster SJ, Fraaije BA, Mccartney HA (2004) Plant pathogen diagnostics: immunological and nucleic acid based approaches. Ann Appl Biol 145:1–16

Wei Y, Shi H, Xia Z, Tie W, Ding Z, Yan Y, Wang W, Hu W, Li K (2016) Genome-wide identification and expression analysis of the WRKY gene family in Cassava. Front Plant Sci 7:25

White TJ, Bruns T, Lee S, Taylor JW (1990) Amplification and direct sequencing of fungal ribosomal RNA genes for phylogenetics. In: Innis MA, Gelfand DH, Sninsky JJ, White TJ (eds) PCR protocols: a guide to methods and applications. Academic Press, New York, pp 315–322

Wolfe LM (2002) Why alien invaders succeed: support for the escape-from-enemy hypothesis. Am Nat 160:705–711

Yadav A, Kon K, Kratosova G, Duran N et al (2015) Fungi as an efficient mycosystem for the synthesis of metal nanoparticles: progress and key aspects of research. Biotechnol Lett 37:2099–2120

Yang H, Li H, Jiang X (2008) Detection of foodborne pathogens using bioconjugated nanomaterials. Microfluid Nanofluid 5(5):571–583

Yang C, Lu X, Ma B, Chen SY, Zhang JS (2015) Ethylene signaling in rice and *Arabidopsis*: conserved and diverged aspects. Mol Plant 8:495–505

Yao KS, Li SJ, Tzeng KC, Cheng TC, Chang CY, Chiu CY, Liao CY, Hsu JJ, Lin ZP (2009) Fluorescence silica nanoprobe as a biomarker for a rapid detection of plant pathogens. Adv Mater Res 79:513–516

Yuen GY, Craig ML, Avila F (1993) Detection of *Pythium ultimum* with a species-specific monoclonal antibody. Plant Dis 77:692–698

Zamioudis C, Pieterse CM (2012) Modulation of host immunity by beneficial microbes. Mol Plant Microbe Interact 25:139–150

Zeng C, Huang X, Xu J, Li G, Ma J, Ji HF, Zhu S, Chen H (2013) Rapid and sensitive detection of maize chlorotic mottle virus using surface plasmon resonance-based biosensor. Anal Biochem 440:18–22

Zhang M, Qin Z, Liu X, Ustin SL (2003) Detection of stress in tomatoes induced by late blight disease in California, USA, using hyperspectral remote sensing. Int J Appl Earth Observ Geoinf 4:295–310

Zhao J, Wang XJ, Chen CQ, Huang LL, Kang ZS (2007) A PCR-based assay for detection of *Puccinia striiformis* f. sp tritici in wheat. Plant Dis 91(12):1969–1674

Zheng Z, Qamar SA, Chen Z, Mengiste T (2006) Arabidopsis WRKY33 transcription factor is required for resistance to necrotrophic fungal pathogens. Plant J 48:592–605

Zhu M, Zhang WN, Tian JY, Zhao WY, Chen ZQ, Sun LH, Liu FQ (2016) Development of a lateral-flow assay (LFA) for rapid detection of soybean mosaic virus. J Virol Methods 235:51–57

Chapter 17
Nanocrystalline Cellulose: Production and Applications

Sai Swaroop Dalli, Bijaya Kumar Uprety, Mahdieh Samavi, Radhika Singh, and Sudip Kumar Rakshit

17.1 Introduction

Cellulose is an abundant renewable resource which can be used for a number of applications. It is mainly found in plant cell wall and plays a vital role in maintaining its structure. Besides plants, cellulose is also found in fungi, bacteria, and some tunicates (George and Sabapathi 2015). Cellulose is a fibrous and linear natural polymer consisting of glucose units bound by β-1,4-linkages with a degree of polymerization ranging from 10,000 to 15,000 units depending on the type of biomass (Sjostrom 1993). For more than a century, several applications of cellulose in food, pharmaceuticals, polymer, pulp, and composite industries have been studied (Coffey et al. 1995; de Souza Lima et al. 2003). Agricultural waste such as rice straw, sugarcane bagasse, sawdust, cotton stables, and woody forest residues are the main source of cellulose for various applications in industries. Production of Nano-Crystalline Cellulose (NCC) is one of the products being evaluated for a number of applications in recent years. NCC produced from

S. S. Dalli
Department of Chemistry and Material Sciences, Lakehead University,
Thunder Bay, ON, Canada

B. K. Uprety · M. Samavi
Department of Biotechnology, Lakehead University, Thunder Bay, ON, Canada

R. Singh
Department of Chemistry, Dayalbagh Educational Institute, Agra, Uttar Pradesh, India

S. K. Rakshit (✉)
Department of Chemistry and Material Sciences, Lakehead University,
Thunder Bay, ON, Canada

Department of Biotechnology, Lakehead University, Thunder Bay, ON, Canada
e-mail: srakshit@lakeheadu.ca

© Springer Nature Switzerland AG 2018
R. Prasad et al. (eds.), *Exploring the Realms of Nature for Nanosynthesis*,
Nanotechnology in the Life Sciences, https://doi.org/10.1007/978-3-319-99570-0_17

cellulose fibers are regarded as nanobiomaterial with variety of applications in chemicals, food, pharmaceuticals, etc. (Habibi et al. 2010).

NCC is produced by breaking the natural polymer of cellulose and separating the crystalline section. The general steps involved in the production of NCC from various sources of biomass were illustrated in Fig. 17.1. Nanocrystalline celluloses are typically 5–70 nm in width and 100 nm to several micrometers in length. NCCs (referred to as whiskers) can be classified based on their dimensions, functions, and preparation methods. NCC has many useful characteristics compared to cellulose such as its physicochemical properties including high surface area, specific strength, and optical properties (Peng et al. 2011). Various studies in literature have reported the chemical structure, physical and mechanical properties of NCC (de Souza Lima and Borsali 2004; Peng et al. 2011; Huq et al. 2012). The biocompatible and biodegradable nature of NCC makes it ideal for several applications and is thus the focus of considerable research. Developing commercial scale methods to produce nanocrystalline cellulose from the forest-based biomass can certainly contribute to the advancement of biobased industries around the world. In this chapter, we have highlighted the important production methods, modification, applications of NCCs, and the economics of its production.

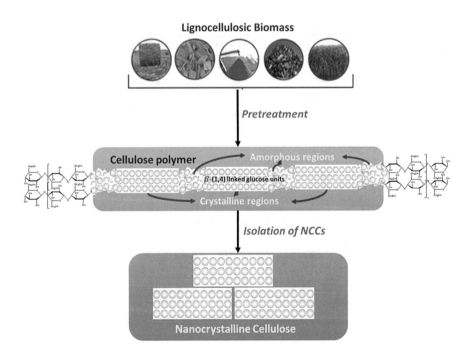

Fig. 17.1 Schematic representation of nanocrystalline cellulose production

17.2   Production of Nano-Crystalline Cellulose (NCC)

Traditionally, NCC can be isolated from any cellulose obtained from plants, animals, bacteria, and algae. However, due to high abundance of cellulose in wood and agricultural residues, these biomass sources have become major substrates for the extraction of NCC. Wood-based NCC extraction has also gained importance as these fibers are thinner than the bacterial cellulose (George and Sabapathi 2015). Abraham et al. (2011) have reported that pineapple leaf fiber produce fine NCC than banana and jute (Abraham et al. 2011). However, jute fiber is one of the cheapest materials to make the nanocellulose production cost-effective. Fortunati et al. (2012) have investigated the extraction of micro- and nanocellulose from okra fibers. Nonetheless, the production process of NCC is similar for all cellulosic materials (Fortunati et al. 2012). There are two stages involved in the isolation of NCC for any type of biomass.

First, the raw material is pretreated to separate cellulose from other constituents (lignin and hemicellulose of lignocellulose biomass). Second, the cellulose is treated to break the amorphous regions of the long glycosidic chain polymer. This is usually done by mechanical, chemical, and enzyme methods. Ranby (1951) reported the synthesis of cellulosic nanofibers for the first time using sulfuric acid. When cellulosic material is subjected to either mechanical or chemical treatments, the amorphous regions disintegrate and leave the crystalline regions intact in the form of short crystals. A number of industries are focusing on commercialization of different forms of NCC by evaluating various methods for economically feasible production (Antonio 2014).

17.3   Pretreatment of Biomass

There are different types of pretreatment techniques reported in literature that are used to separate biomass into two major streams, a solid stream containing cellulose and lignin and a liquid stream containing majorly hemicellulose (Dalli and Rakshit 2015). The pretreatment techniques can be categorized into thermal, chemical, or physical. In most cases, the pretreatment has to be optimized and controlled to avoid unwanted by-products. Some of the effective and most studied pretreatment techniques are reported here.

17.3.1   Hydrothermal Pretreatment

This is a widely used method to separate water soluble and insoluble polymers from the biomass (Ma et al. 2014). As the name indicates, water and heat are involved in this process. Mineral acids such as sulfuric acid are used to enhance the efficiency of the process. Though this process is effective using mineral acids like H_2SO_4, HCl, and H_3PO_4 (Hendriks and Zeeman 2009), waste disposal is a problem as the acid waste stream can cause environmental pollution. In order to avoid acid treatment, the process can be carried out under pressure (Saha et al. 2013). Hydrolysis does not occur below 100 °C (Abatzoglou et al. 1992). Autohydrolysis, steam explosion, steam extrusion are some of the examples of hydrothermal treatment (Dalli and Rakshit 2015) which are considered to be useful depending on the type of biomass used.

17.3.2 Alkaline Hydrolysis

In the alkaline pretreatment process, biomass is treated with aqueous ammonia, sodium carbonate and hydroxides of sodium, calcium or potassium. Generally, lime and sodium hydroxide are used to hydrolyze the biomass at moderate temperature (Park and Kim 2012). Other types of alkaline treatments include the Ammonia Fiber Explosion (AFEX) and the Ammonia Recycling Percolation (ARP). Alkaline peroxide is another type of alkaline pretreatment where oxygen or hydrogen peroxide (1–3%) is added to biomass and catalyzed by the addition of lime or NaOH (Carvalheiro et al. 2008). These processes give better results by removing the lignin from the polysaccharides. In the AFEX process, liquid ammonia is added to biomass at a moderate temperature ranging from 40 to 140 °C and under high pressure (250–300 psi) and the reaction is carried out for a shorter time (Teymouri et al. 2004; Keshwani and Cheng 2009). In the ARP process, ammonia is circulated through the biomass in a column reactor. This is more effective with hardwoods and corn stovers than in softwoods and results in high delignification and moderate hemicellulose solubilization of approximately 40–60% (Carvalheiro et al. 2008). In cases where most of the lignin and hemicellulose are separated, cellulose becomes the major component in the solid fraction.

17.3.3 Organosolvent Pretreatment

In this process, organic solvents (e.g. methanol, ethanol, ethylene glycol etc.) with or without acid catalysts (HCl, H_2SO_4) are used to extract most of the lignin from biomass. Organic solvents dissolve lignin in the presence of acid catalyst and some of the hemicellulosic sugar (Lee et al. 2014). If this pretreatment process is performed under high temperatures (185–210 °C), addition of acids is not necessary because deacetylation from the sugars make the medium acidic. However, when the acids are added externally, this process is more effective in solubilizing lignin and hemicellulose leaving solid cellulose residue, (Zhao et al. 2009). According to Lee et al. (2014), hydroxyl ions from the alcoholic solvents break the bonds of phenolic and polysaccharide linkages in lignin and hemicellulose to dissolve them.

This process is has its benefits as it requires low energy to recover the components (Lee et al. 2014). However, it was also observed that organic solvents swell the cellulose fibers and reduce their crystallinity (McDonough 1993). Another demerit of this process is the formation of clumps of lignin while washing the pretreated biomass with water. Therefore, recovery of cellulose often becomes cumbersome and costly process.

17.3.4 Pretreatment Using Ionic Liquids

Pretreatment of lignocellulosics using ionic liquids is a relatively recent technique developed for isolating the cellulosic components from the lignocellulose matrix. Ionic liquids target the β-glycosidic bonds on cellulose polymer and make it soluble by hydrolyzing the linkages (Xiong et al. 2014). However, it is possible to control the concentration of ionic liquids to make cellulose and hemicellulose dissolve without disturbing the chain's structure (Lee et al. 2014). Ionic liquids are generally composed of an inorganic anion and an organic cation, which can be modified depending on the target compounds. Major benefit involved in this process is the recyclability of ionic liquids (Lee et al. 2014). Some examples of ionic liquids reported in literature are 1-alkyl-3-methylimidazolium [mim]$^+$; 1-alkyl-2,3-dimethylimidazolium [mmim]$^+$; 1-allyl-3-methylimidazolium [Amim]$^+$; 1-allyl-2,3-dimethylimidazolium [Ammim]$^+$; 1-butyl-3-methylpyridinium [C$_4$mP$_y$]$^+$; and tetrabutylphosphonium [Bu4P]$^+$ with n = number of carbons in the alkyl chain (Zavrel et al. 2009; Tadesse and Luque 2011; Lee et al. 2014). Ideally, ionic liquid should possess high dissolution capacity, low melting point, low viscosity, low toxicity and high stability. However, the drawback of using ionic liquids is the cellulose crystallinity would be reduced and become more amorphous. This is because of the tendency of ionic liquids to hydrolyze the hydrogen bonds in the cellulose polymer resulting in bond breaking and subsequent dissolution (Lee et al. 2014). The high costs of the ionic liquids are also an important factor that needs to be taken into account.

17.4 Isolation of Nanocrystalline Cellulose

Though the pretreatment techniques do not produce pure cellulose, most of the lignin and hemicellulose content need to be removed. The solid fraction obtained from the above pretreatment processes contains cellulose and some lignin. This material is then further treated to produce NCC. The purity, physical and mechanical properties of the product depend on the type of method used for the production. Some of the reported methods in literature are mentioned in Table 17.1 and described in detail as follows.

Table 17.1 Isolation methods and processes involved in NCC production

Treatment method	Processes	References
Mechanical	Milling, grinding, cutting, ultrosonication, microfluidization, cryocrushing	Ng et al. (2015), Marimuthu and Atmakuru (2015) and Frone et al. (2011)
Physical	Steam explosion	Cherian et al. (2010)
Chemical	Acid hydrolysis TEMPO-oxidation	Sacui et al. (2014) and Börjesson and Westman (2015)
Enzymatic	Exoglucanases: A and B type cellulases Endoglucanases: C and D type cellulases	George et al. (2011), Pääkkö et al. (2007), Lee et al. (2014) and Khalil et al. (2014)

17.4.1 Mechanical Processes

The mechanical process used for nanocrystalline cellulose production from the pretreated cellulose include milling, grinding, cutting, high pressure homogenization (steam explosion), ultrosonication, microfluidization, cryocrushing, etc.(Ng et al. 2015). The shear forces applied in mechanical treatment make the cellulose disintegrate and help in extracting the crystalline cellulose micrfibrils in the form of a uniform powder (de Souza Lima et al. 2003). Common treatments like milling, cutting or grinding are done in Wiley mill or Fritish Pulverisette mills or grinding machines (Ng et al. 2015). Ribbon-like cellulose nanocrystals are usually obtained in this process. These fibers are then sieved in a vibratory sieve to separate fine particulate fibers. The pore size of the mesh used in the vibratory sieves is usually in the range of 50–250 μm. The smaller the size of the fine fibers, higher the activity in subsequent chemical treatments due to the higher availability of the active groups of cellulose to react with the chemical reagents (Ng et al. 2015). The finely ground fibers are washed with water to remove impurities and to make the fibers softer (Frone et al. 2011; Marimuthu and Atmakuru 2015; Ng et al. 2015). Rosa et al. (2012) suggested dewaxing the fibers in a solvent mixture of toluene/ethanol using a soxhlet type extraction (Rosa et al. 2012).

The other components of the lignocellulosic material, hemicellulose and lignin, often make the NCC fibers impure and reduce its crystallinity. Therefore, it is necessary to separate these materials from cellulose fibers. Commonly used techniques for purification include alkali bleaching treatment or mercerization using sodium hydroxide or potassium hydroxide, sodium chlorite, and acetic acid (Ng et al. 2015). Alkali solutions potentially dissolve the other components except cellulose fibers which can be easily filtered out (Acharya et al. 2011; Faruk et al. 2012). For most applications, the fiber concentration should be limited to the optimal range of 4–6% (w/w) during the alkali treatment because the low fiber to alkali ratio lead to chemical degradation, whereas high fiber to alkali ratio might result in inefficient modification by reducing the active sites of reaction (Ng et al. 2015).

17.4.2   Ultrasonication

It has been reported that ultrasonication processes enhance the efficiency of biomass acid hydrolysis (Brinchi et al. 2013). Production of NCC using ultrasonication in water or an organic acid has been reported in the literature (Filson et al. 2009). Although low yields were obtained, it has been proven that ultrasonication helps to increase the NCC yields. Ultrasonication is an advanced technique to isolate micro- and nanocellulose from lignocellulosic material. Ultrasonication associated grinding and homogenization is an effective and efficient method to produce NCC in large volumes (Hielscher Ultrasonics GMBH 2017). It was reported that oxidation of cotton linter pulp using 2,2,6,6-Tetramethylpiperidine-1-oxyl (TEMPO)-NaBr-NaClO assisted with ultrasonication led to the production of carboxylated NCC (Qin et al. 2011). Leung et al. (2011) suggested a simple procedure to produce carboxylated NCC using ammonium persulfate instead of TEMPO, at 60 °C (Leung et al. 2011). This method showed relatively high yields of NCC and has the potential for scale up (Brinchi et al. 2013). However, this type of oxidation process is not efficient for the production of pure NCC. Besides producing low yields, use of TEMPO is relatively expensive and toxic to dispose. An advantage in this process is that the raw material can be used directly without any pretreatment steps for the isolation of cellulose.

17.4.3   Chemical Treatment

In the chemical treatment of lignocellulosics for NCC production, acid hydrolysis of cellulose is considered as an efficient method as it consumes less energy and produce rod-like crystal structures of cellulose nanofibers (Sacui et al. 2014). The strong acids easily disintegrate amorphous regions of the cellulose fibers, leaving the crystalline regions unaffected. However, boiling with strong sulfuric acid has its affect only up to a certain period, beyond which it slowly reduces the degree of polymerization of the polymer (George and Sabapathi 2015). The threshold level-off the degree of polymerization (LODP) varies with the type of biomass used to produce NCC. According to George and Sabapathi (2015), crystalline nanocellulose obtained from acid hydrolysis exhibit high polydispersity in molecular weight. This was assumed to be due to the lack of regular distribution of amorphous regions. Several other factors such as temperature, reaction time etc., also affect the quality of NCCs. In case of shorter reaction time is used, amorphous regions of the cellulose are retained in the solution and reduce the quality of the NCC. On the other hand, longer reaction time leads to depolymerization of cellulose crystals and reduces the aspect ratio of nanocrystals. The treatment of biomass at high temperature results in short chains of nanocellulose (Elazzouzi-Hafraoui et al. 2008). Bai et al. (2009) reported the synthesis of NCC whiskers from microcrystalline cellulose (MCC) using sulfuric acid. They have reported that the relative centrifugal force (RCF) also affects the length of NCCs (Bai et al. 2009).

Use of other acids like HCl (Börjesson and Westman 2015), HBr, H_3PO_4 (Lee et al. 2009) to isolate NCCs from the cellulosic material have been reported in literature. A mixture of acetic acid and nitric acid was used to enhance the production of nanocellulose by Zhang et al. (2014). Aqueous solution of NCCs was synthesized using HCl by grafting technique. It has been reported that grafting on nanocrystals using 2,2,6,6-tetramethylpiperidine-1-oxyl (TEMPO) in a procedure known as oxidative carboxylation-amidation prevents the particle aggregation and induces the stability of the colloidal NCC solution.

17.4.4 Enzymatic Treatment

Cellulose treated with enzymes along with mechanical shearing and high pressure homogenization is a relatively new method being developed for NCCs production. Commercial cellulase enzyme is used in these methods to obtain NCCs with better physical properties. Enzymatic treatment is considered better than acid hydrolysis as less waste is produced after the process and its ecofriendly nature (George et al. 2011). Pääkkö et al. (2007) have developed a facile method for the usage of enzymes along with mechanical shearing. They reported that the incorporation of enzymatic hydrolysis reduced the energy consumption and has a drastic effect on the mechanical properties of the NCCs produced. Usually, the synergistic action of endoglucanases and exoglucanases or cellobiohydrolases on biomass leads to disintegration of cellulose at the solid-liquid interface (Lee et al. 2014). This results in a shorter chain length of nanocellulose fibers. Although, the enzymatic treatment provides high yields with high selectivity, the limiting factor is the cost of the enzymes. Cellulose degradation using these enzymes take longer time durations to produce the desired high yields of NCC. The slow rate of enzyme hydrolysis also eventually increases the cost of production.

17.5 Chemical Modifications of NCC

One of the important applications of NCC is its use in engineering products as composite material for reinforcement. The chemical modification of nanocellulose is very helpful in improving the physical properties of polymer matrix in biocomposites (Börjesson and Westman 2015). Some of the chemical modification methods along with the functional agents used are listed in the Table 17.2.

Strengthening of nanocellulose-reinforced composites can be done by modifying the NCCs using peroxide. The hydroxyl groups on glucose molecules in a nanocellulose chain are of great interest to synthetic chemists as they can be modified modify to produce NCCs with different properties. Usually, chemical modification

Table 17.2 Chemical modification methods of NCC using various functional agents

Methods	Mechanism	Functional agents	References
Oxidation	Conversion of the hydroxyl groups of cellulose backbone	Maleic acid, lignin, polyvinyl alcohol, polyvinyl acetate	Börjesson and Westman (2015)
Coupling	Formation of covalent bonds between the hydroxyl groups and coupling agents	Organofunctional silanes, polypropylene etc.	Majeed et al. (2013)
Acetylation	Addition of acetyl group on the nanocellulose surface	Acetic acid, propionic acid, alkenyl succinic anhydride (ASA), acetic anhydride and acetyl chloride	Panaitescu et al. (2013) and Majeed et al. (2013)
Grafting	Transforming nanocellulose into a long aliphatic chains	Polycaprolactone, polyurethane, thermos-responsive polymers	Rebouillat and Pla (2013) and Dufresne (2010)

is done to improve the interface between two incompatible phases (Ng et al. 2015). Various compatibilizing agents such as maleic acid (Majeed et al. 2013), lignin, polyvinyl alcohol, polyvinyl acetate have been studied in literature (Ng et al. 2015) to strengthen the polymer containing NCCs. Lignin treatment includes the coating of nanocellulose with kraft lignin which exhibited the compatibility with both hydrophobic and hydrophilic matrices in a composite (Alemdar and Sain 2008). The use of these agents can enhance the interfacial adhesion between nanocrystalline cellulose and the polymer matrix.

Chemical modifications on the surface of NCC can be done by the formation of covalent bonds with the hydroxyl groups on glucose molecules. Several coupling agents have been studied to induce the covalent bonding in the nanocellulose and composite matrix (Majeed et al. 2013). Esterification is beneficial as it requires no solvent and is highly efficient in the conversion of hydroxyl groups and does not affect the crystalline structure of nanocellulose (Ng et al. 2015). The functional properties of the surface of nanocellulose can be altered by esterification. The polar hydroxyl groups are replaced by non-polar carbonyl groups (Panaitescu et al. 2013) which results in altered characteristics.

Acetylation of nanocellulose also one of the commonly employed method to modify the properties of NCC. Acetic acid, propionic acid, alkenyl succinic anhydride (ASA), acetic anhydride and acetyl chloride have been reported as major carriers of acetyl groups which can be easily transferred on to the nanocellulose surface (Majeed et al. 2013; Panaitescu et al. 2013). Dufresne (2010) has reported the studies on a novel modification of nanocellulose by transforming it into a long aliphatic chain. Two approaches such as "grafting-onto" and grafting-from" were used to graft the polymer onto the surface of NCC. In both the approaches, the nanocellulose was suspended in the solvent during grafting (Rebouillat and Pla 2013). In the "grafting-onto" approach, a previously synthesized polymer is directly attached to the available hydroxyl groups on the NCC chain using a coupling agent (e.g. isocyanate or peptide) (Ng et al. 2015). The polymers involved in this type of modification are poly-caprolactone, polyurethane, thermos-responsive polymers etc. (Rebouillat and Pla 2013). The advantage of such methods is that the polymers can be fully characterized before attaching to the NCC chain. However, the reaction time was found to be too long for such modification due to the steric hindrance from the polymer molecules (Rebouillat and Pla 2013).

On the other hand, "grafting-from" is useful as it helps in obtaining long chains of polymers. Unlike grafting-onto, the polymers are built on the surface of the NCC using immobilized initiators via the atom transfer radical polymerization (ATRP) (Rebouillat and Pla 2013). The major bottleneck of this method is that the polymer cannot be characterized before the grafting method. However, the useful factors in this type of surface grafting are the high reaction rates, production of no by-products and high chemical stability (Espino-Perez et al. 2013).

Extensive research is taking place to alter the functional properties of the nanocellulose by reacting it with various reagents having derivative functional groups. Recent reports in literature indicate rapid advancement in improving binding ability and hydrophilicity of NCC (Ng et al. 2015). However, some of the upstream processes for the production of NCCs consume high amount of energy. Some of the recent studies on the energy consumption are discussed below.

17.6 Energy Consumption in Preparing NCC

The energy consumption for NCC production depends mainly on the source of cellulosic fibers and the isolation processes. Different mechanical methods are used to isolate the nanofibrillated cellulose including high pressure homogenization, microfluidization, grinding, cryocrushing and high intensity ultrasonication (Khalil et al. 2014). However, one of the main challenges of all these approaches in nanocellulose production and application results in high energy consumption. Increasing applications of nanocellulose in various field including biomedical uses, food packaging, coating and etc. (George and Sabapathi 2015) requires energy-efficient approaches its production in an industrial scale. There are only a few studies available an energy consumption for a specific pretreatment or property. Further investigation on nanocellulose preparation, modification, production and developing economical approaches are necessary before it can be applied successfully in a large scale.

Production of microfibrillated cellulose is one of the main steps in the NCC isolation. It was reported that, over 25,000 kWh per ton is required for the production of microfibrillated cellulose (MFC) as a result of the multiple passes through the homogenizers (Klemm et al. 2011).

Pretreatments are reported to reduce the energy consumption from 20,000 to 30,000 kWh/ton to 1000 kWh/ton for cellulosic fibers, while improving the production conditions and fiber swelling properties (Zhu et al. 2011). Mechanical and chemical pretreatments are regarded as effective approaches which significantly reduce energy consumption (Stelte and Sanadi 2009). As an example, TEMPO-mediated oxidation of cellulose fibers achieved some level of success in efficiently producing

NFC (Saito et al. 2007). Further, it is reported that enzymatic treatment (Henriksson et al. 2007; Janardhnan and Sain 2007) of cellulose prior to the defibrillation facilitate disintegration requires lower energy for producing MFC. Ankerfors (2012) studied three alternative processes for producing microfibrillated cellulose in which pulp fibers where first pretreated and then homogenized. In the first process sulfite pulp were put through two refining steps and an enzymatic pretreatment. Then high pressure (1600 bar) homogenization with 8 passes was used. 33 and 90 kWh/ton pulp was measured for the first and second refining stage and the total energy required was calculated to be 2344 kWh/ton. The paper claimed 91% reduction in energy use, as the energy consumption reported earlier was 27,000 kWh/ton without pretreatments (Klemm et al. 2011). In the second process, chemical pretreatment such carboxy-methylation was applied prior to high pressure (1650 bar) homogenizing. Mechanical energy consumption was calculated to be 2221 kWh/ton in this way. The third process associated with combined enzymatic and mechanical pretreatment to facilitate disintegration. Approximately the same amount of energy as in the second process was calculated. It was reported that reduction of required energy to 500 and 1500 kWh/ton for the second and the third process respectively can be obtained by optimization of the parameters of reaction and treatment such as concentration and pressure. In terms of low energy consumption, it is reported that these three processes have the potential for industrial scale production.

An overall assumption is that MFC can be produced with an energy consumption of 500–2300 kWh/ton by these methods. Also, the characterizations of the produced materials have been investigated in different ways and it has been demonstrated that the produced MFC fibrils were approximately 5–30 nm wide and up to several microns long. The number of homogenization steps does matter as well. Even though an increased number of steps are found to improve the quality of the product, each homogenization step costs 2200 kWh/ton. Therefore, the nanocellulose isolation process conditions should be taking into this account (Ankerfors 2012).

Another study by Spence et al. (2011), studied the effect of processing on microfibril and film properties, relative to energy consumption on bleached and unbleached hardwood pulp samples by homogenization, microfluidization, and micro-grinding. Film densities of samples in all three approaches were reported approximately to be 900 kg/m^3. However, higher toughness values were reported for microfluidization and micro-grinding with less required energy compare to homogenization.

In fibers processing energy consumption per pass was 3940 kJ/kg by homogenizer and the 620 kJ/kg by micro-grinder. The microfluidizer energy measurements was 200 kJ/kg with operating pressure of 69 MPa, 390 kJ/kg under 138 MPa pressure, and 630 kJ/kg with 207 MPa.

Energy is highly influenced by parameters such as number of passes and flow rate in each processing methods. Homogenization, for instance, had a much slower processing flow rate than the microfluidizer and the micro-grinder, which notably increased the amount of required energy. The processing rates of the suspension were 2 kg/min for micro-grinding, 1 kg/min for microfluidization, and 0.2 kg/min for homogenization. Higher pressures of the microfluidizer also raise the energy required. Generally, the total energy of approximate 9180 kJ/kg for the microfluidizer with pretreatment, 9090 kJ/kg for the grinder with pretreatment, and 5580 kJ/kg for the grinder without pretreatment and 31,520 kJ/kg for the homogenizer is required in processing of MFC films with maximum accessible properties (Table 17.3).

Studies have showed marked increase in tensile index of the MFCs from bleached hardwood fiber after microfluidization and micro-grinding, while homogenization was not considered as an energy-efficient process. However, processing of the unbleached hardwood fibers was more efficient. With overall lower energy consumption, microfluidizer process leads to the highest tensile index (Spence et al. 2011).

From a review of literature, it seems that the following assumptions can be useful for optimizing the processes. Firstly, in microfluidization, even though the increased pressure leads to significant increase in the energy consumption, toughness and tensile index decreases as a result of the microfibrils damage at such high temperature. Secondly, increasing the number of passes would increase the energy consumption, with only a small improvement in properties. It is recommended that 5 steps should be the maximum number of passes in microfluidizing. In terms of tensile strength and toughness, maximum 8 passes are suggested in homogenization including pretreatment (Shahbaz and Lean 2012).

Furthermore, similarity of tensile properties of pretreated and non-pretreated samples shows that no pretreatment is required for micro-grinder processing which leads to great reduction in energy consumption. In general, among the three methods including homogenizer, microfluidization and micro-grinding, nano or micro crystalline cellulose with superior mechanical, optical and physical properties and less energy consumption are produced through microfluidization with a refining pretreatment and the micro-grinding of wood fibers. Finally, unless energy consumption is reduced sufficiently, NCC use at an industrial scale will be limited to some high value applications.

17.7 Applications of Nanocrystalline Cellulose

Celluloses are biocompatible, nontoxic and stable in nature. Even though nanocrystalline cellulose (NCC) are nanometers in dimension, the innate properties of cellulose are retained. Besides the cellulosic characteristics, they possess some unique optical and

Table 17.3 Comparison of energy consumption for homogenization, microfluidization, and microgrinding during NCC production

Methods	Rate flow (kg/min)	Required energy per pass (kJ/kg)	Total energy required (kJ/kg)
Homogenization	0.2	3940	31,520 (with pretreatment)
Microgrinding	2	620	9090 (with pretreatment)
			5580 (without pretreatment)
Microfluidization	1	200/69 MPa 390/138 MPa 630/207 MPa	9180 (with pretreatment)

mechanical properties. Furthermore, nanocrystalline cellulose possess high surface area, better rheological properties, crystallinity, alignment and orientation, etc. Due to these characteristics nanocrystalline cellulose has potential applications in different sectors such as health care, food and beverages, cosmetics, electronics, etc. Its use as a reinforcing agent for nanocomposite materials (Lin and Dufresne 2014) and in the areas of nanomedicine (Dufresne 2013) has been extensively studied.

17.7.1   Use of Nanocomposite Films Production

One of the major applications of NCC is its use as a reinforcing material (or as a filler) during the production of nanocomposite materials such as nanocomposite films. Polymer nanocomposites are multiphasic material made up of polymer and nanomaterials (Jeon and Baek 2010; George and Sabapathi 2015). Because of their nanometric size and large surface area (due to smaller size of the reinforcing materials), they exhibit unique properties. Nanocomposite films obtained using NCC possess characteristics such as low permeability to moisture, gases, aroma, and oil. Due to this, they are commonly used for packaging in food and biomedical field (George and Sabapathi 2015). The properties of nanocomposite films mainly depend on two factors (1) morphology and dimensions of NCC and polymeric matrix and (2) processing method applied. Geometric aspect ratio is one of the factors that dictates the mechanical properties of nanocomposites (Peng et al. 2011). It is defined as the length to diameter (L/d) of the fillers used. Aspect ratio in turn depends on the types of cellulose fibers used and the production conditions involved. Since NCCs have high aspect ratio, they are considered to provide best reinforcing effects (Peng et al. 2011; Dufresne 2013; George and Sabapathi 2015). Good dispersibility of nanocrystals in the polymer matrices are required to enhance the mechanical properties of nanocomposite material. Since, NCCs disperse well on hydrophilic system, they are best suited for water dispersible polymers such as latexes (Hubbe et al. 2008). However, NCCs can also be modified to improve their dispersibility in hydrophobic systems (George and Sabapathi 2015). Thus, some of the commonly used processing methods to produce nanocomposite films includes casting evaporation, electrospinning, extrusion, and impregnation, monolayer films, and multilayer films (Peng et al. 2011).

17.7.2   Biomedical Applications

Due to the safe and natural form of nanocrystalline cellulose, it is gaining increasing attention in biomedical applications. Different scientists have explored its use for the production of biomedical materials suitable for practical clinical applications (Lin and Dufresne 2014). Some of the possible uses of NCCs in biomedical applications have been discussed below.

17.7.2.1   Carrier for Drug Delivery

Toxicity assessment of NCC on human cell lines, insect cells, and aquatic species has been done by various authors (Roman et al. 2009; Kovacs et al. 2010; Male et al. 2012). Most of the results obtained in these studies showed that NCCs were non-toxic to all the samples studied. This makes NCC a potential carrier in targeted drug delivery systems. On the one hand, use of NCC as an excipient in pharmaceutical industries holds considerable potential due to reduced size, hydrophilic nature, and biocompatibility. On the other, with the use of NCC as a carrier for drug delivery, optimal control of dosing can be obtained and large amounts of drugs can be bound to its surface. This is due to its large surface area and possibility of acquiring negative charge during hydrolysis. Surface modification of NCC can also be done to bind nonionized and hydrophobic drugs which normally do not bind to cellulose and its derivatives (George and Sabapathi 2015; Taheri and Mohammadi 2015).

Common forms of nanocellulose-based drug carriers include microspheres (or microparticles), hydrogels (or gels), and membranes (or films) (Lin and Dufresne 2014). Shi et al. (2003) examined the morphology of drug nanoparticles coated onto the cellulosic beads. Similarly, incorporation of the NCC particles into hydrogels has the potential of using such nanocomposite hydrogels as a controlled drug delivery vehicle (Zhang et al. 2010). Some studies carried out by (Clift et al. 2011) have shown that cellulose nanocrystals may slightly induce some dose-dependent cytotoxic and inflammatory effects on human lung cells. Thus, further studies on risk assessment of NCC however should be done before its application in drug delivery system (Peng et al. 2011; Lin and Dufresne 2014).

17.7.2.2   Bioimaging and Biosensor

Nanocellulose can be a very useful material for use in biosensor technologies. It possesses biocompatible transducer surfaces which makes it useful in sensor applications (Edwards et al. 2013). NCC can be derivatized at the hydroxyl regions to make it covalently bound to bioactive molecules. They have chiral neumatic structures which are characteristic for sensitive optical properties (Shopsowitz et al. 2010). Different types of nanocellulose such as nanocomposites, surface grafted molds, and microdialysis membranes were found to show excellent properties at bio-interface of a probe (Edwards et al. 2013). It was reported that cellulose can be used as stimuli responsive mediators by modifying it through radical polymerization (Kang et al. 2013). Such modified celluloses can be used in drug delivery and for bioimaging application (Dong and Roman 2007). These authors reported a method to label NCC with fluorescein-5'-isothiocyanate (FITC). The interaction between labeled NCC with cells where then evaluated using fluorescence techniques.

NCC can be used in biosensing by conjugated it with different biological moieties (such as nucleic acids) or metallic nanoparticles (Lam et al. 2012). Moss et al. (1981) reported a method for producing a DNA-cellulose hybrid suitable for purifying complementary mRNA from total poly(A)-enriched RNA by affinity chromatography.

NCC can also be used to produce robust, porous electrodes and sensors. It can be used to replace the cellulose which has been combined with TiO_2 nanoparticles (Pang et al. 2016) and chitosan for their potential use as biosensors (Manan et al. 2016).

17.7.3   Enzyme Immobilization

NCC has large surface area and is non-porous in nature which makes it suitable for immobilizing proteins or enzymes (Lam et al. 2012). Yang et al. (2008) reported a model system to remove chlorinated phenolic compounds in aqueous solution by immobilizing peroxidase onto NCC. The obtained immobilized peroxidase showed improved activity compared to its soluble counterpart. The immobilized samples had enzyme activity 594 U/g and stable for 3 months at 5 °C. Furthermore, NCC functionalized with gold nanoparticles has been characterized as a support for immobilization of enzymes (Mahmoud et al. 2009; Lam et al. 2012). For instance, cyclodextrin glycosyl transferase (CGTase) and alcohol oxidase can be immobilized on NCC with high enzyme loading capacity. Activity yield and CGTase loading of 70% and 165 mg/g were, respectively, obtained when such NCC was used as matrices for immobilization. This value is higher than those obtained with other commonly used matrices (Ivanova 2010). Recently, detection of phenol using tyrosinase immobilized in the NCC/Chitosan nanomaterial film has been reported (Manan et al. 2016).

17.7.4   Antimicrobial Application

Use of NCC for antimicrobial applications has been demonstrated by Drogat et al. (2011). The author reported the method used to produce silver colloidal suspension using NCC. The obtained Ag-NCC solutions showed inhibiting activities against *E. coli* and *S. aureus*. The inhibiting effect was attributed to the interaction of NCC with the bacterial cell wall. Such Ag nanoparticle NCC suspensions have the potential to be used in antiseptic solutions or wound healing gels. However, detail study involving the long-term toxicity of Ag nanoparticles is yet to be done (Lam et al. 2012). This will confirm the expectations of the possible use of NCC as antimicrobial agents.

17.7.5   Other Applications

In addition to aforementioned applications, use of NCCs as iridescent pigments and biomolecular NMR contrast agent has also been explored (Fleming et al. 2001; Peng et al. 2011). Its use as a reinforcing agent for the production of low thickness polymer electrolytes used in lithium battery has also been reported by some authors

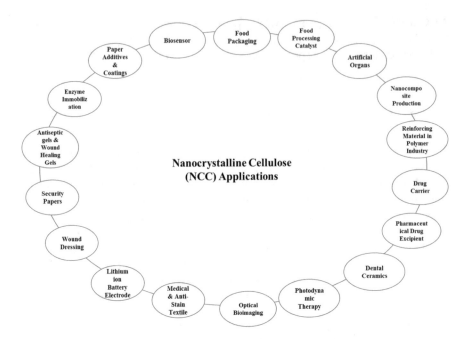

Fig. 17.2 Application of nanocrystalline cellulose in various sectors

(Schroers et al. 2004; Samir et al. 2005). Due to improved crystallinity and interfacial interaction, incorporation of NCCs in polymer nanocomposites can improve the mechanical performance, thermal stability, and optical properties of the same. Thus, biodegradable nanocomposite films with lower permeability to moisture, gases, aromas, and oil can be produced by inclusion of nanocrystalline cellulose. Such obtained films have numerous application in food and biomedical packaging areas (George and Sabapathi 2015). NCCs are biocompatible and their mechanical properties are like natural tissues. Thus, studies involving the use of NCCs as a scaffolding to grow tissue have also been carried out. The use of NCC is being studied in many areas as detailed here. The possible applications of nanocrystalline cellulose in different areas have also been depicted in Fig. 17.2.

17.8   Limitations of NCC Usage

Though NCC was found to be a potential material for several applications, it has several limitations in its upstream processing. For instance, oxidizing the cellulose make the amorphous fibers in NCCs disperse in the aqueous suspension of crystalline cellulose. Separation of these fine fibers is a major hurdle in producing pure NCC. Its usage is limited by factors like its hydrophobic nature and water swelling characteristics in the amorphous regions (Hubbe et al. 2008). Using high concentration sulfuric acid 63.5% (w/w) to produce NCC (Bondeson et al. 2006) also has a major drawback in the perspective of environmental aspects. Despite the use severe conditions, NCC yields have been found to be low (20–40%) (Bondeson et al. 2006; Hamad and Hu 2010). An efficient and reliable process is necessary to maintain the uniform size, aspect ratio, and surface chemistry. It should provide more control in NCC suspensions (Brinchi et al. 2013). Therefore, standardization of the process is necessary to obtain fine NCC fibers. Technological advancements are necessary to develop a method to produce nanocellulose with controlled size, length, and surface properties at low costs. Drying of NCC is necessary to produce in its powder form. It was not listed as one of the limitations by many of the literature reports. However, the conventional drying methods like centrifugation and high temperatures have significant effect on NCC and induce agglomeration (Brinchi et al. 2013). According to Lu and Hsieh (2012), freeze drying and supercritical drying produced a matrix of agglomerates with various sizes of nanocellulose. Therefore, it is important to develop an efficient drying method to produce NCC without altering its structural or functional properties. A number of research studies indicated that the costs and commercial scale production are also the major drawbacks in the production of NCC (Brinchi et al. 2013). However, with the amount of research ongoing on the production, modification, and application of NCC, one can be optimistic about its use in a number of ways in the near future.

References

Abatzoglou N, Chornet E, Belkacemi K, Overend RP (1992) Phenomenological kinetics of complex-systems—the development of a generalized severity parameter and its application to lignocellulosics fractionation. Chem Eng Sci 47:1109–1122

Abraham E, Deepa B, Pothan LA, Jacob M, Thomas S, Cvelbar U, Anandjiwala R (2011) Extraction of nanocellulose fibrils from lignocellulosic fibres: a novel approach. Carbohydr Polym 86:1468–1475

Acharya SK, Mishra P, Mehar SK (2011) Effect of surface treatment on the mechanical properties of bagasse fiber reinforced polymer composite. Bioresources 6(3):3155–3165

Alemdar A, Sain M (2008) Biocomposites from wheat straw nanofibers: morphology, thermal and mechanical properties. Compos Sci Technol 68:557–565

Ankerfors M (2012) Microfibrillated cellulose: energy-efficient preparation techniques and key properties. Innventia, Stockholm

Antonio M (2014) Nanocrystalline cellulose: production, opportunities, and challenges. Available from http://blog.luxresearchinc.com/blog/2014/11/nanocrystalline-cellulose-production-opportunities-and-challenges/. Accessed 12 Jan 2017

Bai W, Holbery J, Li K (2009) A technique for production of nanocrystalline cellulose with a narrow size distribution. Cellulose 16:455–465

Bondeson D, Mathew A, Oksman K (2006) Optimization of the isolation of nanocrystals from microcrystalline cellulose by acid hydrolysis. Cellulose 13:171

Börjesson M, Westman G (2015) Crystalline nanocellulose-preparation, modification, and properties. In: Poletto DM (ed) Cellulose—fundamental aspects and current trends. InTech, Rijeka

Brinchi L, Cotana F, Fortunati E, Kenny JM (2013) Production of nanocrystalline cellulose from lignocellulosic biomass: technology and applications. Carbohydr Polym 94:154–169

Carvalheiro F, Duarte LC, Girio FM (2008) Hemicellulose biorefineries: a review on biomass pretreatments. J Sci Ind Res 67:849–864

Cherian BM, Leão AL, De Souza SF, Thomas S, Pothan LA, Kottaisamy M (2010) Isolation of nanocellulose from pineapple leaf fibres by steam explosion. Carbohydr Polym 81:720–725

Clift MJD, Foster EJ, Vanhecke D, Studer D, Wick P, Gehr P, Rothen-Rutishauser B, Weder C (2011) Investigating the interaction of cellulose nanofibers derived from cotton with a sophisticated 3D human lung cell coculture. Biomacromolecules 12:3666–3673

Coffey DG, Bell DA, Henderson A (1995) Cellulose and cellulose derivatives. In: Stephen AM (ed) Food polysaccharides and their applications. Marcel Dekker, New York

Dalli SS, Rakshit SK (2015) Utilization of hemicelluloses from lignocellulosic biomass-potential products. In: Pittman KL (ed) Lignocellulose. Nova, New York, pp 85–113

De Souza Lima MM, Borsali R (2004) Rodlike cellulose microcrystals: structure, properties, and applications. Macromol Rapid Commun 25:771–787

De Souza Lima MM, Wong JT, Paillet M, Borsali R, Pecora R (2003) Translational and rotational dynamics of rodlike cellulose whiskers. Langmuir 19:24–29

Dong S, Roman M (2007) Fluorescently labeled cellulose nanocrystals for bioimaging applications. J Am Chem Soc 129:13810–13811

Drogat N, Granet R, Sol V, Memmi A, Saad N, Klein Koerkamp C, Bressollier P, Krausz P (2011) Antimicrobial silver nanoparticles generated on cellulose nanocrystals. J Nanopart Res 13:1557–1562

Dufresne A (2010) Processing of polymer nanocomposites reinforced with polysaccharide nanocrystals. Molecules 15:4111

Dufresne A (2013) Nanocellulose: a new ageless bionanomaterial. Mater Today 16:220–227

Edwards J, Prevost N, French A, Concha M, Delucca A, Wu Q (2013) Nanocellulose-based biosensors: design, preparation, and activity of peptide-linked cotton cellulose nanocrystals having Fluorimetric and colorimetric elastase detection sensitivity. Engineering 5:20–28

Elazzouzi-Hafraoui S, Nishiyama Y, Putaux J-L, Heux L, Dubreuil F, Rochas C (2008) The shape and size distribution of crystalline nanoparticles prepared by acid hydrolysis of native cellulose. Biomacromolecules 9:57–65

Espino-Perez E, Bras J, Ducruet V, Guinault A, Dufresne A, Domenek S (2013) Influence of chemical surface modification of cellulose nanowhiskers on thermal, mechanical, and barrier properties of poly(lactide) based bionanocomposites. Eur Polym J 49:3144–3154

Faruk O, Bledzki AK, Fink H-P, Sain M (2012) Biocomposites reinforced with natural fibers: 2000–2010. Prog Polym Sci 37:1552–1596

Filson PB, Dawson-Andoh BE, Schwegler-Berry D (2009) Enzymatic-mediated production of cellulose nanocrystals from recycled pulp. Green Chem 11:1808–1814

Fleming K, Gray DG, Matthews S (2001) Cellulose crystallites. Chem Eur J 7:1831–1836

Fortunati E, Armentano I, Zhou Q, Iannoni A, Saino E, Visai L, Berglund LA, Kenny JM (2012) Multifunctional bionanocomposite films of poly(lactic acid), cellulose nanocrystals and silver nanoparticles. Carbohydr Polym 87:1596–1605

Frone AN, Panaitescu DM, Donescu D, Spataru CI, Radovici C, Trusca R, Somoghi R (2011) Preparation and characterization of Pva composites with cellulose nanofibers obtained by ultrasonication. Bioresources 6(1):487–512

George J, Sabapathi SN (2015) Cellulose nanocrystals: synthesis, functional properties, and applications. Nanotechnol Sci Appl 8:45–54

George J, Ramana KV, Bawa AS, Siddaramaiah (2011) Bacterial cellulose nanocrystals exhibiting high thermal stability and their polymer nanocomposites. Int J Biol Macromol 48:50–57

Habibi Y, Lucia LA, Rojas OJ (2010) Cellulose nanocrystals: chemistry, self-assembly, and applications. Chem Rev 110:3479–3500

Hamad WY, Hu TQ (2010) Structure–process–yield interrelations in nanocrystalline cellulose extraction. Can J Chem Eng 88:392–402

Hendriks AT, Zeeman G (2009) Pretreatments to enhance the digestibility of lignocellulosic biomass. Bioresour Technol 100:10–18

Henriksson M, Henriksson G, Berglund L, Lindström T (2007) An environmentally friendly method for enzyme-assisted preparation of microfibrillated cellulose (MFC) nanofibers. Eur Polym J 43:3434–3441

Hielscher Ultrasonics GMBH (2017) Ultrasonic production of nano-structured cellulose [online]. Hielscher ultrasonics GMBH. Available from https://www.hielscher.com/ultrasonic-production-of-nano-structured-cellulose.htm#62347. Accessed 12 Jan 2017

Hubbe MA, Rojas OJ, Lucia LA, Sain M (2008) Cellulosic nanocomposites: a review. Bioresources 3(3):929–980

Huq T, Salmieri S, Khan A, Khan RA, Le Tien C, Riedl B, Fraschini C, Bouchard J, Uribe-Calderon J, Kamal MR (2012) Nanocrystalline cellulose (NCC) reinforced alginate based biodegradable nanocomposite film. Carbohydr Polym 90:1757–1763

Ivanova V (2010) Immobilization of cyclodextrin glucanotransferase from *Paenibacillus Macerans* ATCC 8244 on magnetic carriers and production of Cyclodextrins. Biotechnol Biotechnol Equip 24:516–528

Janardhnan S, Sain MM (2007) Isolation of cellulose microfibrils–an enzymatic approach. Bioresources 1:176–188

Jeon IY, Baek JB (2010) Nanocomposites derived from polymers and inorganic nanoparticles. Materials 3:3654

Kang H, Liu R, Huang Y (2013) Cellulose derivatives and graft copolymers as blocks for functional materials. Polym Int 62:338–344

Keshwani DR, Cheng JJ (2009) Switchgrass for bioethanol and other value-added applications: a review. Bioresour Technol 100:1515–1523

Khalil HA, Davoudpour Y, Islam MN, Mustapha A, Sudesh K, Dungani R, Jawaid M (2014) Production and modification of nanofibrillated cellulose using various mechanical processes: a review. Carbohydr Polym 99:649–665

Klemm D, Kramer F, Moritz S, Lindström T, Ankerfors M, Gray D, Dorris A (2011) Nanocelluloses: a new family of nature-based materials. Angew Chem Int Ed 50:5438–5466

Kovacs T, Naish V, O'connor B, Blaise C, Gagne F, Hall L, Trudeau V, Martel P (2010) An eco-toxicological characterization of nanocrystalline cellulose (NCC). Nanotoxicology 4:255–270

Lam E, Male KB, Chong JH, Leung AC, Luong JH (2012) Applications of functionalized and nanoparticle-modified nanocrystalline cellulose. Trends Biotechnol 30:283–290

Lee SY, Mohan DJ, Kang IA, Doh GH, Lee S, Han SO (2009) Nanocellulose reinforced PVA composite films: effects of acid treatment and filler loading. Fibers Polym 10:77–82

Lee HV, Hamid SBA, Zain SK (2014) Conversion of lignocellulosic biomass to nanocellulose: structure and chemical process. Sci World J 2014:1–20

Leung ACW, Hrapovic S, Lam E, Liu Y, Male KB, Mahmoud KA, Luong JHT (2011) Characteristics and properties of carboxylated cellulose nanocrystals prepared from a novel one-step procedure. Small 7:302–305

Lin N, Dufresne A (2014) Nanocellulose in biomedicine: current status and future prospect. Eur Polym J 59:302–325

Lu P, Hsieh YL (2012) Preparation and characterization of cellulose nanocrystals from rice straw. Carbohydr Polym 87:564–573

Ma XJ, Yang XF, Zheng X, Lin L, Chen LH, Huang LL, Cao SL (2014) Degradation and dissolution of hemicelluloses during bamboo hydrothermal pretreatment. Bioresour Technol 161:215–220

Mahmoud KA, Male KB, Hrapovic S, Luong JHT (2009) Cellulose nanocrystal/gold nanoparticle composite as a matrix for enzyme immobilization. ACS Appl Mater Interfaces 1:1383–1386

Majeed K, Jawaid M, Hassan A, Abu Bakar A, Abdul Khalil HPS, Salema AA, Inuwa I (2013) Potential materials for food packaging from nanoclay/natural fibres filled hybrid composites. Mater Des 46:391–410

Male KB, Leung AC, Montes J, Kamen A, Luong JH (2012) Probing inhibitory effects of nanocrystalline cellulose: inhibition versus surface charge. Nanoscale 4:1373–1379

Manan FAA, Abdullah J, Nazri NN, Malik INA, Yusof NA, Ahmad I (2016) Immobilization of tyrosinase in nanocrystalline cellulose/chitosan composite film for amperometric detection of phenol. Malay J Anal Sci 20:978–985

Marimuthu TS, Atmakuru R (2015) Isolation and characterization of cellulose nanofibers from the aquatic weed Water Hyacinth: Eichhornia crassipes. In: Pandey JK, Takagi H, Nakagaito AN, Kim HJ (eds) Handbook of polymer nanocomposites. Processing, performance and application, Polymer nanocomposites of cellulose nanoparticles, vol C. Springer, Berlin

McDonough TJ (1993) The chemistry of organosolv delignification. TAPPI J 76(8):186

Moss LG, Moore JP, Chan L (1981) A simple, efficient method for coupling DNA to cellulose. Development of the method and application to mRNA purification. J Biol Chem 256:12655–12658

Ng HM, Sin LT, Tee TT, Bee ST, Hui D, Low CY, Rahmat A (2015) Extraction of cellulose nanocrystals from plant sources for application as reinforcing agent in polymers. Compos Part B 75:176–200

Pääkkö M, Ankerfors M, Kosonen H, Nykänen A, Ahola S, Österberg M, Ruokolainen J, Laine J, Larsson PT, Ikkala O, Lindström T (2007) Enzymatic hydrolysis combined with mechanical shearing and high-pressure homogenization for nanoscale cellulose fibrils and strong gels. Biomacromolecules 8:1934–1941

Panaitescu DM, Frone AN, Nicolae C (2013) Micro- and nano-mechanical characterization of polyamide 11 and its composites containing cellulose nanofibers. Eur Polym J 49:3857–3866

Pang Z, Yang Z, Chen Y, Zhang J, Wang Q, Huang F, Wei Q (2016) A room temperature ammonia gas sensor based on cellulose/TiO2/PANI composite nanofibers. Colloids Surf A Physicochem Eng Aspects 494:248–255

Park YC, Kim JS (2012) Comparison of various alkaline pretreatment methods of lignocellulosic biomass. Energy 47:31–35

Peng BL, Dhar N, Liu HL, Tam KC (2011) Chemistry and applications of nanocrystalline cellulose and its derivatives: a nanotechnology perspective. Can J Chem Eng 89:1191–1206

Qin ZY, Tong G, Chin YCF, Zhou JC (2011) Preparation of ultrasonic-assisted high carboxylate content cellulose nanocrystals by tempo oxidation. Bioresources 6(2):1136–1146

Ranby BG (1951) Fibrous macromolecular systems. Cellulose and muscle. The colloidal properties of cellulose micelles. Discuss Faraday Soc 11:158–164

Rebouillat S, Pla F (2013) State of the art manufacturing and engineering of nanocellulose: a review of available data and industrial applications. J Biomater Nanobiotechnol 4:165–188

Roman M, Dong S, Hirani A, Lee YW (2009) Cellulose nanocrystals for drug delivery. polysaccharide materials: performance by design. American Chemical Society, Washington, DC, pp 81–91

Rosa SML, Rehman N, De Miranda MIG, Nachtigall SMB, Bica CID (2012) Chlorine-free extraction of cellulose from rice husk and whisker isolation. Carbohydr Polym 87:1131–1138

Sacui IA, Nieuwendaal RC, Burnett DJ, Stranick SJ, Jorfi M, Weder C, Foster EJ, Olsson RT, Gilman JW (2014) Comparison of the properties of cellulose nanocrystals and cellulose nanofibrils isolated from bacteria, tunicate, and wood processed using acid, enzymatic, mechanical, and oxidative methods. ACS Appl Mater Interfaces 6:6127–6138

Saha BC, Yoshida T, Cotta MA, Sonomoto K (2013) Hydrothermal pretreatment and enzymatic saccharification of corn stover for efficient ethanol production. Ind Crop Prod 44:367–372

Saito T, Kimura S, Nishiyama Y, Isogai A (2007) Cellulose nanofibers prepared by TEMPO-mediated oxidation of native cellulose. Biomacromolecules 8:2485–2491

Samir MASA, Alloin F, Sanchez JY, Dufresne A (2005) Nanocomposite polymer electrolytes based on poly(oxyethylene) and cellulose whiskers. Polímeros 15:109–113

Schroers M, Kokil A, Weder C (2004) Solid polymer electrolytes based on nanocomposites of ethylene oxide–epichlorohydrin copolymers and cellulose whiskers. J Appl Polym Sci 93:2883–2888

Shahbaz M, Lean HH (2012) Does financial development increase energy consumption the role of industrialization and urbanization in Tunisia. Energy Policy 40:473–479

Shi HG, Farber L, Michaels JN, Dickey A, Thompson KC, Shelukar SD, Hurter PN, Reynolds SD, Kaufman MJ (2003) Characterization of crystalline drug nanoparticles using atomic force microscopy and complementary techniques. Pharm Res 20:479–484

Shopsowitz KE, Qi H, Hamad WY, Maclachlan MJ (2010) Free-standing mesoporous silica films with tunable chiral nematic structures. Nature 468:422–425

Sjostrom E (1993) Wood chemistry: fundamentals and applications. Academic Press, San Diego, CA

Spence KL, Venditti RA, Rojas OJ, Habibi Y, Pawlak JJ (2011) A comparative study of energy consumption and physical properties of microfibrillated cellulose produced by different processing methods. Cellulose 18:1097–1111

Stelte W, Sanadi AR (2009) Preparation and characterization of cellulose nanofibers from two commercial hardwood and softwood pulps. Ind Eng Chem Res 48:11211–11219

Tadesse H, Luque R (2011) Advances on biomass pretreatment using ionic liquids: an overview. Energy Environ Sci 4:3913–3929

Taheri A, Mohammadi M (2015) The use of cellulose nanocrystals for potential application in topical delivery of hydroquinone. Chem Biol Drug Des 86:102–106

Teymouri F, Laureano-Perez L, Alizadeh H, Dale BE (2004) Ammonia fiber explosion treatment of corn stover. Appl Biochem Biotechnol 113-116:951–963

Xiong Y, Zhang Z, Wang X, Liu B, Lin J (2014) Hydrolysis of cellulose in ionic liquids catalyzed by a magnetically-recoverable solid acid catalyst. Chem Eng J 235:349–355

Yang R, Tan H, Wei F, Wang S (2008) Peroxidase conjugate of cellulose nanocrystals for the removal of chlorinated phenolic compounds in aqueous solution. Biotechnology 7:233–241

Zavrel M, Bross D, Funke M, Buchs J, Spiess AC (2009) High-throughput screening for ionic liquids dissolving (ligno-)cellulose. Bioresour Technol 100:2580–2587

Zhang X, Huang J, Chang PR, Li J, Chen Y, Wang D, Yu J, Chen J (2010) Structure and properties of polysaccharide nanocrystal-doped supramolecular hydrogels based on Cyclodextrin inclusion. Polymer 51:4398–4407

Zhang PP, Tong DS, Lin CX, Yang HM, Zhong ZK, Yu WH, Wang H, Zhou CH (2014) Effects of acid treatments on bamboo cellulose nanocrystals. Asia Pac J Chem Eng 9:686–695

Zhao X, Cheng K, Liu D (2009) Organosolv pretreatment of lignocellulosic biomass for enzymatic hydrolysis. Appl Microbiol Biotechnol 82(5):815

Zhu JY, Sabo R, Luo X (2011) Integrated production of nano-fibrillated cellulose and cellulosic biofuel (ethanol) by enzymatic fractionation of wood fibers. Green Chem 13:1339–1344

Index

© Springer Nature Switzerland AG 2018
R. Prasad et al. (eds.), *Exploring the Realms of Nature for Nanosynthesis*,
Nanotechnology in the Life Sciences, https://doi.org/10.1007/978-3-319-99570-0

Printed in the United States
By Bookmasters